D0204086

ASTROBIOLOGY
An Introduction

Series in Astronomy and Astrophysics

The Series in Astronomy and Astrophysics includes books on all aspects of theoretical and experimental astronomy and astrophysics. Books in the series range in level from textbooks and handbooks to more advanced expositions of current research.

Series Editors:
M Birkinshaw, University of Bristol, UK
J Silk, University of Oxford, UK
G Fuller, University of Manchester, UK

Recent books in the series

Nuclear Processing of Heavy Elements
Hermann Beer

An Introduction to the Physics of Interstellar Dust
Endrik Krugel

Numerical Methods in Astrophysics: An Introduction
P Bodenheimer, G P Laughlin, M Rózyczka, H W Yorke

Very High Energy Gamma-Ray Astronomy
T C Weekes

The Physics of Interstellar Dust
E Krügel

Dust in the Galactic Environment, 2nd Edition
D C B Whittet

Dark Sky, Dark Matter
J M Overduin and P S Wesson

An Introduction to the Science of Cosmology
D J Raine and E G Thomas

The Origin and Evolution of the Solar System
M M Woolfson

The Physics of the Interstellar Medium, 2nd Edition
J E Dyson and D A Williams

Optical Astronomical Spectroscopy
C R Kitchin

Dust and Chemistry in Astronomy
T J Millar and D A Williams (eds)

Series in Astronomy and Astrophysics

ASTROBIOLOGY
An Introduction

Alan Longstaff
Associate Lecturer in Astronomy and Earth Sciences
Open University, UK

CRC Press
Taylor & Francis Group
Boca Raton London New York

CRC Press is an imprint of the
Taylor & Francis Group, an **informa** business

Cover credit: Carter Roberts, Eastbay Astronomical Society, CA.

CRC Press
Taylor & Francis Group
6000 Broken Sound Parkway NW, Suite 300
Boca Raton, FL 33487-2742

© 2015 by Taylor & Francis Group, LLC
CRC Press is an imprint of Taylor & Francis Group, an Informa business

No claim to original U.S. Government works

Printed on acid-free paper
Version Date: 20141014

International Standard Book Number-13: 978-1-4398-7576-6 (Hardback)

Visit the Taylor & Francis Web site at
http://www.taylorandfrancis.com

and the CRC Press Web site at
http://www.crcpress.com

Contents

Constants

Fundamental constants

Speed of light in vacuum	c	2.998×10^8 m s^{-1}
Faraday constant	F	9.65×10^4 C mol^{-1}
Gravitational constant	G	6.673×10^{-11} N m^2 kg^{-2}
Planck constant	h	6.626×10^{-34} J s
Boltzmann constant	k	1.381×10^{-23} J K^{-1}
Avagadros number	N_A	6.022×10^{23} mol^{-1}
Universal gas constant	R	68.3 J mol^{-1} K^{-1}
Stefan–Boltzmann constant	σ	5.671×10^{-8} J m^2 K^{-4} s^{-1}

Particle constants

Charge of proton	e	1.602×10^{-19} C
Electron rest mass	m_e	9.109×10^{-31} kg
Neutron rest mass	m_n	1.675×10^{-27} kg
Proton rest mass	m_p	1.673×10^{-27} kg
Atomic mass unit	u	1.661×10^{-27} kg

Astronomical quantities

Luminosity of the sun	$L_\odot$	3.83×10^{26} J s^{-1}
Mass of the Earth	$M_\oplus$	5.98×10^{24} kg
Mass of the sun	$M_\odot$	1.99×10^{30} kg
Mass of Jupiter	M_J	1.90×10^{27} kg
Radius of the Earth	$R_\oplus$	6.378×10^6 m
Radius of the sun	$R_\odot$	6.96×10^8 m
Radius of Jupiter	R_J	7.14×10^7 m

Equivalent quantities

Energy

$$1 \text{ eV} = 1.602 \times 10^{-19} \text{ J}$$

Distance

$$1 \text{ AU} = 1.496 \times 10^{11} \text{ m}$$
$$1 \text{ ly} = 9.461 \times 10^{15} \text{ m}$$
$$1 \text{ pc} = 3.086 \times 10^{16} \text{ m} = 3.261 \text{ ly}$$

Acknowledgments

Iain Nicholson (University of Hertfordshire) and Radmila Topalovic (Royal Observatory, Greenwich) provided very helpful comments and suggestions on several chapters. Patricia Revest (Queen Mary University of London) heroically worked through the entire text, giving invaluable critique. I am extremely grateful to them for their time and expertise. There are almost certainly errors remaining despite our best efforts and of course I take full responsibility for them. My thanks are due to Rachel Holt, Francesca McGowan, Amber Donley, Kari Budyk, and Robert Sims at CRC Press, Sarfraz Khan, and to Karthick Parthasarathy of Techset Composition for their unfailing guidance and patience, and to John Navas, who commissioned the book.

Author

Alan Longstaff, BSc (Lond), BSc (Open), PhD, FRAS, originally trained as a biochemist, and after a Senior Lectureship in the Biosciences Department at the University of Hertfordshire for a number of years he became a university student once again to study astronomy and planetary science. He now divides his time between teaching and writing. Since 2003, he has worked part-time for The Royal Observatory, Greenwich, as an astronomy tutor and planetarium presenter, and held part-time teaching posts at Queen Mary University of London, Waldegrave Science School for Girls, London, and the Open University. He has lectured widely to astronomical and geological societies, written and coauthored several textbooks, and is a regular contributor to the UK magazine *Astronomy Now*.

Introduction

The idea that there is life on other worlds has a long history that stretches back at least as far as the Greek philosophers' writing in 400 BCE. More recently, Giordano Bruno (1548–1600 CE) formulated a cosmology that was remarkably ahead of his time. He subscribed to the Copernican heliocentric description of the solar system, considered that the stars were other suns, and would be orbited by planets just as the sun is. Moreover, he reasoned that there would be an infinite number of inhabited worlds. We had to wait almost 400 years after Bruno's dreadful execution for the first exoplanet to be discovered. In the meantime, we have seen some fantastic claims, including life on the moon, and even the sun! For the first half of the twentieth century, the notion that Mars was inhabited was widely entertained; a view encouraged by illusory canals, surface features that darken and lighten with the Martian seasons attributed to the growth of vegetation and even, in 1956, the apparent detection of spectroscopic evidence for plant life.

As a research enterprise, astrobiology—the study of the origin, nature, distribution, and future of life in the universe—is young. The term exobiology, the study of life elsewhere than on Earth, was first coined by the Nobel Prize-winning molecular biologist Joshua Lederberg, who became a powerful advocate for the new field. It emerged in the 1950s and 1960s from a convergence of planetary science, the search for exoplanets, origin of life studies, and the Search for Extraterrestrial Intelligence (SETI), four disciplines that use very different research methods and techniques and started out with very different goals. Unlike exobiology, described disparagingly in 1964 by an evolutionary biologist as a science without a subject, astrobiology—its successor—encompasses life on Earth. NASA has become its most important patron, spending $1 billion on the 1977 Viking missions to search for life on Mars, and funding much else, including biology programs.

Astrobiology is a multidisciplinary pursuit that in various guises encompasses astronomy, chemistry, planetary and Earth sciences, and biology. It relies on mathematical, statistical, and computer modeling for theory, and space science, engineering, and computing to implement observational and experimental work. Consequently, when studying astrobiology we need a broad scientific canvas. For example, it is now clear that the Earth operates as a system; it is no longer appropriate to think in terms of geology, oceans, atmosphere, and life as being separate. This textbook reflects this multiscience approach, so if you are an astronomy major, there is some planetary science and biology to get to grips with, if you are a biologist, be prepared to learn some planetary science and astronomy, and if you are doing primarily Earth sciences— well, you get the picture.

This book is introductory in the sense that there is no calculus or curly arrow chemistry, nor details of modeling, and I have been sparing in areas that experience shows students find hard, such as isotope geochemistry and molecular spectroscopy. Also I provide reminders at appropriate places so that your study will not be compromised by lack of a bit of physics, chemistry, or whatever. Despite that, some of the material is not straightforward because there are uncertainties and controversies which is sometimes instructive to follow.

Inevitably, it is not possible to cover everything. The astonishingly rapid rate of exoplanet discoveries means that I have not included statistical data on exoplanets, since anything I include now will be out-of-date within a few months. There are several websites and computer applications that will allow you to mine the exoplanet database and construct any number of graphical representations using the latest exoplanet count if you wish. My choice of material has been dictated to some extent by the emphasis of recent conferences and by ideas that seem as if they have a future. Thus, my coverage of how life might have originated on Earth has focused (some might say too much) on alkaline vents. My future defense will be that it seemed like a good idea at the time! I have also taken the traditional line that life everywhere will be carbon- and water-based (for good reasons). There is often considerable speculation in the media on matters astrobiological; just a "sniff" of water is seemingly enough to populate a world. Yet NASA's mantra, "follow the water," has to be interpreted carefully: water is the second most common molecule in the universe (after molecular hydrogen) and most of the places where it is found would be inimical to life. Nonetheless, some astrobiologists do consider not just whether worlds such as Mars, Titan, or Europa are habitable, but what sort of organisms might make a living there—that is their job after all—and although this is entirely speculative, where it is more plausible than fanciful I have included it. But this material comes with a warning: we must surely demand extraordinary evidence for any extraordinary claim of extraterrestrial life.

No textbook author is likely to have an unbiased opinion, especially on something as exciting as astrobiology. I have tried not to let my biases get too much of an upper hand, but just so that you know: the number of planets out there makes me pretty sure that there is life elsewhere in the universe, so Earth is not unique. But my feeling is that it is rare, and that complex life is extremely rare indeed. Overall, I remain to be convinced by arguments about the habitability of Mars or Europa, but I would be delighted to be proved wrong. What I am convinced about is that we must continue the quest.

1

Origin of the Elements

1.1 Elements for Life

For the first 100 million years or so the universe contained only three elements: hydrogen, helium, and trace amounts of lithium. The other elements were forged inside the stars. As far as we know, all life in the universe is based on carbon chemistry, mainly involving carbon, hydrogen, oxygen, nitrogen, phosphorus, and sulfur. Some other elements such as calcium, magnesium, and iron, while being only minor components of living materials, serve critical biological functions. Consequently, until the first stars had been born, died, and dissipated their carbon (and other essential elements) into the clouds that would spawn the next generation of stars, there could be no life. This means that we live now in a special time. There could have been no life in the early universe. Of course, heavy elements such as silicon and iron are also required to build potentially habitable real estate such as rocky planets.

In this chapter, we examine how the elements came into existence and discuss the properties of the stars that gave birth to them.

1.2 The Universe Started from a Hot and Dense State

The hot Big Bang hypothesis favored by most cosmologists is that the universe originated about 13.82 Gyr ago from an extraordinarily hot, small, and dense state. It has expanded and cooled ever since.

1.2.1 Abundances of Primordial Elements Are Predicted by the Big Bang Hypothesis

Theory predicts that all hydrogen and most of helium seen in the current universe were created in the first few minutes. Indeed, these give mass abundances of about 75% of $_1^1H$, about 25% of $_2^4He$, about 0.01% of deuterium ($_1^2H$) and helium-3 ($_2^3He$), and trace amounts of lithium-7 ($_3^7Li$) (Box 1.1). That observed abundances are largely consistent with these predicted abundances is a strong evidence for the Big Bang hypothesis.

1.2.1.1 The Early Universe Was Dominated by Radiation

There is considerable speculation about what happened at the earliest times after the Big Bang, but by about 1 µs, we think that standard physics is a reasonable guide to the state of the universe, even though conditions were extreme! By this time the universe still had the density of nuclear matter despite having expanded to 0.025 light

---

BOX 1.1   ATOMIC NUCLEI

The number of protons, the atomic number, Z, uniquely specify an element. The mass number, A, of a nucleus is the number of protons plus the number of neutrons. Many elements have several isotopes that differ in mass number because they have different numbers of neutrons. For example, the helium nucleus has two protons, but one isotope of helium has two neutrons ($^4_2$He) while the other has one neutron ($^3_2$He). (Note that A is shown as a back superscript, Z as a back subscript.) The hydrogen nucleus generally contains just a single proton ($^1_1$H), but the deuterium nucleus or "heavy hydrogen" has a single neutron ($^2_1$H) as well, and the tritium nucleus has two neutrons ($^3_1$H).

It is common practice not to show Z explicitly, in which case the isotope is defined by A alone, for example, ^{3}He or helium-3.

Many isotopes are radioactive. For example, the tritium nucleus is unstable and undergoes β decay in which one its neutrons emits an electron (a β particle) and an anti-neutrino (ν) to become a proton:

$$n \rightarrow p + e^- + \nu.$$

Consequently,

$$^3_1H \rightarrow {}^3_2He + e^- + \nu. \tag{1.1}$$

Note how the result of this type of radioactive decay is the synthesis of a new element with atomic number $Z + 1$. While the decay of any individual nucleus is probabilistic and unpredictable, given a large number of nuclei, half will decay in a time (half-life) that is predictable and characteristic of the isotope. For example, the half-life of tritium is 12.3 years.

---

years across (about 20-fold the size of the solar system). It had cooled considerably but was still a searing 10^{13} K. At this time, particles and antiparticles formed that rapidly annihilated with the release of radiation in the form of gamma rays. In the case of the electron and its antiparticle, the positron:

$$e + e^- \rightarrow 2\gamma. \tag{1.2}$$

For reasons that are unclear, a small excess of particles over antiparticles were created initially and these remained after the annihilation. This matter (essentially free protons, neutrons, electrons, and neutrinos) and the huge amounts of radiation produced by the annihilation were in thermal equilibrium.

1.2.1.2 Primordial Hydrogen and Helium Were Synthesized in the First 20 min

Big Bang nucleosynthesis (BBN) produced all three isotopes of hydrogen, helium-3, and helium-4, and isotopes of lithium and beryllium. By far the most abundant were $^1_1$H and $^4_2$He. Moreover, tritium and beryllium isotopes are radioactive, so these decayed long ago and do not contribute to the current inventory of primordial elements.

Work on BBN began with Ralph Alpher, a graduate student, and his supervisor, George Gamov, in 1948. Gamov, an extrovert Russian with a huge sense of fun, invited the chemist Hans Bethe to be a coauthor of the paper reporting the work, even though he made no contribution, because Alpher, Bethe and Gamov made a pun on the first three letters of the Greek alphabet ($\alpha\beta\gamma$)! In fact, Hans Bethe subsequently went on to do significant work on BBN.

There was only a very limited time window for nucleosynthesis to happen for two reasons. The first reason is that the free neutrons are not stable but have a mean lifetime of ~15 min, so any neutrons that had not combined with protons in an atomic nucleus soon decayed:

$$n \rightarrow e^- + p + \nu. \tag{1.3}$$

The second reason is temperature. Stable nuclei could only form when the temperature had fallen below 10^9 K, at about 100 s after the Big Bang. Many nucleosynthesis reactions were happening at this time. However, it is sufficient for our purpose to look at just the one that generated helium-4:

$$n + p \rightarrow {}^2_1H + \nu, \tag{1.4}$$

$${}^2_1H + {}^2_1H \rightarrow {}^4_2He + \gamma. \tag{1.5}$$

These reactions can run in reverse directions provided that the γ radiation is energetic enough to drive the reverse reaction. This was the case at first and meant that deuterium was destroyed as fast as it was made. Since deuterium is a mandatory intermediary for all reactions generating helium-4, helium synthesis did not take off until the temperature fell below about 10^9 K and the γ radiation become too weak to break up the deuterium nucleus. This is referred to as the deuterium bottleneck. By ~1000 s the temperature had fallen below 4×10^8 K, and there was not enough energy for any further nucleosynthesis. Hence, the abundances of the various nuclei (except the radioactive isotopes that continued to decay) became fixed.

By this time the ratio of neutrons to protons was 1 to 7. So out of every 16 baryons (2 neutrons and 14 protons), 4 (25%) combined to form one helium-4 nucleus. Although tiny amounts of lithium and beryllium were made, BBN virtually stopped with ${}^4_2He$. This is because helium-4 is very strongly bound; in other words, it is very stable. In addition there is no nucleus of mass number 5; neither is there a stable nucleus with $A = 8$ (${}^8Li$ and ${}^8Be$ decay with half-lives of 0.8 s and 3×10^{-16} s (!), respectively) to act as intermediates for the synthesis of heavier elements.

How do these BBN predictions measure up to the abundances of primordial elements found by astronomers? The difficulty in determining the primordial helium mass fraction is that all stars produce helium by hydrogen fusion, and some stars consume helium in nuclear fusion reactions late in their lives. To get around this, astronomers concentrated their attention on nearby dwarf galaxies containing stars and gas low in oxygen and nitrogen. Such galaxies are thought to have changed little over cosmic time. Spectral lines of hydrogen and helium emitted from regions containing very young stars (HII regions) in these galaxies give a primordial helium mass fraction of 24% (with uncertainty of 1%), in excellent agreement with predictions made by BBN theory.

1.2.1.3 Lithium Abundance Is Problematic

Astronomers have sought to confirm the lithium abundance predicted by BBN in extremely old, cool, dwarf stars. Although lithium is destroyed by nuclear fusion reactions in the cores of the stars, their atmospheres should preserve the primordial lithium abundance. Being so old implies that the material from which they were formed has not experienced much nucleosynthesis in earlier generations of stars. In the early 1980s, astronomers identified a population of such stars in the halo of the Milky Way, all with the same low lithium abundance. These were termed lithium plateau stars and their "plateau" lithium content was assumed to be primordial. Unfortunately, BBN theory in combination with a recent determination of the ratio of baryons to photons in the universe predicts that lithium abundance should be four-fold higher than that in the lithium plateau stars. This mismatch is termed the "lithium problem."

1.3 The Message of Light

1.3.1 Atoms and Molecules Process Electromagnetic Radiation

Because the stars lie at such remote distances the only way in which we can discover their basic properties (temperature, composition, etc.) or that of planets beyond our solar system is by examining the light and other electromagnetic radiation that comes from them (Box 1.2).

BOX 1.2   ELECTROMAGNETIC RADIATION

At the end of the nineteenth century, Thomas Young showed that two sets of water ripples would interfere: where two crests (troughs) coincided, a larger crest (deeper trough) resulted; where the crest and the trough coincided, they cancelled out. The discovery that light passing through two slits produced comparable interference patterns on a screen was proof that light was a wave.

In the 1860s, the Scottish physicist James Clerk Maxwell derived a set of equations that explained the mutual relationships between electricity and magnetism that had been revealed by nineteenth century experimental physics. He showed that light could be described as self-sustaining oscillations of electric and magnetic fields. Remarkably, his equations predicted the speed at which these electromagnetic waves are propagated. It implied that the speed of light is a constant, presumably with respect to a medium through which it was postulated to travel; after all, there cannot be water waves without water or sound waves without air!

But there can be light waves without a medium. This surprising result was demonstrated experimentally by Albert Michelson and Edward Morley in 1887. These scientists showed that the speed of light was unchanged regardless of the direction of the Earth through space. This ruled out the existence of the luminiferous ether through which light had been assumed to propagate. The conclusion that light travels at a constant speed in vacuum, independent of the velocity of the light source with respect to the observer who is performing the measurement, became an axiom for the special theory of relativity.

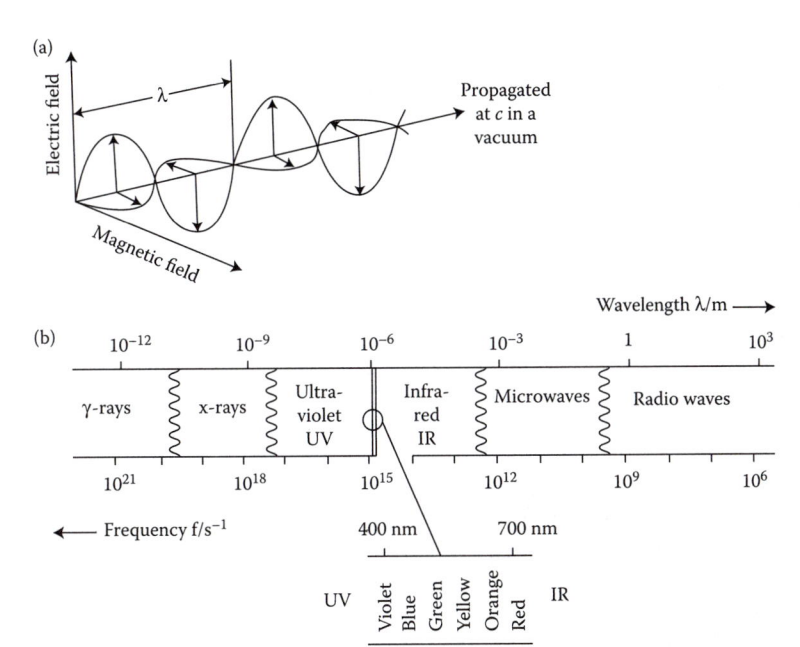

FIGURE 1.1 (a) Light as a wave. (b) The electromagnetic spectrum.

Earlier, in 1800, the English astronomer William Herschel had demonstrated the existence of radiation beyond the visible region by observing a rise in temperature registered by a thermometer placed beyond the red end of the sun's spectrum. Over the succeeding 150 years, the entire electromagnetic (EM) spectrum was revealed (Figure 1.1).

The distance from any point on a sine wave to the corresponding point on the next cycle is the wavelength, λ. If the wave is travelling at constant speed, c, then the number of wavelengths that will pass a fixed point is the frequency, f, or ν (Greek "nu"). We see that there is a reciprocal relationship between the wavelength and frequency, $\lambda \propto 1/f$. Indeed:

$$c = f\lambda, \quad \text{or} \quad c = \nu\lambda. \tag{1.6}$$

The EM radiation runs from short wavelength, high frequency, and high energy gamma radiation at one end to long wavelength, low frequency, and low energy radio waves at the other. Visible light is just a very narrow rainbow spanning wavelengths from 380 to 740 nm. The boundaries between the types of EM radiation are generally rather loosely defined. For example, radiation with $\lambda = 10^{-11}$ m can be described as hard x-rays or soft γ-rays.

EXERCISE 1.1

What is the frequency range of the visible region of the electromagnetic spectrum?

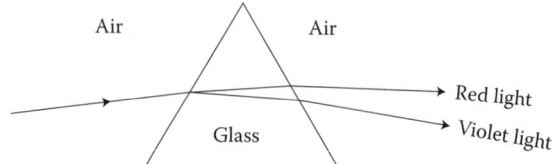

FIGURE 1.2 The dispersion of white light by a prism.

Remarkably, in 1835 the French philosopher Auguste Comte asserted that the chemical composition of celestial bodies was unknowable. But by 1859 chemists Robert Bunsen and Gustav Kirchhoff had succeeded in obtaining pure samples of a few metals, including sodium, and had developed a gas burner (the Bunsen burner) that produced a colorless flame. Thus armed, they were able to examine with a spectroscope the light produced by sodium (and other metals) heated in the flame. In this instrument, light is passed via a narrow slit through a glass prism. The light is dispersed because the refraction of light through the glass depends on the wavelength of the light. Violet light is slowed down more than red light as it hits the interface between air and glass and, consequently, is bent through a larger angle (Figure 1.2).

When sodium was heated, Bunsen and Kirchhoff observed two emission lines close together in the yellow part of the spectrum. They then showed that these lines corresponded to 2 of the 570 or so dark lines (the D lines) that had been identified by Joseph Fraunhofer in 1814 in the solar spectrum. Bunsen and Kirchhoff concluded that the atmosphere of the sun contained sodium atoms that *absorbed* the yellow light. Hence the dark lines in the solar spectrum became known as absorption lines. These studies heralded the field of spectroscopy—a technique that lies at the heart of observational astronomy.

Spectroscopy works because photons interact with matter. Atoms absorb and emit light at specific and characteristic wavelengths as their electrons jump between discrete energy levels determined by quantum mechanics. This insight would emerge from the failure of classical physics.

1.3.2 Electronic Transitions Are Quantized

The radiation from a source at thermal equilibrium has the characteristic black body spectrum (Figure 1.3) that has a very different shape compared to that predicted by physics at the end of the nineteenth century. According to classical electromagnetism, the spectrum of black body radiation should increase continuously without limit at ever higher frequencies. This disagreement between theory and experiment was called the "ultraviolet catastrophe."

Experimental observations showed that as the temperature of a black body source increased its spectrum shifted to shorter wavelengths. Wilhelm Wien showed in 1893 that the temperature (in Kelvin) and the peak wavelength (in meter) are directly related:

$$\lambda_{\text{peak}} = \frac{2.90 \times 10^{-3}}{T}. \tag{1.7}$$

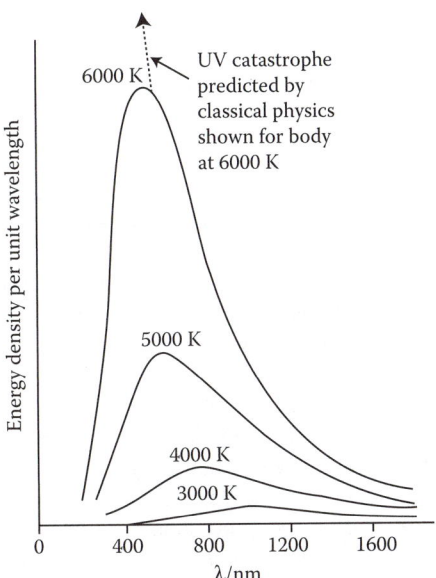

FIGURE 1.3 Planck curves of black body radiation emitted by sources at different temperatures.

As radiation from the stars can be approximated to that of a black body, this relation (termed the Wien displacement law) allows us to calculate the effective temperature of the stars if we can determine the wavelength at which they radiate maximally. The effective temperature, T_e, of a star is the temperature of a black body that has the same luminosity (power output) and the surface area of the star.

EXERCISE 1.2

The peak wavelength for the sun's radiation is 502 nm. What is the effective temperature of the sun?

After a couple of false starts, by 1900 the German physicist Max Planck had derived an empirical expression for black body radiation that agreed with experimental observations. This was the starting point for quantum mechanics because it was based on the postulate that energy and angular momentum are not continuously distributed at the atomic scale and that only discrete values are allowed. Planck showed that the energy radiated is proportional to multiples of the frequency (ν or f) of the radiation:

$$E = h\nu \quad \text{or equivalently} \quad E = hf, \tag{1.8}$$

where h is the Planck constant. Since h is a very small number (6.626×10^{-34} J s) we see that a "quantum leap" is extremely tiny!

That light appeared to be emitted in packets suggests that electromagnetic radiation behaves as if it were a stream of particles (photons) rather than waves. In support of this quantum theory of light is the photoelectric effect in which electrons are knocked out of a metal target when light is shone on it. The energy of the electrons depends not on the intensity of the light but on its frequency.

In quantum mechanics light can behave like a wave or a particle depending on the nature of the experiment. This wave–particle duality is a deep mystery.

EXERCISE 1.3

Calculate the energy range of visible light photons in joules.

On a more prosaic level quantum mechanics tells us that electron orbitals in atoms have fixed discrete energies. An electron can be made to jump to another orbital with a higher energy if a photon with exactly the right energy collides with the atom. The atom is said to be excited because it has absorbed the energy of the photon. If large numbers of identical atoms are excited in the same way enough photons with the same wavelength are removed from the light emitted by the gas to produce a dark absorption line in the spectrum. Because the energies of electron orbitals are unique to a given atom or ion, the absorption lines are characteristic signatures of specific elements and are therefore diagnostic. When atoms de-excite, electrons jump to lower energy orbitals and emit photons resulting in emission lines that are once again characteristic of a given atom or ion. This is because the photons have precisely the same energy that is equal to the difference between the two energy levels. We can examine this in more detail by looking at the most abundant atom in the universe, hydrogen.

1.3.3 Energy Levels Govern Electronic Transitions in the Hydrogen Atom

According to quantum theory electrons in an atom are permitted to exist in a number of states, each associated with a fixed amount of energy. In general, these states are completely defined by four quantum numbers. However, in the case of neutral hydrogen atom having just one electron, the energy levels can be designated by just the principal quantum number, n, that takes integer values 1, 2, 3, ... In the model of the hydrogen atom derived by Niels Bohr in 1913, the principal quantum number is related to the radius, r, of the electron's orbit (actually $r = n^2 a_0$, where $a_0 = 0.0529$ nm, the Bohr radius), giving rise to the notion of electron shells (K, L, M, ...) in the atom. The energy level diagram of the hydrogen atom is shown in Figure 1.4.

The hydrogen atom with its electron in $n = 1$ quantum state is said to be in its ground state. The atom has an energy $E_1 = -13.60$ eV (Box 1.3). When the electron is in higher energy states, E_2, E_3, ... (i.e., with $n = 2, 3, ...$) the atom is said to be in excited states. The energy of the atom is given by

$$E(\text{eV}) = -\frac{13.60}{n^2}. \tag{1.9}$$

A hydrogen atom can be boosted to a higher energy level by absorbing a photon that has exactly the same energy as the difference between the initial and final energy levels ($E_f - E_i$). From Equation 1.10, the frequency of such a photon is given by

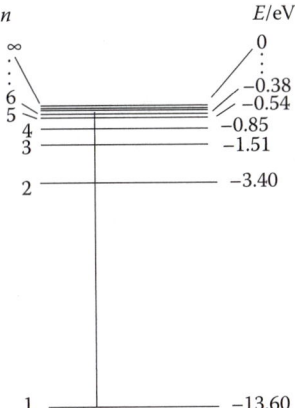

FIGURE 1.4 Energy levels in the hydrogen atom. As n increases, energy levels get ever closer together.

$$f_{fi} = \frac{E_f - E_i}{h}. \tag{1.10}$$

As we have seen, for a given electronic transition the photon energy must be exactly right. So, what happens if a photon with energy equal to or greater than 13.60 eV hits a hydrogen atom? This will knock the electron out of the hydrogen atom altogether, a process called ionization. Any energy in addition to 13.60 eV goes to increase the kinetic energy of the ejected electron.

Atoms usually do not remain excited for long but instead de-excite, either spontaneously or as a result of collisions with other atoms. (Collisional de-excitation needs higher gas densities than spontaneous de-excitation.) Electrons make transitions to lower energy states by emitting photons with allowable frequencies given by Equation 1.8. This will entail a drop from E_f to E_i. If these energy levels are not adjacent (i.e., if there are other energy levels intercalated between them), then the

---

BOX 1.3   THE ELECTRON VOLT

Photon energy is often expressed in electron volts (eV). 1 eV is the work done in moving a unit charge (e, the magnitude of the charge on an electron or a proton) through a potential difference of one volt. The unit charge is 1.602×10^{-19} C and the volt is a joule per coulomb (J/C) so 1 eV = 1.602×10^{-19} J.

EXERCISE 1.4

What is the energy range of visible light photons in eV?

Typical molecular bond energies are ~1 eV, while the rest mass of the electron and the proton are 0.5 MeV and 938.3 MeV, respectively.

de-excitation can occur in steps, each of which results in the emission of a photon with appropriate frequency. Hence, de-excitation gives bright emission lines in the spectrum.

Since a gas can emit photons at the same frequency as it absorbs them, how do we ever get the line spectra? The answer is that it all depends on how we view the gas (Figure 1.5).

Three series of electronic transitions of the hydrogen atom were observed by early spectroscopists. Although the wavelengths of these transitions could be predicted at the start of the twentieth century by empirical formulae, they could not be explained until Niels Bohr developed his model of the hydrogen atom. The Balmer series (see Figure 1.6) are transitions to and from $n \geq 2$ energy levels and as seen in visible light. For example, the H_α transition between $n = 3$ and $n = 2$ is responsible for the red color of the emission of the nebulae and, closer to home, the red glow of the chromosphere (the lowest part of the atmosphere) of the sun. The Lyman series, the transitions to and from the ground state ($n = 1$) are of much higher energies than the Balmer series; so, they are seen in the ultraviolet region. The Lyman alpha (Ly_α) transition between $n = 2$ and $n = 1$ is particularly strong and easily recognized; so it has proved important for astronomers probing distant galaxies. The Paschen series are transitions to and from $n > 3$ and $n = 3$.

EXERCISE 1.5

(i) Calculate the frequency of the Paschen α photon absorbed when the hydrogen atom is excited from $n = 3$ to $n = 4$ electronic level. (ii) What is the wavelength of the Paschen α photon? (iii) In which region of the electromagnetic spectrum does the Paschen α photon lie?

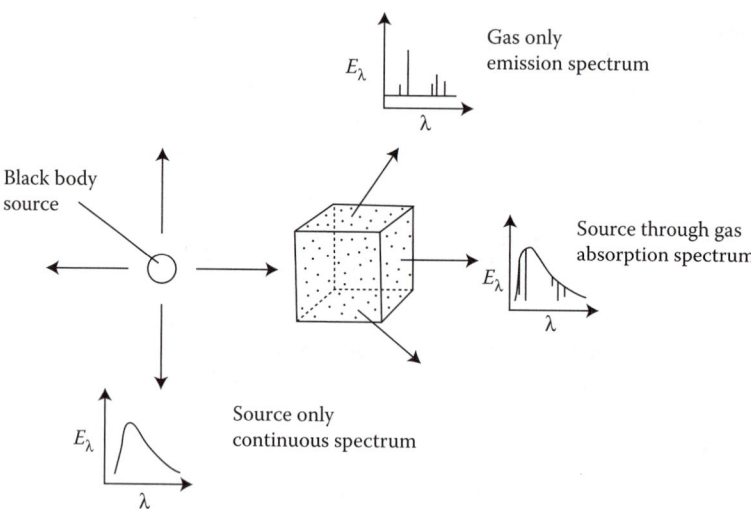

FIGURE 1.5 Different spectra. Viewing a black body source directly we see a continuous spectrum. Looking at such a source through interposed gas gives an absorption spectrum. The gas itself yields an emission spectrum.

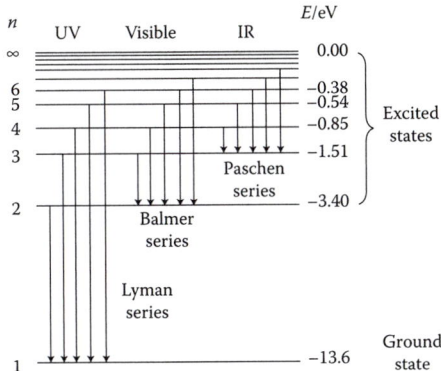

FIGURE 1.6 Spectral line series for the hydrogen atom.

1.3.4 Spectrographs Are Used to Capture Spectra

A spectrograph is used to disperse light from distant sources. Light captured by a telescope is brought to focus as it passes through a narrow slit. It is reflected by a collimating mirror onto a diffraction grating. The grating has thousands of closely etched lines that disperse light by diffraction, the bending of light around obstructions. This effect can be reproduced by allowing a narrow beam of bright sunlight to illuminate a compact disc (CD). When the angle between the beam and observer is just right a continuous spectrum (rainbow) is seen. The light is dispersed into its colors because the long wavelength light is diffracted more than the short wavelength light. The dispersed light (the spectrum) is captured by a charge-coupled device (CCD) camera. Astronomical CCD cameras are not unlike those found in digital cameras except that they have larger CCD chips, are cooled to low temperature to reduce thermal noise, are built to higher specification to maximize efficiency, and are much more expensive!

Spectroscopy is not confined to visible light. Spectra can be obtained in x-ray, UV, IR, and radio frequencies. Molecules undergo electronic transitions that are seen in the visible and UV parts of the spectrum that indicate changes in the strengths of chemical bonds and hence the reactions they are engaged in. However, molecules also produce lines in infrared and microwave parts of the spectrum.

1.3.5 Stellar Spectra Encode Temperature and Elemental Abundances

By the nineteenth century, astronomers realized that the different colors of stars indicate their surface temperature, just as an iron poker heated in a fire goes from dull red through orange, yellow, and white as it gets hotter. Remarkably, between 1863 and 1867, which was before the advent of astronomical photography, Angelo Secchi examined the spectra of 400 stars and classified them into five groups based on color, the strength, number, and the distribution of spectral lines. It was the expansion and refinement of this scheme by Henry Draper and Annie Jump Cannon that gave us the Harvard spectral classification we use today.

1.3.5.1 Harvard Spectral Classification Is Based on the Strength of Absorption Lines

The spectra of stars fall into well-established categories based on the appearance and strength of particular absorption lines (see Figure 1.7 and Plate 1.1). Spectral classes are listed in decreasing order of surface temperature: O, B, A, F, G, K, and M. This was traditionally remembered with the mnemonic: "**O**h **Be A**Fine **G**irl **K**iss **M**e." (A modern "politically correct" version replaces "Girl" with "Girl/Guy" thereby rendering it gender and sexual orientation neutral!) Cannon subdivided each letter into 10 subclasses, 0–9, with 0 being the hottest, except for O, which is subdivided into 5–9 with 5 being the hottest. More recently, the discovery of cooler stars ($T < 3000$ K), which give out most of their radiation in the infrared, has added classes L, T, and Y to the end of the Harvard classification. Some L and all T and Y are brown dwarfs (see Section 1.4.1.1).

From Figure 1.7 we can see clear trends in the strengths of key spectral lines. The sparse spectra of O and B stars are dominated by helium lines. Hydrogen lines are strongest for A type and weaker for hotter and cooler stars. The cool K and M stars have abundant spectral lines of neutral metals (e.g., sodium, NaI) and molecules (e.g., titanium oxide, TiO).

Spectral lines encode information about temperature. To see how let us consider the H_α line. If a star is very hot, all its hydrogen will be ionized; so transitions from $n = 2$ to $n = 3$ cannot occur as there are no electrons available to make the transition! By contrast, if a star is too cool there will be no photons with energy high enough to excite the atom to make the transition. In between these two extremes, the H_α line will have intensities that reflect the distribution of photon energies and the proportion of hydrogen that is ionized. For A type stars the temperature is such that maximum number of hydrogen atoms are able to make the H_α transition. A single spectral line cannot give an unambiguous temperature because, except at the maximum, a given line strength will correspond to two temperatures. But with several spectral lines the ambiguity is removed (Figure 1.8).

The intensity of spectral lines is also sensitive to the abundance of an element, but care is needed in interpreting this. The sun has neutral Na and Ca lines that are more intense than the hydrogen lines but this does not mean that the sun has more Na and

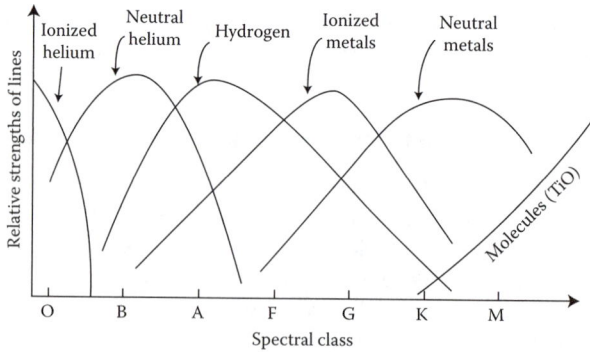

FIGURE 1.7 Stellar spectra. The strength of absorption lines depends on the temperature and hence the spectral class.

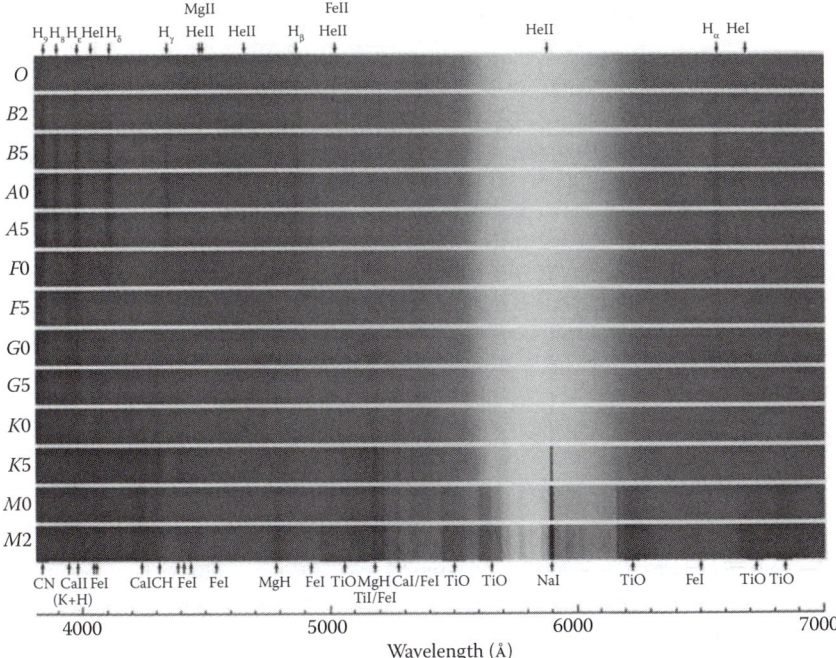

PLATE 1.1 (**See color insert.**) Stellar spectra for spectral classes of the star. (Photo credit: Spectral class/absorption feature diagram © Michael Briley.)

Ca in its atmosphere than hydrogen! It simply means that the temperature of the sun (5770 K) favors the transitions that generate the Na and Ca lines more than the H lines. In deducing stellar temperatures, astrophysicists use the lines of elements that have similar abundances from star to star.

The intensity and precise wavelengths of spectral lines are sensitive to other properties apart from temperature and composition. We explore these as the need arises.

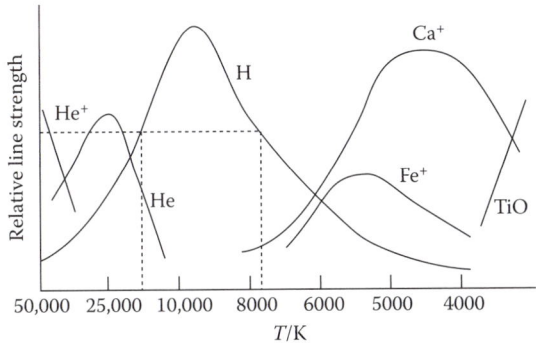

FIGURE 1.8 The strengths of absorption lines vary with temperature. Note that the same line strength for H corresponds to two temperature lines.

1.3.5.2 Luminosity Classes Are Based on the Sizes of Stars

The luminosity of a star is its power output, and it is usual to express this in relation to the luminosity of the sun to avoid cumbersomely large numbers. So, with the luminosity of the sun given by $L_\odot \sim 4 \times 10^{26}$ W, a star with a luminosity $L = 8 \times 10^{26}$ W would be described as having $L = 2L_\odot$. Astronomers use similar normalizations for stellar masses and radii.

The luminosity of a star depends on two factors. The first is its surface area, A_s, which varies with the square of its radius, R; namely, $A_s = 4\pi R^2$. The second is the power per unit area, the flux F, which is given by σT^4, where σ is the Stefan–Boltzman constant $(5.67 \times 10^{-8}$ J m^{-2} K^{-1} s$^{-1})$ and T is the effective temperature of the star; that is, the temperature of a black body that has the same luminosity and surface area as the star. Hence, we can write,

$$L = 4\pi R^2 \sigma T^4. \qquad (1.11)$$

An increase in radius or surface temperature increases a star's luminosity. That luminosity goes with T^4 might suggest that temperature is the most important determinant of changes to the intrinsic brightness of a star as it evolves, but this is not so. Toward the end of their lives stars undergo huge changes in radius but only relatively small alterations in surface temperature. Consequently, it is the change in the size of a star that has the most important effect on its luminosity.

Stars of the same spectral type can differ greatly in their luminosity. Since spectral class defines the temperature, Equation 1.8 tells us that this must be because they vary in size. Fortunately, spectral lines are broader in small stars than in the big ones. This is termed "pressure broadening" and provides the basis for luminosity classes, which subdivide stars on the basis of size (Table 1.1).

Pressure broadening happens because smaller stars have a higher gas density and pressure in their atmosphere. This increases the frequency of interactions between atoms in smaller stars compared to that in larger stars. Interactions distort the atomic energy levels causing random fluctuations in the energies of electronic transitions, and hence in the wavelengths of absorbed photons. This broadens the spectral lines for smaller, less luminous stars.

It is worth noting the effect that radius can have. Supergiants can have luminosities of the order of 10^6-fold that of the solar luminosity, despite being cool M-type

TABLE 1.1

Luminosity Classes

Luminosity Class	Description
Ia	Bright supergiants
Ib	Supergiants
II	Bright giants
III	Giants
IV	Subgiants
V	Dwarfs (main sequence)
VI	Subdwarfs
VII	White dwarfs

stars, while white dwarfs (which do not fall neatly into the Harvard classification) are extraordinarily faint ($<10^{-3}L_\odot$) because they are only about the size of the Earth, despite having surface temperatures of 10,000–25,000 K.

1.3.5.3 A Plot of Luminosity against Temperature Is a Hertzsprung–Russell Diagram

By about 1910 the distances to a number of stars had been determined allowing their absolute magnitudes (the brightness a star would have if 10 pc away) to be found. Moreover, it was becoming clear that stars of identical spectral class differed in the widths of their spectral lines, which we now know to be due to pressure broadening. The question arose as to whether there was any systematic relationship between the absolute magnitude and the spectral class. This was studied independently by Ejnar Hertzsprung and Henry Norris Russell, who in 1912 and 1913, respectively, plotted graphs of absolute magnitude versus spectral class for several hundred stars. They discovered that stars fell into two bands: the main sequence—so called because it contained most stars—sloped diagonally from hot luminous stars in the top left to cool faint stars in the bottom right, while a giant branch across the top comprised very luminous stars in all spectral classes.

Plots of absolute magnitude versus spectral class, or its theoretical equivalent, luminosity versus temperature, are now the mainstay of astrophysics research and termed Hertzsprung–Russell (H–R) diagrams (Figure 1.9).

H–R diagrams are a snapshot of stellar populations. They show that at any given instant of time sampled by the plot, 90% of stars fall into the main sequence while relatively few are found in the giant branch or in the bottom left where hot, but very dim white dwarfs lurk. The implication is that stars spend most of their time on the main sequence and relatively little time as giants. The paucity of white dwarfs is due to a selection bias (they are

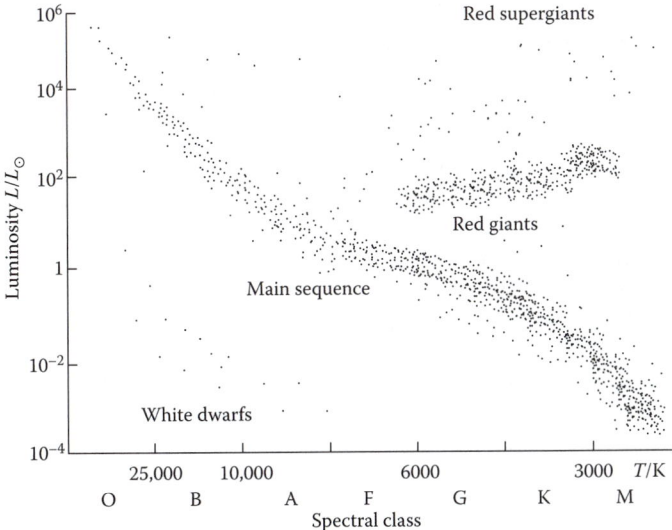

FIGURE 1.9 Hertzsprung–Russell diagram. Note the unconventional direction of the *x*-axis in these plots, with temperature increasing to the left!

extremely faint and so very hard to detect, and the majority of stars have yet to become white dwarfs) and not because stars spend little time as white dwarfs. H–R diagrams are very powerful. They can be used to track the evolution of stars, estimate stellar ages, and estimate luminosities so as to allow distances to be calculated.

1.4 Stellar Evolution

Stars are born, live their lives, and die. Crucially, stars maintain their power output by nuclear fusion, transmuting hydrogen nuclei to heavier nuclei. Indeed, all the elements other than those produced by Big Bang nucleosynthesis have been created inside stars. The surface temperature of a star, how long it will live, and how it will die depend on just one variable—its mass at birth as it joins the main sequence. Massive stars are hot, live life in the fast lane, and die young, whereas low mass stars are cool and can have lifetimes longer than the current age of the universe. Stars with masses $>8M_\odot$ end their days in catastrophic explosions termed "core collapse supernovae." Lower mass stars end their lives as white dwarfs. But whatever their fate, all stars lose much of their mass to the interstellar medium (ISM), the thin gruel of gas and dust between the stars. In this way, the ISM is enriched by elements made in stars. Moreover, the ISM is the raw material from which new stars and planets are formed. This means that over billions of years successive generations of stars have become richer in elements other than hydrogen and helium. Such elements (i.e., those with $Z > 2$) are referred to by astronomers as metals. (This use should not be confused with the way in which metals are defined by physicists or chemists.) The abundance of metals in the stars (Box 1.4) is an important variable in astrobiology as we shall see.

BOX 1.4   METALLICITY

The chemical composition of any region of the interstellar medium (ISM) or a star can be summarized by three parameters: the mass fraction of hydrogen (X), the mass fraction of helium (Y), and the mass fraction of all the elements heavier than helium (Z). A mass fraction of a component is simply the mass of the component divided by the total mass. Clearly $X + Y + Z = 1$. For most astrophysical locations except the deep interior of stars, $X = 0.74$ and $Y = 0.24$. So, this leaves the mass fraction of all the elements in the Periodic Table from lithium through uranium, $Z = 0.02$.

As we have seen astronomers, in defiance of the usual rules of chemistry, regard all elements other than H and He as metals. Although they make up only about 2% of the mass of baryonic matter, they have important effects on the properties of stars, and of course, without them there would be no planets or life! Hence, it is often useful to know Z. This is termed metallicity.

However, it is almost impossible to know the complete inventory of metals in a star or a region of the ISM, so the definition of Z given above is not practical. Consequently, astronomers resort to defining metallicity on the basis of a single metal, the abundance of which can be determined by spectroscopy. Because of its high abundance and ease of measurement iron is often the metal of choice. Then,

$$Z = \left[\frac{Fe}{H}\right] = \log_{10}\left[\frac{(Fe/H)_*}{(Fe/H)_\odot}\right] = \log_{10}\left[\frac{Fe}{H}\right]_* - \log_{10}\left[\frac{Fe}{H}\right]_\odot. \qquad (1.12)$$

The square brackets in [Fe/H] flags that this is a log value. It is the ratio of iron to hydrogen in a star normalized to the iron/hydrogen ratio in the sun. For the sun, $Fe/H = 1/20,000 = 4 \times 10^{-5}$. So $[Fe/H]_\odot = \log_{10}(4 \times 10^{-5}) = -4.3$.

EXAMPLE 1.1

If a star has $Fe/H = 1/30,000 = 3.3 \times 10^{-5}$, then it has a metallicity $Z = [Fe/H] = \log_{10}(3.3 \times 10^{-5}) - (-4.3) = -4.8 + 4.3 = -0.18$.

EXAMPLE 1.2

What is the metallicity of a star that has one iron nucleus for every 15,000 hydrogen nuclei? Now, $Fe/H = 1/15,000 = 6.67 \times 10^{-5}$. So, $[Fe/H] = \log_{10}(6.67 \times 10^{-5}) - (-4.3) = -4.176 + 4.3 = +0.124$.

Hence, we see that stars that have a lower metallicity than the sun have $Z < 1$, while stars that are metal-rich by comparison with the sun have $Z > 1$.

1.4.1 The Properties of Main Sequence Stars Are Determined by Their Masses

A star joins the main sequence (MS) when its temperature is high enough that it can fuse hydrogen nuclei to helium nuclei in its core. This is the defining characteristic feature of an MS star. A star will remain on the main sequence for most of its lifetime.

Some properties of MS stars are summarized in Table 1.2. With the exception of the proportions of stars in each spectral class, all the other properties are determined by the initial mass of the star. Let us see how. Protostars grow by accreting gas from their surroundings. As the gas falls inward, its gravitational potential energy (E_G) is converted to kinetic energy (E_K) and it heats up. Eventually, the temperature in the core of the star is high enough for nuclear fusion to happen, and this maintains its energy output. At this stage, the star settles into hydrostatic equilibrium in which inwardly directed gravitational forces acting to collapse it are balanced by the gas and radiation pressure acting outward. This determines the radius. The core temperature at which this balance is achieved depends on the mass; the greater the mass, the more extreme the compression of the star and hence the higher its central temperature and pressure. So, the more massive a star is, the hotter it is, and its surface (effective) temperature determines its color. Its temperature and radius together determine its luminosity.

Nuclear reaction rates increase with temperature and density, so the gas and radiation pressures are correspondingly higher for more massive stars. This staves off stellar collapse, but at the expense of a higher rate of hydrogen fusion. Although more massive stars start out with more core hydrogen, the increase in fusion rate rises with mass much faster than can be compensated for by the additional fuel. This is because nuclear reactions are very sensitive to temperature. Even a small increase in core

TABLE 1.2

Harvard Spectral Classification for Main Sequence Stars

Class	Average Temperature (K)[a]	Apparent Color	Mass ($M_\odot$)	Radius ($M_\odot$)	Luminosity ($L_\odot$)	Proportion (%)[b]
O	30,000	Blue	>16	>6.6	>30,000	0.0003
B	17,000	Blue-white	2.1–16	1.8–6.6	25–30,000	0.13
A	9000	White to blue-white	1.4–2.1	1.4–1.8	5–25	0.6
F	7000	White	1.04–1.4	1.15–1.4	1.5–5	3
G	5500	Yellowish-white	0.8–1.04	0.96–1.15	0.6–1.5	7.6
K	4500	Yellow-orange	0.45–0.8	0.7–0.96	0.08–0.6	12.1
M	3000	Orange-red	<0.45	<7	<0.08	76.45

[a] These are approximate surface (photosphere) temperatures for the middle of each class.
[b] Percentage of main sequence stars that fall in the class.

temperature produces a dramatic boost in the reaction rate. The result is that the more massive a star, the shorter its lifetime.

1.4.1.1 Brown Dwarfs Are Not Massive Enough to Become Stars

The lowest mass a star can have is well defined by quantum physics. It is $0.08M_\odot$ (about 80-fold the mass of Jupiter, $80M_J$). Below this mass the core temperature is not high enough to initiate hydrogen-1 fusion. However, objects below this mass, named brown dwarfs by Jill Tartar, can form in a way similar to stars. For a short period they can sustain themselves against gravitational collapse by deuterium fusion, which happens at lower temperatures and density than hydrogen-1 fusion. But the amount of deuterium is so low that this cannot continue for more than a million years or so, after which these "failed stars" slowly contract, generating feeble amounts of energy by converting gravitational potential energy to kinetic energy.

Brown dwarfs range from $80M_J$ to $13M_J$ and fall into spectral classes L, T, and Y. L-dwarfs have temperatures around 2000 K, cool enough that dust condenses in their atmospheres. This dust scatters the outgoing light, making them appear very red. The most massive L-dwarfs can sustain hydrogen fusion so are very cool main sequence red dwarfs. Lower mass L-dwarfs are brown dwarfs. The very cool ($T \sim 1000$ K) T-dwarfs have prominent absorption lines of methane, which makes them look blue-green in color like Uranus and Neptune (which have CH_4 in their atmospheres). This is because methane absorbs red light strongly. Y-dwarfs have $T < 600$ K and have absorption lines due to ammonia. The coolest Y-dwarfs have a surface temperature of just 300 K (25°C) and modeling suggests that rain may fall in their atmospheres.

1.4.1.2 The Boundary between Brown Dwarf and Planet Is Hard to Define

The $13M_J$ lower limit for brown dwarfs comes about because the standard story is that below this mass no core fusion, even of deuterium, is possible. So we ought to regard a ball of gas with $M < 13M_J$ as a planet. This is the view of the International Astronomical Union (IAU), the committee that has the task of defining and naming astronomical bodies. But things are not that simple. One reason is that the mass limit for core fusion is influenced by a star's metallicity, [Fe/H]. It could be as low as $11M_J$ for stars with zero or negative [Fe/H] and up to $16.3M_J$ for metal-rich stars with [Fe/H] up to 3-fold solar. Some researchers argue for an upper limit between planets and brown dwarfs as high as $25M_J$.

Another reason is that planets are supposed to form from an accretion disc around a star, whereas stars arise by collapse and fractionation of a giant molecular cloud (see below). However, the discovery of sub-brown dwarfs, objects with masses between $13M_J$ and $1M_J$ (i.e., giant gas masses) that form *like stars*, has muddied the waters. We know that planets can be ejected from stellar systems, and sub-brown dwarfs could be captured by stars, so the formation process cannot always be known. Not surprisingly, astronomers argue about whether the formation process is a useful way to define what is meant by a planet.

1.4.1.3 The Upper Limit for Stellar Masses Is Not Well Defined

Observational evidence suggests an upper limit to stellar mass of $\sim150M_\odot$.

1.4.2 Stars Form by the Collapse of Giant Molecular Clouds

Giant molecular clouds (GMCs) make up the coldest and densest clumps of the interstellar medium. They are located in the thin discs of spiral galaxies, generally associated with spiral arms. The "giant" epithet of GMCs is well earned given that they can be from 1 pc up to 100 pc across and have masses ranging from 10 to $10^6 M_\odot$. They are mostly hydrogen and helium gases, but about 2% by mass are metals (elements with $Z > 2$). Because GMCs are extraordinarily cold, 10–20 K, much of the hydrogen is present as H_2 molecules, and the metals are in the form of dust grains, largely silicates and elemental carbon. Low melting point compounds (e.g., H_2O, CO, CO_2, CH_4, and NH_3) are frozen around the dust grain.

Giant molecular clouds have a hierarchical density structure. While most of their volume are diffuse clouds transparent to visible light, they harbor dense clouds, which themselves contain even denser dense cores, that are opaque to visible and UV wavelengths because of their high density of dust grains. As a result, dense clouds in GMCs obscure the light from distant stars and so appear in the sky as dark nebulae. Dense clouds are potential stellar nurseries.

Once the first stars have been born in a GMC, their UV light ionizes the hydrogen gas in their neighborhood forming HII regions (Box 1.5). As electrons recombine with protons and cascade down the Balmer series, the hydrogen atoms emit visible light photons. This processing of UV radiation into visible light is called fluorescence, and it makes HII regions emission nebulae. The most spectacular emission nebula associated with a GMC because of its closeness to us is the Orion nebula (M42). Small (~1 pc), cold (10 K), dense molecular clouds that resist the effects of UV can be seen silhouetted against HII regions. These surviving remnants of the GMC are called Bok globules, and there is evidence that star formation can occur within them.

BOX 1.5 ASTRONOMICAL NOMENCLATURE FOR IONIZATION STATES

In chemistry, a neutral hydrogen atom is depicted by its chemical symbol, H. This is the case for all elements. Thus Fe signifies an iron atom. The hydrogen ion is H^+, while singly and doubly ionized helium are He^+ and He^{2+}, respectively. This rule works for any other ions; so singly ionized oxygen is O^+, while iron that has lost 13 of its normal complement of 26 electrons is Fe^{13+}.

Astronomers often use an alternative nomenclature in which neutral atoms are designated with a Roman numeral I, for example, HI, HeI or OI. In this system, ionized hydrogen is HII and singly ionized helium and oxygen are HeII and OII, respectively. The Fe^{13+} ion would be written as FeXIV.

In this book, we use both systems as appropriate.

1.4.2.1 Cloud Collapse Can Be Modeled by Simple Physics

A dense cloud will be in equilibrium if the gas pressure exerted by the thermal energy of its constituent particles is balanced by the gravitational force exerted by its mass. If the gravitational force were to gain the upper hand, the cloud would contract.

EXERCISE 1.6

(i) What would you expect to happen to the temperature of a gas in a self-gravitating cloud? (ii) Consequently, what will happen to the volume of the gas?

Clearly, for a cloud to collapse under self-gravity it must be able to radiate so that it does not heat up and/or be so massive that gravity overcomes thermal pressure. In the early 1900s, James Jeans modeled this balance of forces in a spherical, uniform, nonrotating cloud so as to quantify the conditions under which it would contract. He showed that for a given temperature and number density of particles, n, of average mass, m, there was a critical cloud mass above which contraction would occur. This is called the Jeans mass, M_j:

$$M_j = \frac{9}{4} \left(\frac{1}{2\pi n} \right)^{1/2} \times \frac{1}{m^2} \left(\frac{kT}{G} \right)^{3/2}. \tag{1.13}$$

So we can see that cloud contraction, in other words star formation, is favored by high particle number density and low temperature. Dense clouds and dense cores that exceed M_j are ripe for star formation.

Elegant though this model is, it is not the whole story, because the rate at which stars form is much lower than expected, given the number of molecular clouds around that exceeds the Jeans mass.

EXERCISE 1.7

What might account for the observation that star formation rate is lower than that predicted by the Jean's model?

1.4.2.2 Any Plausible Model of Star Formation Must Be Able to Explain Several Key Facts

Observations that have to be explained by any model of star formation include:

1. Stars form in clusters rather than in isolation.
2. More young stars exist in binary systems than older ones. This implies that binary systems get disrupted.
3. More high mass stars than low mass ones form binaries.
4. The majority of stars are low mass, dim, cool stars, while high mass, bright, hot stars are rare. Indeed recent observations suggest there may be as many brown dwarfs (L, T, and Y spectral classes) as there are stars (classes O–M). We can see this highly skewed distribution of stellar masses in Table 1.2. This distribution of stellar masses is termed the initial mass function (IMF), and one of the fundamental questions in astrophysics is what processes shape it.

We need to account for how dense clouds with masses up to a million solar masses collapse to form stars that mostly have masses $\sim 0.5 M_\odot$ and less, yet still generate some intermediate mass stars and a few very massive stars.

1.4.2.3 Fragmentation and Accretion Are Key Processes in Star Formation

Two processes seem to be at work in star formation. The first is fragmentation. When dense clouds collapse they fragment hierarchically; that is, large structures break up into smaller ones. At each stage, the fragment must equal or exceed the Jean mass if it is to continue to contract.

The second is accretion. A protostar forms from the inside-out as low angular momentum material in a dense core falls to the center. This makes the heart of the would-be star. High angular momentum material outside forms a flattened circumstellar accretion disc fed by an extended envelope. The protostar grows as material falls onto it from the accretion disc. Something must bring this accretion to a halt and, whatever it is, presumably regulates stellar masses and so the IMF.

High mass stars form in the core of the cluster as the gravitational potential of the entire system funnels gas toward its center. The effect is that accretion onto the stars there is faster and a few become very massive in a "winner-takes-all" process. The gas velocity near the center is high as a result of its acceleration by the gravitational field. This high velocity gas cannot be captured by low mass protostars but it can by the high mass ones. Collisions between stars at the center of star-forming clusters may account for the most massive stars. Low mass stars and brown dwarfs probably arise because low mass fragments are easily ejected from gas-rich regions by gravitational interactions, halting further accretion.

EXERCISE 1.8

If stars form by the competitive accretion mechanism outlined above what does this model predict for the distribution of stars in star-forming regions.

Several factors are thought to trigger star formation including shock waves from exploding high mass stars (supernovae).

1.4.3 Protostars Contract Down Onto the Main Sequence

1.4.3.1 Initial Protostar Contraction Is Isothermal

Initially, a dense core must contract isothermally if it is to become a protostar. If it contracts adiabatically (i.e., not exchanging energy with its surroundings), it will warm up and hence expand. This means it has to process the gravitational potential energy, E_G, in such a way that it does not heat up, or must lose heat on the same timescale as it collapses.

An isothermally contracting dense core uses much of its gravitational potential energy to dissociate hydrogen molecules and then to ionize the resulting hydrogen atoms:

$$H_2 \rightarrow H + H \quad \text{(requires 4.5 eV)}, \tag{1.14}$$

$$H \rightarrow H^+ + e^- \quad \text{(requires 13.6 eV)}. \tag{1.15}$$

Energy also goes to excite molecules (to higher rotation and vibration levels: see Chapter 2) and their emission carries away radiant energy. This is termed molecular line cooling. Although H_2 is not effective in molecular line cooling below 1000 K, CO, H_2O, HCN, and NH_3 are all capable of cooling efficiently at temperatures down to ~50 K. Another cooling process operates at lower temperatures in dense cores. Dense cores are opaque to UV and optical photons, but the gas within them can be heated by cosmic rays that do penetrate the dense cores. The warmed hydrogen molecules collide with cooler dust grains warming these so that they radiate in the infrared, to which the dense core is transparent. This is the only cooling channel for dense core gas that is effective below 50 K.

It is a combination of these mechanisms that allows isothermal collapse and fragmentation of cold dense cores ~1 pc across to protostellar cores.

1.4.3.2 Contraction onto the Main Sequence

Once a protostar has become opaque to its own radiation, gravitational collapse is opposed by thermal pressure, and the rate of contraction slows dramatically. The protostar collapses adiabatically and as it converts gravitational potential energy to kinetic energy and then to thermal energy, its surface temperature rises from ~100 to 4000 K. Although its surface area is getting smaller, the rise in temperature increases its luminosity so it tracks up and to the left on the H–R diagram. One motivation for studying protostars (young stellar objects, YSOs) is that later stages of star formation coincide with planet formation (Figure 1.10).

1.4.4 Main Sequence Stars Fuse Hydrogen to Helium

Once the core temperature is high enough, hydrogen nuclei will fuse to form helium nuclei. This defines a main sequence star. High densities and temperatures are required to ensure a high frequency of collisions with sufficient energy to overcome the mutual electrical repulsion of the positively charged protons. Nuclear fusion reactions in main sequence stars are exothermic. The energy generated maintains the core temperature and supplies radiation at a rate to match the loss from the surface. This maintains hydrostatic equilibrium.

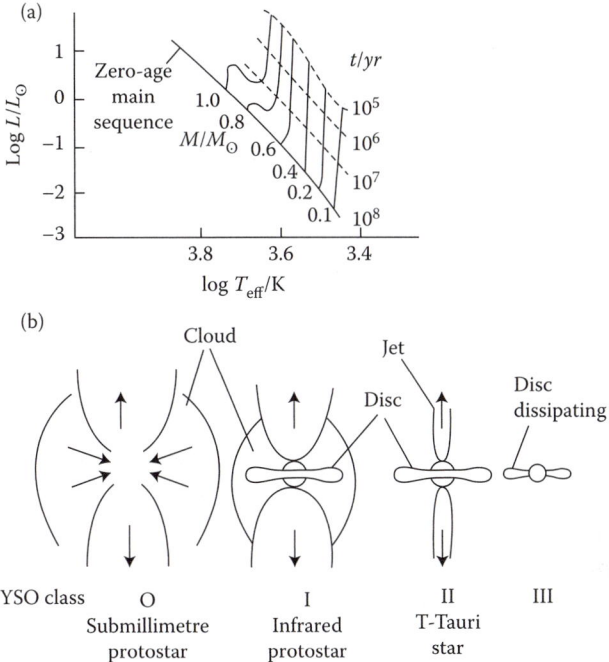

FIGURE 1.10 Star birth. (a) Structures of YSOs. (b) Track of protostar on the H–R diagram.

1.4.4.1 Proton–Proton (PP) Chains Are the Main Energy Yielding Fusion Reactions in Main Sequence Stars

In stars with $M < 1.5M_\odot$, the bulk of the energy is produced by fusion reactions called proton–proton chains. These start when the core reaches about 5×10^6 K. The overall reaction for the major PP interaction (PPI) chain is

$$4\,{}^1_1\text{H} \rightarrow {}^4_2\text{He} + 2e^+ + 2\nu_e + 2\gamma. \tag{1.16}$$

Note how the reaction conserves mass number and charge. It also conserves energy. The mass of four protons is slightly greater than that of the helium nucleus, positrons, and electron neutrinos together. This mass difference is converted to energy according to

$$E = mc^2. \tag{1.17}$$

The energy is released as γ photons. The positrons almost immediately annihilate with electrons with the release of further γ photons. The total energy liberated is 26.7 MeV (4.28×10^{-12} J) per helium nucleus formed.

The neutrinos have an extremely small mass and barely interact with matter so they leave the star unimpeded at nearly the speed of light.

The PPI chain occurs in a series of steps rather than as a single reaction as implied by Equation 1.23. (The probability of four protons with enough energy colliding simultaneously is vanishingly small!). Two further PPI chains also operate but they are quantitatively less important. A cycle of fusion reactions that fuses hydrogen nuclei to helium nuclei occurs in stars with core temperatures exceeding 15×10^6 K (comparable to that of the sun) and so is important in relatively few stars. It is termed the CNO cycle after the nuclei that catalyze the fusion.

EXERCISE 1.9

i. Calculate the amount of energy released during the main sequence lifetime, t_{MS}, of the sun (take $t_{MS} = 10$ Gyr and $L_\odot = 3.84 \times 10^{26}$ J s^{-1}).

ii. The mass of the sun is 2×10^{30} kg. Assuming all of this could be used in the PPI chain, what is the total amount of energy available in the sun during its main sequence?

iii. Estimate the efficiency of the PPI chain.

iv. Is the value you obtain higher or lower than you would expect given the assumptions in (ii)? Briefly explain your answer.

v. Use your answer to (i) to confirm the amount of energy produced by one PPI event. Take the mass of a $^4_2$He nucleus as 6.645×10^{-27} kg, the mass of a proton as 1.673×10^{-27} kg, and the mass of an electron/positron as 9.110×10^{-31} kg.

vi. What is the rate at which the PPI chain occurs in the sun assuming that the sun's power output is provided exclusively by the PPI chain.

vii. What mass of hydrogen is consumed each year by the PPI chain?

1.4.4.2 Stars Process γ Radiation to Lower Energy Radiation

The core of the star is opaque to the γ radiation released by hydrogen fusion. This means that photons are scattered by electrons and absorbed/emitted by nuclei many times. Consequently, it takes thousands of years for radiation generated in the core of a star to emerge from its surface. In addition the energy of the original γ rays is partitioned among the particles in the star to produce many more low energy photons. The result is that stars radiate almost all of their energy in UV, visible or IR wavelengths, depending on their surface temperature.

1.4.5 Many Low Mass Stars Become Red Giants

Stars leave the main sequence when they run out of hydrogen in their core. If they have masses exceeding $\sim 0.5 M_\odot$, they swell in size so their luminosity increases dramatically, despite a reduction in surface temperature, becoming red giants. A star will typically remain a red giant for about 10% of its main sequence lifetime (Figure 1.11).

1.4.5.1 Hydrogen Fusion Moves from Core to Shell

The stars slowly become more luminous while they are on the main sequence. This is a consequence of the change in the composition of the core as hydrogen is converted to helium. This has the effect that the core gradually contracts and heats up. Modeling this process tells us that the sun has increased its luminosity by 25% since it was formed 4.6 Gyr ago. Toward the end of the star's main sequence lifetime, increased radiation from the shrinking core triggers hydrogen fusion in a thin shell surrounding it.

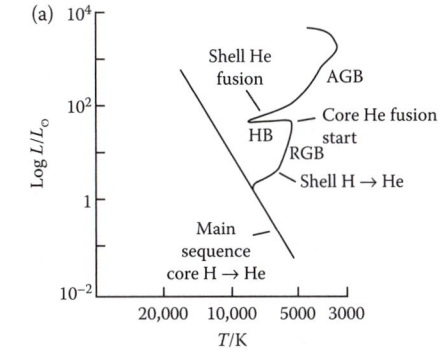

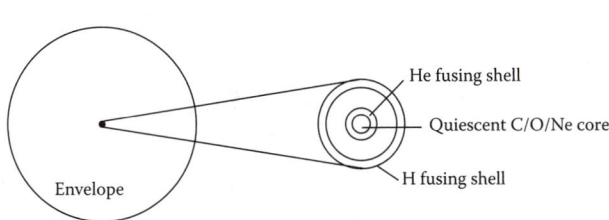

FIGURE 1.11   (a) Trajectories of $1M_\odot$ post–main sequence star. (b) Structure of low mass red giant.

EXERCISE 1.10

Why does the core of a star slowly contract while it is on the main sequence?

Eventually core hydrogen is completely exhausted, marking the star's departure from the main sequence. Failure of core fusion means that there is no longer an energy supply to oppose gravity. The core now contracts rapidly, increasing its temperature. This keeps shell hydrogen fusion going, and the helium it generates is dumped into the core, increasing its mass, thereby increasing its gravitational potential energy and its temperature. Consequently, the hydrogen burning shell marches outward, causing the outer layers of the star to expand so that its radius increases 10-fold. The surface temperature of the star falls to about 3500 K and its color becomes red; it has become a red giant. This takes the star up the red giant branch (RGB) on the H–R diagram. (Note that it is commonplace for astrophysicists to describe fusion as "burning" even though this is usually taken to mean combustion, which of course is a chemical reaction, not a nuclear reaction.)

1.4.5.2 *Core Fusion Reactions Generate Carbon and Oxygen in Red Giant Stars*

The core continues to contract until it reaches 10^8 K whereupon helium fusion to carbon detonates at its center. This is termed the triple alpha (3α) process:

$$\begin{aligned} {}_2^4\text{He} + {}_2^4\text{He} &\rightleftharpoons {}_4^8\text{Be} \\ {}_4^8\text{Be} + {}_2^4\text{He} &\rightarrow {}_6^{12}\text{C} + 2\gamma \end{aligned} \tag{1.18}$$

This liberates only about 10% of the energy generated by hydrogen fusion to helium, which is why a star's red giant phase is so much shorter than its main sequence life. Another key nuclear reaction in all red giants is the synthesis of oxygen:

$$^{12}_{6}C + {}^{4}_{2}He \rightarrow {}^{16}_{8}O + \gamma. \tag{1.19}$$

The 3α process and nucleosynthesis of oxygen in red giants are the major sources of carbon and oxygen in the universe, and hence is crucial for life.

Core helium fusion raises the surface temperature of the star so it moves along the horizontal branch (HB) on the H–R diagram.

Subsequently, helium burning migrates to a shell around the new carbon core. This shell moves progressively further out as helium is consumed, causing the outer layers of the star to expand again, taking it to 100 times its original main sequence radius. The star now tracks up the asymptotic giant branch as its luminosity increases, and it cools.

1.4.5.3 Asymptotic Giant Branch (AGB) Stars Build Elements by Slow Neutron Capture

Except for the first stars to be born—rather paradoxically called population III stars—all stars contain tiny amounts of elements heavier than helium (see Box 1.4). These include the iron group elements ($_{24}Cr$, $_{25}Mn$, $_{26}Fe$, $_{27}Co$, and $_{28}Ni$) that act as the seeds for the nucleosynthesis of most of the elements beyond iron by neutron capture. There are two types of neutron capture, slow (the s-process) and rapid (the r-process). The s-process occurs in AGB stars, while the r-process happens during the explosions that end the lives of the most massive stars.

How does neutron capture work? Iron group elements have a fairly large cross section (i.e., they are a sizable target) for neutrons, which having no charge, do not have a large energy barrier to overcome. All this makes neutron capture by a nucleus rather easy. What happens next depends on the nucleus that has captured the neutron. If the new nucleus is stable then,

$$^{A}_{Z}X + n \rightarrow {}^{A+1}_{Z}X + \gamma. \tag{1.20}$$

Note that this has made a new isotope, not a new element. However, if the neutron capture results in an unstable nucleus it will decay radioactively, emitting a β particle (e^-):

$$^{A}_{Z}X \rightarrow {}^{A}_{Z+1}X + e^- + \underline{\nu}_e. \tag{1.21}$$

Here, $\underline{\nu}_e$ is an electron anti-neutrino. Essentially, the neutron has spat out an electron to become a proton and the result is a new element. Decay continues until a stable isotope is reached. Further neutron capture may then occur.

Slow neutron capture (s-process) happens where the number density of neutrons is relatively low (10^{11} m^{-3}) and at intermediate temperatures. This is the situation in the helium-burning shells of AGB stars. The s-process can synthesize about half of the isotopes heavier than iron ($^{56}_{26}Fe$), but cannot produce elements heavier than bismuth ($^{209}_{83}Bi$). This is because neutron capture times are much longer than β decay times; in other words, neutron capture is slow enough that radioactive decay occurs before there has been time to build ever more massive isotopes by capturing more neutrons. The starting point for the s-process is $^{56}_{26}$Fe and the timescale for individual s-process steps varies widely, from minutes to 10^6 years. The main source of neutrons in AGB stars are

$$^{13}_{6}C + {}^{4}_{2}He \rightarrow {}^{16}_{8}O + n, \qquad (1.22)$$

$$^{22}_{10}Ne + {}^{4}_{2}He \rightarrow {}^{25}_{12}Mg + n. \qquad (1.23)$$

Although the main site of the s-process is in AGB stars, high mass stars also produce light s-process elements during helium burning on the red giant branch.

1.4.5.4 AGB Stars Distribute Metals into the Interstellar Medium

The increased energy production in the interior of AGB stars has the effect of increasing the temperature gradient in the stellar envelope, making it convective. This brings carbon, oxygen, and s-process elements made in the thermal pulses to the surface.

AGB stars are large, so they have a low surface gravity and thus only a tenuous hold on their atmospheres. They are also unstable, undergoing long-period pulsations, and have powerful stellar winds. Consequently, the star can lose over half of its mass into an extensive circumstellar envelope (CSE) that can be several parsecs across. These are very high density regions of the interstellar medium (with number densities $\sim 10^{17}$ m^{-3} in their inner regions), but differ from dense molecular clouds in being warm rather than cold. They contain material dredged up from deep in the stellar interior.

At a sufficiently large distance from the star, the CSE is cool enough for refractory elements and minerals to condense as solids. AGB stars are the main sites for the production of interstellar dust. Because it is warm, the dust can be seen in infrared, but it obscures the light from late AGB stars.

The effective temperature of the AGB star increases as it contracts down. The last of its stellar envelope is detached by stellar winds, expands out to about 0.3 pc, and is turned into an emission nebula by UV radiation from the star's core. These are called planetary nebulae (PNe) because of their resemblance to gas giants, such as Uranus, seen through a modest size telescope. They are beautiful objects shaped by successive episodes of mass loss and magnetic fields, but are ethereal, dissipating into the interstellar medium (ISM) in just a few thousand years, and enriching it with the elements forged in the star. The ISM is the raw material from which the next generation of stars and planets may be made.

Finally, all that remains is the star's hot (100,000 K), dense, degenerate core made of some combination of neon, oxygen, carbon, and helium (depending on the original mass of the star), with (typically) a radius about the size of the Earth and a mass of $\sim 0.6 M_\odot$. This is a white dwarf, which gradually cools over billions of years.

1.4.5.5 Very Low Mass Stars Do Not Alter the Interstellar Medium

Stars less than half the mass of the sun (red dwarfs) cannot burn helium. When they run out of core hydrogen, their cores contract but they never get hot enough to ignite helium fusion. They may ignite shell hydrogen burning to produce a modest increase in size and luminosity; however, this is short lived. These stars just fade away.

1.4.6 High Mass Stars Make High Mass Elements

Stars greater than $1.5M_\odot$ are able to carry nuclear fusion further than we have explored so far. Moreover, stars greater than about eight solar masses end their lives not as planetary nebulae and white dwarfs, but in catastrophic supernova explosions during which the heaviest elements are forged and most of a star's mass is flung into the interstellar medium.

1.4.6.1 Red Supergiants Can Forge Elements up to Iron

The greater the mass of a star, the more nucleosynthesis reactions it is able to do. It does this in a series of steps. As the core exhausts one fuel, it contracts while gravity is winning the battle against radiation and thermal pressure, only to stabilize temporarily at a higher temperature when another set of fusion reactions become possible. How far along the sequence a star can get depends on its mass. We list of few key reactions below. In general the energy yield for successive steps gets smaller and the reaction rates are faster because of the progressively increasing temperature. The outcome is that each burn is shorter than the previous one.

Once a red supergiant has exhausted its core helium, the core contracts and its temperature rises to 5×10^8 K, igniting carbon burning that creates elements with mass number around 20:

$$^{12}_{6}\text{C} + {}^{12}_{6}\text{C} \rightarrow {}^{23}_{11}\text{Na} + \text{p}, \tag{1.24}$$

$$^{12}_{6}\text{C} + {}^{12}_{6}\text{C} \rightarrow {}^{20}_{10}\text{Ne} + {}^{4}_{2}\text{He}. \tag{1.25}$$

The proton and the helium nucleus participate in further fusion reactions. For example, the fusion of helium with carbon makes oxygen (see Equation 1.19).

With core carbon used up after about 600 years the core undergoes another phase of contraction and heating, attaining a temperature of ~10^9 K at which neon burning takes place:

$$^{20}_{10}\text{Ne} + {}^{4}_{2}\text{He} \rightarrow {}^{24}_{12}\text{Mg} + \gamma. \tag{1.26}$$

After neon burning (which lasts a year) and a further round of core contraction and heating, oxygen burning occurs:

$$^{16}_{8}\text{O} + {}^{16}_{8}\text{O} \rightarrow {}^{28}_{14}\text{Si} + {}^{4}_{2}\text{He}. \tag{1.27}$$

Oxygen is consumed in about 180 days and core contraction takes the temperature to 3×10^9 K, at which the γ photon energies are so high that photodisintegration reactions become significant. These yield helium nuclei (α particles):

$$^{28}_{14}\text{Si} + {}^{4}_{2}\text{He} \rightarrow {}^{32}_{16}\text{S} + \gamma, \tag{1.28}$$

or in abbreviated form $^{28}_{14}\text{Si}(\alpha,\gamma)^{32}_{16}\text{S}$.

The importance of the helium nuclei is that they are able to participate in a series of fusion reactions, which build elements up to the iron group that cluster around $^{56}_{26}\text{Fe}$

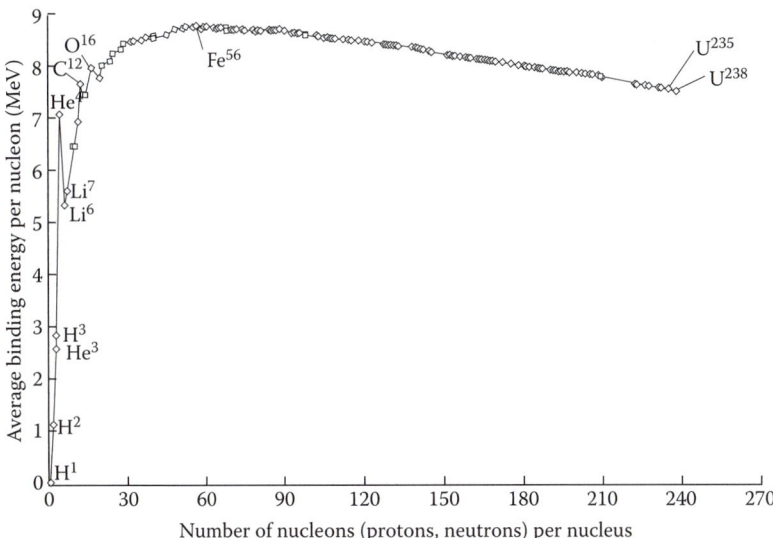

FIGURE 1.12 Nuclear-binding energy. The energy released per nucleon when a nucleus is formed as a function of mass number, A. Note the plateau of iron group elements around $A = 56$ that are the most tightly bound.

in the Periodic Table. Continuing from Equation 1.28 this goes (linking successive reactions):

$$^{32}_{16}S(\alpha,\gamma)^{36}_{18}Ar(\alpha,\gamma)^{40}_{20}Ca(\alpha,\gamma)^{44}_{22}Ti(\alpha,\gamma)^{48}_{24}Cr(\alpha,\gamma)^{52}_{26}Fe(\alpha,\gamma)^{56}_{28}Ni. \qquad (1.29)$$

This is the end of the line for energy production. For nucleosynthesis up to this point the mass of the resulting nucleus is slightly lighter than the combined masses of the constituent protons and neutrons. The difference is the nuclear binding energy that is released as γ photons during fusion (Figure 1.12). This is what makes nuclear fusion exergonic (energy releasing).

Nuclear fusion beyond nickel is endergonic—it *requires* energy—because for nuclei heavier than nickel nuclear binding energy (which is short-range) is overcome by the long-range electrical repulsion between positively charged protons. (Beyond nickel it is *fission* reactions that are exergonic.) Hence, once a massive star has fused silicon to iron and nickel in its core, further fusion is impossible without the input of energy. This is unstable because it removes radiation and thermal pressure support against gravitational collapse; the star is now seriously out of hydrostatic equilibrium. Silicon burning is only reached in stars of more than eight solar masses and lasts just a day. It is the star's last day (Figure 1.13).

1.4.6.2 Core Collapse Supernovae Forms r-Process Elements

Silicon burning makes an iron–nickel core with no further prospect of generating energy by fusion. Unable to resist its own gravity the core collapses in a timescale of minutes, with spectacular results.

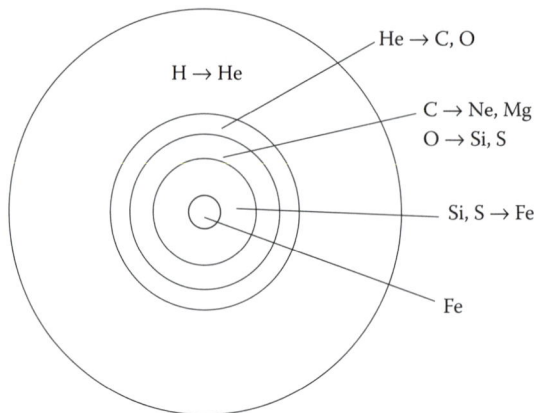

FIGURE 1.13 Structure of the center of a red supergiant during silicon burning.

As the core collapses the temperature rises to 10^{10} K; extremely energetic γ photons photodissociate the iron into individual protons, neutrons, and electrons. Because the core is larger than the critical mass called the Chandrasekhar limit (1.4 solar masses), its collapse cannot be halted by electron pressure that holds up white dwarfs. In fact, the electrons in the core combine with protons to form neutrons: a process that generates huge numbers of neutrinos. Collapse is only halted when the core has become a solid ball of neutrons. At this stage the outer layers of the star are falling inward at velocities of around 70,000 km/s. When they crash into the neutron core they rebound, setting up a shock wave that races back into the in-falling material (most of it hydrogen and helium), heating it to 10^{10} K. This, plus the high neutron flux, detonates an orgy of nuclear reactions, including rapid neutron capture (r-process) reactions. This synthesizes elements up to uranium, and blasts the star apart in a stupendous explosion, a type II or Ib/c core collapse supernova (SN).

The r-process occurs in environments where the number density of neutrons is high (10^{22}–10^{30} m^{-3}), the flux of neutrons is high (10^{26} m^{-2} s^{-1}), and the temperature is high (~10^{9} K): typically core collapse supernovae (SNe) or colliding neutron stars. It is responsible for synthesizing about half of all isotopes heavier than $^{56}_{26}$Fe. Neutron capture times are shorter (10^{-4}–10^{-3} s) than the time for β decay; that is, neutron capture is so rapid that there is no time for radioactive decay between successive capture events, and this allows elements to be built beyond $^{209}_{83}$Bi. The starting point is $^{56}_{28}$Ni. The r-process lasts only a few seconds and is terminated when $A \sim 270$, when elements undergo spontaneous fission to less massive isotopes. The heaviest element on the Earth, with a half-life (4.5 Gyr) about the current lifetime of the solar system, is $^{238}_{92}$U. Heavier elements on the Earth, for example, plutonium ($_{94}$Pu), are synthesized in nuclear reactors, since their isotopes have lifetimes far too short for any to have survived from stellar nucleosynthesis.

The abundance of r-process elements is low. It is currently thought that some supernovae are not efficient at distributing r-process elements, perhaps because they are swallowed up by the compact objects—neutron stars (pulsars) or black holes—that are the end products of core collapse.

1.4.6.3 Supernova Remnants Seed the Interstellar Medium with Elements

Most of the stellar envelope ($5–10M_\odot$) is ejected as a supernova remnant (SNR), expanding shells of gas that slam into the surrounding interstellar medium at speeds of up to 10^4 km s^{-1}. This creates shock fronts where the gas is compressed and heated to 10^5 K. Light from this shocked gas dominates the peak of the light curve of the SN.

Substantial amounts of iron group elements are made during a core collapse supernova. In fact, it is the radioactive decay of ^{56}Ni to ^{56}Co and finally to stable ^{56}Fe that dominates the later light output from supernovae. Core collapse supernovae are the most important source of elements heavier than neon in the ISM.

2

The Chemistry of Space

2.1 From Elements to Molecules

Elements created by stellar nucleosynthesis can combine to form molecules in a variety of locations: the circumstellar envelopes and planetary nebulae around old, cool stars; the interstellar medium (ISM); and the circumstellar discs around protostars out of which planets can form. These locations are not independent. Dying stars shed their envelopes into the ISM, and this becomes the raw ingredients for the next generation of stars and protoplanetary systems, but at each stage, chemistry forges many compounds anew.

The study of this chemistry in space is variously described as astrochemistry, cosmochemistry, or molecular astrophysics. In this chapter, we explore the conditions under which chemistry can work in astrophysical environments and discuss some of the most important reactions. Along the way we see how molecules can be detected in space. Cosmochemistry can provide many of the small molecules that act as reactants for more complex biochemistry.

2.2 Astrochemical Environments

2.2.1 Cool Stars Have Molecular Absorption Lines

The atmospheres of many stars are sufficiently cool that a variety of the molecules can survive there. These are revealed by characteristic absorption lines in stellar spectra.

Surprisingly, molecules have been found in the photosphere of the sun. For example, H_2 molecules have been detected in cool sunspots, which have temperatures as low as 3000 K. But it is the cool M-type stars that have by far the largest inventory of molecules, as the richness of their optical absorption spectra shows. They include metal hydrides (MgH, CaH, CrH, CoH, and NiH), metal oxides (mainly TiO), water and hydroxyl (OH) molecules, N_2, CO, CO_2, and simple organic molecules (e.g., HCN, C_2H_2).

2.2.2 The Interstellar Medium Is Extremely Tenuous

The interstellar medium is the matter between the stars. In terms of number density it is about 90% hydrogen, 9% helium, and 1% heavy elements. Compared with the densities we are familiar with on the Earth, the densities encountered in the ISM are extraordinarily low, and the temperature ranges from close to absolute zero to several million degrees (Figure 2.1).

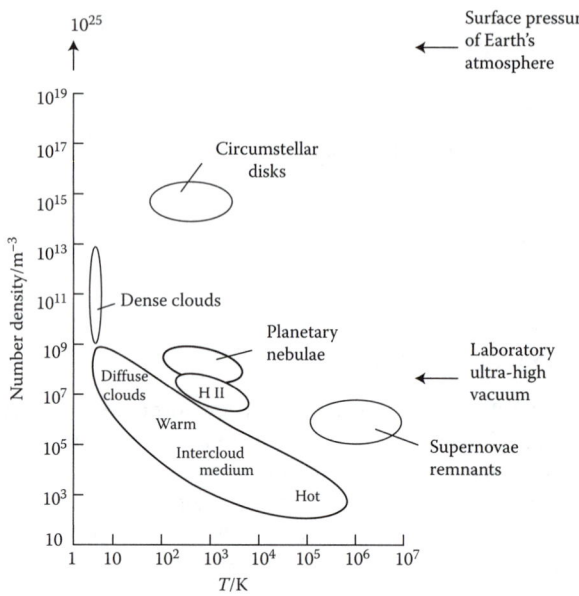

FIGURE 2.1   The number density of particles, n, and the temperature characterizes the interstellar medium.

2.2.2.1 Warm–Hot ISM Pervades Most of the Space between the Stars

Not surprisingly, the warmest fractions of interstellar medium (the warm–hot inter-stellar medium, WHIM) have the lowest density, so they make up by far the greatest volume of the ISM even though it has only a modest amount of mass (see Table 2.1). Its low density makes it transparent so it does not obscure anything lying within or

TABLE 2.1

The Make-Up of the ISM

Type of Region	Fraction of the ISM (%)		Typical Size (pc)	Typical Mass (M)	Predominant Form of Hydrogen	Abundance of Molecules
	By Volume	By Mass				
Hot intercloud medium	~60	≤0.1	–	–	H^+	Very low
Warm intercloud medium	~30	~20	–	–	H^+ or H	Very low
Diffuse clouds	~3	~30	~3 to ~10	1–100	H or H_2	Diatomic molecules common
Dense clouds	≤1	~45	~0.1 to ~20	$1–10^4$	H_2	Molecules common, even large ones
HII regions	~10	~1	~1 to ~20	$10–10^4$	H^+	Very low

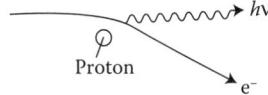

FIGURE 2.2   The hot ISM converts electron kinetic energy into x-rays by free–free radiation, also referred to as bremsstrahlung or "braking radiation."

beyond it. However, we can still detect it. The hydrogen in coronal gas (hot inter-cloud medium) is fully ionized, but the UV/x-ray spectrum has lines which come from energy transitions of highly ionized species such as O^{5+}, N^{4+}, C^{3+}, and Si^{3+}. Coronal gas is also seen by its emission of diffuse x-rays. This is produced in a variety of ways by virtue of its million degree temperatures. One of them is when high energy electrons are decelerated as a result of deflection by the positive charge of an atomic nucleus; some of the kinetic energy of the electrons is converted into x-ray photons. This free–free radiation (so called because both particles remain free after the emission) has a continuous spectrum (Figure 2.2).

The warm intercloud medium has both HI and HII (H^+), so can be seen by emissions as electrons recombine with H^+.

2.2.2.2 Atomic and Molecular Hydrogen Dominates Diffuse and Dense Clouds

Embedded within the WHIM are diffuse clouds and the dense clouds nested within them, which together make up three-quarters of the mass of the ISM. These are the cold giant molecular clouds out of which new stars and planetary systems are born, and they trace out the spiral arms of the Galaxy (Plate 2.1).

The hydrogen in these regions is either atomic (H) or molecular (H_2). Cold atomic hydrogen can be detected because it emits radiation at 21 cm. This comes about because a hydrogen atom can have its electron and proton with their spins in the same direction (parallel) or in opposite directions (antiparallel). The parallel configuration has slightly more energy than the antiparallel one (the energy difference between the two states, $\Delta E = 9.4 \times 10^{-25}$ J), so if the atom goes from parallel to antiparallel state it emits a photon with that energy.

EXERCISE 2.1

Show that a photon with energy $E = 9.4 \times 10^{-25}$ J would produce radiation at 21 cm (take $h = 6.626 \times 10^{-34}$ J s).

The transition, formally termed the hyperfine atomic transition of neutral hydrogen, has a very low probability. In other words, the excited state has a long lifetime: a single isolated hydrogen atom will take ~11 Myr on average before it flips. The more energetic parallel spin state is regenerated by the collision of the hydrogen atom with another atom: collisional excitation. Despite the low transition probability, the huge number of hydrogen atoms in the ISM means that the 21 cm emission line is easily seen. The transition was predicted in 1944 and first detected in 1951. It is by system-atically mapping Doppler shifts in the 21 cm radiation that the spiral arm structure of the Milky Way has been revealed (Figure 2.3).

PLATE 2.1 (**See color insert.**) The spiral galaxy M74. The dark dust lanes are giant molecular clouds. Together with diffuse clouds they are the site of much chemistry. The overall blue color of the spiral arms is caused by young, hot stars. HII regions are seen in red H_α light. (Photo credit: NASA, ESA, and the Hubble Heritage (STScI/AURA)-ESA/Hubble Collaboration.)

The 21 cm line of neutral hydrogen is an example of a forbidden line. The transition takes so long that in high density environments such as a laboratory, collisions with other atoms or the walls of a container will excite or de-excite the atom long before it can make the transition. Thus, forbidden lines are only seen from the low densities encountered in the ISM. They are seen quite commonly in astrophysical settings and are denoted by square brackets notation; for example, [OII]. Forbidden lines are so called because, unlike permitted lines, they do not obey all of the quantum mechanics selection rules and this is what makes them such low probability events.

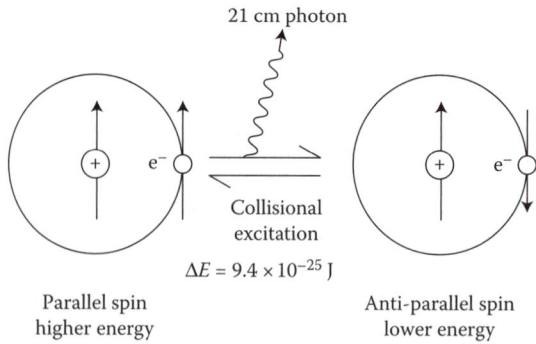

FIGURE 2.3 The spin-flip transition of HI allows it to be seen by emission (or absorption) at 21 cm.

EXERCISE 2.2

Calculate the frequency (in megahertz) of the hydrogen HI line.

Hydrogen molecules are impossible to detect in dense clouds, the coldest regions of the ISM, even though they are dominated by H_2. This is because it is too cold for the molecule to make any transitions. To map giant molecular clouds astronomers rely on markers that are codistributed with hydrogen. The most common proxy is carbon monoxide (CO), which emits at characteristic microwave frequencies when excited by collisions with hydrogen molecules. For example, after collisional excitation, ^{12}CO emits at 115 GHz ($\lambda = 2.6$ mm). This was first discovered in 1970 by Robert Wilson, Keith Jefferts, and Arno Penzias, the same Wilson and Penzias who discovered the cosmic microwave background radiation.

EXERCISE 2.3

What assumption must be made if H_2 in the ISM is mapped using CO as a marker?

2.2.3 The ISM Contains Dust Grains

About 1% by mass of the ISM is heavier elements (metals) and their compounds, usually in the form of dust grains that are mainly silicates or carbonaceous grains. Dust is responsible for extinction (scattering and absorption) of starlight over a range of wavelengths, and the size distribution of dust grains can be deduced from the extinction. Grain size ranges from 0.001 μm to a sharp cut-off at ~1 μm. However, many are about 0.2 or 0.4 μm across, comparable in size to the wavelength of blue light. Hence they scatter blue light more effectively than red light. This means that light passing through ISM dust is reddened. Dust grains also absorb UV light. This combination of scattering, reddening, and absorption it attenuates the transmission of short wavelength radiation. In fact, the densest regions of the ISM are so large they are opaque to UV and visible light, which is why they appear as dark nebulae at these wavelengths. Longer wavelengths in the infrared, microwave, and radio regions are transmitted much more readily than visible light, so the astronomers use IR and radio telescopes to see through dense ISM (Figure 2.4).

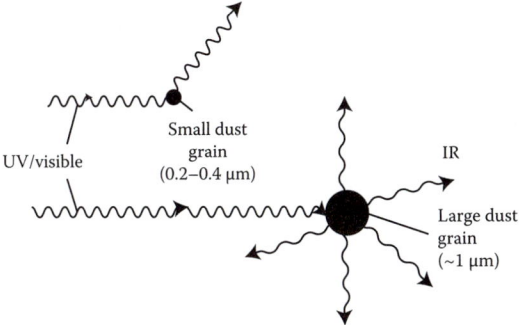

FIGURE 2.4 Extinction of starlight by dust grains depends on grain size distribution and composition.

EXERCISE 2.4

What color might you expect a cloud of gas and dust to be if seen by reflected starlight?

Dust grains that end up in diffuse and dense clouds ($T < 100$ K) acquire a mantle of ices. These include H_2O, CO, CO_2, NH_3 (ammonia), CH_3OH (methanol), and HCHO (formaldehyde). Some form in gas phase and condense out onto the grain because of the low temperature while others (e.g., methanol and formaldehyde) are actually synthesized on the surface of the grain and remain there as solids because sublimation temperature is higher than that of the cloud. At the mantle surface, further chemical reactions between the ices, driven by UV, are thought to generate a coating of complex organic molecules such as polycyclic aromatic hydrocarbons (PAHs). These are planar carbon compounds with three or more linked aromatic rings and no substituents. The simplest is anthracene. PAHs are lipophilic molecules (water solubility decreases with the number of rings), found in oil and tar and, since they are formed by the incomplete combustion of fossil fuels and fats, most of us have direct exposure to PAHs if we eat smoked fish or barbequed meat. When they were discovered in meteorites and comets, PAHs reached the attention of astrobiologists, but it is important to be aware that this is not evidence for biology; abiogenic synthesis of PAHs is a much more likely explanation than cooked organisms. About 300 weak, broad absorption lines in the IR, visible, and UV discovered in 1922 termed diffuse interstellar bands (DIBs) are probably due to PAHs (Figure 2.5).

Dust grains are critically important components in the ISM because:

- Their surfaces allow crucial chemical reactions to happen that would not work in gas phase.
- They shield molecules from destruction by absorbing the damaging UV light and reradiating it as IR.

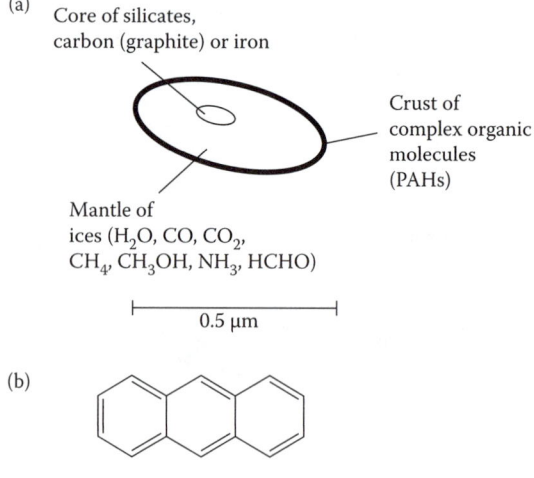

FIGURE 2.5   (a) Dust grains in diffuse and dense clouds are covered by a carapace of ice. (b) The polyaromatic hydrocarbon, anthracene.

- Diffuse and dense clouds are relatively transparent to IR radiation, so dust grains provide a cooling mechanism for the clouds as a whole.
- Warm dust (up to 100 K) heats molecules producing a small degree of ionization. This allows some ion–molecule chemistry and modest coupling of the cloud to ambient magnetic fields.

EXERCISE 2.5

Why is it important for giant molecular clouds to be able to radiate heat energy?

Dust grains can be eroded to produce molecules in gas phase by evaporation, sputtering (molecules ejected by high speed particles hitting the grain), and shattering (grain–grain collisions). The energy of the collisions is radiated away in the infrared.

Dust grains have distinctive spectral signatures in the infrared. Silicates have strong absorption lines at 9.7 and 18 μm, while carbonaceous grains have a strong line at 3.4 μm due to the vibration of the C–H bonds in organic molecules. The ice on the grains in dense clouds can also be detected; for example, lines due to the O–H in water and to CO_2 occur at 3.1 μm and 15.7 μm, respectively. These signatures have been used to detect and study protoplanetary discs, and so provide insights into the early stages of planet formation around other stars.

2.2.4 AGB Stars Have Either Oxygen- or Carbon-Rich Atmospheres

From a chemical perspective asymptotic giant branch (AGB) stars (those with main sequence masses between 1 and $8M_\odot$) either have oxygen-rich atmospheres or carbon-rich atmospheres (carbon stars). Convection in the stellar envelope brings the oxygen and carbon into the atmosphere.

2.2.4.1 Fusion Products Are Convected into the Atmospheres of AGB Stars

There are three brief periods when material is brought from the interior of a post–main sequence star to the surface (dredge-ups), by temporary deepening of convection in the stellar envelope. The first dredge-up happens when core hydrogen fusion stops and the star begins to expand up the red giant branch. The second dredge-up occurs at the end of core helium fusion as the star leaves the horizontal branch of the H–R diagram to become an AGB star. The third dredge-up is actually a series of episodes of convective instability that bring up products of H shell and He shell burning. Since the deep interior of the star has a higher abundance of ^{12}C than ^{16}O, each third dredge-up enriches the surface with carbon. Hence, although AGB stars start out with atmospheric C/O < 1 if they undergo a sufficient number of third dredge-ups, the C/O may become greater than 1 so they morph into carbon stars. Why one AGB star develops an atmosphere that is oxygen-rich while another becomes carbon-rich is a complex issue and not fully worked out. Most stars, including the sun, are oxygen-rich.

2.2.4.2 C/O Ratio Is the Key to Stellar Chemistry

The important issue for us is that the chemistry of the circumstellar discs around AGB stars is determined by the C/O ratio. Once the temperature is low enough, carbon will react readily with oxygen to form carbon monoxide (CO) which is stable. Oxygen-rich stars (confusingly described as M-stars) have C/O < 1. Formation of CO will have locked up all of the carbon and the chemistry will be dominated by the remaining oxygen. This results in the formation of oxygen-rich molecules, such as hydroxyl (OH), water, and oxides (e.g., TiO, ZrO, and VO), all of which stay in the gas phase, and silicates (formed when metal ions bond with SiO_4^{2-} ions) which condense to form dust particles. The first minerals to form in the dust grains are those with high melting temperatures (e.g., corundum, Al_2O_3, and calcium aluminum silicate, $Ca_2Al_2SiO_7$). These act as nuclei for the lower temperature condensation of Mg-olivine (forsterite, Mg_2SiO_4) and pyroxene (enstatite, $MgSiO_3$). At temperatures below 1100 K the forsterite and enstatite react with gas phase iron generating modest amounts of Fe-containing minerals.

For carbon stars (C-stars) C/O > 1, so all the oxygen is sequestered as CO and carbon chemistry then dominates. The gas phase molecules formed are carbon compounds (e.g., C_2, CN, CH, C_2H_2, C_3, and HCN). These act as raw ingredients from which more complex organic molecules are built in the ISM. The dust particles from carbon stars contain a variety of carbides (SiC, SiC_2) and carbon as graphite, diamond, or in amorphous (noncrystalline) form, that is, soot.

We might imagine that all of the molecules brewed up in stars survive intact in the ISM but this is not the case. Most of the volume of the ISM is hot and awash with photons and cosmic rays which have more than enough energy to destroy chemical bonds. Hence, *de novo* chemical synthesis must occur in some relatively shielded part of the ISM: it is the dense and diffuse clouds that are the chemical factories.

2.2.5 Doing Chemistry in Space

The Earth's surface is a cool, high density chemical environment.

Chemistry close to the surface of the Earth that occurs naturally (as opposed to in a laboratory) happens at cool temperatures—the mean temperature at the Earth's surface is 288 K (15°C)—and high density. The number of oxygen and nitrogen molecules packed into each cubic meter of the atmosphere (the number density) at the surface of the Earth is around 10^{25}. So, although atoms and molecules do not have very high kinetic energies, the chance of a collision between any two of them is extremely high. Indeed the time between collisions in this sort of environment is only about 1 ns. So, even if very few collisions succeed in triggering a chemical reaction, the huge numbers of collisions in any given time mean that reaction rates are fast. This is particularly true of collisions between atoms that lead to the formation of a molecule. For two atoms to come together to make a molecule the kinetic energy of the atoms has to be just right. If the velocities of the atoms are too low, there is not enough energy to overcome the repulsion of their respective electrons. On the other hand, if the velocities are too high the molecule has only a fleeting existence before shaking itself apart. But in a high density situation, such as the Earth's atmosphere, there is a high probability of three atoms colliding simultaneously, in which the third atom has its kinetic energy boosted by the collision and so carries away excess energy, allowing the two other atoms to form a stable molecule.

2.2.5.1 *Astrochemistry Happens Over a Wide Temperature Range*

As we have seen, astrophysical environments cover a large range of temperatures, from 2.7 K (the temperature of the cosmic microwave background radiation now) to billions of degrees in the maelstrom of a supernova explosion. We can dismiss the higher end of the temperature scale for the purposes of chemistry on the grounds that molecules cannot exist much above 3000 K. Hence, the temperature window for astrochemistry extends from about 3 to 3000 K. The rate of chemical reactions increases with temperature. Hence, in the coolest regions we expect chemistry to be very slow or not to happen at all.

2.2.5.2 Astrochemistry Happens at Extraordinarily Low Densities

Most astrophysical environments have extraordinarily low densities compared to that of the surface of the Earth. This means that the probability of atoms and/or molecules colliding is much less, and reactions go correspondingly slowly. In particular, the chance of three atoms colliding at the same time, so that one can carry away excess energy, allowing the other two to form a molecule, is vanishingly small. In the most tenuous material, that between the galaxies, the time between collisions is about one million years! For the bloated stellar atmospheres thought to be where some of the interesting cosmochemistry happens, the collisional timescale is rather better, perhaps around 10 μs. But it is worth noting that this is still 10,000-fold slower than terrestrial chemistry.

The conditions under which much astrochemistry happens cannot yet be replicated in the laboratory. Temperatures of 10 K can be obtained using liquid helium routinely but the hardest vacuum that can currently be achieved has a number density of 10^8 m^{-3}, matching that in dense clouds but not diffuse clouds. Given the difficulties and expense this entails, laboratories often work at higher pressure; typically 10^{-10} mbar, which corresponds to a number density of 10^{12} m^{-3}. This is at the very high end of that estimated for dense clouds. This makes it hard to study ISM reactions and, in particular, to get accurate values for reaction rate constants for astrochemical reactions. This produces uncertainty in the relative contributions of particular reaction pathways to the network of chemical reactions that are possible. In practice, what happens is that theoretical models of chemical networks are derived. These are grounded on the basis of laboratory experiments, including estimates of rate constants at low temperature and density. They are also informed by measuring the abundances of chemical species in the ISM, although some are hard to detect, because their concentrations are low or their spectral signatures are weak. In summary, while our understanding of the chemistry of space is likely to be broadly correct, much important detailed work is in progress.

2.2.5.3 Reaction Rate Constants Describe Reaction Rates

The rate at which a chemical reaction happens is given by its rate constant. For a simple bimolecular reaction between two molecules A and B:

$$A + B \rightarrow \text{products},$$

$$\text{Rate} = k[A][B],$$

(2.1)

where k is the reaction rate constant, and square brackets denote concentrations. Doubling the concentration of A (or B) doubles the number of collisions which will double the rate. Doubling the concentration of both A *and* B will increase the rate fourfold.

The rate constant depends partly on the activation energy. This is the minimum energy reactants must have in order for a reaction to go. It can be thought of as a potential energy barrier that must be overcome for the reaction to occur and is related to the strength of the chemical bonds that must be broken and made during the progress of the reaction. In Figure 2.6 the combined energy of the products is less than the combined energy of the reactants. This is an exergonic reaction since energy has been lost (usually as heat) from the system to the surroundings. Endergonic reactions in which the combined energy of the products exceeds that of the reactants are virtually impossible at the low temperatures of diffuse and dense clouds; there is simply not enough energy.

If two atoms, ions, or molecules collide, the probability that they react is essentially down to two factors. One is that their mutual orientation has to be just right for the reaction to happen. This allows the reactive parts of the molecules to be brought sufficiently close together. Since particles hit each other with random orientations only a small fraction of collisions result in a proper alignment. More obvious perhaps is that the energy of the collision—the translational kinetic energies of the atoms and their internal energies (vibrational and rotational kinetic energies)—has to be large enough. For a population of atoms this depends on the Maxwell–Boltzmann distribution of particle speeds and hence the temperature (Box 3.3). The rate constant, k, is empirically described by the Arrhenius equation:

$$k = Ae^{-E_a/RT}, \tag{2.2}$$

where E_a is the activation energy (J mol^{-1}). The exponential term is the fraction of reactants with energies equal to or greater than the activation energy at a particular temperature. This is because RT is the average kinetic energy per mole. So E_a/RT is the ratio of the activation energy to the average kinetic energy available, as given by the Maxwell–Boltzmann distribution. The larger this ratio is the lower the rate constant (note the minus sign). In other words, low activation energy favors a high

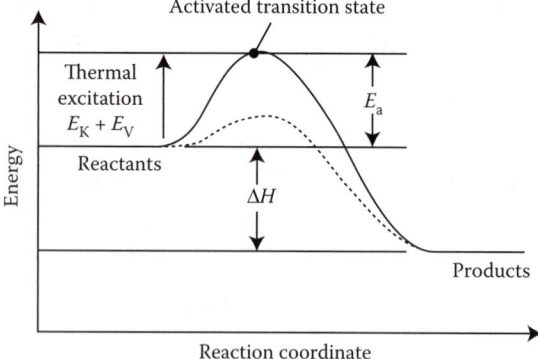

FIGURE 2.6 Activation energy must be achieved for a reaction to happen. The reaction coordinate plots the progress of the reaction. ΔH is the change in internal energy of the system and depends on the thermodynamics, so is independent of the reaction pathway. As we shall see later catalysts such as enzymes (dotted line) lower the activation energy.

rate constant and hence a faster reaction. Reactions between ions or highly reactive radicals (species with unpaired electrons) often have very low or even zero activation energies.

EXERCISE 2.6

a. How does the rate constant depend on temperature?
b. How likely is it that a reaction with a high activation energy will happen in diffuse and dense clouds?

A is the experimentally determined frequency factor, the number of collisions per second with the appropriate orientation to lead to the relation:

$$A = Z\rho, \tag{2.3}$$

where Z is the total number of collisions per second and ρ is a steric factor, the fraction of collisions that have the correct orientation to allow a reaction to occur (provided they are sufficiently energetic).

Hence k is simply a measure of the number of collisions that result in a reaction per second.

EXERCISE 2.7

a. What exactly is decreasing exponentially in the Arrhenius equation?
b. The activation energy for a hydrogen molecule to bind to a graphite grain surface has been measured to be 17.4 kJ mol^{-1}. Plot $\exp(E_a/RT)$ against T from 0 to 100 K for this process. What is the significance of this plot for chemistry at ultra low temperatures?
c. For a simple bimolecular reaction A has units of m^3 mol^{-1} s^{-1}. What are the units of k?
d. Under what circumstance will $k \simeq A$?
e. Activation energy can be found from the rate constant. Show that a plot of ln$k$ versus $1/T$ has a slope $-E_a/R$.

2.2.6 Different Chemistry Operates in Dense Clouds and Diffuse Clouds

2.2.6.1 Dense Clouds Provide Safe Haven for Molecules

There are two features of dense clouds that at face value make them seem unlikely venues for chemistry: (1) They have a high opacity to UV and visible radiation, so chemistry cannot be driven by photons to any appreciable extent; that is, photoionization and photodissociation are unimportant; and (2) The frigid 10 K temperature means that whatever chemistry there is happens very slowly. However, it turns out that dense clouds are a hotbed (or perhaps "coldbed" would be more appropriate) of interesting chemistry. This is because

- Cosmic rays can penetrate and these can ionize atoms and molecules.
- Dust grain surfaces promote much chemistry that cannot happen in gas phase at low temperature.
- Molecules that do get made are much less likely to be destroyed by photons than in diffuse clouds.

In diffuse and dense clouds (where $T < 100$ K), reactions with appreciable activation energies are not likely to happen. This rules out any gas phase reactions between neutral atoms or molecules which have significant activation energies. However, neutral chemistry is still possible on the surface of dust grains, as we shall see. By contrast, a rich network of reactions between ions and molecules is possible in the cold ISM. Ion–molecule reactions tend to have lower activation energies than neutral chemistry because the charge on an ion distorts the electron orbitals of a molecule. This electrostatic attraction increases the probability of a correctly oriented collision and hence of a reaction.

2.3 Molecular Spectroscopy

2.3.1 Molecules Are Detected Mostly by Vibrational and Rotational Spectra

Molecules can be detected in visible and ultraviolet light spectra as a result of electronic transitions, but under the low temperature and density in which chemistry occurs in space these transitions do not happen. But molecules vibrate and rotate under these conditions and these vibrational and rotational transitions occur in the infrared and microwave parts of the spectrum, respectively. Transitions between vibrational states have energies ~0.1 eV, while transitions between rotation levels have energies ~10^{-4} eV, so correspond to absorption or emission of photons with frequencies of order GHz.

Although the Earth's atmosphere is opaque to much of the radiation over this range, there are windows of transparency through which spectra can be obtained by ground-based telescopes. These provide information on vibrational transitions (1–5 μm and 8–20 μm) and on rotational transitions (0.35–1.3 mm). Orbiting IR observatories (e.g., Spitzer) can, of course, see the entire gamut, but have operational lifetimes limited by the period that their liquid helium (used to keep the detector cold) can last; typically just a few years.

2.3.1.1 Molecular Vibrations Are Quantized

Vibration is the periodic oscillation of atoms in a molecule. Each single frequency of oscillation is a normal mode of vibration, so can be thought of as a resonance. As with electronic energy levels, a molecule can vibrate only at discrete frequencies. In other words, vibrational modes are quantized and only certain transitions are allowed. These are defined by selection rules of quantum physics.

Transitions between vibrational levels occur by the absorption or emission of infrared photons with appropriate energy, $E = h\nu$. At a higher vibrational level a molecule vibrates more vigorously (i.e., higher amplitude) than at a lower level. If enough energy is absorbed by a molecule, it will dissociate.

2.3.1.2 Molecules Are Similar to Springs

To a first approximation, molecular vibrations are an example of simple harmonic motion. In the simplest case of a diatomic molecule, this is analogous to a spring obeying Hooke's Law in which the force needed to stretch a spring is proportional to the

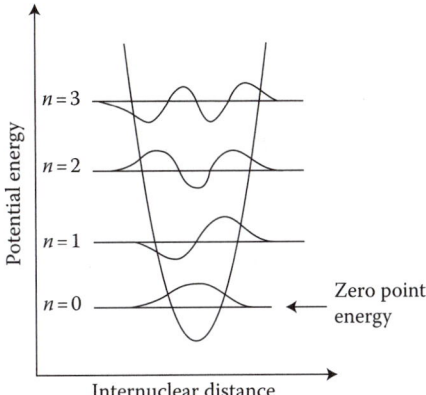

FIGURE 2.7 The potential well of a simple harmonic oscillator is a parabola (see Equation 2.3). Each vibrational level is an overtone of the fundamental frequency of the molecule in its ground state, $v = 0$.

extension. Harmonic oscillators continually transform energy back and forth between potential and kinetic energy. The potential energy, E_P, of the oscillator has a minimum value (zero) at the equilibrium position x_0, and increases to a maximum at the extremes of its excursion. A standard result from mechanics shows that E_P changes with position, x, as

$$E_P = (1/2)k(x - x_0)^2, \tag{2.4}$$

where k is the force constant or stiffness of the spring (with SI units N m^{-1}), a measure of the extent to which it resists deformation by an applied force (Figure 2.7).

For molecules k is the strength of the bond, and the vibration frequency is related to bond strength and the masses of the atoms forming the bond.

In reality, however, molecular vibrations are anharmonic oscillations. This distorts the shape of the potential well, and the energy levels are not equally spaced. The first overtone has a frequency that is slightly lower than twice that of the fundamental. In fact, excitation of ever higher overtones involves progressively less and less additional energy, although the total energy is increasing.

EXERCISE 2.8

What happens to the interatomic distance, r, at the top of the potential well of the anharmonically oscillating diatomic molecule in Figure 2.8? What is the physical interpretation? Is this outcome predicted by the simple harmonic motion model?

Note that at $v = 0$ molecules have vibrational energy. This is called the zero point energy and is a direct consequence of Heisenberg's uncertainty principle which forbids exact knowledge of both the position and momentum of a particle. It means that molecules vibrate even at their lowest vibrational energy level and have internal energy even at 0 K. In fact, most molecules are in the $v = 0$ vibrational state at the temperatures of diffuse and dense clouds.

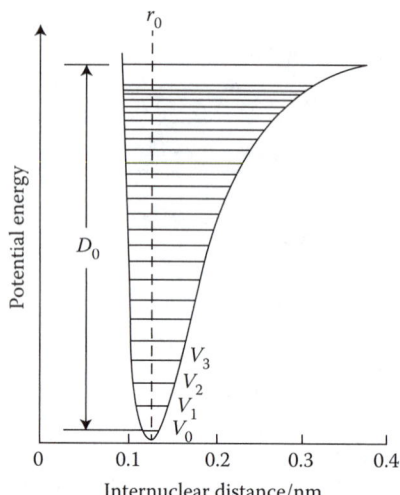

FIGURE 2.8 Real diatomic molecules are anharmonic oscillators. r_0 is the equilibrium bond length D_0 is the dissociation energy.

Diatomic molecules have just one normal mode of vibration, the stretching and relaxing of the chemical bond between its two atoms. Figure 2.8 shows that the vibrations oscillate about r_0, the bond length. The narrow shape of the potential energy surface at the vibrational ground state, $v = 0$, means the atoms are vibrating rapidly over a short distance, so the vibration frequency is highest there and drops for each successive vibrational energy state higher up the potential well.

2.3.1.3 Not All Vibrations Are Infrared Active

For a molecule to absorb and emit strongly in the infrared (i.e., to be infrared active) a vibrational mode must change the electric dipole moment of the molecule. (The electrical dipole moment is a measure of the separation of positive and negative charges in a molecule; i.e., how polarized it is.) This requires either that a molecule is asymmetric or it vibrate in such a way that its symmetry changes.

A diatomic molecule is asymmetric if it is heteronuclear and if the two nuclei have different electron withdrawing powers so that partial charges exist on the atoms; such a molecule is said to be a polar molecule. This is the case for carbon monoxide that has the electronic structure $C^{\delta+}O^{\delta-}$: the partial charges, though opposite in sign, are equal in magnitude.

Homonuclear diatomic molecules (e.g., H_2, N_2, or O_2) are largely IR inactive. This is because they are symmetrical—their two electrons are "shared" equally between the two identical nuclei—so they have no permanent electric dipole and cannot vibrate in any way that alters their symmetry. Consequently, in cold environments where electronic transitions are impossible, homonuclear molecules can be extremely hard to detect.

Polyatomic molecules (those with three or more atoms) vibrate in a variety of interesting ways. Each mode is independent and different bits of a molecule can all vibrate in a given mode at the same time (Figure 2.9).

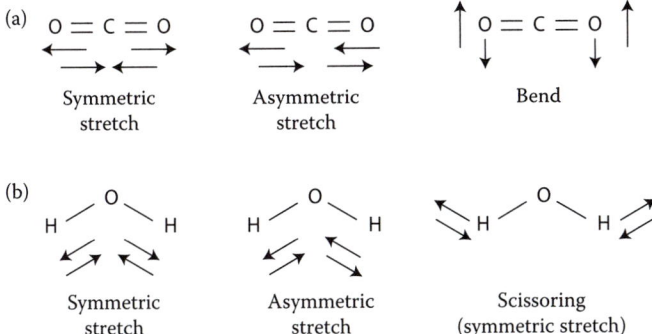

FIGURE 2.9 Vibration modes of (a) CO_2 and (b) water. All modes are IR active, except for the symmetric stretch of the linear CO_2 molecule which does not alter its electric dipole.

2.3.1.4 Molecular Rotations Are Quantized

A molecule in the gas phase is free to rotate relative to a set of mutually orthogonal axes, with their origin at the center of mass of the molecule. Free rotation of molecules is prevented in liquid and solid phases by intermolecular forces. Rotation about each unique axis is associated with a set of quantized energy levels that is dependent on the moment of inertia about that axis (Box 2.1) and a quantum number. Each vibrational level has several rotational levels associated with it.

BOX 2.1   MOMENT OF INERTIA

The moment of inertia, I, is a measure of the shape of a body, its mass and how this mass is distributed within the body. It is defined in terms of the turning force (torque, τ) that needs to be applied to a body to effect a given change in its angular velocity about an axis of rotation. To change angular velocity, an angular acceleration, α, is needed, so

$$\tau = I\alpha. \tag{2.5}$$

This is equivalent to Newton's Second Law for linear motion:

$$F = ma, \tag{2.6}$$

that is, moment of inertia is the rotational analog of mass in linear motion. Alternatively mass can be thought of as linear inertia. In general,

$$I = kmr^2, \tag{2.7}$$

where r is the distance of a point mass from an axis of rotation and k is the inertial constant, which depends on the shape of the object and how the mass is distributed within it. For a uniform sphere $k = 0.4$.

Any three-dimensional body has three mutually orthogonal principal axes of rotation. Each of these will have a parallel angular momentum vector and hence a principal moment of inertia which is constant, provided the body is rigid.

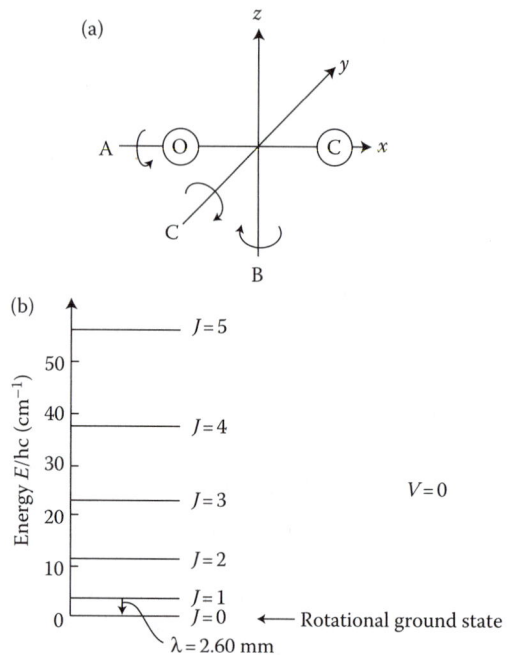

FIGURE 2.10 (a) Principle axes of rotation of the diatomic molecule. (b) First five rotational energy levels of a diatomic molecule.

To a first approximation molecules can be thought of rigid rotors. A diatomic molecule resembles a dumbbell rotating around the center of the rod. The more rotational energy it has the faster it spins. In this model, the quantized rotational energy levels are all evenly spaced. But molecules depart from rigid behavior as their rotational energy gets larger and the average distance between atoms increases, reducing the spacing between rotational energy levels (Figure 2.10).

For a homomeric diatomic molecule (those with two identical atoms, mass m, bond length r), rotating like a propeller, $I = 2mr^2$. More generally, for diatomic molecules where the center of rotation is not equidistant between the two atoms because they are of different masses, m_1 and m_2, $I = \mu r^2$, where μ is a weighted average of the masses.

2.3.1.5 Rotational–Vibrational Spectra

Vibrational transitions in a diatomic molecule changes the bond length (see Figure 2.8). This will modify the moment of inertia of a molecule changing its rotational energy thereby triggering rotational transitions. In this way molecules generate rotational-vibrational (rovibrational) spectra. For a diatomic molecule undergoing a vibrational transition $\Delta v = 1$ the rotational quantum number J can change by ∓ 1. Transitions for which $\Delta J = -1$ generate the P-branch of the spectrum, those for which $\Delta J = +1$ generate the R branch and $\Delta J = 0$ gives the Q branch which should be a single line at the fundamental frequency corresponding to the pure vibrational transition, but is split by the effects of anharmonicity and centrifugal distortion (Figure 2.11).

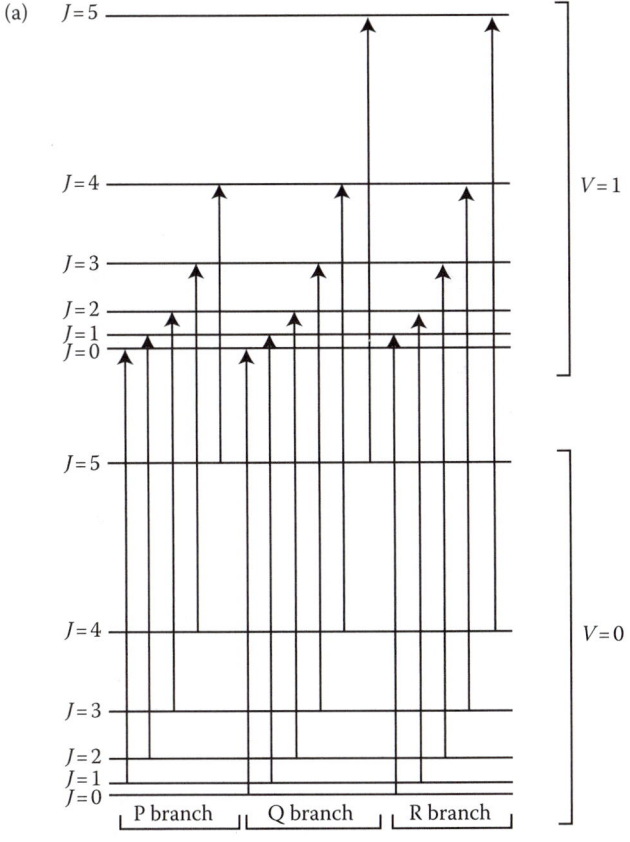

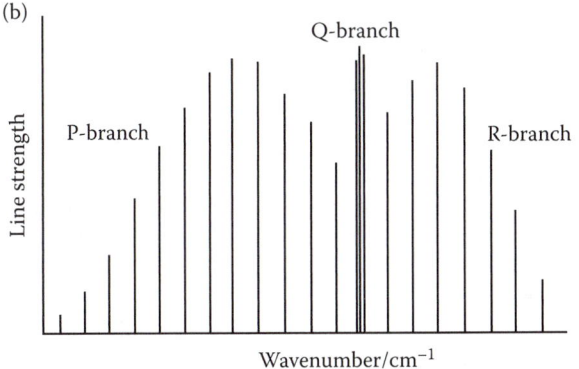

FIGURE 2.11 (a) The rovibrational energy level diagram for a linear molecule (b). Idealized rovibrational spectrum for a heteronuclear diatomic molecule. Wavenumber = $1/\lambda$. Note how P and R branches correspond to lines with frequencies below and above the fundamental frequency, respectively.

2.3.1.6 Isotopes Can Be Distinguished by Rotational Spectra

The moment of inertia of a molecule can be calculated from a rotational spectrum. Since the moment of inertia of a molecule depends on its mass, it can reveal its isotopic composition. Hence, the isotopic make up of molecules in space can be amenable to observation.

2.3.1.7 Temperature Can Be Deduced from Rotational Spectra

The rotational energy levels a molecule has will depend on the kinetic energy imparted to it in collisions, and hence by the temperature of its local environment; for example, the ISM or photospheres of cool stars. This means that rotational energy levels can be good thermometers provided that the molecule is in local thermal equilibrium with its surroundings. For this the density must be high enough to allow sufficient collisions to partition energy between the molecule and its surroundings. In practice, it is possible to calculate the temperature by comparing the intensities of spectral lines corresponding to two rotational energy levels.

2.3.1.8 Polyatomic Molecules

Polyatomic molecules are problematic because with each additional atom the rotational and vibrational spectra get much more complicated. Rotational spectra get harder to interpret because usually each of the three moments of inertia that give rise to the spectrum is different. This is true even for very simple molecules such as that of water! In general, all that can be done is to attempt to fit a given spectrum to different moments of inertia. If the molecular formula is known, moments of inertia can be calculated from guesstimates of the possible structure, and if the calculated moments of inertia agree well with those that fit the spectra, then the guessed structure is probably the correct one. Libraries of individual molecular spectra are available to help identify molecules.

Groups in organic compounds can vibrate in numerous ways. For example, methylene ($>CH_2$) group hydrogen atoms can vibrate in six different modes with respect to each other and their associated carbon atom. The molecule of which the methylene group is a part will also show other vibration modes (e.g., C–C stretching, C–H stretching, H–C–H bending, etc.). Each of these can occur in any combination.

In the laboratory where spectra can be studied in detail, identification can be far easier than with lower resolution astronomical spectra. Moreover, even simple molecules with just a dozen or so carbon atoms have microwave spectra so complex as to make identification hard.

2.3.1.9 Electronic States of Molecules

Most molecules have dissociation energies less than the lowest electronic excitation state and so are destroyed rather than excited. However, some molecules have electronic transitions in the ISM by which they can be detected. In fact, the first molecules detected in the ISM (CH, CN, and CH^+ discovered in 1937, 1940, and 1941, respectively) were revealed by their electronic transitions in visible light. Molecules containing transition metals (TiO, FeH, and VO) seen in cool M-dwarfs have numerous low energy electronic states that can be occupied at the temperatures of these stars.

2.4 Building Molecules

2.4.1 Molecules in the ISM

The first molecule detected by radio telescope was OH in 1963, soon to be joined in 1968 and 1969 by NH_3, H_2O, and HCHO (formaldehyde). Since then over 160 molecules have been found in the ISM and some have even been seen in galaxies with appreciable redshifts. Most are identified by emission rather than absorption.

2.4.2 Reaction Mechanisms

To appreciate the chemistry that can happen in various astrophysical environments we briefly examine some of the main reaction mechanisms at play. Those involving UV and visible light photons occur mostly in diffuse clouds that are transparent to these wavelengths. Cosmic rays are energetic enough to penetrate dense clouds.

2.4.2.1 Ionization

Electromagnetic radiation (i.e., photons) or cosmic ray particles (mostly protons) can knock electrons from atoms and molecules to form atomic and molecular ions: positively charged particles. For each electron ejected the ion acquires a single positive charge. For example, if one electron is ejected from an oxygen atom (which astronomers by tradition depict as OI), it becomes O^+ (astronomers write this as OII), whereas if two electrons are kicked out it is O^{2+} (or OIII). Ionization happens a great deal in diffuse clouds where it is a key driver of chemistry. This is because gas phase reactions involving ions are much more likely than those between neutral species at the low temperature ($T < 100$ K) of diffuse clouds.

2.4.2.2 Atomic Photoionization

A photon with enough energy, on striking an atom, will completely eject an electron from the atom so that it becomes an ion. For example, a hydrogen atom in its ground state ($n = 1$) will be ionized by a ultraviolet photon with energy, $E = 13.6$ eV or more. A photon with this energy has a wavelength of

$$\lambda = ch/v = (3 \times 10^8 \times 6.626 \times 10^{-34})/(13.6 \times 1.602 \times 10^{-19}) = 9.124 \times 10^{-8} \text{ m or 91.2 nm.}$$

Of course, if a hydrogen atom is already in an excited state (i.e., if it has an electron in an energy level $n = 2$ or higher), a photon with less energy (longer wavelength) will ionize it. We can depict the ionization of hydrogen as follows:

$$H + h\nu \rightarrow H^+ + e^-, \tag{2.8}$$

where $h\nu$ is the shorthand for the energy of the incoming photon.

Since hydrogen has only one electron, no further ionization is possible. Regions dominated by UV photon emission in which most or all of the hydrogen is ionized are called HII regions. They are found in the spiral arms of disc galaxies and are places where young, massive stars of early spectral types are born. These hot stars emit the bulk of their radiation in the ultraviolet.

The amount of energy needed to eject a given electron varies with the atom and with each electron. For example, the energy needed to form HeII and HeIII is higher than the ionization energy for hydrogen, and sufficiently energetic photons have wavelengths in the far UV.

2.4.2.3 Molecular Photoionization

Molecules too can be ionized by radiation. The electronic levels in a molecule are not the same as the atoms from which it is made because the formation of the chemical bonds alters the distribution of the electrons. Hence, the energy needed to ionize a molecule is different from those of its constituent atoms. An example is the photoionization of the hydroxyl molecule:

$$OH + h\nu \rightarrow OH^+ + e^-. \qquad (2.9)$$

Molecular ions are, generally, more reactive than their parent molecules.

2.4.2.4 Cosmic Ray Ionization

Cosmic ray particles are atomic nuclei and a smattering of electrons with high energies. The nuclei have the same abundance as the ISM; namely, 90% protons, 9% helium, and 1% heavy elements. At least some cosmic rays originate from protons that have been accelerated to velocities close to the speed of light by supernovae both within and beyond our galaxy, though there may be other sources. Although the majority are not particularly energetic, their sheer numbers mean that ionization by cosmic rays is important. Both hydrogen and helium atoms can be ionized by cosmic rays:

$$H + cr \rightarrow H^+ + e^- + cr, \qquad (2.10)$$

$$He + cr \rightarrow He^+ + e^- + cr. \qquad (2.11)$$

Molecules can also be ionized by cosmic rays. In the case of the hydrogen molecule this is depicted as

$$H_2 + cr \rightarrow H_2^+ + e^- + cr. \qquad (2.12)$$

All these cosmic ray ionizations have important consequences for the chemistry of the interstellar medium, which we will explore later.

2.4.2.5 Photodissociation

As well as ionizing atoms and molecules, photons can also break molecules up into neutral atoms. This is termed photodissociation and is a major way in which molecules are destroyed in interstellar environments. Usually, it requires the energy of UV photons to split molecules in this way. Hence, if the energy of the incoming photon is sufficient, the hydroxyl molecule is split apart:

$$OH + h\nu \rightarrow OH^* \rightarrow O + H. \qquad (2.13)$$

Here the OH* represents an unstable excited intermediate state.

2.4.2.6 Dissociative Recombination

Ionization generates as many free electrons as there are positively charged species. (Overall, the interstellar gas is electrically neutral.) The attraction between positively charged atoms and molecules and electrons ensures that recombination occurs, provided the radiation field is not too energetic. However, the resulting neutral molecule is frequently unstable and so dissociates. This sequence of events is called dissociative recombination. The H_3O^+ ion plays a key part in carbon and oxygen chemistry. On recombining with an electron it dissociates to give either water or the hydroxyl molecule:

$$H_3O^+ + e^- \rightarrow H_3O \rightarrow H_2O + H, \qquad (2.14a)$$

$$H_3O^+ + e^- \rightarrow H_3O \rightarrow OH + H_2. \qquad (2.14b)$$

2.4.2.7 Dissociative Recombination of Molecules Is Fast and Common

Electrons also recombine with atomic ions, but this often results in an excited, unstable intermediate with enough energy to dissociate once again. In such cases, a neutral atom results only if the energetic intermediate immediately emits a photon. In the case of carbon, for example,

$$C^+ + e^- \leftrightarrow C^* \rightarrow C + h\nu. \qquad (2.15)$$

Because the intermediate usually dissociates rather than emitting a photon, neutralization of atomic ions is slow.

2.4.2.8 Neutral Chemistry

The conditions for two uncharged species, atoms or molecules, to react on colliding are fairly unfavorable in most astronomical settings. This makes neutral chemistry slow.

Two atoms colliding usually effectively just bounce off one another, because they will not have the right amount of kinetic energy to form a stable molecule. In high density environments, reactions between atoms can occur with high frequency via three-atom collisions, in which one atom removes excess kinetic energy so that the other two can combine. But in low density astronomical environments the chance of this is very small. Hence, throughout regions of space where the interstellar gas is a highly dilute soup of neutral atoms, very little chemistry happens at all.

Interactions between an atom and a molecule may or may not result in a chemical reaction, depending on the temperature. There are two reasons for the relative unreactivity of neutral species. One is that neutral reactions can have quite large activation energies. Thus, at the temperature of a cold molecular cloud the reaction $O + OH \rightarrow O_2 + H$ is fast, whereas $N + NO \rightarrow N_2 + O$ is extremely slow. This is because of the difference in the size of the activation energies of the two reactions. At 10 K the kinetic energy of the nitrogen atom and nitric oxide is barely able to supply the large activation energy needed to form the triple bond of the nitrogen molecule, whereas the *same* kinetic energy of the oxygen atom and hydroxyl molecule is easily enough to provide the lower activation energy to make the oxygen molecule double bond.

The second reason for neutral reactions being slow is if the reactants have a total internal energy less than the products. The reaction will not go unless the energy deficit is supplied. This is the case for the reaction

$$C + H_2 \rightarrow CH + H. \tag{2.16}$$

Here the energy required to split the hydrogen molecule is greater than the energy released by forming the CH molecule. Consequently, this reaction needs high temperatures (several thousand Kelvin) to proceed.

2.4.2.9 Ion–Molecule Reactions

A reaction is much more likely when an ion collides with a molecule than if two neutral species collide. This is because the charge on the ion polarizes the molecule, attracting electrons in the chemical bond of the molecule so that it breaks, permitting a new bond to be made with the incoming ion. For example, in the reaction between a singly ionized oxygen ion and molecular hydrogen, the ion distorts the covalent bond of the hydrogen molecule so that one of its electrons becomes associated with the oxygen to form a hydroxyl molecule, liberating a hydrogen atom:

$$O^+ + H_2 \rightarrow OH^+ + H. \tag{2.17}$$

2.5 Chemical Networks

Chemistry happens in the denser and cooler regions of the ISM. In gas phase reactions, the energy is supplied by visible and ultraviolet radiation, x-rays, and cosmic rays (mostly high energy protons and helium nuclei). This radiation heats dust grains, raising the temperature of the gas. In general, cosmic rays can penetrate more deeply than x-rays which in turn penetrate to greater depths than UV radiation. All radiation is attenuated as it goes through the ISM, so dense cores that are furthest from the sources of radiation will receive the lowest fluxes and will therefore be the coolest and densest regions. Consequently, although some gas phase chemistry can be energized by cosmic radiation in dense cores, much chemistry there is catalyzed on the surface of dust grains. Regions of the ISM in which most of the flux is in the far ultraviolet (FUV) are described as photon-dominated regions (PDRs) while those that are close to x-ray sources are x-ray-dominated regions (XDRs). Considerable chemistry occurs in both PDRs and XDRs. The prevailing drivers in both cases are photodissociation and ionization, and the details depend on the flux of radiation and the density of hydrogen, and sometimes that of other species, in the locality.

2.5.1 Dust Grain Surfaces Catalyze Synthesis of Hydrogen Molecules

The starting point for ISM chemistry is the formation of hydrogen molecules in dense clouds. Conversion of H into H_2 must occur in dense clouds since the gas here is overwhelmingly molecular. But there is a problem. If two H atoms collide with each other in the gas, they almost always separate shortly afterwards; their time together is only $\sim 10^{-12}$ s! Formation of a hydrogen molecule is extremely unlikely because the system cannot rid itself of excess energy to stabilize the H_2 bond, even though the

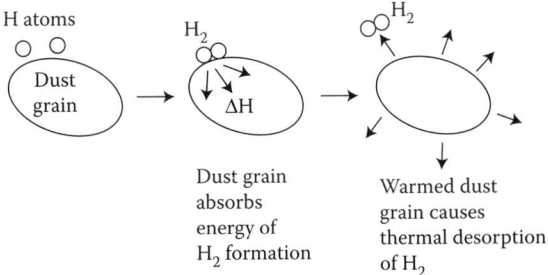

H atoms

Dust grain

Dust grain absorbs energy of H_2 formation

Warmed dust grain causes thermal desorption of H_2

FIGURE 2.12 The activation energy of a reaction is reduced by a catalyst such as the surface of an interstellar dust grain.

reaction is exothermic. There are two ways to dump the excess energy. The first one, radiative association, requires that the collision complex lasts long enough to emit the energy as radiation. But this is not the case for hydrogen under interstellar conditions. The other is for a third hydrogen atom to collide with the newly formed molecule and carry away its excess energy before it dissociates. But the density is so low that such three-body collisions virtually never happen.

Instead, hydrogen molecules are synthesized on the surface of dust grains:

$$H + H \rightarrow H_2. \tag{2.18}$$

Hydrogen atoms are adsorbed onto the surface and remain for long enough to encounter each other and react to form a hydrogen molecule. The dust grain acts as a third body, absorbing the heat of formation (4.7 eV). The warmed surface of the grain imparts enough kinetic energy to the hydrogen molecule to eject it into the gas phase, a process called thermal desorption.

In this molecular cookery, the surface of the dust grain is acting as a catalyst, increasing the rate at which the reaction will go by providing an environment in which reactants are brought into a favorable orientation thereby reducing the activation energy (Figure 2.12).

As we shall see, other reactions are catalyzed by dust grains, particularly the hydrogenation of O, C, and N, but the formation of H_2 is the most common because hydrogen is the most abundant element.

2.5.1.1 Molecular Hydrogen Is Protected by Self-Shielding

Molecular hydrogen in dense clouds is protected from ionization by UV photons by a phenomenon termed self-shielding. Photons with energies greater than the ionization potential of atomic hydrogen (13.65 eV) will drive the reaction, $H \rightarrow H^+ + e^-$. There is enough H around that this protects the bulk of molecules within the cloud from ionization. This is because the ionization potential of H_2 is 15.43 eV, but photons with this energy have been absorbed at the edges of the cloud to ionize atomic hydrogen. Despite this, dense clouds are not entirely neutral. Some chemical species, particularly atomic carbon, have ionization potentials lower than 13.65 eV: for $C \rightarrow C^+$ it is only 11.3 eV. In addition, some cosmic rays can penetrate dense clouds. Hence, hydrogen molecules are not completely safe from ionization.

2.5.1.2 Grain Surfaces Are Important for Oxygen Chemistry

Molecular oxygen is in much lower abundance in dense clouds than predicted by chemical models. The most likely explanation is that O_2 is mostly adsorbed onto grain surfaces where it undergoes hydrogenation. This is supported by the recent identification of OH_2 and H_2O_2 in the gas phase of a warm molecular cloud where O_2 has also been found.

2.5.1.3 Methanol Synthesis Involves Grain Surfaces

Methanol is synthesized by the hydrogenation of CO on grain surfaces:

$$CO + H \rightarrow \underset{\text{formyl radical}}{HCO}, \tag{2.19a}$$

$$HCO + H \rightarrow \underset{\text{formaldehyde}}{H_2CO}, \tag{2.19b}$$

$$H_2CO + H \rightarrow H_3CO, \tag{2.19c}$$

$$H_3CO + H \rightarrow CH_3OH. \tag{2.19d}$$

Formaldehyde is potentially an important precursor in prebiotic chemistry, since it may be involved in the synthesis of sugars, amino acids, and RNA.

2.5.1.4 Diffuse Clouds Are Photon-Dominated Regions

Diffuse clouds are predominantly neutral regions of the ISM but have a low enough density that they allow appreciable amounts of radiation to penetrate; that is, they have a much lower opacity than dense clouds. The number density of the most abundant species, hydrogen, is around 3×10^7 m^{-3}. A typical diffuse cloud contains 100-fold the mass of the sun in a space 3 pc across which, given its opacity, means it absorbs about a quarter of the visible starlight which irradiates it. However, the opacity of these clouds varies quite a lot; optically thin clouds greatly outnumber their thicker brethren. Diffuse clouds are flooded with UV and visible light photons, and these are crucial drivers of the chemistry that occurs inside them; hence they are prototypical PDRs. UV and light photons are absorbed by dust grains. This is the main factor limiting radiation penetration into the PDR. Thus warmed, the grains radiate in the infrared heating the gas in the cloud to 80 K or so. A small fraction of the far ultraviolet (FUV) heats the gas by knocking electrons off the grain surfaces: the photoelectric effect.

EXERCISE 2.9

What would happen qualitatively to photoelectric heating in a PDR if the dust grains became very positively charged because the FUV flux rate exceeded the rate at which electrons recombined on the grain surface?

Diffuse cloud PDRs are the transition region between HII regions that are dominated by ionization and dense clouds that contain mostly neutral molecules, but

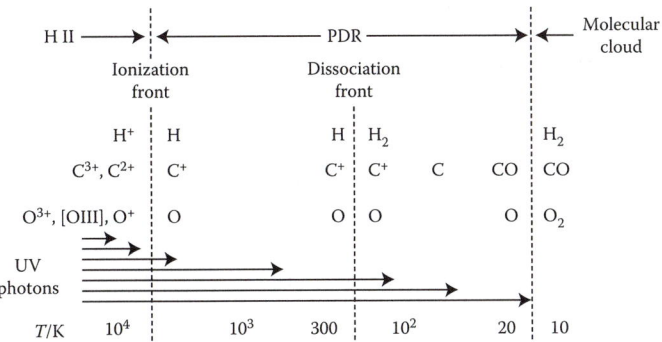

FIGURE 2.13 The abundances of ion and neutral species inside PDRs are determined by the extent to which far ultraviolet light can penetrate.

reflection nebulae, AGB circumstellar discs, the neutral envelopes of planetary nebulae, and protoplanetary disc atmospheres are also PDRs.

The chemistry of PDRs changes with distance into the region as radiation is increasingly attenuated by scattering and absorption (extinction).

On the outskirts the radiation flux is high enough for photodissociation to produce atomic hydrogen, oxygen, and carbon. Here, typically, the temperature is relatively high (10^2–10^3 K), because the high radiation flux heats the dust grains effectively, and so the gas density is low. The overwhelming abundance of hydrogen mops up almost all the 13.6 eV photons in the hydrogen atom ionization:

$$H \rightarrow H^+ + e^-, \tag{2.20}$$

and consequently oxygen, which has the same ionization potential, is shielded and remains as neutral atomic oxygen. However, atomic carbon has a lower ionization potential (11.26 eV) and hence is ionized to C^+. This favors carbon chemistry.

The center is dominated by H_2 and CO because very little radiation can penetrate. Consequently, there is little heating of dust grains and so the temperature is low (10–10^2 K). The H/H_2 transition occurs where the opacity due to dust and H_2 self-shielding conspire to choke-off photodissociation; it is therefore partly determined by how efficiently H_2 is synthesized on the dust grains. Deeper in are the transitions from C^+ to C to CO and S^+ to S to SO. Hence, PDRs are zoned structures, and each zone favors particular chemical reactions (Figure 2.13).

2.5.1.5 X-Ray-Dominated Regions

Some astrophysical environments, such as protoplanetary discs around young stellar objects (T-Tauri stars), can have x-ray fluxes greater than FUV fluxes. These are x-ray-dominated regions. Just as for PDRs ionization, photodissociation and heating are key processes, except that x-rays have a larger penetration depth and greater heating efficiency. X-rays preferentially ionize heavy elements and it is the low abundance of metals that enables x-rays to penetrate more deeply than UV. But the electrons produced by heavy element ionization collide with atomic and molecular hydrogen, ionizing these.

2.5.1.6 H_3^+ Is the Starting Point for a Great Deal of Chemistry

Both cosmic radiation and FUV photons have enough energy to ionize hydrogen molecules in PDRs and XDRs:

$$H_2 \rightarrow H_2^+ + e^-. \tag{2.21}$$

This is followed by

$$H_2^+ + H_2 \rightarrow H_3^+ + H. \tag{2.22}$$

The discovery of the trihydrogen cation, H_3^+, was both a surprise and a blessing: a surprise because initially it was not possible to account for how three hydrogen nuclei could form a stable equilateral triangle held together by just two electrons; and a blessing because the range of molecules found in the ISM could only be explained by having H_3^+ as a key intermediate in the chemistry. H_3^+ is a powerful protonating agent and takes part in reactions of the type

$$H_3^+ + X \rightarrow XH^+ + H_2, \tag{2.23}$$

where X is an atom such as C and O or a molecule such as CO.

Reaction 2.22 is very fast because it has a very small activation energy. This means that H_2^+ is quickly removed from the ISM so its abundance is very low. Similarly H_3^+ is rapidly snapped up in a variety of Equation 2.7-like protonation reactions. Consequently, both H_2^+ and H_3^+ are hard to detect in the ISM. A network of key reactions centered on H_3^+ is depicted in Figure 2.14. Note how the protonation of oxygen leads to the formation of water.

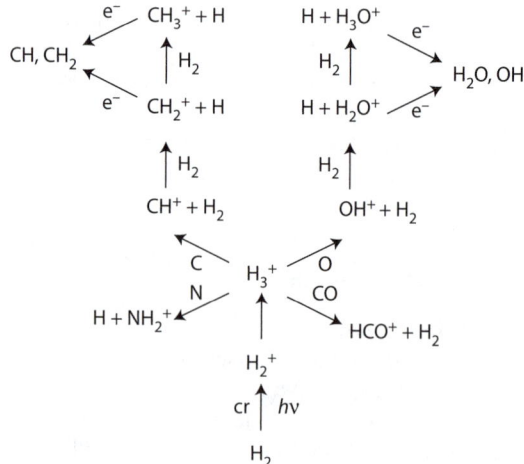

FIGURE 2.14   The trihydrogen cation incorporates carbon and oxygen into chemical networks as this highly simplified scheme shows.

2.5.1.7 Hydrocarbon Synthesis from H_3^+ Gives Methane

The trihydrogen cation also acts as a starting point for the synthesis of hydrocarbons:

$$H_3^+ + C \rightarrow CH^+ + H_2, \tag{2.24a}$$

$$H_2 + CH^+ \rightarrow CH_2^+ + H, \tag{2.24b}$$

$$H_2 + CH_2^+ \rightarrow CH_3^+ + H, \tag{2.24c}$$

$$H_2 + CH_3^+ \rightarrow CH_5^+ + h\nu, \tag{2.24d}$$

$$CH_5^+ + e^- \rightarrow CH_4 + H. \tag{2.24e}$$

The route to more complex hydrocarbons is via C^+, for example,

$$CH_4 + C^+ \rightarrow C_2H_3^+ + H, \tag{2.25a}$$

$$C_2H_3^+ + e^- \rightarrow C_2H_2 + H. \tag{2.25b}$$

2.5.1.8 The C^+ Ion Is a Starting Point for PDR Chemistry

We have seen that the environment in photon-dominated regions is not harsh enough to ionize oxygen atoms but carbon is ionized to C^+. This eventually leads to the formation of carbon monoxide as can be seen in Figure 2.15.

We have seen that CO is involved in neutral chemistry on dust grains in cold dense cores, but in warm PDRs, it reacts with hydrocarbon molecular ions to produce a number of oxygen-containing molecules, for example,

$$CO + C_2H_2^+ \rightarrow H_2C_3O^+. \tag{2.26}$$

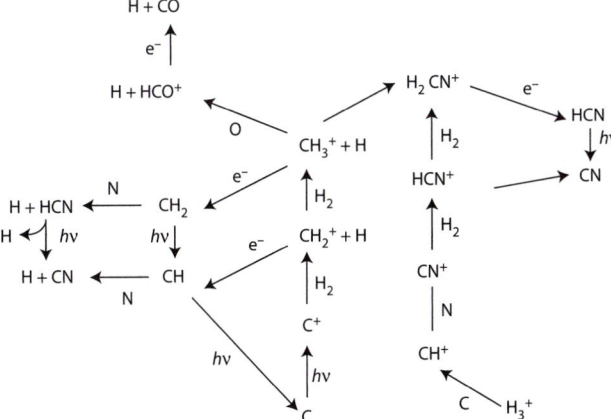

FIGURE 2.15 C^+ in PDRs is an alternative to the trihydrogen cation in generating backbone oxygen and nitrogen chemistry.

2.5.1.9 Nitrogen Chemistry Yields Ammonia

The synthesis of ammonia proceeds via protonation in dense clouds. It starts by breaking the N_2 triple bond. This is a very strong bond, and the energy is effectively provided by cosmic rays ionizing helium:

$$He + cr \rightarrow He^+. \tag{2.11}$$

Reaction of HeII with the nitrogen molecule leads to the formation of the N^+ ion:

$$N_2 + He^+ \rightarrow N + N^+ + He. \tag{2.27}$$

Sequential additions of molecular hydrogen then occur:

$$N^+ + H_2 \rightarrow NH^+ + H, \tag{2.28a}$$

$$NH^+ + H_2 \rightarrow NH_2^+ + H, \tag{2.28b}$$

$$NH_2^+ + H_2 \rightarrow NH_3^+ + H, \tag{2.28c}$$

$$NH_3^+ + H_2 \rightarrow NH_4^+ + H. \tag{2.28d}$$

Finally,

$$NH_4^+ + e^- \rightarrow NH_3 + H. \tag{2.29}$$

2.5.2 Chemical Species Can Trace ISM Conditions and Processes

The chemical networks outlined above are just some of the possibilities, and this simple account downplays the degree of interconnectedness. In practice, the networks are worked out by kinetic modeling. The temperature and extinction of the ISM and initial relative abundances are assumed and put into the model at the start.

But how do we know what these conditions are? In fact, over the years a variety of chemical species have become important probes of the conditions in the ISM as our understanding of the reactions in which they are formed and destroyed has improved.

There are several examples of chemical tracers to choose from. We finish our exploration of astrochemistry by looking at how three nitrogen compounds serve this role: CN, HCN (hydrogen cyanide), and its tautomer HNC (hydrogen isocyanide). The motivation for studying these is that they are found in comets which may have been important in delivering organic molecules to planetary surfaces and may have been key reactants in early biochemistry.

2.5.2.1 CN, HCN, and NHC Are Intimately Related

The synthesis of CN and HCN starts with radiative association of C^+:

$$C^+ + H_2 \rightarrow CH_2^+ + h\nu. \tag{2.30}$$

This is followed by

$$CH_2^+ + H_2 \rightarrow CH_3^+ + H, \tag{2.31a}$$

and

$$CH_3^+ + e^- \rightarrow CH + H_2 \rightarrow CH_2 + H. \qquad (2.31b)$$

The CH and CH_2 radicals then react with abundant N atoms via neutral–neutral reactions:

$$CH + N \rightarrow CN + H, \qquad (2.32)$$

$$CH_2 + N \rightarrow HCN + H. \qquad (2.33)$$

Actually, rather more HCN is produced by dissociative recombination:

$$H_2CN^+ + e^- \rightarrow HCN + H, \qquad (2.34a)$$

and the tautomer of HCN hydrogen isocyanide (HNC) can be formed in the same way:

$$H_2CN^+ + e^- \rightarrow HNC + H, \qquad (2.34b)$$

whereas dissociative recombination with H_2NC^+ (the tautomer of H_2CN^+) produces HNC exclusively:

$$H_2NC^+ + e^- \rightarrow HNC + H. \qquad (2.35)$$

HCN is destroyed in the ISM in a variety of ways depending on the location in the cloud. In PDRs, photodissociation occurs either by FUV or (at greater depth) cosmic rays, that is,

$$HCN \xrightarrow{h\nu \text{ or cr}} CN + H. \qquad (2.36)$$

In the dense core, two competing mechanisms destroy it:

$$HCN + H^+ \rightarrow HCN^+ + H, \qquad (2.37)$$

$$HCN + HCO^+ \rightarrow H_2CN^+ + CO. \qquad (2.38)$$

The reaction with HCO^+ dominates by a factor of ~3.5.
HNC is destroyed mainly through ion–neutral reactions with C^+ and H_3^+:

$$C^+ + HNC \rightarrow C_2N^+, \qquad (2.39)$$

$$H_3^+ + HNC \rightarrow H_2CN^+ + H_2. \qquad (2.40)$$

The abundances of CN, HCN, HNC, and HCO^+ in any part of the ISM depend in each case on the balance between the reactions which form the molecules and those that destroy them, providing insights into the conditions within the ISM such as density, temperature, radiation fluxes, etc. These molecules have large dipole moments making them easy to detect by their rotational transitions, some of which are in atmospheric windows allowing them to be observed using ground-based telescopes.

2.5.2.2 HCN and Friends Provide Clues to the State of the ISM

CN and HCN formation and destruction sequences vary with location within a PDR or cosmic ray-dominated dense core. Modeling shows that CN/HCN abundance ratios can reveal a great deal about the environment over a wide range of conditions. Both

CN and HCN form at the C^+/C/CO transition region, but HCN is more vulnerable to photodissociation than CN, so it is rapidly destroyed in these outer parts of the PDR.

EXERCISE 2.10

   a. Qualitatively what would the CN/HCN ratio be at the outer regions of a PDR?
   b. What do you predict would happen to the CN/HCN ratio deeper in the PDR? Explain your answer.

HNC/HCN ratios have also proved instructive. HNC has a higher energy than HCN so that at equilibrium at $T < 100$ K, kinetic calculations suggest we would expect [HNC]/[HCN] $< 10^{-25}$. In other words, the tautomerization reaction HNC $\rightleftarrows$ HNC should dramatically favor the forward reaction. However, observation shows that the ratio is actually close to unity! This is because there is a huge activation energy needed to "kick start" the tautomerization which corresponds to a temperature at which HNC would be destroyed. Hence, the HNC/HCN ratio is not determined by equilibration between these species but by the reactions that form and destroy each of them.

It turns out that there is a strong temperature dependence on the HNC/HCN ratio, so determining the ratio allows the temperature to be calculated. The HNC/HCN ratio has been shown to be an excellent method for distinguishing between PDRs and XDRs since it is about unity in PDRs but >1 in XDRs.

Finally, the HCO^+/HNC line ratio is a good tracer of the density of the ISM.

3

Habitable Earth

3.1 Earth in Context

Earth is the only planet in the universe we know has life. From this one example we have to construct astrobiology. The principle of mediocrity tells us we should regard the Earth as an unremarkable planet orbiting an "average" star in a spiral galaxy that is indistinguishable from countless billion others. Taken at face value this means that life in the universe is common. However, that is not how it feels from our perspective. Is the apparent uniqueness of Earth a consequence of our not having explored far enough for long enough? Perhaps, but it may be that the principle of mediocrity is a poor guide and that Earth is indeed special. Maybe there is actually something unusual about the sun or the circumstances in which the solar system formed. One way to try to answer this question is to hunt for habitable planets. However, it makes sense to first fathom what makes Earth habitable. To do this it is useful to compare the Earth to two other planets, Venus and Mars, because they share much in common with us, and being our nearest neighbors, are relatively easy to study.

Understanding why Earth is habitable is the main goal of this chapter. First we look at the family to which Earth belongs.

3.1.1 There Are Eight Major Planets

Earth is a planet, one of the eight in orbit around the sun, a G2V star.

3.1.1.1 Terrestrial Planets Have Low Masses

Although it has a dense atmosphere, and oceans of liquid water, most of the Earth is made of iron and rock (Figure 3.1a). Terrestrial planets have a central core of iron and nickel, a mantle of silicate rock overlain by a thin crust of low-density silicate rock. For a sense of the relative proportions of these components a peach is a fair model! There are three other terrestrial planets, Mercury, Venus, and Mars. The terrestrial planets are all relatively small and, being closest to the sun, they are known as the inner planets.

3.1.1.2 Gas and Ice Giants Have High Masses

The four outer planets (Jupiter, Saturn, Uranus, and Neptune) are large, massive, and have extensive atmospheres of hydrogen and helium (Figure 3.1b). Jupiter and Saturn are gas giants because most of their bulk is composed of H and He, although deep inside their worlds these elements are not gases because of the high temperatures and pressures. The gas giants have metal/rock cores with masses of 8–15 Earth masses ($M_\oplus$).

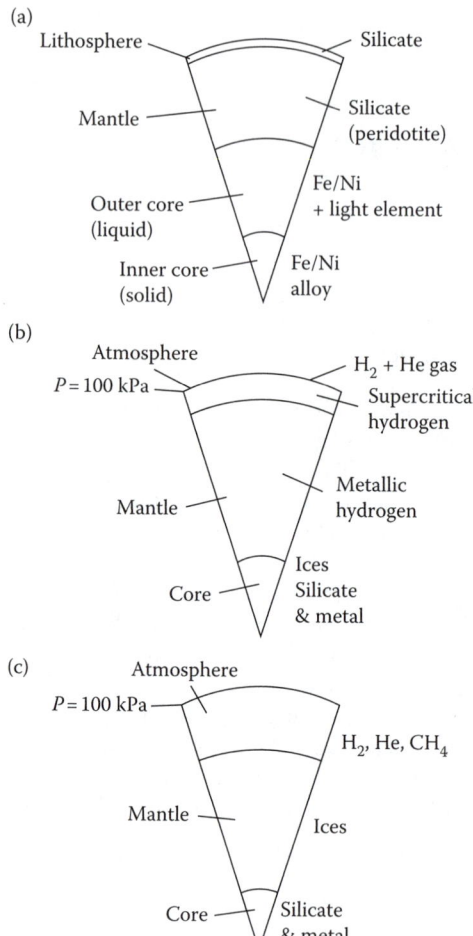

FIGURE 3.1 The internal structures of the planets: (a) Terrestrial planet, Earth. (b) Gas giant, Jupiter. (c) Ice giant, Uranus.

Uranus and Neptune are considerably smaller and lighter than the gas giants and much of their bulk is composed of volatile compounds that are ices at 0°C and one atmosphere pressure—water, methane, and ammonia—so they are described as ice giants. This is a misnomer because these volatiles are mostly at high temperatures inside these planets and are gases in the atmospheres of the ice giants.

All the giant planets have rings. Rings consist of billions of lumps of water ice in orbit around the planet. The ring particles are laced with silicate particles and tholins, large organic molecules cooked up by UV light chemistry.

3.1.1.3 Most Planets Have Moons

All of the planets except Mercury and Venus have moons, natural satellites, in orbit around them. Ganymede, the largest moon of Jupiter, is larger than the planet Mercury.

However, some moons are just a few kilometers across. Regular moons tend to be relatively large and close to their host planet. They are formed from a circumplanetary disc of gas and dust. Irregular moons tend to be small and orbit relatively far from their planet, often in a retrograde direction (i.e., in the opposite sense to the orbit of the planets around the sun): these are thought to have been captured. Planetary rings were produced either when icy moons strayed too close to the giants and were broken up by tidal forces, or were shattered by giant impacts.

3.1.1.4 Many Solar System Bodies Are Differentiated

All of the planets, many moons (and at least one asteroid, Vesta) have a layered internal structure. This differentiation came about because they are massive enough that early in their formation they heated up enough to undergo partial melting. This allowed high-density material (metals) to sink to the center and low-density material (silicates) to rise to the surface. Heat sources include radioactive isotopes and impacts. Once differentiation started, core formation itself would generate heat as gravitational potential energy is converted into kinetic energy.

3.1.2 The Planets Were Condensed from a Spinning Disc

All major planets orbit anticlockwise around the sun as viewed from above the north pole of the Earth. This is evidence that the solar system started out as a rotating disc. Anticlockwise orbits are said to be prograde. We might then expect all the planets to spin in a prograde sense around their rotational axes but Venus and Uranus have an axial tilt (obliquity) (Box 3.1) greater than 90° and hence spin in a retrograde direction. This is usually attributed to destabilizing impacts in the early solar system.

3.1.3 The Solar System Contains Numerous Small Bodies

All objects that orbit the sun that are not planets are termed small solar system bodies. The great majority lie beyond the orbit of Neptune and are termed Trans-Neptunian objects (TNOs).

BOX 3.1   OBLIQUITY

Obliquity (axial tilt, axial inclination) is defined as the angle between the equatorial plane and orbital plane of a planetary body. Equivalently it is the angle between the rotation axis and orbital axis of a body.

The relationship between orbital direction and obliquity is given by the right-hand rule. With the fingers of the right hand curled in the direction of a planet's rotation, the thumb points towards its geographical north pole. Using the right-hand rule on Venus shows that it has almost completely flipped over. In fact it has an obliquity of 177° (Figure 3.2).

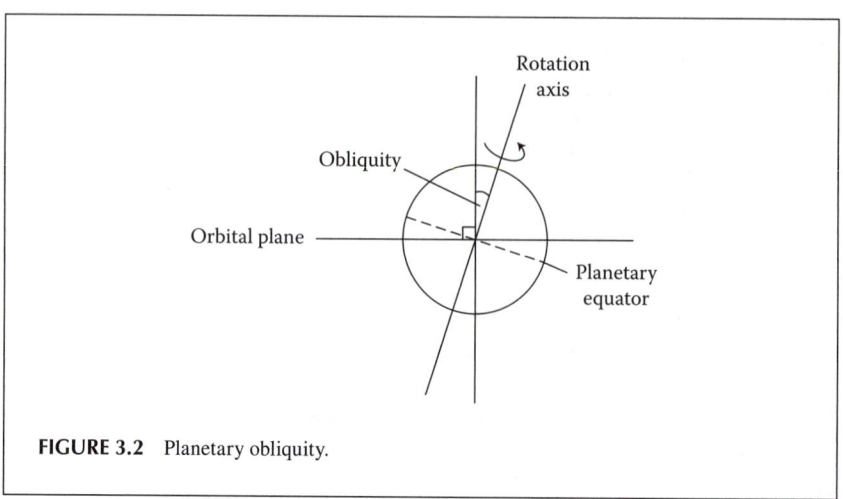

FIGURE 3.2 Planetary obliquity.

3.1.3.1 The Scattered Kuiper Belt Is the Source of Jupiter Family Short-Period Comets

A torus (doughnut-shape region), the Edgeworth–Kuiper belt or classical Kuiper belt, extends from about 30 to 52 AU and contains perhaps as many as a billion ice/rock bodies. The first Kuiper belt object (KBO), Pluto, was discovered by Clyde Tombaugh in 1930, and for three quarters of a century was regarded as a planet. We had to wait until 1992 for the second KBO to be found. The orbital inclinations (i) and eccentricities (ε) of classical KBOs are fairly modest, but there are numerous KBOs in high i and high ε orbits in a scattered disc. The combined mass of the Kuiper belt and scattered disc is only of order $0.1M_{\oplus}$, surprisingly low, but this has been a crucial clue in fathoming the evolution of the early solar system. The scattered disc is a reservoir of short-period comets.

A comet is an icy small solar system body, a few hundreds of meters to tens of kilometers across, that gets close enough to the sun (~3 AU) for some of its ices sublime to form an atmosphere (coma) and often ion tails and dust tails as well. About 500 short-period comets are known, defined as having orbits with periods <200 years.

3.1.3.2 The Oort Cloud Is the Source of Halley Family Short-Period and Long-Period Comets

The orbits of long-period comets show they originate from a distant reservoir of trillions of icy bodies termed the Oort cloud. It is also the original source of Halley family short period comets (Figure 3.3). Its inner region (2000–20,000 AU) is a torus but its outer region is a spherical shell that may extend to 200,000 AU. The total mass of the Oort cloud is reckoned to be ~$5M_{\oplus}$. Modeling suggests that the scattered disc was the source for the Oort cloud and may still be supplying the cloud now. The sun's gravitational hold on the outer edge of the Oort cloud is very weak and is located at a large fraction of the distance to Alpha Centauri, our nearest star system. It is possible then that the solar system could be contaminated by comets from other stellar systems.

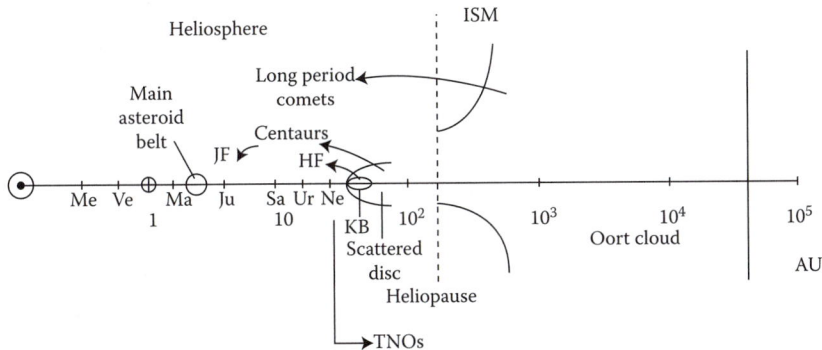

FIGURE 3.3 Large scale anatomy of the solar system and the origin of comets. There are two populations of short period comet. The Halley family (HF) are Oort cloud bodies captured by the gas giants. Objects perturbed out of the scattered disc become Centaurs and then Jupiter family (JF) comets. KB, Kuiper belt.

3.1.3.3 Asteroids Sample Very Early Solar System Material

Between Mars and Jupiter (2.2–3.3 AU) lies the main asteroid belt. This contains several million bodies with rocky and/or metal compositions, some of which are parent bodies of meteorites. Low mass asteroids are far more numerous than high mass ones (c.f. the initial mass function of stars) so the total mass of the main belt <0.01% Earth mass. Asteroids are thought to be early solar system material that failed to form a planet because of the disruptive gravitational field of Jupiter. The outer part of the asteroid belt contains cometary nuclei.

Meteorites are fragments of rock and metal that enter the atmosphere and survive frictional ablation to land on the surface. Almost all meteorites are derived from asteroids, so we can learn much about the early solar system from them. Vesta is the fourth largest asteroid and is the parent body of over 450 meteorites.

3.1.4 What Is a Planet?

In 2006, the International Astronomical Union (IAU), the organization responsible for naming celestial bodies, voted to redefine what is meant by a planet.

3.1.4.1 A Planet Must Satisfy Three Criteria

The IAU's current definition of a planet has three points:

1. It is a celestial body in orbit around the sun. (A planet-like body around a star other than the sun is an exoplanet.)
2. It has sufficient mass for its self-gravity to overcome rigid body forces so that it approaches hydrostatic equilibrium and/or assumes a nearly round shape. Rigidity depends on composition, so for silicate or water ice bodies, hydrostatic equilibrium is achieved at masses (diameters) of 3.4×10^{20} kg (600 km) and 1×10^{19} kg (320 km), respectively.

3. It has cleared its neighborhood. This means that its gravitational field is strong enough to scatter other bodies in its vicinity, collide with them, capture them so that they become satellites (moons), or lock them into orbital resonances.

A body, such as Pluto, that has not cleared its neighborhood but which satisfies the other two criteria, and is not a satellite, is a dwarf planet.

3.2 Habitability Is an Attribute of the Entire Earth System

To understand how the Earth came to be habitable we need to know how the Earth system works as a whole. This is because the solid Earth, oceans, atmosphere, and life are all intimately coupled.

The mantle is the ultimate source of the Earth's atmosphere and oceans and helps determine their composition. Plate tectonics, a continual recycling of the ocean floor and mantle, stabilizes the Earth's surface temperature by regulating the CO_2 concentration of the atmosphere. Plate tectonics requires water and much of the Earth's water is in the mantle. Thus, the oceans are recycled through the mantle. The Earth's magnetic field (magnetosphere) protects the atmosphere from being blasted away by the solar wind. The geodynamo that generates the magnetosphere lies in the core yet may need plate tectonics to operate. Life itself has terraformed the Earth, altering its atmosphere and oceans, and creating hundreds of new minerals in the solid Earth.

Therefore, it is the interplay between all the components of the Earth system that contributes to the Earth's habitability. Moreover, geology played a critical role in the origin of life itself.

To set the scene for our investigation of Earth as a habitable planet, we look first at the internal composition and structure of the Earth.

3.2.1 The Structure and Composition of the Solid Earth

Rocks much below the Earth's surface are largely inaccessible so we have not been able to *systematically* sample the interior of the Earth.

We are even worse off with the other terrestrial planets. We currently have 69 meteorites from Mars, about 180 from the moon, and about 382 kg of lunar rock returned by the Apollo missions, and these are not representative samples. Remote sensing data from orbiting spacecraft can tell us about the composition of terrestrial planets down to depths exposed by impact craters.

Hence, we have been forced to use indirect methods to work out the nature (i.e., composition, state) of materials inside terrestrial planets, including laboratory experiments on tiny rock samples that mimic high-temperature, high-pressure conditions deep in the Earth and theoretical modeling. We look at two other approaches in the following sections.

3.2.2 The Chondritic Earth Model Provides a First Approximation to Its Bulk Composition

One approach to understanding the make-up of the Earth and terrestrial planets is to assume that they accreted from a class of primitive meteorites, chondrites, the

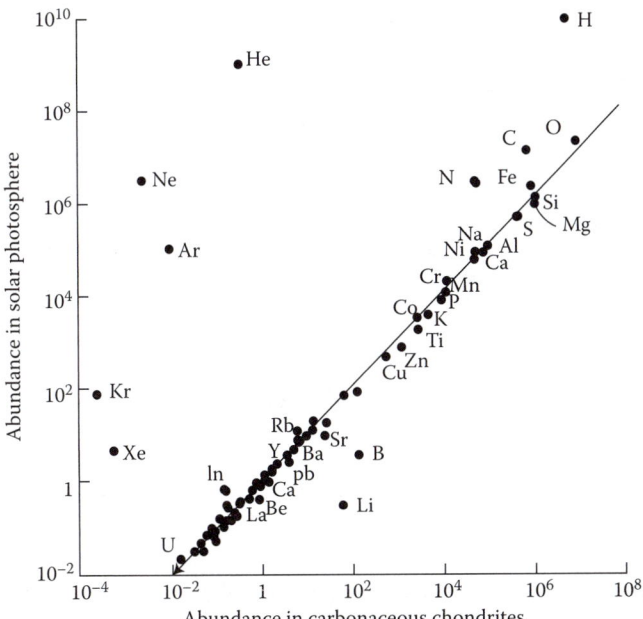

FIGURE 3.4 Relative abundances of elements in solar photosphere and carbonaceous chondrites. Abundances are normalized to Si = 10^6 atoms.

composition of which is known. This chondritic Earth model is based on the high correlation between abundances of elements in the sun with their abundances in chondrite meteorites (Figure 3.4), with the exception of the most volatile elements (H, He, C, N, O, and noble gases). Solar abundances for most elements have been determined by spectroscopy. Chondrites are thought to be fragments of the first kilometer-sized bodies (planetesimals) to accrete. Thus, they are assumed to sample the nonvolatile material out of which the planets were built. Chondrites are undifferentiated: their parent bodies never heated enough for their metal and silicate fractions to separate.

The difficulty with the chondritic Earth model is that mass balance calculations show that no class of chondrite or mixture can account for any plausible model Earth composition.

3.2.3 Seismology Provides a Picture of the Earth's Interior

Seismic (earthquake) waves are refracted or reflected at boundaries between rocks of different stiffness and density, altering their direction in predicable ways in a manner analogous to the refraction and reflection of light rays. Seismometers scattered around the globe detect the arrival of seismic waves. From seismometer records, geologists can piece together the locations of major boundaries inside the Earth from changes in the velocity of seismic waves. The boundaries or discontinuities occur where the mechanical properties of the medium alter because of a phase change. This could be a switch between solid and liquid, a change in chemical composition, or a transformation to a different molecular structure (e.g., at high pressures minerals morph

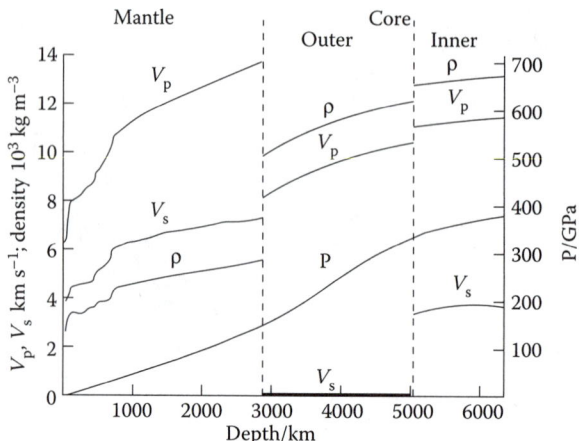

FIGURE 3.5 The interior of the Earth revealed by seismic studies. v_p, p-wave velocity; v_s, s-wave velocity; ρ, density.

into higher density forms that take up less volume, thereby minimizing their internal energy).

Two types of seismic waves propagate deeply enough to probe the Earth's interior: primary or P waves and secondary or S waves. P waves are the first to arrive at a seismometer. They are acoustic (sound) waves: longitudinal waves that propagate at the speed of sound as compressions and rarefactions of the rock. S waves, sometimes called shear waves, are transverse waves that oscillate at right angles to their direction of travel. P waves will go through liquids and solids, but S waves cannot travel through liquids, because liquids cannot be sheared. This distinction has proved crucial. Generally, seismic wave velocity is higher in dense, cooler rock and lower in less dense, hotter rock.

By the mid-twentieth century, seismology had pretty much given us the modern picture of our planet's interior. By deriving a differential equation that describes self-compression, mathematical models for how density, gravity, pressure, and temperature vary with depth were derived and matched with predictions of seismology (Figure 3.5).

3.3 What Makes Earth Habitable?

The Earth has been inhabited for most of its 4.5 Gyr history. What is remarkable about this is that early Earth was very different to what it is now and would have been very inhospitable to most modern organisms. Covered almost entirely by water, there would have been little dry land. The ocean temperature *may* have been warmer than today and tides would have been much higher because of the proximity of the moon. There was no free oxygen in the atmosphere until 2.4 Gyr ago so the surface would have been bathed in high UV flux, and the atmospheric pressure may have been several times greater than now. What this tells us is that conditions for habitability have changed over time. A number of factors that contribute to Earth's habitability have been proposed.

3.3.1 Temperature at the Earth's Surface Is Largely Determined by the Sun

Life has an absolute requirement for free energy, as we shall see in Chapter 6. Crucial energy sources for terrestrial life are the sun and geothermal energy.

Three factors contribute to the surface temperature of the Earth. By far the most important is solar radiation. The others are the composition of the atmosphere and internal heat sources. We can calculate the temperature of the Earth assuming that solar radiation is the only energy source and that the Earth is a black body radiator. This is termed the effective temperature, T_{eff}.

We start by working out the flux of solar radiation (across all wavelengths) per unit area at 1 AU from the sun. This is the total solar irradiance or solar constant (I). As the sun's power is distributed over the surface of a sphere of area $4\pi a^2$, where a is the average Earth–sun distance, the total solar irradiance is given by

$$I = \frac{L_\odot}{4\pi a^2}. \tag{3.1}$$

At the top of the atmosphere, I averages 1370 W m^{-2}, but it varies by 0.1% during the solar cycle (greatest at solar maximum) and by about 6.9% during the year as the Earth–sun distance changes due to the eccentricity of the Earth's orbit.

The cross section of the Earth (the area of a disc with Earth's radius, R_E) is πR_E^2 so it intersects 1370 W m^{-2} × πR_E^2 m^2 of the solar power. This is the power incident on the whole Earth:

$$P_{in} = \frac{L_\odot \pi R_E^2}{4\pi a^2}.$$

Not all of the incident solar radiation is absorbed; a fraction is reflected. This is the albedo. If all the power is absorbed $A = 0$; if all is reflected $A = 1$. Hence, the power absorbed by the Earth, P_{abs}, is given by

$$P_{abs} = \frac{L_\odot \pi R_E^2 (1 - A)}{4\pi a^2} = \frac{L_\odot R_E^2 (1 - A)}{4a^2}.$$

Treating the Earth as a black body means that the power it radiates, P_{rad}, is given by the Stefan–Boltzmann relation

$$P_{rad} = 4\pi R_E^2 \sigma T_{eff}^4.$$

For a black body, P_{abs} and P_{rad} are the same; so,

$$\frac{L_\odot R_E^2 (1 - A)}{4a^2} = 4\pi R_E^2 \sigma T_{eff}^4,$$

$$T_{eff} = \left(\frac{L_\odot (1 - A)}{16\pi \sigma a^2} \right)^{1/4}. \tag{3.2}$$

The Earth reflects about 30% of incident radiation back into space, so has an albedo, $A = 0.3$. With this we find that $T_{eff} = 255$ K. But the average annual global surface temperature of our home world is 15°C (288 K) *not* −18°C! The difference arises because while the Earth absorbs mostly short-wave (visible) radiation from the sun it radiates long-wave (IR) radiation. This is absorbed and reradiated by greenhouse gases in the atmosphere, notably water, carbon dioxide, and methane. This warms the atmosphere and the surface. While greenhouse gases get a bad press we see that without them there could be no liquid water on the surface of the Earth.

Because the Earth is spherical, over a period longer than a day the solar irradiance intercepted by a disc of area πR_E^2 will be spread over an area $4\pi R_E^2$; that is, fourfold larger. So the effective solar flux at the top of the atmosphere will be $I/4 = 1370/4 = 343$ W m^{-2}. With an albedo of 0.3, then $0.7 \times 343 = 240$ W m^{-2} will be absorbed by the atmosphere and surface to warm the planet.

Geothermal heat flowing from the Earth's interior to the surface is negligible by comparison to solar irradiance. The average geothermal heat flow is ~0.086 W m^{-2}; only 0.036% of the heat that is supplied by sunlight.

3.3.2 Liquid Water Exists on the Earth's Surface

All life on Earth depends on water. The combination of sea level pressure of 101.3 kPa and global average surface temperature of 288 K allow liquid water to exist over much of the surface of the Earth. Indeed the surface temperature is close to the triple point at which all three phases of water exist together at equilibrium. Moreover, the atmospheric pressure is 100-fold that of the triple point pressure, so, on the Earth's surface ice always melts to form liquid water if the temperature is high enough. This is in contrast to Mars. Here the surface pressure is lower than the triple point pressure for water, so ice on the Martian surface goes straight to water vapor. In other words, liquid water cannot exist on the surface of Mars (Figure 3.6).

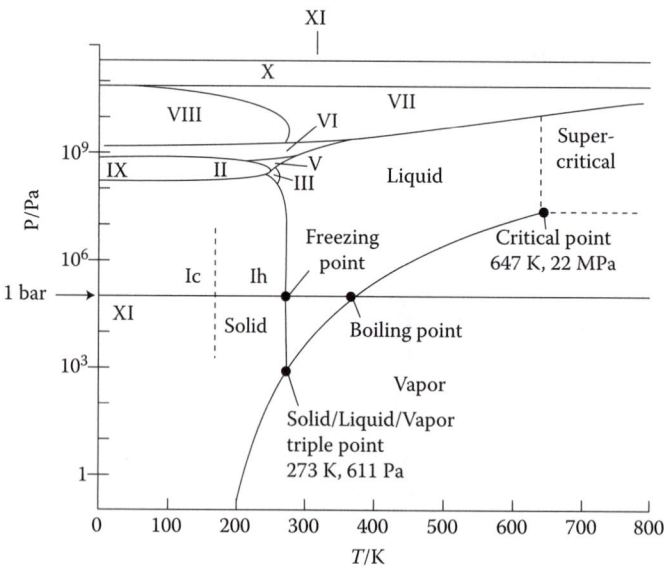

FIGURE 3.6 Phase diagram for water. Note that there are numerous solid phases; ice Ih is the one we use to cool our drinks!

BOX 3.2   UNITS OF PRESSURE

The SI unit of pressure is N m^{-2} or the pascal (Pa) and is convenient for expressing pressures in planetary interiors that range from MPa to GPa depending on depth. Pressures of planetary atmospheres are often given in bar, where 1 bar = 0.1 MPa. Since the mean pressure of the Earth's atmosphere at sea level is 101.3 kPa or 0.1013 MPa, a bar is almost the same as one atmosphere (atm).

Above the critical point ($T_c = 647$ K, $P_c = 22$ MPa) water is a supercritical fluid where there is no distinction between liquid and gas phases. The atmosphere of Venus consists of ~96.5% CO_2 and ~3.5% N_2. At the surface pressure (9.5 MPa or 93 bar) (Box 3.2) and temperature (735 K) of the Venusian atmosphere both of these molecules are above their critical points so the atmosphere of Venus is a supercritical fluid.

3.3.3 Earth Is in a Stable Orbit in the Habitable Zone

As we have seen, the Earth's orbit is the correct distance from the sun for the solar flux to allow liquid water to exist at the surface, given that the Earth has a dense nitrogen atmosphere containing the greenhouse gases, water, and carbon dioxide (Plate 3.1). The range of distance over which this is possible is termed the circumstellar

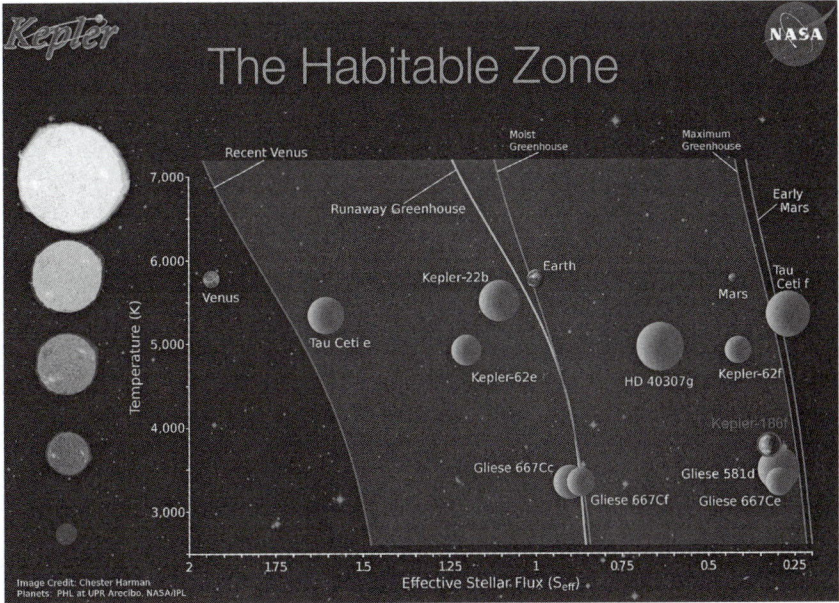

PLATE 3.1 **(See color insert.)** Habitable zones for stars. For definitions of the inner and outer edges of the zone, see text. Kepler 186f is the first Earth-sized planet discovered in a habitable zone. (Photo credit: NASA/Chester Harman.)

habitable zone (HZ) or "Goldilocks zone" (after the nursery rhyme "Goldilocks and the Three Bears" about a girl who tasted three bowls of porridge, finding one too hot, one too cold, and one "just right"). It is important to note that the HZ is defined not just by the distance from the sun, but by the atmosphere as well.

3.3.3.1 The Inner Edge of the Habitable Zone Is Defined by the Moist Greenhouse Effect

A number of estimates of the width of the current solar system HZ have been made. Most are similar to an influential 1993 climate model of Jim Kasting, Daniel Whitmire, and Ray Reynolds. This is a one-dimensional (1-D) model, so called because it treats the atmosphere as a column. It is also a radiative–convective model. Its goal is to calculate the temperature of a planetary surface. Its starting point is that the temperature is a balance between radiation gain from the sun and radiation loss, the basic physics of which is encapsulated in Equation 3.4. This is combined with data on how effectively greenhouse gases absorb and emit radiation and the spectrum of the atmosphere, which shows quantitatively how greenhouse gases are distributed through the atmosphere. The radiative part of the model works out how radiation is absorbed and emitted at different heights. The convective part models transfer heat by convection. This must consider the exchange of latent heat as water condenses and evaporates. The result is a temperature profile of the atmosphere.

Kasting's model places the inner boundary at 0.95 AU where the solar flux is 10% higher than at 1 AU. Modeling shows that if the Earth were to be transported to 0.95 AU, the radiation gain would still be balanced by radiation loss although at a higher surface temperature. This increases the rate of water evaporation from the ocean to the extent that it would saturate the stratosphere. Here it can be photodissociated by solar UV radiation:

$$H_2O \rightarrow H_2 + O. \qquad (3.3)$$

The resulting hydrogen would escape into space. This represents an irrevocable loss of water and it defines the water loss or moist greenhouse limit. But the surface temperature, while higher than it is at 1 AU, is constant.

If instead the Earth is moved even closer, to 0.84 AU, the loss of radiation by the Stefan–Boltzmann process can no longer balance the insolation: the Earth no longer behaves like a black body. The Earth cannot attain radiative balance so the temperature keeps rising because the greenhouse effect of the evaporating water retains heat more effectively than the planet can radiate it. This is a positive feedback effect: increasing temperature boosts water evaporation adding to the temperature rise. The end result is that the oceans boil away. Some authorities regard this runaway greenhouse effect as marking the inner limit on the HZ.

A 2013 reworking of Kasting's climate model by Ravi Kumar Kopparapu and colleagues, which uses the revised values for the absorption of radiation by CO_2 and water, locates the water loss limit at 0.99 AU and the runaway greenhouse limit at 0.97 AU. This puts the Earth perilously close to the moist greenhouse limit. However, the model does not include clouds. Yet a warmer Earth would have greater cloud cover and hence a higher albedo. The model also fully saturates the troposphere with water, maximizing the greenhouse effect, but this probably would not be the case. For these reasons, the inner edge of the HZ would likely be a bit closer than the model implies. For the Earth the moist greenhouse limit occurs at a surface temperature of 340 K. To achieve this at 1 AU would require a 10-fold increase in atmospheric CO_2.

Note that Venus, at 0.7 AU is too hot to retain water even with the 1993 model. If it had oceans early in its history they would have evaporated, driving a runaway greenhouse effect. There is no sink on Venus, as there is on Earth, to sequester most of the CO_2 in its atmosphere. Hence Venus is now a dry CO_2 greenhouse.

3.3.3.2 The Outer Boundary of the Habitable Zone Is Defined by Maximum CO_2 Greenhouse

The outer boundary is determined by the physics of carbon dioxide in the atmosphere. It is assumed that habitable planets will have volcanoes that outgas CO_2. At the outer boundary, lower insolation or greenhouse gases reduce the temperature so much that surface water freezes, raising the albedo and so reflecting more sunlight. This is a positive feedback process. The lower water vapor pressure also means less of a water greenhouse effect. Together, both of these processes drop the temperature to the point where CO_2 condenses, thereby removing the CO_2 greenhouse. This is defined as the outer edge of the HZ in the 1993 climate model. However, it is now known that CO_2 clouds have a warming effect, so the 2013 model defines it as the maximum CO_2 greenhouse effect. This works in the following way: CO_2 is a very efficient Rayleigh scatterer, so a dense carbon dioxide atmosphere has a high albedo that works in opposition to its greenhouse effect. The outer edge of the HZ is where this scattering begins to outweigh the greenhouse effect. Hence, it is the point where the CO_2 greenhouse is at its greatest. It depends on distance from the sun and the amount of CO_2 in the atmosphere. For the Earth it lies at about 1.7 AU, although this may be a conservative estimate because it does not include CO_2 clouds that produce radiative warming.

Mars, at 1.5 AU, is currently outside the HZ mainly because it lacks a dense atmosphere. Almost all of its CO_2 has condensed at the polar caps or in high altitude clouds. Whether Mars was in the HZ early in its life is a controversial point we take up in Chapter 10.

3.3.3.3 The Habitable Zone Changes with Time

The sun was about 30% less luminous 4.6 Gyr ago than today (see Chapter 1, Section 1.3.5.1). This means that the HZ must have been closer to the sun then than now. The 1993 model has the outer edge at 1.4 AU. Hence the region between 0.95 and 1.4 AU has been habitable for the entire history of the solar system. This is termed the continuously habitable zone (CHZ). The Earth has always been inside it because its orbit has been stable over the lifetime of the solar system and it has a low eccentricity. None of the above should be taken to mean that the Earth's climate has not changed over geological time; it has, and the Earth's orbit has been an important forcer.

3.3.3.4 Habitable Zones Are Defined for Exoplanet Systems

Assuming that water is a universal requirement for life, the concept of circumstellar habitable zones can be applied to planets around other stars, including those of different spectral type (Plate 3.1). Consequently, we shall return to the topic of habitable zones again in Chapter 13.

It is worth noting that life may thrive in places that are not within a circumstellar habitable zone. We will look at these in detail later, but for now have a go at Exercise 3.1.

EXERCISE 3.1

Suggest locations where water may be liquid outside a habitable zone around a star.

3.3.4 Earth's Dense Atmosphere Contributes to Habitability

The Earth has a dense atmosphere. As we have seen, the high surface pressure contributes to the presence of liquid water at the surface. But a dense atmosphere can contribute to habitability in other ways. It protects the surface from high-energy particles such as cosmic ray particles, which are energetic enough to destroy biomolecules. It distributes heat energy around the planet, minimizing temperature variations across latitudes, and between the day and night sides.

3.3.4.1 A Planet's Mass Determines What Gases It Can Retain

The column mass of the Earth's atmosphere is 10^4 kg. The force this mass exerts is approximated by $F = mg = 10^4$ kg $\times 9.81$ m s$^{-2} \sim 10^5$ kg m s^{-2}. Since this is over a unit area of the Earth's surface this is a pressure of 10^5 kg m^{-1} s^{-2} or 10^5 Pa. Clearly the pressure depends on the mass and radius of the Earth since this determines the acceleration due to gravity, g, at the surface. But whether a planet holds onto its atmosphere also depends on its temperature. We explore this next by looking at a parameter termed the escape velocity.

The escape velocity, v_e, is the speed a particle launched into a ballistic trajectory (the path a body will take under gravity and without any propulsion after launch) requires for it to escape from a particular location within a gravitational field. A rocket need not achieve the escape velocity at the Earth's surface to escape from the Earth's gravity, because it can continue to gain kinetic energy for as long as its engines generate thrust.

Escape velocity is defined as the speed at which the kinetic energy plus gravitational potential energy of a body is zero. Consider a gas atom with mass m in a planetary atmosphere. At time, t, it has a kinetic energy, $E_{K,t}$, and a gravitational potential energy, $E_{G,t}$, by virtue of its distance, r, from the center of gravity of the planet, which has mass M. At some later time, t_0, it has escaped from the gravitational field. By the conservation of energy,

$$(E_G + E_K)_t = (E_G + E_K)_{t0}.$$

But if the particle has escaped from the gravitational field, then it will be infinitely far from the planet so will have zero gravitational potential energy and if it only just escaped it will have zero kinetic energy as well. Hence,

$$(E_G + E_K)_{t0} = (0 + 0).$$

So with $E_K = \frac{1}{2}mv^2$ and $E_G = -GMm/r$:

$$\frac{1}{2}mv_e^2 + \left(-\frac{GMm}{r}\right) = (0 + 0),$$

$$v_e = \left(\frac{2GM}{r}\right)^{1/2}.$$

(3.4)

EXERCISE 3.2

Calculate v_e for the Earth.

A major route by which atoms and molecules escape from the atmosphere is Jeans escape. This is thermal escape from the high-energy tail of the Maxwell–Boltzmann velocity distribution (Box 3.3), where velocities are much higher than the average value. As a rule of thumb, there are significant numbers of particles up to about six times the typical velocity, so a planet is expected to hold onto a gas for a timescale comparable to the age of the solar system if its atoms or molecules have velocities less than one-sixth escape velocity.

Of course molecules are lost from the top of a planetary atmosphere. To see what governs this we need to take a brief look at the structure of the Earth's atmosphere (Figure 3.8). In the bottom 100 km of the atmosphere, below the homopause, turbulent mixing (eddy diffusion) ensures that the composition of the atmosphere is homogeneous, but in the thermosphere the composition varies with height. The thermosphere contains significant numbers of atoms of oxygen and nitrogen generated by photodissociation of O_2 and N_2 by solar ultraviolet radiation. The number of heavier molecules of nitrogen and oxygen falls off more rapidly with increasing altitude than does the

---

BOX 3.3   MAXWELL–BOLTZMANN DISTRIBUTION

Except at absolute zero, particles—electrons, atoms, ions, and molecules—have kinetic energy. For an ideal gas, particle speeds (or equivalently kinetic energies) have a characteristic Maxwell–Boltzmann distribution. Most gas atoms or molecules will have speeds close to the most probable value, but some speeds will be slower, and there will be a substantial number with appreciably higher speeds (Figure 3.7).

The peak of the distribution is the most probable speed v_p. This is given by

$$V_p = \left(\frac{2kT}{m}\right)^{1/2}.$$

(3.5)

where m is the mass of the particle. The average speed is higher than this because the distribution is not symmetrical.

EXERCISE 3.3

Show that Equation 3.5 has units of velocity.

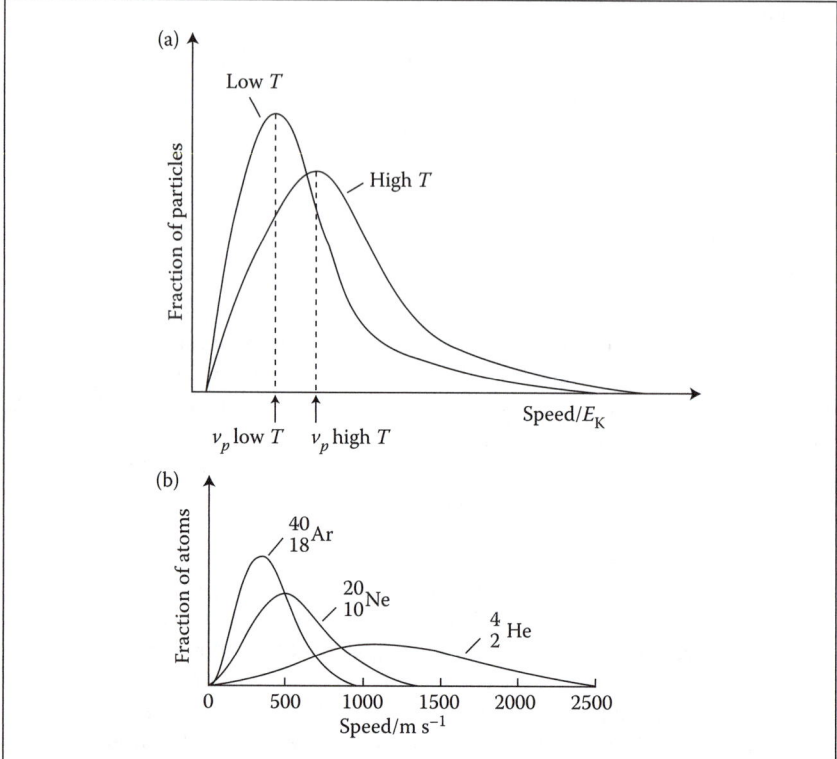

FIGURE 3.7 Maxwell–Boltzmann distribution for a gas at thermal equilibrium. The area under the curve is proportional to the mass of the gas. (a) As the gas is heated, the most probable speed and the average kinetic energy (temperature) increase, as does the fraction of particles in the high speed tail. (b) The lower the mass number, the higher is v_p.

number of the lighter atoms of nitrogen and oxygen. Hydrogen and helium, which are even lighter, become relatively abundant in the upper thermosphere. This stratification occurs because without mixing, the various gas species separate by molecular diffusion. Lighter particles diffuse faster than heavier ones.

At the top of the thermosphere is the exobase. Here the density is so low that the chance of an atom or molecule colliding and being scattered from an outgoing trajectory is very low. For the Earth, the exobase is the top of the thermosphere at an average altitude of about 600 km (though it varies between 500 and 1000 km). Here v_e is a bit less than at the surface because of the lower gravitational potential; about 10.8 m s^{-1}. Because this region is rich in oxygen (a EUV absorber) and poor in CO_2 (an IR radiator), it is hot. The temperature ranges between ~1000 K (solar minimum) and ~2500 K (solar maximum).

EXERCISE 3.4

The exobase is virtually a collisionless region. How does this lead to thermal escape for particles with speeds greater than escape speed?

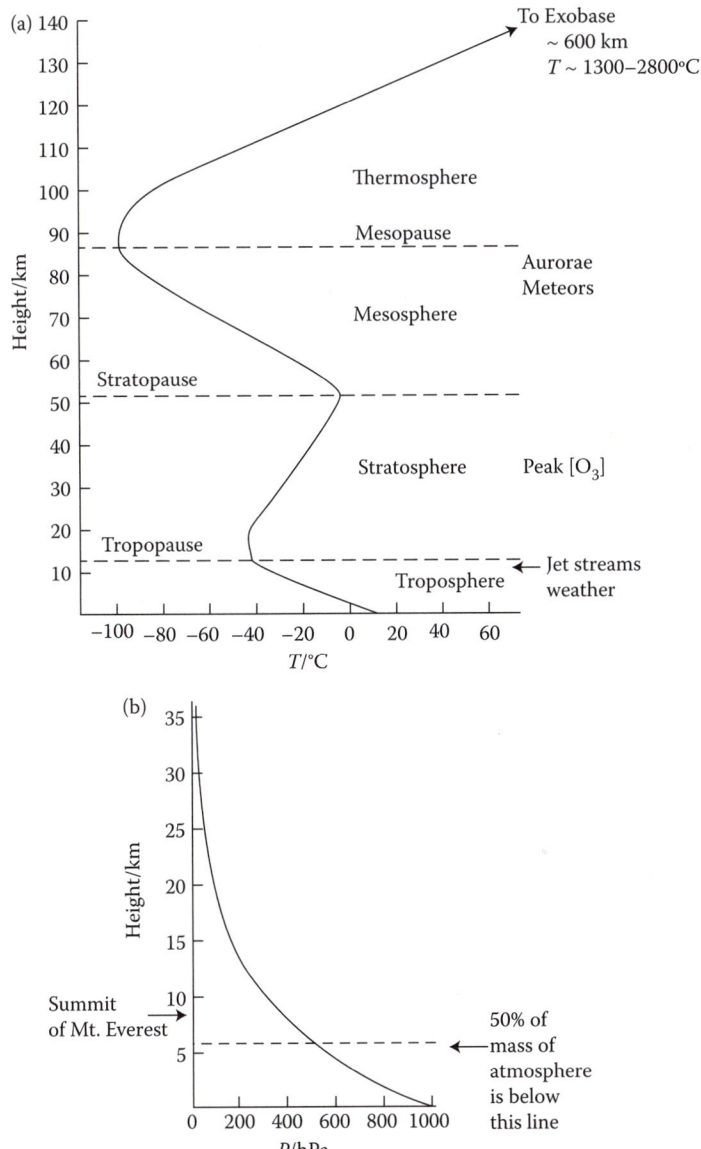

FIGURE 3.8 The structure of the Earth's atmosphere. (a) temperature structure (b) pressure. The summit of Mt. Everest (8848 m), at over 8000 m, is in the death zone where the partial pressure of oxygen is too low to sustain human life for longer than a few hours.

EXERCISE 3.5

 a. What is the most probable speed of a nitrogen molecule (composed of two $^{14}_{7}\text{N}$ atoms) at the Earth's exobase around solar maximum? (1 amu = 1.661×10^{-27} kg)

b. Is there any risk that in the long term the Earth will lose its nitrogen atmosphere?

EXERCISE 3.6

Is the Earth likely to lose hydrogen atoms from its atmosphere?

You should have found that the most probable speed for hydrogen atoms at the exobase is not so different from the exobase escape velocity of ~10.8 km s^{-1}, so appreciable hydrogen could be lost from the Earth's atmosphere even at the lowest temperature. We would expect hydrogen loss to be even greater at solar maximum. In fact, there is a barrier to hydrogen diffusion at the homopause that can slow hydrogen loss from the top of the atmosphere. Figure 3.9 shows how V_e and temperature determine the atmospheric composition of planets.

EXERCISE 3.7

When might be the rate-limiting boundary for hydrogen escape from the atmosphere at (a) the homopause, and (b) the exobase?

3.3.5 A Global Magnetic Field May Be Required for Habitability

The Earth has a global magnetic field. This acts as a shield against high energy particles in the solar wind that might otherwise erode the atmosphere. Since a dense atmosphere contributes to Earth's habitability, maybe its magnetic field does too. Here it is instructive to compare our planet with Venus and Mars, neither of which has global magnetic fields.

3.3.5.1 *Earth's Magnetic Field Helps It Retain Its Atmosphere*

Cosmic rays and solar wind particles are mostly protons and electrons with sufficient energy to break chemical bonds in biological molecules. Earth's field offers little

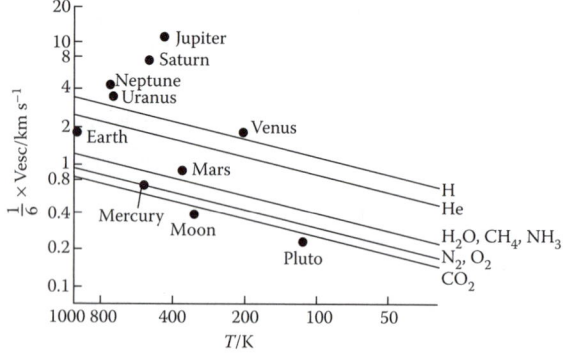

FIGURE 3.9 Gases that solar system bodies can retain. If the body lies above the line, the gas is retained. For bodies with an atmosphere, T is exobase temperature. For bodies without an atmosphere, $T = T_{eff}$.

protection against high energy galactic or extragalactic cosmic rays. However, it does shield us from solar wind and this protects the atmosphere from being eroded. In this process, termed sputtering, high velocity solar wind particles collide with gas molecules, atoms, and ions in the upper atmosphere giving them enough kinetic energy to exceed escape velocity.

3.3.5.2 Earth's Atmosphere Has Been Shielded by a Magnetic Field for at Least 3 Gyr

Measurements of remnant palaeomagnetism in a "fossil" Archaean magma reservoir in Montana, USA show that the Earth's magnetic field strength was about the same 3 Gyr ago as it is now. We have reason to think that the Earth had a global magnetic field at earlier times. However, the early sun had higher x-ray and UV fluxes than now. This would have heated the upper atmosphere so that it expanded beyond the magnetopause making it vulnerable to the solar wind.

3.3.5.3 Mars Lacks a Magnetic Field and Has Lost Much of Its Atmosphere

Mars, which has lacked a magnetic field for the last 4 Gyr, could have lost much of its early atmosphere to sputtering. Indeed the Aspera instrument on Mars Express shows the red planet is still losing its atmosphere in this fashion. The lower escape velocity of Mars makes its atmosphere more vulnerable than that of Earth or Venus.

3.3.5.4 Venus Lacks a Magnetic Field yet Has a Dense Atmosphere

Venus poses something of a puzzle. It lacks a magnetic field and yet has an atmosphere almost 100-fold denser than that of Earth. Of course, Venus and Earth have similar masses and therefore comparable escape velocities, which make them less vulnerable to sputtering than Mars. The survival of the dense Venusian atmosphere may be because, being almost entirely carbon dioxide, it is less susceptible to sputtering than Earth's atmosphere, which is mostly nitrogen. This is because CO_2 is a greenhouse gas whereas N_2 is not. CO_2 at the top of an atmosphere radiates heat, so cools and is less likely to be accelerated beyond escape velocity by solar wind particles than N_2. The implication is that a magnetic field may not be important for the habitability of a planet that is massive enough to retain a dense atmosphere of CO_2.

3.3.5.5 A Global Magnetic Field Requires a Liquid Core

Generation of a global magnetic field in a terrestrial planet requires a liquid core or partly liquid core capable of convection and (generally) planetary rotation. Electric currents in the molten metal produce magnetic fields and electromagnetic induction results in a self-sustaining dynamo. A global magnetic field is not possible for a terrestrial planet with a completely solid core since the metal would be above its Curie temperature.

The heat source that sustained a (partly) liquid core for the first few hundred million years was the core formation. As dense metal fell toward the center, gravitational potential energy was turned into kinetic energy and hence heat. This is insignificant now. The dominant *core* heat source for most of Earth's history has probably been inner core freeze-out. Secular cooling of the entire Earth means that the solid inner core is gradually growing, at about 1 mm per year, at the expense of the liquid outer core. As new inner core freezes, it releases the latent heat of fusion that helps drive thermal convection in the outer core.

3.3.5.6 The Core Has a Light Element That Is Crucial for the Geodynamo

Freeze-out of the inner core is thought to play an important role in the operation of the geodynamo. The reason is that the Earth's core contains one or more light elements (S, O, Si, etc.) which have partition coefficients (Box 3.4), $D = [i]_{liquid}/[i]_{solid} > 1$, where i is any element; in other words, they preferentially go into the liquid outer core.

The outcome is that Earth's outer core is about 8 wt% of light elements. This has two effects. First, it lowers the melting point of the outer core, keeping it liquid despite the crushing pressure. Second, light element-enriched material at the inner core/outer core boundary has a lower density and buoyantly rises, adding compositional convection to thermal convection.

Thermal convection will occur in the core if its temperature gradient is sufficiently steep. This requires the mantle heat flow to be above a critical value. If the mantle removes heat faster than the critical rate, convection occurs and the dynamo is activated. If the mantle is too hot or fails to cool, there will be no convection and no dynamo. If we add compositional convection, we can reduce the thermal constraints: outer core convection may still happen even if heat flow through the core/mantle boundary is less than the critical value.

The nature of the light element in the Earth's core is uncertain. Candidates include O, S, Si, C, and H.

3.3.5.7 Why do Mars and Venus Lack Magnetic Fields?

The melting points of plausible composition planetary cores are rather low, making *total* freezing of the core difficult. Indeed, currently we think all the terrestrial planets have at least partially liquid cores. So, why do Venus and Mars lack a global magnetic field? We cannot be sure.

Three models have been proposed to account for why the Martian dynamo failed. The most plausible is that the Martian core has a high sulfur content (perhaps 10–17

BOX 3.4   PARTITION OF ELEMENTS BETWEEN LIQUID AND SOLID PHASES

The extent to which an element distributes itself between different phases is its distribution coefficient or partition coefficient, D. A given element, i, will distribute between phase-a and phase-b so that

$$D_i = [i]_a/[i]_b$$

If one of the phases is liquid and the other solid, the convention is that phase-a is the liquid phase.

wt%) which keeps it entirely liquid. The cooling rate of the core would have declined very early (as small Mars lost its heat rapidly), and once it got too low, the dynamo would have shut off. This model predicts that Mars will not have a solid inner core; the other models require one. Unfortunately, moment of inertia and tidal deformation measurements by Mars orbiters has not been able to distinguish between a partial or entirely liquid core. Laboratory experiments that recreate the pressures (up to 40 GPa) and temperatures (up to 2200 K) of the Martian core show that a Fe–Ni alloy with 10.6% sulfur is entirely liquid. Intriguingly, as Mars continues to cool, the solid inner core would be expected to freeze out which might eventually reactivate the magnetic field of Mars!

The lack of a global magnetic field on Venus is often attributed to its slow rotation rate. However, rotation is not an *absolute* requirement for a planetary dynamo, though it is much easier for geodynamo models to generate a magnetic field if there is rotation. But there is an alternative explanation. Venus, lacking plate tectonics, has retained much more of its internal heat than Earth, despite being of comparable size. This could mean that the Venusian core is entirely liquid, assuming it contains light elements. Without the freeze-out of an inner core Venus and with presumably only a modest temperature gradient across the core/mantle boundary, there is no core convection and so no magnetic field.

3.4 Earth Seems Unique in Having Plate Tectonics

Plate tectonics is a continuous recycling of Earth's surface with the mantle below. It acts like a thermostat to regulate the CO_2 concentration of the atmosphere and hence the surface temperature. This it does by means of a feedback mechanism that may have ensured that the Earth remained within the continuously habitable zone over its 4.5 billion year existence, even though the sun's luminosity was lower in the past. Plate tectonics may also be necessary to maintain a global magnetic field over geological timescales. If so, it has helped prevent the loss of the atmosphere by solar wind sputtering. Water is needed for plate tectonics, so it is tempting to think that the two may go hand-in-hand on other worlds too.

3.4.1 Plate Tectonics Depends on a Weak Mantle Layer

Plate tectonics emerged from geological evidence that continents now a long way apart were once attached (continental drift) and the hypothesis that the mantle convected. Where hot mantle rose it ruptured the overlying lithosphere. The convecting mantle then carried the rigid lithosphere, piggyback, until it cooled and descended.

The notion of mantle convection may seem very strange given that the mantle is (mostly) solid. However, hot crystalline mantle undergoes solid state creep in which it behaves like a very viscous liquid. This is convection, although it happens many orders of magnitude slower than conventional fluid convection.

There is a layer in the upper mantle in which the seismic wave velocities are anomalously low called, appropriately, the low velocity zone (LVZ) or asthenosphere. Here the mantle is mechanically weak and behaves plastically (that is it deforms nonelastically) because its dynamic viscosity is low: in fact, recent work shows that

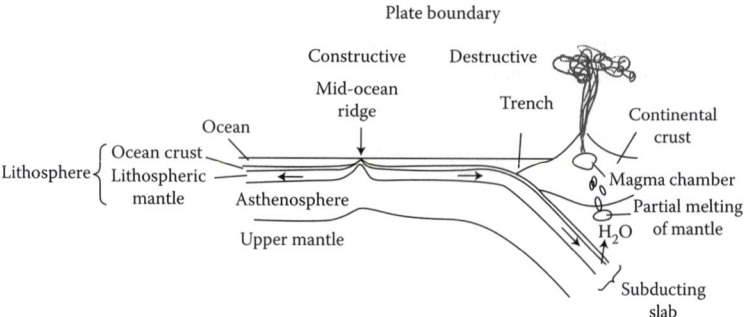

FIGURE 3.10 Plate tectonics allows Earth to lose internal heat.

in places the asthenosphere is partially molten. It is on this low velocity zone that the rigid lithosphere floats. The lithosphere consists of the crust and an underlying layer of solid mantle, and it is broken into a dozen large plates and a number of smaller ones. Oceanic plates are about 60 km thick. Most of this is mantle because oceanic crust is only about 7 km thick. Oceanic crust consists mostly of basaltic composition rock though it acquires a sedimentary covering. Continental plates are about 100 km thick. Continental crust varies greatly in thickness, 30–90 km, though averages about 35 km. The higher the topography, the thicker the crust; mountains have "roots" so this is where continental crust is thickest. Continental crust is made from many different rock types, but its bulk composition has higher silica (SiO_2) content than basalt and importantly has a lower density than oceanic crust (Figure 3.10).

3.4.2 New Ocean Crust Is Made at Constructive Plate Boundaries

Mid-ocean ridges (MORs) are where new ocean lithosphere is being made so the plate boundary they define is called a constructive (or divergent) boundary. Here convection brings hot mantle up and as it ascends it decompresses, and this causes it to partially melt to form basalt magma. This is injected between the plates, driving seafloor spreading at rates between 20 and 200 mm yr^{-1}. Constructive boundaries are born when rising magma thins overlying lithosphere so that it eventually rifts, heralding the opening of an ocean. The Red Sea is a site of recent rifting.

3.4.3 Ocean Crust Is Destroyed at Destructive Plate Boundaries

Clearly if new ocean crust is being made continuously some process must remove it at a comparable rate since the Earth is not getting bigger! We know that ocean crust is being continuously destroyed because none is older than 180 Myr.

New ocean lithosphere is hot but as it is carried away from the MOR it cools and becomes denser. Eventually, it is so dense that its leading edge sinks deep into the mantle. The region where the slab descends, dragged down by a negative buoyancy slab-pull force, is marked by a deep ocean trench and the site of numerous earthquakes called a subduction zone. These are the features of a destructive (or convergent) plate boundary because here ocean lithosphere is being destroyed. The basalt in the descending slab and overlying sediment contains water locked up in hydrous minerals. As the slab reaches a depth of about 120 km dehydration reactions release this water into the mantle above. Even a low concentration of water (0.5–1%) causes the mantle to undergo partial melting. Therefore, the injection of water into a wedge of mantle above the slab causes it to partially melt to form basalt. This ascends and often undergoes further processing to form magmas with higher SiO_2 content such as andesites and granites, forming new continental crust.

Where ocean lithosphere subducts under other ocean lithosphere the eruption produces arcs of islands. Indonesia is the result of island arc volcanism. Where ocean lithosphere is subducted under continental lithosphere, the eruptions generate long chains of volcanic mountains about 100 km inland from the coast. The most obvious example is the Andes. If subduction happens faster than seafloor spreading, then intervening ocean closes and this causes continents to collide. Such monumental collisions uplift mountains. A classic example is when India collided with the underbelly of Asia to make the Himalayas and the Tibetan plateau. This started about 50 Myr ago and has almost ground to a halt now. The formation of high mountain ranges, such as the Himalayas, has profound effects on the Earth's climate.

3.4.4 Hot Spot Volcanism: Evidence for Mantle Convection?

There are a number of places on the Earth where volcanism occurs a long way from plate boundaries. This is termed hot spot volcanism and it builds shallow-sided shield volcanoes. Hot spot magmas are low viscosity basalts and their geochemistry suggests they have a very deep mantle source. This and other evidence are taken to mean that hot spot volcanism happens where mantle is rising in the ascending limb of a convection cell that originates at the core–mantle boundary. These are called mantle plumes.

Both Mars and Venus have shield volcanoes, from which we infer that deep mantle convection is a feature of terrestrial planets and operates independently of plate tectonics (Figure 3.11).

3.4.5 Plate Tectonics Is Self-Regulating

For plate tectonics the lithosphere has to be rigid and has sufficient strength not to fail under extensional and compressional forces applied over long times. This depends on

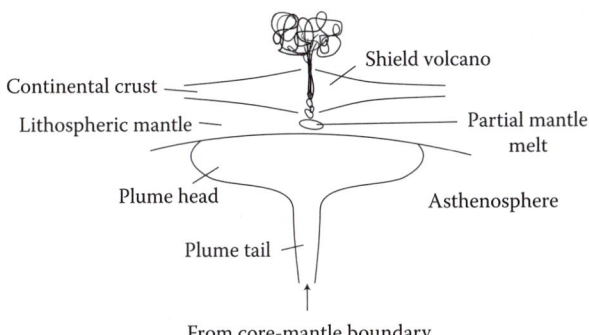

FIGURE 3.11 Mantle plumes are responsible for hot spot volcanism.

the composition, including water content, temperature, and thickness. Dry, cold lithosphere is stronger than wet, warm lithosphere. The plasticity of the asthenosphere on which the plates ride depends on the formation of hydrous minerals at a depth of about 100 km, and this depends on mantle water abundance. Even 50 ppmv of water reduces the dynamic viscosity of the mantle by two or three orders of magnitude. This allows the mantle to deform much more readily in response to loading and ensures that the mantle undergoes solid-flow convection. If the water content is too low, the viscosity will be too high and so convection and plate motion will cease. The lithosphere now becomes a stagnant lid.

In fact, plate tectonics is regulated by both positive and negative feedback that controls the water content of the mantle. As mantle viscosity is reduced by water, subduction gets easier and descending plates carry more water into the mantle. This is positive feedback. But faster subduction means greater slab-pull forces which results in faster seafloor spreading rate. Melting at mid-ocean ridges dehydrates the mantle locally, so its viscosity increases, and this acts as a force that opposes ridge-push. This applies a negative feedback brake on plate movement. In the long term, melting at MORs reduces the water content of the mantle.

EXERCISE 3.8

What is likely to happen to plate tectonics when the sun reaches the end of its main sequence lifetime?

3.4.5.1 The Carbonate–Silicate Cycle Is a Planetary Thermostat

There are a number of linked carbon cycles on the Earth. All but one involves biomass reservoirs, which is hardly surprising given the pivotal role of carbon in life. The exception is a geological cycle that could in theory operate on any planet with plate tectonics, even if it were a sterile world. It is termed the carbonate–silicate cycle and it regulates the surface temperature of the Earth by controlling the amount of CO_2 in the atmosphere (Figure 3.12).

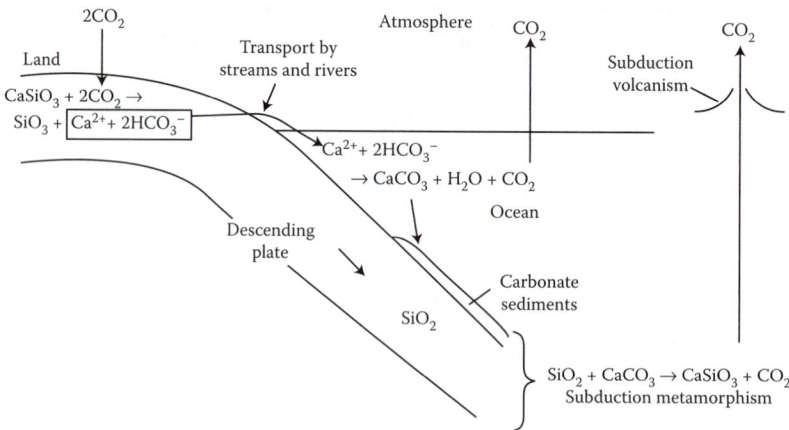

FIGURE 3.12   The geological carbon cycle.

3.4.5.2 Silicate Weathering Reduces Atmospheric CO_2

As we have seen, plate tectonics builds mountains. These are formed from the continental crust and contain silicate minerals that react with weak acids. For example, granites are common continental rocks composed of three major minerals: quartz, feldspars, and micas. Quartz is silica (SiO_2) and is resistant to acids, but feldspars and micas are not. A weak acid is produced when CO_2 in the atmosphere dissolves in water:

$$H_2O + CO_2 \rightleftharpoons \underset{\text{carbonic acid}}{H_2CO_3} \rightleftharpoons H^+ + \underset{\text{bicarbonate}}{HCO_3^-}, \tag{3.6a}$$

$$HCO_3^- \rightleftharpoons H^+ + \underset{\text{carbonate}}{CO_3^{2-}}. \tag{3.6b}$$

The higher the pH the more these reactions are pushed to the right. The pH of rainwater is 5.6 or less, which favors the carbonic acid–bicarbonate equilibrium (Equation 3.6a). Seawater is a little on the alkaline side of neutral (the pH ranges from 7.2 to 8.2), which drives the equilibria further toward the right so that some carbonate forms (Equation 3.6b). Note that these are acid–base reactions involving the transfer of protons, not oxidation–reduction reactions in which electrons are exchanged. (The oxidation state of carbon remains +4 throughout.)

The reaction of acidic rainwater with silicate minerals is called silicate weathering. Taking the simple case of the weathering of the calcium silicate mineral wollastonite, we have:

$$\underset{\text{Wollastonite}}{CaSiO_3} + 2CO_2 + H_2O \rightarrow Ca^{2+} + 2HCO_3^- + \underset{\text{Silica}}{SiO_2}. \tag{3.7}$$

This is the rate-limiting step in the removal of CO_2 from the atmosphere. The calcium and bicarbonate ions, and some of the silica, are dissolved in the water that runs off the mountains into streams and rivers and ends up in the oceans. Here, where the pH is higher, and carbonates can form:

$$Ca^{2+} + 2HCO_3^- \rightarrow \underset{\text{calcium carbonate}}{CaCO_3} + CO_2 + H_2O. \tag{3.8}$$

Note that only one of the two original CO_2 molecules has been locked up as carbonate, the other is in solution in the ocean. This means that the overall weathering reaction is:

$$CaSiO_3 + CO_2 \rightarrow CaCO_3 + SiO_2. \tag{3.9}$$

Silicate weathering reactions are generally more complicated than those above because most silicate minerals have a more complex chemistry than wollastonite, but all draw down atmospheric carbon dioxide and sequester it as carbonate. Some of this carbonate ends up on the ocean floor as sediments and is eventually subducted into the mantle.

EXERCISE 3.9

A more realistic weathering reaction involves a feldspar mineral found in granite, albite. Complete the silicate weathering reaction of albite by rainwater:

$$2NaAlSi_3O_8 + 2CO_2 + 3H_2O \rightarrow Al_2Si_2O_5(OH)_4 +$$

albite   clay mineral

3.4.5.3 Subduction Volcanism Increases Atmospheric CO_2

As carbonate sediments are subducted they are subjected to high pressure and temperature, and this produces metamorphic decarbonation reactions. Sticking with our simple example,

$$CaCO_3 + SiO_2 \rightarrow CaSiO_3 + CO_2. \tag{3.10}$$

Carbon dioxide ends up dissolved in the partial mantle melt that buoyantly rises to supply the magma for subduction volcanism. As the lava erupts, CO_2 is outgassed. Carbon dioxide and water are the most common gases vented by volcanoes.

EXERCISE 3.10

Why does the magma give off its dissolved CO_2 when it is erupted as lava?

3.4.5.4 The Carbonate–Silicate Cycle Regulates Temperature by Negative Feedback

Imagine a time when the CO_2 concentration in the Earth's atmosphere is high. The greenhouse effect boosts the temperature. Under warmer conditions water evaporates from oceans faster, and this makes precipitation rates greater. (Although warmer air can hold more moisture than cold air, this effect is not enough to prevent it raining more on a greenhouse Earth.) Higher precipitation means faster rates of silicate weathering. This results in global cooling, providing that CO_2 sequestration by weathering exceeds CO_2 outgassing by volcanism.

EXERCISE 3.11

Describe what would happen at a time when the CO_2 concentration in the Earth's atmosphere is low.

It is easy to see how this cycle could, in principle, maintain a constant temperature. Of course this is a rather simple negative feedback model (geologists call it the GEOCARB model) for temperature regulation by the carbonate–silicate cycle. It assumes that in the long term (of order 100 Myr), weathering rates and volcanism are matched.

EXERCISE 3.12

What does the GEOCARB model predict for the relationship between silicate weathering rates and temperature?

However, the planet need not be in a steady state over periods of a few tens of millions of years, and the imbalance can have a dramatic effect on climate. Currently, the carbonate–silicate cycle is thought to be locking up 0.03×10^{12} kg C from the atmosphere each year while volcanic outgassing is only returning 0.02×10^{12} kg C yr^{-1}. The discrepancy is thought to be due to the recent uplift of the Himalayas. This has increased the silicate weathering over the past 50 Myr and has resulted in a fall in global deep ocean temperature of 15°C. So, in the short term, mountain-building seriously disturbs the steady-state assumption. At times of extensive mountain-building, high rates of silicate weathering correspond to a cooling planet.

3.4.6 Plate Tectonics May Be Required for Habitability

There are several ways in which plate tectonics could contribute to planet habitability.

1. It operates a carbon cycle over geological timescale that exerts negative feedback control of the temperature of the Earth's surface. In other words, plate tectonics acts as a planetary thermostat. This may have kept the Earth's temperature in the habitable range over the past 4.6 Gyr despite the 25% increase in the luminosity of the sun over this time.

2. By allowing the Earth to lose its internal heat more or less continually on geological timescales, plate tectonics has provided a temperature gradient across the mantle large enough to have allowed the geodynamo to work. The magnetic field has protected the Earth's atmosphere from sputtering by the solar wind. The atmosphere is needed for surface water; this also provides protection from cascades of cosmic radiation potentially lethal to surface organisms.

3. Plate tectonics builds continents and, over periods of hundreds of millions of years cycles of supercontinent rifting and break up followed by reassembly, has produced dramatic changes in sea level that has altered the amount of exposed continental shelf. These changes in the Earth's geography have been crucial drivers of evolution and biodiversity of complex life.

None of these arguments applies to life deep below the surface. A meter of solid rock is all that it takes to provide protection from cosmic rays, and there is almost certain to be some depth in the crust where the temperature is in the habitable range even for a planet with a tenuous atmosphere and a global annual mean surface temperature of −63°C, such as Mars.

3.4.7 Plate Tectonics Seems Not to Operate on Mars or Venus

Plate tectonics on the Earth requires surface water, and since Earth is the only planet to have surface water now, it seems to be the only planet in the solar system to have plate tectonics.

Liquid water has flowed on Mars, though probably not for long or often, over the last 3.5 Gyr. There are claims that Mars had plate tectonics earlier but the evidence is not persuasive.

Venus does not have plate tectonics, presumably because it has experienced a runaway greenhouse hike in temperature, losing any surface water it had early on. The volcanism that accompanies plate tectonics is a major route by which the Earth loses its internal heat. Crater counts show Venus appears to have been resurfaced by catastrophic volcanic eruptions sometime between 500 million and one billion years ago. This could be because Venus was not able to lose internal heat in the absence of plate tectonics. Consequently, radiogenic heating raises the temperature of the upper mantle and crust until eventually widespread melting and volcanism are triggered. This allowed the planet to cool.

EXERCISE 3.13

What might allow a planet to be habitable in the absence of plate tectonics?

4

Building the Solar System

4.1 Planet Formation Is Contingent

Most of what we know about how planets form comes from studies of the terrestrial planets, the moon and meteorites, and modeling the dynamics of the early solar system. This is now being heavily influenced by the make-up of other stellar systems, which are turning out to be very diverse. It seems that the solar system is not a model for how other planetary systems must look. Indeed, even in the solar system, of the eight major planets, several dwarf planets, asteroids, and more than 160 moons that have been studied in any detail, no two are alike. It seems that building planetary bodies is a matter of contingency.

The way a planetary system evolves early on is governed by numerous accidents of birth, random dynamical events such as near encounters or collisions between growing protoplanets, the outcome of which determines the number, location, and composition of the planets that a system finally ends up with. This makes the question about how habitable planets emerge a difficult and interesting one to try to answer.

4.2 Planets Formed by Accretion from the Solar Nebula

The notion that a cloud of gas and dust, the solar nebula, collapsed to form the sun and a rotating flat disc from which planets formed was first advanced in the eighteenth century. It was seen as a natural explanation of the fact that all the planets orbit the sun in the same direction and in the same plane. However, at the time it was impossible to explain why the planets orbit so rapidly. How did the nebula collapse so that the sun got 99.9% of the mass but only 1% of the angular momentum? This problem meant that for a long while the idea languished in favor of other proposals.

4.2.1 The Solar Nebula Was a Dynamic Environment

Our prevailing view of planet formation is a reincarnation of the solar nebula disc model based on the work in the early 1970s by the Russian astronomer Viktor Safronov, which solved several problems with the theory, including the angular momentum question. At first, the accretion process was modeled assuming that the planets were formed by clearing out material from a ring-shaped feeding zone in the solar nebula. Hence, the planets were made from local ingredients in the solar nebula. It explained why inner planets are rock worlds while the outer planets are gas giants.

The discovery of circumstellar discs of cool gas and dust around young stars such as β-Pictoris, and infrared excess from protostars, which is explained by nebula dust, provided observational evidence that brought the solar nebula model back into favor by the 1980s.

George Wetherill was stimulated by Safronov's work to try to understand how the terrestrial planets originated. He developed a new numerical method to calculate how orbits evolve and hence was able to simulate the physical and orbital properties of terrestrial planets in good agreement with observations. This showed that the early solar system was far more dynamic than Safronov had modeled. Wetherill's calculations started out with numerous planetesimals and included the mutual gravitational effects between them as they grew. He discovered that planetesimals were perturbed to move large radial distances in the solar nebula. Planets are therefore composed of materials from different orbital distances in the solar nebula. Specifically, water could be delivered to the inner solar system by planetesimals from beyond the snow line, where it is cold enough for water vapor to freeze. Wetherill's model predicted that the final round of assembly would involve really large impacts.

4.2.2 The Solar Nebula Formed a Spinning Disc

The solar nebula started out as a dense molecular core perhaps 20,000 AU across. It consisted mostly of hydrogen and helium with just 2% being all the heavier elements. Initially, given its temperature of just 10–20 K, many of the most abundant elements such as carbon, nitrogen, and oxygen were present as ices (e.g., H_2O, CO, CO_2, CH_4, and NH_3), coating grains of silicates, and metals present mostly as sulfides or oxides. Discs around stars in nearby star-forming regions can be up to 2000 AU across, although one model suggests that the solar nebula was only 100 AU across by the time planets were forming.

As the solar nebula contracted it span up to conserve angular momentum, flattening into a disc because centrifugal force due to rotation opposed the inwardly directed gravitational force in the plane of the disc. Material fell onto the disc from above and below the plane of the disc, since here it would be opposed only by gas pressure. This material lost its angular momentum and was transported inward to fall on the growing protosun at the center. In this way, the protosun acquired most of the mass of the system while the surrounding disc retained most of the angular momentum.

4.2.3 Planet and Star Formation Occur Together

Planets began forming from the disc when the protosun was a T-Tauri star. A high proportion of T-Tauri stars are surrounded by a circumstellar disc (Figure 4.1). These are also termed protoplanetary discs (proplyds) because they are assumed to be the birthplace of planets. But as T-Tauri stars begin core fusion (deuterium to helium), they generate fierce stellar winds. These T-Tauri winds blow away the circumstellar disc on a timescale of just 3–6 million years, including all the gas and dust. Initially in our solar system the effect was to sweep rocky bodies smaller than about 10 m across from the inner nebula and to concentrate them with the ices beyond 2.7 AU from the sun. This provided sufficient raw ingredients for the gas/ice giants to form by the time the solar nebula dissipated. Rocky planets could continue to accrete in the inner solar system after this event.

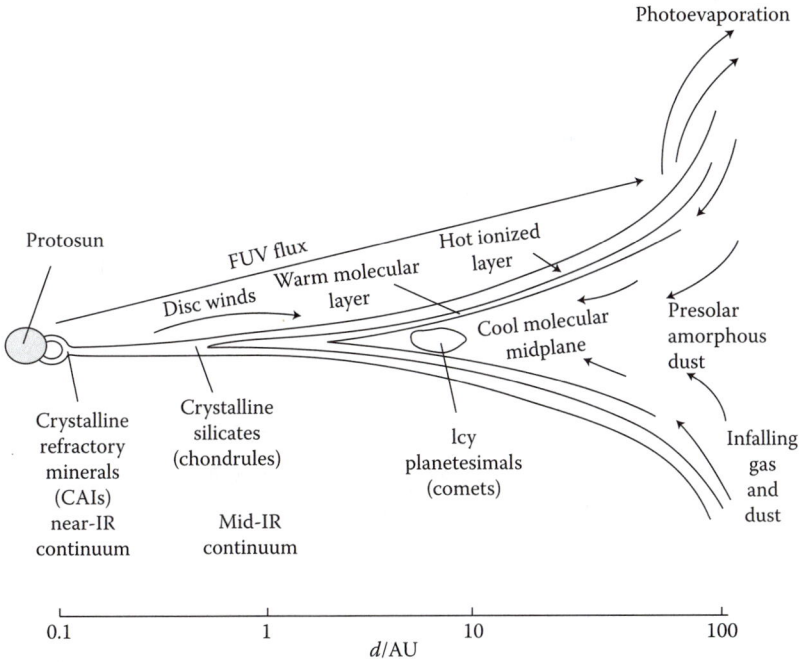

FIGURE 4.1 Infall of material onto a circumstellar disc continues until stellar winds choke it off.

Stellar evolution models suggest that the sun did not join the main sequence as a fully fledged, core hydrogen-burning star until 50 million years after the first solids condensed in the solar nebula, by which time the planets had almost finished accreting.

4.2.4 Condensation of Solids from the Solar Nebula Depended on Temperature

The solar nebula was heated by solar radiation and by gas and dust falling onto the disc and being transported through the disc. Gravitational potential energy is converted into kinetic energy and collisions (friction, turbulence) convert this ordered motion into the disordered motion of thermal energy. The end result, at the height of the collapse, is a radial temperature gradient across the disc, from ~1700 K at about 0.1 AU to ~20 K at the outer edge of the nebula located perhaps 80 AU from the sun. There would also have been a vertical temperature gradient through the disc, warmest at the midplane (Figure 4.2).

John Lewis of the University of Arizona has modeled the physics and chemistry of the solar nebula and worked out the order in which elements and compounds (minerals, ices) would condense with distance from the sun (Table 4.1). This depends on the temperature and how refractory or volatile the material is. This is defined by the T_{50}, the temperature at which 50% of the element has condensed at a given pressure; the lower its T_{50}, the more volatile it is. Materials with $T_{50} > 1100$ K are regarded as refractory, those with condensation temperatures between 1100 K and 400 K are moderately volatile, and those with $T_{50} < 400$ K are volatile. At 1700 K, all but the most

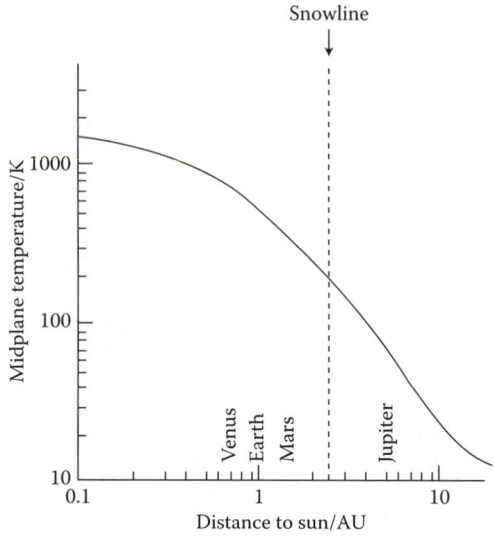

FIGURE 4.2   The solar nebula radial temperature gradient.

refractory materials would be in the gas phase. At ~2.5–5 AU where the temperature is 145–170 K at nebula pressures of ~10–100 Pa, highly volatile materials (e.g., H_2O) can condense as ices at the snow line and beyond. The snow line probably migrated early on and this will have influenced how the terrestrial planets acquired water.

The temperature gradient varied with time, and there is evidence for substantial radial mixing in the nebula as a result of turbulence, so there is little reason to expect a simple relationship between distance from the sun and planetary composition. That said, we would expect to find planets to have a higher inventory of refractory

TABLE 4.1

T_{50} Values (at $P = 100$ Pa) for Some Important Compounds in the Solar Nebula

Compound	Chemical Formula	T_{50} (K)
Corundum	Al_2O_3	1758
Perovskite	$CaTiO_3$	1647
Iron-nickel alloy	Fe, Ni	1471
Pyroxene (diopside)	$CaMgSi_2O_6$	1450
Olivine (forsterite)	Mg_2SiO_4	1444
Troilite	FeS	700
Hydrated silicates	$(OH)_x$-containing	550–330
Water	H_2O	170
Ammonia	$NH_3 \cdot H_2O$	120
Methane	$CH_4 \cdot 6H_2O$	70
Nitrogen	$N_2 \cdot 6H_2O$	70

Those at the top condensed earliest and closest to the protosun. The four highly volatile compounds at the bottom condensed as ices at and beyond the snow line.

materials and a lower inventory of volatile materials the closer they are to the sun, and this is broadly the case.

4.2.4.1 The First Solids Condensed in the Solar Nebula 4.567 Gyr ago

The first solids to condense were calcium aluminum inclusions (CAIs), visible as white specks in some chondritic meteorites. CAIs contain the most refractory minerals and froze out of a region of the solar nebula where the temperature was >1300 K and the pressure was low (10^{-4} bar) over just a few hundred thousand years. At this time the sun was a Class 0 protostar. The age of CAIs has been measured by radiometric dating schemes as 4567.3 ± 0.16 Myr. This is usually taken to be the age of the solar system.

4.2.5 Accretion Involves Several Distinct Mechanisms

Making terrestrial planets happens in four stages:

1. Cohesion and settling of dust in the midplane of the nebula
2. Growth of planetesimals up to ~1–10 km
3. Runaway growth of oligarchs to ~1000 km
4. Oligarch growth to planetary embryos planets as a result of collisions

We look at these in turn.

4.2.5.1 Cohesion Builds Particles a Few Millimeters in Size

The first stage in accretion is cohesion of solid grains about 0.1–1 μm in size by electrostatic and weak molecular forces. This is thought to have made centimeter-sized particles at 1 AU from the sun in a couple of thousand years. Such particles would have been massive enough to decouple from the gas motions and sink to the midplane of the disc. On the face of it we might expect the column density (mass per square meter perpendicular to the nebula midplane) to decrease radially outward. If so, cohesion should be slower at greater distances from the sun because collisions between grains would happen less often. However, as we have seen, it is thought that T-Tauri winds swept material from the inner solar system, increasing the column density at and beyond the snow line, thereby favoring rapid assembly of the rocky kernels of the gas giants.

This stage saw the formation of chondrules. These are millimeter-sized blebs, mostly of the silicate minerals olivine and pyroxene, which have undergone one or several episodes of melting at temperatures in excess of 1000 K followed by cooling over several hours. CAIs and chondrules were then incorporated into a fine-grained matrix of unmelted material (anhydrous and hydrated silicates, FeS, Fe-Ni alloy, and sometimes carbonaceous material) to form chondritic asteroids and meteorites (Box 4.1).

4.2.5.2 Chondrules Formed in Episodes of Intense Heating

Exactly how chondrules formed is not certain. It seems that most started out as coalesced aggregates of silicate dust grains that were rapidly heated to ~1000 K at an ambient pressure of 10^{-3} bar.

BOX 4.1 METEORITES

Over 25,000 meteorites exist in collections around the world. They are traditionally divided into irons, stony-irons and stones. Irons and stony-irons are from differentiated worlds. The majority of stones are undifferentiated chondrites, but the achondrites—which include lunar, Martian, and Vesta meteorites—are from differentiated bodies.

Iron meteorites are Fe–Ni alloys that were once cores of planetesimals. Radiometric dating reveals that some separated from silicates as early as 1.5 Myr after the first solids condensed in the solar system, so differentiation of planetesimals started very early. Simulations suggest that their parent bodies formed in the hot inner solar system (promoting melting and differentiation) and were subsequently ejected out into the asteroid belt.

Stony-irons originally resulted from the collision of two differentiated bodies, a protoplanet with radius about 200 km and a smaller impactor. Liquid metal from the impactor core was injected into the mantle of the protoplanet to form an olivine–metal mix. Tens of millions of years later the protoplanet was fragmented by further collisions. The fragments ended up in the asteroid belt and as sources of stony-iron meteorites. Remnant magnetism in the stony-irons shows that the progenitor protoplanets had magnetic fields.

The stony chondrites contain chondrules and are divided into three major groups on the basis of chemical composition and minerals into ordinary, enstatite,

PLATE 4.1 (**See color insert.**) Ordinary chondrite. Slice through North West Africa 869. Note the glistening flecks of iron–nickel metal and pale circular section chondrules visible most clearly at the top of the meteorite. (Photo credit: Alan Longstaff.)

and carbonaceous chondrites. They are also further classified into six petrologic types (1–6). Types 1 and 2 have been extensively altered by water, type 1 more so. Types 3–6 have been increasingly altered by heat.

Ordinary chondrites (Plate 4.1) are the most common meteorite falls and are subdivided into H, L, and LL categories depending on their total iron content. Category H has the highest amount of iron, and is the most reduced.

Enstatite chondrites are so called because they are rich in the orthopyroxene mineral enstatite ($MgSiO_3$). They are much reduced, having almost all their high iron content present as native metal or iron sulfides.

Carbonaceous chondrites contain 0.25–4 wt% carbon, either in amorphous form or in the form of organic molecules such as hydrocarbons, aldehydes, carboxylic acids, purines, and amino acids. They are divided into eight groups generally named after a type meteorite (Table 4.2). CI chondrites are regarded as the most primitive meteorites because they have elemental abundances close to that of the sun, and show the least signs of alteration by heat. For this reason planetary scientists often normalize elemental abundances of rocks and other meteorites to CI values.

TABLE 4.2

Classification of Carbonaceous Chondrites

Type	Type Meteorite	Main Petrologic Type	Notes
CI	Ivuna	1	Condensed at <50°C, oxidized, rich in water and organic molecules
CM	Mighei (Murchison)	2	Oxidized, rich in organic molecules, parent body 19-Fortuna?
CR	Renazzo	2	Reduced, parent body 2-Pallas?
CH	"High iron"	2	~40 wt% Fe–Ni, reduced, CR-like
CB	Bencubbin	2	>50 wt% Fe–Ni, highly reduced, CR-like
CO	Ormans	3	Chondrite-rich, slightly reduced
CV	Vigarano (Allende)	3	Abundant CAIs, interstellar nanodiamonds, oxidized and reduced subgroups
CK	Karoonda	4 (3–6)	Oxidized, related to CV
C UNGR	Tagish Lake	2	Low density, few chondrules, elemental abundances between CI and CM, many interstellar nanodiamonds, rich in organic molecules

We do know that chondrules must have condensed from a vapor phase with which they were in equilibrium. The reasoning behind this is that despite being depleted in relatively volatile elements (e.g., K and Mg) by comparison with average meteorite abundances—hardly surprising given that they have been heated to such high temperatures—they do not show any preferential loss of light isotopes. Because light isotopes evaporate more readily than heavy ones, we might expect the silicon in chondrules, for example, to be relatively depleted in light isotopes such as ^{28}Si compared with heavy isotopes such as ^{30}Si. That this has not happened suggests that chondrules were immersed in chondrule vapor, and the light isotopes that most readily evaporate from chondrules would just as quickly re-condense back onto them. Calculations show that the pressure of the volatile element vapor must have been 95% of the saturation pressure during the time that the chondrules were forming, and given the low density of the solar nebula this means that the region in which chondrules condensed had to be of order 1000 km across. For this scenario to work the chondrule density must have been high implying they formed close to the midplane.

The scenario above constrains the formation mechanisms that are plausible. Shock waves best fits the bill. A shock wave is produced if a fast-moving gas is decelerated as it ploughs into a slower moving gas. The gas is rapidly compressed and heated. Shock wave could be generated in the nebula in several ways. Recent radio-dating shows that some chondrules formed at the same time as the CAIs. At this stage, the young Class 0 protosun was accreting material at very fast rates ($10^{-5}M_\odot$ per year) for about 10^5 years, so accretion shocks could produce the first chondrules. However, chondrules were formed over about 3 Myr, so other shock mechanisms must have operated later.

4.2.5.3 Assembly of Planetesimals Is Poorly Understood

Accretion next builds planetesimals, rocky bodies ~1–10 km across, from the centimeter-sized particles. How this happens is poorly understood. The prevailing view is that the centimeter-sized particles settled into the midplane of the nebula producing a thin (100 km) dense sheet of chondritic material that became gravitationally unstable and broke up into many billions of planetesimals. Modeling this suggests it could occur in a thousand to a million years.

4.2.5.4 Runaway Growth Builds Oligarchs and Planetary Embryos

Planetesimals have sufficient mass that gravity now becomes important. The more massive ones attract their lighter neighbors (gravitational focusing) resulting initially in low velocity collisions that build successively larger bodies. Gravity, frequency, and velocity of collisions become ever higher and runaway growth occurs, a positive feedback mechanism in which the most massive bodies, termed oligarchs, engage in a "winner takes all" race. After perhaps 10^5 years, this results in about 100 moon- to Mars-sized planetary embryos. These would come to dominate a "feeding zone" within the nebula, a region cleared of planetesimals either through accretion, or because they had collided at high velocities, broken up, and/or been ejected out of the plane of the nebula by the gravity of the embryo.

At this stage, planetary formation depends on the distance from the sun. Beyond the snow line, growth continues because there are plenty of solids in the form of frozen volatiles to build a metal–rock–ice core massive enough to hang onto gases in the solar nebula and hence build ice or gas giants. Closer in, growth ceases once the feeding zone has been depleted, and so terrestrial planet formation proceeds chaotically as planetary embryos perturb each others orbits. This causes embryos to eject remaining planetesimals and collide with each other.

4.2.5.5 Debris Discs Result from End Stages of Planet Building

In the early 1980s, several A-type stars were identified as emitting more IR than their high effective temperatures suggested they should. These stars, β-Pictoris, Vega, and Fomalhaut, are surrounded by gas-depleted debris discs. Particles in the disc absorb sunlight and are warmed, which accounts for the infrared excess. Debris discs have now been identified around over 900 stars including cool K-type stars. In most cases the stars are too old for the discs to be protoplanetary nebulae: both Vega and Fomalhaut are ~450 Myr old. Instead, debris discs are what remain after planet formation has largely finished, and indeed planets have been identified in many systems with debris discs.

What makes β-Pictoris remarkable is its youth, 8–20 Myr old, so its edge-on dusty discs (a smaller secondary disc is inclined at 5° to the primary) represent a relatively early stage in the genesis of a planetary system. They have been probed at mid-infrared wavelengths (2–30 μm) to reveal a series of rings of tiny (<1.5 μm) amorphous olivine (Mg_2SiO_4) grains at radii 6.4, 16, and 30 AU from the star. The small size of the grains means they should be blown out of the system by the radiation pressure from β-Pictoris within just tens of years. So their presence can only be explained if something is continually supplying them, perhaps collision of planetesimals in the rings. Moreover, the lack of dust between the 6.4 and 16 AU rings attests to the presence of β-Pictoris b, a massive planet (~$8M_J$) orbiting at ~8 AU.

4.2.6 Heat Sources Drive Differentiation

When planetesimals got to a critical size in the early solar system, they differentiated into a metallic core and silicate carapace. This required heat, initially from the decay of radioisotopes and impacts. Radiogenic heating was far higher in the early solar system not only because the abundance of long-lived isotopes would have been greater, there were also short-lived isotopes that are now extinct. This produced partial melting of planetesimals and planetary embryos.

The first melt of a mix of Fe–Ni alloy with a lighter element such as S is likely to occur at a lower temperature than that of silicates. (Mixtures of metals or compounds melt over a range of temperatures rather than at a single melting point.) Hence segregation of a liquid metal phase from solid silicate could occur provided that the metal forms an interconnected network so that it can percolate between the silicate grains. Laboratory experiments show that this can occur if the pressure is high enough, as it would be deep inside a terrestrial planet. However, for differentiation of smaller bodies might require 40–50% of the silicates to melt to allow the metal to sink to the center.

The most important heat source in the first 2.5 Myr was the decay of the short-lived radioisotope ^{26}Al. We know this because calculations show that the concentration of ^{26}Al was sufficient to provide heat for partial silicate melting and the timescale of planetesimal differentiation coincides with the decay of radiogenic heating from decay of ^{26}Al.

There is geochemical evidence for extremely deep magma oceans; that is, extensive silicate melting, on the Earth, moon and Mars, which implies that the final stages of planet assembly were extremely violent.

4.2.6.1 Vesta Is an Ancient Differentiated World

The asteroid 4-Vesta appears to be a fully differentiated world, an oligarch that never made it to planetary status. We know this because we have meteorites from Vesta and all are igneous rocks that come from a differentiated crust. These form the Howardite–Eucrite–Diogenite (HED) suite of meteorites.

The Dawn spacecraft which went into orbit around Vesta in July 2011 for 10 months has confirmed earlier findings that the HED meteorites match the surface of the asteroid, has identified craters large enough to be the origin of all the mass of the HED meteorites, and confirms that Vesta has an iron core with a radius of about 110 km (see Plate 4.2).

4.2.7 Differentiation Redistributes Elements

Working out details of the differentiation of the terrestrial planets and other bodies comes partly from knowledge of how elements distribute themselves between different phases: the Goldschmidt classification of elements.

4.2.7.1 Elements Have Distinct Chemical Affinities

If a mixture of molten iron and silicate coexists, as would have been the case in the differentiating Earth, then some elements (siderophile or "iron-loving" elements) will preferentially dissolve into the metal phase while others (lithophile or "rock-loving" elements) will enter the silicate phase. Highly siderophile elements include osmium (Os) and iridium (Ir), while cobalt (Co), nickel (Ni), and tungsten (W) are moderately

PLATE 4.2 (**See color insert.**) Asteroid 4-Vesta imaged by the Dawn spacecraft. The first million years of the solar system would have been dominated by thousands of objects like Vesta. (Photo credit: NASA/JPL-Caltech/UCAL/MPS/DLR/IDA.)

**BOX 4.2 ELECTRONEGATIVITY DETERMINES
THE CHEMICAL AFFINITIES OF ELEMENTS**

Electronegativity, E, is the tendency for an atom to attract electrons toward itself in a covalent bond and so acquires a negative charge. It increases systematically across periods and up groups in the Periodic Table to a maximum value at fluorine. The higher an element's electronegativity, the greater is the ionic quality of the bonds it forms. It increases across a period because the number of protons in the nucleus gets higher and the atomic radius gets smaller. Electronegativity decreases down a group as extra shells of electrons are added, increasing the atomic radius, taking the valence electrons further from the nucleus which provides additional shielding from the positive charge of the nucleus.

In general, elements with low electronegativity ($E < 1.7$) form positive ions. These will readily form ionic bonds with silicate anions (SO_4^{2-}), so are lithophiles. Elements with $1.7 < E < 2.1$ tend to form covalent bonds with sulfur, so are chalcophile. Siderophiles have high electronegativity ($2.1 < E < 2.5$), so form metallic bonds. Elements with $E > 2.5$ form negative ions that make ionic bonds with low electronegativity lithophiles. These include O, N, and the halogens in Group 17.

---

siderophile. Many elements are lithophile. A few elements exist as gases at the Earth's surface (H, C, N, O, and the inert gases) and are described as atmophiles. Elements that bind preferentially to sulfur are termed chalcophiles. These chemical affinities of elements are determined by their electronegativity (Box 4.2). How elements distribute themselves is relative. If no metal phase is present, siderophiles will partition into silicates rather than the atmosphere, behaving as lithophiles. Some elements can be siderophile, lithophile, or chalcophiles, depending on conditions.

The partitioning of an element, i, between metal and silicate is quantified by its distribution coefficient:

$$D_i = [i]_{metal}/[i]_{silicate} \tag{4.1}$$

A siderophile will have $D > 1$, while a lithophile will have $D < 1$. Distribution coefficients vary with pressure and temperature.

4.2.8 Gas Giants Must Have Assembled Within a Few Million Years

Timing constraints mean that the rocky cores of the gas/ice giants had to be in place early. Jupiter and Saturn managed to capture massive envelopes of hydrogen and helium from the primordial nebula before the T-Tauri wind had blasted away all the gas. Uranus and Neptune, being further out, assembled too slowly to do the same trick to anything like the same extent. This can explain the difference in mass and internal structure of the gas and ice giants, which had just a few million years to form.

The worlds now beyond the orbit of Neptune formed too far out to assemble into bodies much larger than Pluto, presumably because the number density of planetesimals there was so low that collisions were rare. This explains the existence of the Kuiper belt. The ice/rock composition of Kuiper belt objects comes about because the temperature of the solar nebula here would have allowed volatiles to condense out.

4.2.9 Accretion of Terrestrial Planets Took Tens of Millions of Years

Provided that accretion had made bodies greater than 10 m across by the time the T-Tauri winds swept out the solar nebula gas and dust, the inner planets could continue to accrete and differentiate. One way we can estimate how long it takes to build terrestrial planets is by measuring how long it takes planetary cores to form. Core formation times have been estimated for Earth from two radioisotope decay schemes: hafnium–tungsten (Hf–W) and uranium–lead (U–Pb). These give 35 and 80 Myr, respectively. The discrepancy is thought to be due to the effect of the massive impact that formed the moon (Chapter 5). In effect, 35 Myr may mark the initial phase of core formation and 80 Myr the late stage. Hf–W dating of core formation for Vesta and Mars is also subject to large uncertainties, but 1–4 Myr for the asteroid and no later than 15 Myr for the red planet are currently in vogue.

4.3 The Solar System Started with a Bang

We suspect that a supernova exploded locally at about the time the sun was forming because it seeded the solar nebula with radiogenic isotopes such as ^{26}Al. This was responsible for the extremely rapid differentiation of the planetesimals and the short timescale of planet formation. The supernova may even have contributed to the Earth's habitability since the amount of ^{26}Al in a protostellar nebula could have a profound influence on the architecture of the resulting stellar system.

4.3.1 ^{26}Mg Traces the Original ^{26}Al

Aluminum-26 is produced in core collapse supernovae and decays to the stable isotope ^{26}Mg with a half life of 717,000 years, generating a great deal of heat in the process. CAIs, the first solids to condense in the solar nebula have very high levels of ^{26}Mg. Although the first chondrules condensed at the same time, chondrule formation continued for longer, so they average somewhat lower levels. This implies that a high amount of ^{26}Al was injected into the nascent solar nebula by a nearby core collapse supernova.

However, as well as ejecting ^{26}Al, type II supernovae also produce ^{60}Fe and ^{58}Fe. Iron-60 decays to ^{60}Ni with a half-life of 2.6 Myr, so ^{60}Ni abundance today is a proxy for how much ^{60}Fe there was in the early solar system, and ^{58}Fe is stable. Measurements of ^{60}Ni and ^{58}Fe in many meteorites (chondrites, achondrites, and irons) show that the initial solar system abundance of ^{60}Fe was very low, consistent with having been derived from the Galactic interstellar medium, and that ^{58}Fe, which would have been distributed the same as ^{60}Fe, was homogenously mixed in the early solar system.

The implication is clear. A type II supernova was not responsible for seeding the solar system with ^{26}Al. Instead ^{26}Al was probably delivered by copious stellar winds ejected by a high mass ($M > 20M_\odot$) Wolf–Rayet star near the end of its life. These stars blast off their entire envelopes to leave behind their exposed scorching massive cores. Within 2 Myr these undergo Type 1b/c supernovae, ejecting ^{60}Fe. It is intriguing that our solar system appears not to have any trace of this later ejection of iron. Be that as it may, this scenario suggests that the sun may have formed as part of a cluster containing extremely massive, hot, bright stars (e.g., an OB association).

4.3.2 Did the Decay of ^{26}Al Make Life on Earth Possible?

The amount of ^{26}Al there must have been in chondrites (as shown by the amount of ^{26}Mg now) is close to the amount of ^{26}Al required to give enough heat to differentiate planetesimals. Struck by this coincidence Jamie Gilmore of University College, London and Ceri Middleton have sought to work out what the consequences of different levels of ^{26}Al would be for planet building in stellar systems.

If there was more ^{26}Al there would be higher temperatures throughout a stellar nebula, so the snowline will form at a greater distance. This means that gas giants would form more slowly, because the density of planetesimals is less at larger distances, and consequently the probability that they will collide and accrete to form planetary embryos is lower. The outcome is that gas giants may not form because the protostar solar wind sweeps away all the nebula gas before the planetary embryos are massive enough to capture it.

By contrast, with less ^{26}Al, gas giants would form more rapidly—by the reverse of the arguments in the paragraph above—and hence could get to be appreciably larger than those in the solar system. Planets migrate (see Section 4.5 below). They can continue to do so as long as there is still gas in the nebula. However, in the low ^{26}Al-cold nebula scenario, because gas giants would have been in place early they would have had time to migrate large distances before their star blasted the nebula gas away. How does this affect habitability? If gas giants have a net protective effect, shielding inner planets from potential impactors (and this has been questioned), then the high ^{26}Al-hot nebula scenario, lacking gas giants, could be problematic. Worse though is the low ^{26}Al-cold nebula scenario, since simulations show that terrestrial planets are ejected from habitable orbits by inwardly migrating gas giants.

EXERCISE 4.1

Assuming that terrestrial planets did form in a low ^{26}Al-cold nebula system, how might their composition differ from the terrestrial planets in our solar system?

Ultimately, the amount of ^{26}Al delivered to a protostellar nebula depends on the proximity and timing of nearby core-collapse supernovae. The early solar system was more than 10-fold richer in ^{26}Al than the interstellar medium. Indeed, calculations show that less than 6% of Galactic material is sufficiently enriched in ^{26}Al to act as precursor material for the solar system, and that the chance of a stellar system forming with a ^{26}Al concentration as high as ours is less than 1 in 3000 if the source were a close supernova. The implication of this is that we live in a rare type of stellar system. It seems likely that most stellar systems form in giant molecular clouds (GMC) that have not been seeded by ^{26}Al from supernovae, perhaps because the clouds are too small to sample nearby supernova remnants, or because they accrete too early. It may be that the solar system formed in a particularly large GMC at just the right time. If the Earth's habitability did indeed depend on a supernova going off at just the right distance and time to give us the "Goldilocks" amount of ^{26}Al, then we are left to ponder that this isotope, though critical to our existence, had long since decayed before ever life emerged on Earth.

4.4 Dating Events in the Early Solar System Relies on Radioactive Isotopes

Crucial to understanding how the solar system evolved has been the ability to determine the ages of its components (e.g., meteorites, terrestrial, and lunar rocks) and to time key events, such as when the Earth's core formed. The key is the use of radioactive isotopes naturally present in rocks for radio-dating.

4.4.1 Radiometric Dating Relies on the Exponential Decay of Radioisotopes

A parent radioisotope, P, decays exponentially into a daughter isotope, D. The rate at which this decay occurs is known (from laboratory experiments), so measuring the amount of D present now gives the age of the rock, provided we know how much P there was originally. The decay rate is often quoted as a half-life, the time taken for the amount of P to drop to half its initial value. Clearly, after two half-lives P will be one-quarter of the initial value, after three half-lives P will be one-eighth the initial value, and so on. Decay rate can also be expressed as a rate constant. This is the time for the quantity of P to drop to 1/e of its initial value (Figure 4.3).

But how do we know how much P there was in a rock to begin with? Fortunately it is possible to calculate this because of the inherent variability in the concentrations of

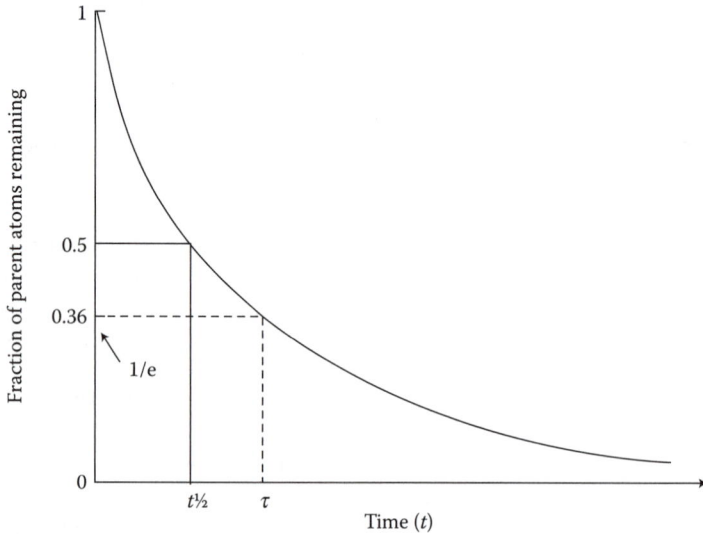

FIGURE 4.3 Radioactive decay rate can be expressed as a half-life or as a rate constant.

elements found even in closely related rocks. To understand how this works we need to derive the equation used for radiometric dating.

Consider the decay of $P \rightarrow D$. At any time, t, the number of daughter atoms is related to the number of parent atoms by

$$D = P(e^{\lambda t} - 1), \qquad (4.2)$$

where λ is the characteristic decay constant for the isotope.

However, there may have been daughter atoms D_0 present initially that did not come from the decay of P. So Equation 4.2 becomes

$$D = D_0 + P(e^{\lambda t} - 1). \qquad (4.3)$$

This is the basic linear equation used for radiometric dating.

The decay constant is directly related to the half-life of the parent isotope, $t_{1/2}$. This is because at the half-life $D = P$ and by definition $t = t_{1/2}$.

EXERCISE 4.2

From Equation 4.2 show that $\lambda = 0.693/t_{1/2}$.

4.4.2 Radiometric Dating Uses Isochron Plots

The radiometric dating Equation 4.3 is used to plot a straight line with a y-axis intercept, c, which defines the initial amount of P and slope, m, which is a function of the age, t. This line is termed an isochron (Figure 4.4).

This method is used for a variety of radioisotope decay schemes. Here we take the decay of samarium-147 to neodymium-143 by emission of an alpha particle as an example:

$$^{147}_{62}\text{Sm} \rightarrow {}^{143}_{60}\text{Nd} + \alpha. \qquad (4.4)$$

This has a half-life, $t_{1/2} = 1.06 \times 10^{11}$ yr.

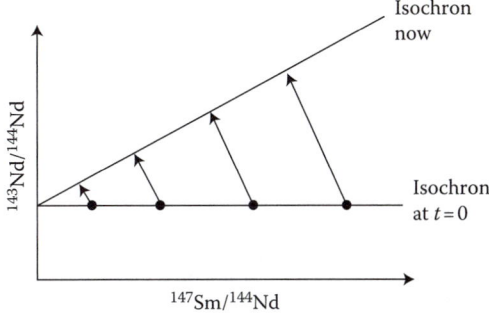

FIGURE 4.4 Isochron plots. Four rock samples from the same place and of the same age will naturally differ in $^{147}\text{Sm}/^{144}\text{Nd}$ ratios. These ratios evolve over time from the $t = 0$ isochron. Each sample moves up and to the left with time as it acquires more ^{143}Nd at the expense of ^{147}Sm.

Writing Equation 4.3 for this scheme, we get

$$^{143}_{60}\text{Nd} = {}^{143}_{60}\text{Nd}_0 + (e^{\lambda t} - 1){}^{147}_{62}\text{Sm}. \tag{4.5}$$

Because mass spectrometers measure isotope ratios to a much higher accuracy than single isotopes, this is rewritten as ratios with respect to a stable neodymium isotope, in this case $^{144}_{60}\text{Nd}$:

$$(^{143}\text{Nd}/^{144}\text{Nd}) = (^{143}\text{Nd}/^{144}\text{Nd})_0 + (e^{\lambda t} - 1)\,(^{147}\text{Sm}/^{144}\text{Nd}), \tag{4.6}$$

$$y \quad = \quad c \quad + \quad mx$$

Hence, a plot of $(^{143}\text{Nd}/^{144}\text{Nd})$ versus $(^{147}\text{Sm}/^{144}\text{Nd})$ will give a straight line with slope $m = (e^{\lambda t} - 1)$. Since λ is easily calculated from $t_{1/2}$, the age, t, can be determined. This line is called an isochron since it joins data points of rocks with the same age (Figure 4.5).

Figure 4.4 shows how this works. It is an Sm–Nd isochron plot of olivine and pyroxene crystals from a lunar Highlands rock (anorthosite) collected by Apollo 16. The radioactive clock started when these crystals froze and there was no further exchange of isotopes with the surroundings. At this time, $t = 0$, the only ^{143}Nd is that derived from exogenous Sm. This is the D_0 component and it fixes the initial $^{143}\text{Nd}/^{144}\text{Nd}$ ratio. Although all of the crystals come from anorthosite that solidified at the same time they will naturally differ in $^{147}\text{Sm}/^{144}\text{Nd}$ because of the details of crystallization. This gives the range of $^{147}\text{Sm}/^{144}\text{Nd}$ that lie along the x-axis and the horizontal $t = 0$ isochron. Over time the ^{147}Sm decays. The greater the $^{147}\text{Sm}/^{144}\text{Nd}$ ratio of the rock the higher, proportionally, is the resulting $^{143}\text{Nd}/^{144}\text{Nd}$ ratio. With

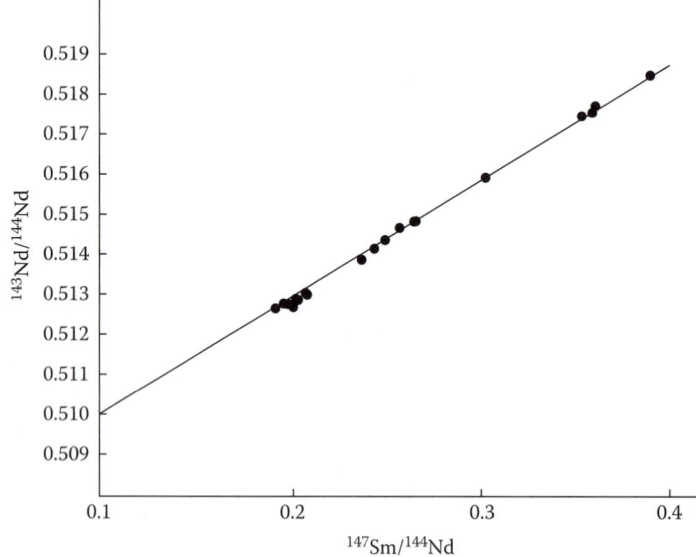

FIGURE 4.5 Sm–Nd isochron plot of olivine and pyroxene crystals from a lunar Highlands anorthosite.

time, each point moves up and to the left. At any given time these points lie on an isochron that pass through the y-axis intercept and as the rocks get older the slope of the isochron increases.

Isochron plots will only yield a reliable age if the rocks have the same initial $^{143}Nd/^{144}Nd$ ratio, if they are a genuinely closed system (i.e., there has been no exchange of isotopes with the surroundings since they crystallized) and if the rocks are of the same age. This last point sounds obvious but it is not trivial; if the geology is not well understood it can be hard to pick samples that have the same birthday. If a suite of rocks fails to meet any one of these three conditions, the data will not fit a single straight line.

EXERCISE 4.3

Calculate the age of the olivine and pyroxene of lunar anorthosite from the Sm–Nd isochron plot in Figure 4.4.

4.5 Planetary Migration Is Required to Resolve Several Paradoxes

It is easy to imagine that the planets in the solar system have always been in their current locations. However, it has now become clear that the solar system and other stellar systems (exoplanets going round other stars) are very dynamic places in which planets can form in one location and migrate long distances toward or away from their parent star. It turns out that planetary migration could be a solution to a number of problems in understanding the architectures of the solar system and exoplanet systems. Here, after a general introduction, we confine our attention to the solar system. In Chapter 13, we explore planetary migration in exoplanet systems.

4.5.1 Theories of Planetary Migration Have Been Derived

Planetary migration requires the presence of a stellar nebula of gas and dust, so is only possible in the first few million years of a stellar system. Planetary migration will not occur in the solar system now.

The mathematics of planetary migration is quite hard, so here we take a qualitative approach. As we have seen, planets form by accretion from a disc of gas and dust swirling around a protostar. As the growing planet orbits, it creates spiral density waves in the disc. These are formed by gravitational interaction between the planet and the disc and in turn lead to gravitational torques that cause planets to migrate.

To see how this works we note that disc particles inside a planet's orbit move faster than the planet, whereas those outside a planet's orbit have a lower velocity than the planet. Hence, as the planet interacts with the *inner* disc, it will tend to *gain* angular momentum, thereby driving it into a larger orbit. As it interacts with the *outer* disc, however, it will tend to *lose* angular momentum so that it relaxes into a smaller orbit. If these two effects balance, the orbit will not change. However, if one effect is larger than the other, then the planet will move in or out.

4.5.1.1 Planets Embedded in a Gas Disc Experience Type I Migration

Simulations of how planets embedded in protoplanetary discs migrate show that torques induced by the outer disc generally dominate and, so the planet will lose angular momentum and migrate inward. The timescale for the migration is directly proportional to the mass of the planet: a $10M_\oplus$ planet spiral inward 10 times faster than a $1M_\oplus$ planet. The timescale is short compared to the few million year lifetime of the disc; when account is taken of how the disc cools, which is important in determining the behavior of the spiral density waves, it is typically between 10,000 and 100,000 years for planets starting a few astronomical units out. This is termed type I migration and can lead to a planet falling into its host star.

Alternatively, type I migration is brought to a halt when the planet gets massive enough, and its gravity high enough, that its accretion of gas and dust from the disc is faster than the rate at which this gas and dust can be supplied to its feeding zone. This leads to a gap in the disc. Now the planet can no longer experience tidal torques from neighboring gas, so it stops migrating.

4.5.1.2 Planets in Gaps Experience Type II Migration

Even when type I migration has ceased, over timescales comparable to the disc lifetime gas in the disc drift into the gap so as to produce torques on the planet, thereby modifying its orbit. This process is known as type II migration. As with type I migration, detailed simulations show that inward migration is more common than outward. Hence, in general, both the planet and the gap it has cleared out get closer to the star. Current thinking is that this is how hot Jupiters arise.

4.5.1.3 Halting Migration

As we have seen, type I migration stops when the planet reaches a large enough mass, but there is no obvious way to halt type II migration. As the planet gets more massive, the gap in the disc gets wider, and the density of residual gas within it gets lower, but the effect of this on type II migration is proving difficult to model. Simulations do not give simple answers. In some scenarios migration ceases, in others migration actually reverses so that the planet moves *outward*. Sometimes inward migration does not stop at all and the planet falls into the parent star. Ultimately, type II migration ceases when the disc is dispersed by stellar winds and photo-evaporated by the intense UV of the nascent star. Subsequently, exchange of angular momentum can only occur by close interactions between bodies.

4.5.1.4 Gravitational Scattering

Gravitational interactions during close encounters between a planet and another body cause gravitational scattering, the effect of which depends partly on the relative masses of the bodies. An encounter between a planetesimal or an oligarch and a planet will excite the less massive body into a highly eccentric orbit, and in the process, the planet will lose or gain a small amount of angular momentum. The most dramatic effect is likely to be on a planetesimal because of its much smaller mass. Indeed such an encounter may fling a planetesimal into such a highly eccentric orbit that it is lost from the system entirely. However, the interaction will produce a small effect on the planet and numerous such encounters can result in its slow migration.

Possible outcomes of close encounters between two planets are that the orbital separation between them will be increased until a new stable configuration is reached. More violent outcomes include ejection of one of the planets from the system (usually, but not always, the lighter one), or into the sun, or a collision. Collisions between planetary-sized bodies have been invoked to explain several features of the solar system including the origins of the Earth's moon and the moons of Pluto, the northern lowlands of Mars, the extreme obliquities of Venus and Uranus, and the rings of Saturn. The implication is that collisions are commonplace.

Which of these outcomes actually happens depends on the details of the orbital parameters, the masses of the bodies, and so on, and cannot be predicted other than stochastically on the basis of simulations. This sort of planet–planet scattering can potentially cause havoc with the orbits of other planets in the same system, so could be important determinants of stellar system architectures.

4.5.2 Some Features of Solar System Architecture Have Been Hard to Explain

There are a number of puzzling features about the solar system. One is that the eccentricities and inclinations of the giant planets cannot be explained simply based on accretion from the solar nebula. Standard dynamical models show that the gas giants should be in virtually circular and coplanar orbits.

Others have to do with the Kuiper belt (KB), the disc of ice/rock worlds at 40–52 AU. First is the missing mass problem; namely, the KB contains only ~0.1 Earth masses, but it should have been of order 100-fold higher for the density to have been high enough to accrete bodies that are far from the sun. Second, the KB contains two populations of objects, one in low inclination orbits (<4°) assumed to reflect the original state of KB objects and easy to account for, the other in high inclination orbits (>30°) which must have been flung there by some violent dynamic process. Something similar would also be needed to explain the existence of the scattered disc, a region beyond the KB with objects in high inclination and very eccentric orbits.

Another puzzle is what caused the cataclysmic impacts the terrestrial planets experienced early in their history.

4.5.2.1 The Late Heavy Bombardment (LHB) Is One Model of Impact History

The late heavy bombardment is the barrage of asteroids and comets that formed the impact basins (craters >300 km across) that underlie the lunar maria. It is commonly thought to have ended with the creation of the 930-km diameter Orientale basin about 3.8 Gyr ago. One view of the LHB (Figure 4.6) is that it marked the end of a steadily decreasing bombardment of the inner solar system by left-over planetesimals. Another is that it was an intense spike of impacts peaking at about 3.9 Gyr caused by some sort of cataclysmic event superimposed on a background decline. In both models, the LHB consists of particularly large impacts because by this time accretion had built massive planetesimals.

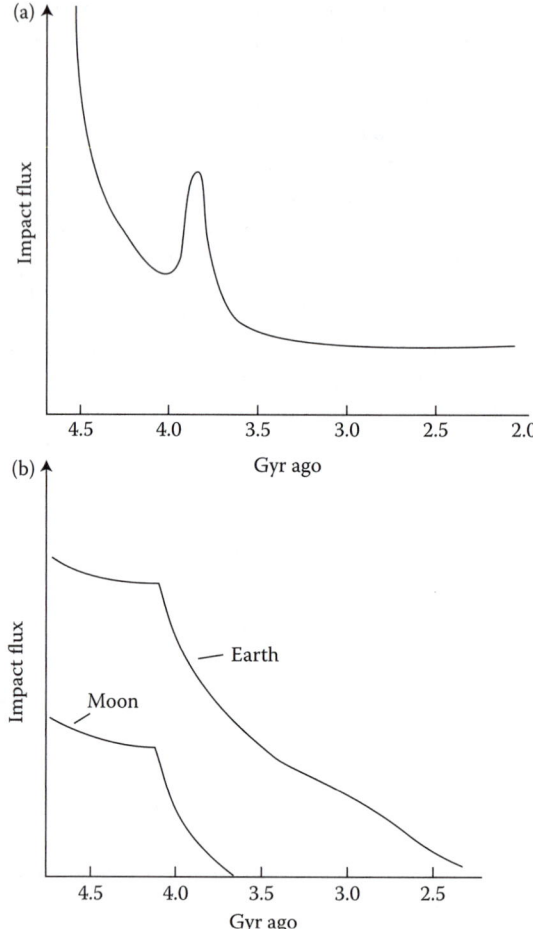

FIGURE 4.6   End-member models for solar system impact history. (a) Classical model. (b) Long-tail LHB model.

The cataclysmic LHB idea is based on the fact that the age of impact melt rocks from four lunar basins (Imbrium, Serenitatis, Crisium, and Nectaris) all cluster around 3.9 billion years ago. Few impact melt rocks brought back by the Apollo astronauts have dates earlier than this. Moreover, crater counts at lunar landing sites with measured ages show a steep decline in cratering rate from about 3.8 to 3.1 Gyr ago, and a constant impact rate (within a factor of ~2) thereafter. However, this idea is not without controversy. Firstly, the clustering of impact melt ages could be a selection effect of Apollo mission sampling. It is argued that the ejecta from the formation of just one impact basin, the Imbrium basin, are likely to be found at all of the Apollo landing sites, so all that has been measured is the date of this youngest lunar impact basin. Secondly, the lack of impact melts older than 4.1 Gyr is an artifact: older rocks existed but subsequent bombardment has reset their radioactive ages. Thirdly, dates of impact melts from lunar meteorites and Apollo glass spherules do not show the expected clustering. It may be that the LHB lasted much longer than we have thought, in which case these difficulties evaporate. We shall explore this option later (Section 4.5.3.5).

4.5.3 The Nice Model Accounts for Solar System Architecture by Planetary Migration

In a remarkable *tour de force,* a group of astronomers working with Alessandro Morbidelli at Observatoire de la Cote d'Azur in Nice, France, developed computer simulations of the early evolution of the solar system that incorporated planetary migration. The model reproduces the orbits of the gas giants and of the Trojan asteroids of Jupiter, shows how the Kuiper belt could come to look the way it is today, and the cause of the LHB. This approach, which has come to be known as the Nice model, was first published in 2005, and continues to be developed.

4.5.3.1 Simulating the Early Solar System

The early evolution of the solar system was driven by several interactions.

1. For the first few million years while the solar nebula gas was present, type I or type II planetary migration occurred. Simulations show that giant planets should get trapped in mean motion resonances (MMR) in which the orbital periods of two planets is an integer ratio. For example, Nice simulation runs typically start with Jupiter and Saturn in a 3:2 resonance in which for every three times Jupiter goes round the sun Saturn orbits twice.
2. Once the gas had dissipated, the outer solar system probably underwent a violent phase in which the giants gravitationally scattered off each other and acquired eccentric orbits.
3. Subsequently, interactions with remaining planetesimals eventually damped the excess orbital energy of the giants, circularizing their orbits and driving them to migrate radially into their present orbits, but at the expense of exciting the planetesimals.

It is the last stage that is captured in the original incarnation of the Nice model. The original motivation was to explain the eccentricities of the giant planet orbits.

4.5.3.2 The Original Nice Model Explains the Eccentricities of the Giants

Simulations typically start with Jupiter slightly further out than it is now while the other giants are closer in; for example, Neptune is at 14 AU. Moreover, simulations initially confine the Kuiper belt to between 15 and 34 AU and give it far more mass ($35M_\oplus$) than it has now. One motivation for starting out with a much more compact outer solar system is that the accretion of Uranus and Neptune at their present locations, about 20 and 30 AU respectively, needs an implausibly long time.

The Nice simulations show that Jupiter moves inward while Saturn, Uranus, and Neptune migrate outward as a result of the exchange of angular momentum with planetesimals in the Kuiper belt. Planetesimals at the inner edge of the disc are first gravitationally scattered by the outermost giant planet, which tends to send them inward, while the planet moves outward. Inward falling planetesimals then scatter off the next planet they encounter, moving its orbit outward. This process continues until the planetesimals encounter Jupiter. So great is Jupiter's mass that the planetesimals are excited into highly elliptical orbits or are ejected from the solar system. This causes Jupiter to move slightly inward.

A few hundred million years of this slow migration eventually brings the distance between Jupiter and Saturn to the point where they go into a 1:2 mean motion resonance. This triggers gravitational mayhem. It drives Saturn and the ice giants rapidly outward and pumps up their orbits, so they become eccentric and inclined. The ice giants plough straight into the Kuiper belt scattering 99% of its planetesimals (see Figure 4.7). This is thought to account for the dynamically excited (high inclination) population of Kuiper belt objects, the scattered disc, the Oort cloud, for the low mass of the KB now, and for the late heavy bombardment.

In 33% of the simulations—each has a slightly different starting configuration—the researchers found that Neptune is actually ejected from the solar system. However, in virtually all other runs, the four giants eventually settle into stable orbits. Simulations of other mean motion resonances between giants during planetary migration failed to account for the observed eccentricity and inclination of Jupiter's orbit.

4.5.3.3 The Nice Model Has an Explanation for the Late Heavy Bombardment

The origin of the LHB has always been a mystery. The Nice model suggests that it was triggered by the disruption of the Kuiper belt. While most of the icy planetesimals ejected from the Kuiper belt probably end up in the Oort cloud, a few were sent into the inner solar system where they hammered the terrestrial planets or perturbed asteroids in the main belt to do likewise. One estimate reckons that about one-tenth of the LHB mass hitting the moon was cometary, but most was asteroidal. The size distribution of lunar impact basins formed in the LHB seem to match the size distribution of main belt asteroids and simulations show how main belt asteroids could indeed have been flung into the inner solar system. In some simulations the main belt loses as much as 95% of its mass to fuel the LHB. This suggests that the bulk of the LHB impactors were asteroids. Confusingly, 3.8-Gyr-old rocks in Isua, Greenland that bear geochemical traces of the LHB imply that the impactors were comets.

4.5.3.4 The Main Asteroid Belt May Provide Clues to the Origin of the LHB

The main asteroid belt has gaps, named for Daniel Kirkwood who reported them in 1867, where the asteroids are in mean motion resonances with Jupiter. For example, there is a gap at 3.3 AU where an asteroid would orbit twice, every time Jupiter orbited once. Mean motion resonances can make asteroid orbits unstable: put an asteroid into a Kirkwood gap and it would soon be flung out.

By running simulations, researchers find that the distribution of asteroids in the main belt cannot be explained simply by the gravitational influence of the gas giants

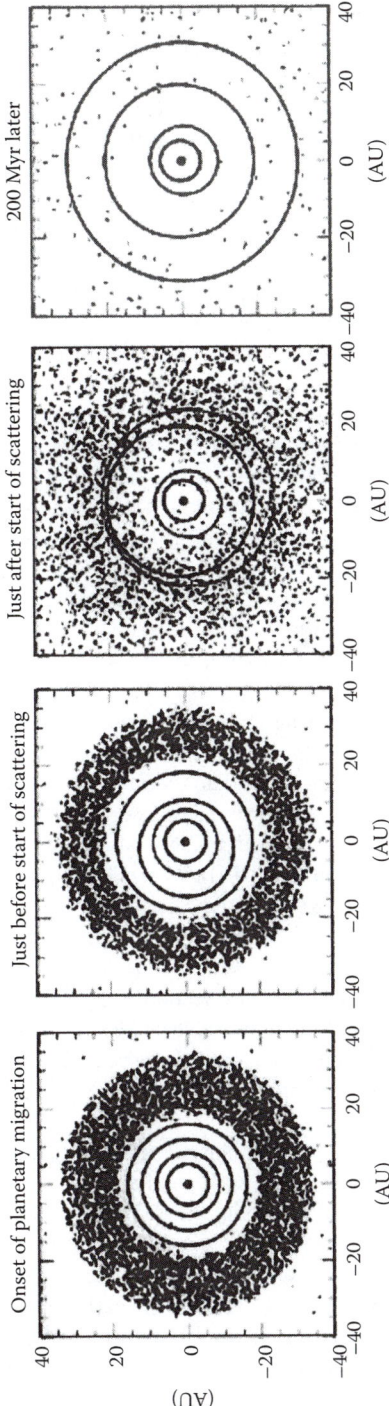

FIGURE 4.7 The Nice model shows how the Kuiper belt might have been completely disrupted by planetary migration early in the history of the solar system. (From Gomes et al., *Nature*, 2005;435:466–469. With permission.)

in their current positions. The Nice model of the LHB has had some success in predicting how the asteroid belt has come to look the way it does. For example, it shows that many icy planetesimals would end up in the outer asteroid belt where they would have been fragmented by collisions. The outer asteroid belt is indeed populated by cometary nuclei (D-type asteroids) with the correct size distribution.

Bill Bottke of the South West Research Institute, Boulder, Colorado, and colleagues have revised the Nice model. They argue that the LHB lasted much longer than previously thought, from 4.1 to 2.1 Gyr ago, with most of the impactors coming from an extended, and now severely depleted region of the inner main belt, termed the E-belt, located between 1.7–2.1 AU. What is the evidence for this?

Firstly, simulations show that left over planetesimals in the inner solar system cannot account for anything like the number of impact basins seen on the moon. This means that some sort of catastrophic event must have contributed to the LHB.

Secondly, when a large impactor strikes the Earth it throws a cone-shaped ejecta curtain that contains numerous tiny melt droplets. Many of these will reach altitudes of tens of kilometers. These cool and fall back to form a global layer of spherules that can be several millimeters thick in the case of a large cratering event. Twelve globally distributed beds of spherules that date between 3.47 and 2.1 Gyr ago attest to large impacts that the Earth suffered throughout the Archaean and into the Proterozoic eons.

A third clue comes from the E-belt. Located there now are the high inclination Hungaria family asteroids. Simulations show these are a likely origin of Near Earth Objects. Scaling up the numbers of E-belt asteroids showed that as the gas giants migrated, as specified by the Nice model, a resonance between Saturn and asteroids swept rapidly inward through the asteroid belt. This pumped up the inclinations and eccentricities of the E-belt asteroids, making them 10 times more likely to be flung from the E-belt into Earth-crossing orbits than their main belt cousins. However, the E-belt is dynamically "sticky" in the sense that it can take a long time for objects to leave it. Consequently, the LHB on Earth could have lasted about 2 billion years, some 10-fold longer than the traditional LHB. The simulations produced 10 lunar basins between 4.1 and 3.7 Gyr ago, and predicted that 15 basins and some 70 large (180 km diameter) craters would have been formed on Earth between 3.7 and 2.5 Gyr ago, enough to explain the spherule beds.

Because the moon has such a small cross section compared to the Earth, calculations show it could easily have avoided being struck by LHB impactors after about 3.8 Gyr, as appears to have been the case, even though the Earth continued to be battered. That is why lunar cratering history has led us to postulate that the LHB ended by then.

The significance of this is that cratering chronologies of other planets (most importantly for astrobiologists, Mars), used to estimate the *relative* ages of planetary surfaces, are calibrated on the basis of the moon, since we have absolute radiometric ages for the moon. If lunar cratering has "missed" much of the impact history of the early solar system, then we may be wrong about the ages of planetary surfaces.

Another consequence of Bottke's model is that the early Earth would have been a violent place for a lot longer than the classical LHB postulates, and that if life did get started by 3.8 Gyr ago, then it clearly survived the battering that persisted throughout the Archaean. Indeed, as we shall see, impacts may actually have been beneficial for life on early Earth.

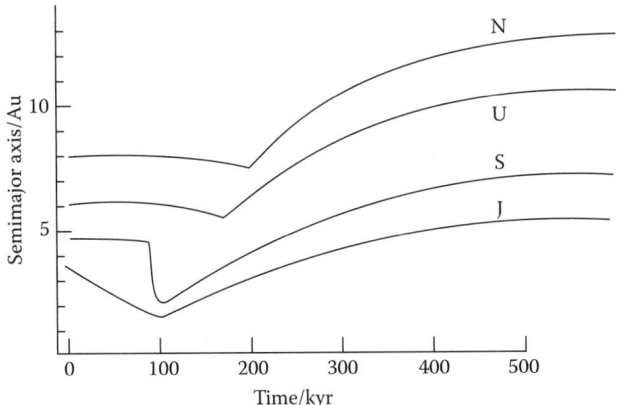

FIGURE 4.8 The grand tack model has Jupiter and Saturn migrate in and out of the inner solar system. (Redrawn from Walsh et al. *Nature* 2011;475:206–9. With permission.)

4.5.3.5 The Grand Tack Accounts for the Inner Solar System

The Nice model provides a plausible account of the architecture of the outer solar system. It has now been extended by Kevin Walsh with Nice researchers to account for the structure of the inner solar system with the grand tack hypothesis. The idea is that Jupiter and Saturn, initially at 3.5 and 4.5 AU, respectively, both migrate inward and become locked into a 2:3 resonance, with Jupiter at 1.5 AU and Saturn at 2 AU (Figure 4.8). Both gas giants then migrate out, still in resonance, capturing Neptune and Uranus in resonance and so driving them out too. During this process Saturn, Uranus, and Neptune finish their accretion.

This gravitational dance drives most of the anhydrous S-type planetesimals in the inner disc inward, while scattering 14% of them outward, while dynamically exciting the wet C-type planetesimals in the outer disc. The result is the formation of an asteroid belt in which the S-type asteroids (parents of the ordinary chondrites) are in the inner belt while the wet C-type asteroids (parents of the carbonaceous chondrites) in the outer belt (Plate 5.1). This explains how objects so different in volatile content could end up close together, even though they must have accreted in different locations. The model shows that the disc gets truncated beyond about 1 AU, and hence why Mars ended up being relatively small. Simulations show $(3–11) \times 10^{-2} M_{\oplus}$ of C-type asteroids are delivered to the region where the terrestrial planets form: 6–22 times the minimum mass required to account for the Earth's water inventory. We return to this issue in Section 5.4.4.2.

5

Early Earth

5.1 Assembly of the Earth Can Be Modeled

5.1.1 To First Approximation Earth Grew at a Decreasing Exponential Rate

Simulations show that a terrestrial planet as big as the Earth grows over tens of millions of years. If it is assumed that planets accrete at a decreasing exponential rate, then the time scale of the accretion of the exponential growth is given by the time constant, τ. The time constant is the mean of an exponential function. In this context, it is the time for 63% of the mass of the Earth to be accreted (Figure 5.1).

EXERCISE 5.1

The curve in Figure 5.1 has an equation of the form:

$$M_t = M_{t0}(1 - e^{-t/\tau}). \tag{5.1}$$

Show that at time t the Earth has accreted 63% of its final mass.

5.1.2 Accretion Was Probably Heterogeneous

The current thinking is that the Earth accreted from raw ingredients that changed in composition and oxidation states over time. This is termed heterogeneous accretion. The first 80–90% of materials were probably iron-rich and reducing in nature and the last 10–20% were more oxidizing. However, melting and differentiation homogenized the Earth's mantle and stripped many of the iron-loving (siderophile) elements into the core, leaving little trace of heterogeneous accretion except for the final 1% which was delivered too late to mix properly throughout the mantle and never had the chance to equilibrate with the core.

5.2 Early Earth Was Shaped by a Moon-Forming Impact

As we have surmised, it is highly likely that the moon was produced as a result of a giant impact. This added almost one-tenth of the Earth's mass. It marks the boundary between the Chaotian and Hadean eons. A final 1% of the Earth's mass was added subsequently by the late heavy bombardment (LHB), which could account for the late veneer, a geochemical signature of material added after the Earth's core had

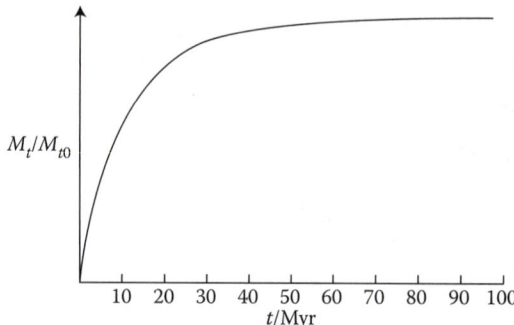

FIGURE 5.1 Exponential model of planetary accretion. M_t/M_{t0} is the cumulative fractional mass of the Earth at time t relative to now, t_0, and τ is the time constant.

formed. These were cataclysmic events and set the context in which Earth spawned life. However, their timing is not well constrained, the exact nature of the moon-forming impact remains uncertain, and there is controversy over whether the late veneer is real or an "artifact" generated in other ways.

James Hutton, the Scottish physician and geologist concluded his 1788 paper, "Theory of the Earth" with the comment "we find no vestige of a beginning—no prospect of an end." Although this is now a somewhat pessimistic view, it captures succinctly the fact that the Hadean is a very data-poor time. For example, we have only two terrestrial rocks that could be Hadean, both Canadian. Acasta gneiss from the Northwest Territories is dated at 4.031 Gyr, and Nuvvuagittuq greenstone belt exposed on the eastern shore of the Hudson Bay has a Sm–Nd age of 4.28 Gyr, though another interpretation of the data suggests 3.8 Gyr is closer to the mark. Hence, almost all of what we think we know about this remote time comes from indirect clues and intelligent guesswork.

5.2.1 The Moon-Forming Impact Is Supported by Theory and Geochemistry

The current view is that the moon formed as a result of a catastrophic collision between the proto-Earth and a second high mass body very early in the history of solar system.

5.2.1.1 *Computer Simulations Show a Moon-Forming Impact Is Dynamically Feasible*

Simulations must capture the essential features of the system they attempt to model, but not be so complex as to become mathematically intractable or take up too much computing time. By making reasonable assumptions some parameters can be fixed. Simulations are then run in which other parameters are varied in a systematic way to see what happens. For many years, the "standard" moon-forming impact has been based on the assumption that the collision did not result in much loss of mass or change the angular momentum of the system very much. This means that the velocity of the impact and angular momentum can be more or less fixed. The simulations then attempt to home-in on the correct Earth/moon mass ratio by tweaking the impactor mass and impact angle. Any impact theory has to account for how the moon gets to be iron deficient, so into the model goes the assumption that both the proto-Earth and impactor are differentiated worlds with chondritic composition: 30 wt% iron core and 70 wt% silicate. The simulations then follow hundreds of thousands of mass particles during the impact, keeping track of the type of material (metal or silicate), its origin, and variables such as velocity and temperature, while gravity, pressure forces, and shock dissipation play out over time.

A leading researcher of moon-forming impacts is Robin Canup of the Southwest Research Institute, Boulder, Colorado. Given the high angular momentum of the Earth–moon system, she has shown that low velocity (barely higher than escape velocity), ~45° angle, grazing impacts are compatible with the mass ratio of Earth and moon, provided the mass ratio of the proto-Earth/impactor is 9:1 and the total mass is $1.05M_\oplus$. In other words, the impact happened between a proto-Earth that had already accreted 90% of its current mass and a Mars-sized planetary embryo, Theia. Many impacts produce two moons, but one either collides with the other or falls to Earth. Off-center grazing impacts generally result directly in a single moon. In such impacts, Theia is almost completely melted and/or vaporized, but condenses within a few hours to spiral in for a second hit. The end result is a disc of solids and vapor around the Earth from which the moon subsequently accretes.

EXERCISE 5.2

1. Briefly describe how you would expect the proto-lunar disc to be oriented with respect to the Earth.
2. Why might this be a problem for a giant impact model?

Simulations of this type of model consistently show that the moon is made mostly from Theia. As for the Earth, the impact probably heats the mantle to an average temperature of about 2000 K, creating a magma ocean and a silicate vapor atmosphere from the crust and upper mantle. The entire moon-forming process takes just 2 days.

5.2.1.2 Much of Theia's Core Becomes Part of Earth's Core

Lunar rocks are highly depleted in siderophile elements such as Ge, Co, Ag, Pd, Au, and Ir. This shows that Theia must have been a differentiated world because it had a metal core that had stripped most of the siderophile elements from its silicate mantle. A reanalysis of seismic data collected by the Apollo missions suggests that the moon has a core, but with a radius of only 330 km it is small in relation to terrestrial planet cores. The implication is that most of Theia's core ended up in the Earth.

5.2.1.3 Lunar Rocks Are Depleted in Volatiles

An impact explains much of the detailed chemistry of the 382 kg of lunar rocks brought to the Earth by Apollo astronauts. It clarifies why the moon is depleted in volatiles (e.g., the alkali metals Na_2O, K_2O, and H_2O) and enriched in refractory elements (e.g., Ti, Th, and U) relative to the Earth and primitive meteorites. This is because the moon is thought to have formed substantially from Theia, which generally experienced higher temperatures than proto-Earth material, it would be expected to have lost more volatiles. Volatile depletion has now been confirmed. Isotopes are not fractionated from each other by magmatism, so erupted melts faithfully preserve isotope ratios of their source material. Lunar samples have $^{66}Zn/^{64}Zn$, which is 1.5 parts per thousand greater than samples from the Earth, Mars, and carbonaceous chondrites; that is, the moon has lost relatively more of the light isotope of zinc. This provides unambiguous evidence that the moon is depleted in volatiles with respect to the Earth.

5.2.1.4 Earth and Moon Have the Same Oxygen and Silicon Isotopes

Remarkably, oxygen and silicon (and many other) isotope ratios are identical for the Earth and the moon. On the face of it, this is extremely puzzling. If the moon originated largely from Theia it requires that Theia formed in the same place as the Earth (because planetary bodies born in different parts of the solar nebula should have their own distinctive isotopic signatures) or that thorough homogenization of proto-Earth and Theia material occurred, which implies that both bodies must have extensively melted and vaporized.

Could Theia have formed in the same part of the solar nebula as the proto-Earth, so started out with the same isotopic composition? Generally, this is not feasible as any small planetary body accreting in an orbit close to the much larger proto-Earth would have been ejected out of its orbit, or collided with the Earth long before it could grow to the size of Mars. (This follows from the dynamical definition of a planet we saw in Chapter 4.) The one potential exception is if Theia had accreted at the L4 or L5 Lagrange points of the sun–Earth system that are located 60° ahead and 60° behind the Earth in its orbit around the sun—like the Trojan asteroids of Jupiter. For all Lagrange points, the vector sum of gravitational and centrifugal forces is zero, and objects at L4 and L5 will tend to remain there. Edward Belbruno and Richard Gott have shown that the accretion of a $0.1M_{\oplus}$ planetary body at L4 or L5 is plausible and that it could be gravitationally perturbed by interactions with remaining planetesimals into a parabolic collision course with the proto-Earth (Figure 5.2).

An alternative explanation for identical oxygen and silicate isotopes is that the impact was so energetic that silicate melt/vapor from the proto-Earth and Theia were well mixed, homogenizing the isotope ratios before the moon accreted. There are two difficulties with this notion.

1. For very refractory elements there may not have been enough time for homogenization. Titanium is a highly refractory element and only likely to

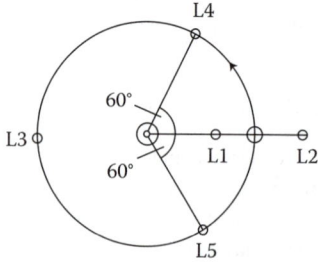

FIGURE 5.2   If Theia accreted at the L4 and L5 Lagrange points of the sun–Earth system, it would have the same isotope composition as the Earth.

vaporize at very high temperatures, and this makes the identical $^{50}Ti/^{47}Ti$ for Earth and moon hard to explain. This is because the more refractory an element is, the longer it takes to homogenize. It would require at least 100 years for the vapor in the disc and the Earth's silicate atmosphere to homogenize the titanium isotopes by which time most estimates have the disc condensing and coalescing to form the moon.

2. Recent isotopic data from the deep mantle shows that the whole mantle was not completely mixed at the end of accretion.

EXERCISE 5.3

What might be a solution to both of these problems?

5.2.1.5 The Moon May Have Originated from the Earth's Mantle after It Had Already Produced a Crust

The Earth's mantle has nonchondritic (much lower) abundances of siderophile elements because these have been stripped by core formation. The accessible mantle of the Earth also has nonchondritic abundances of lithophile elements because when the mantle partially melts, the melt acquires more than its fair share of incompatible elements (those that distort crystal structures and hence partition into a melt fraction) and then freezes to form crust. The residual mantle is thus depleted in incompatible lithophiles, but enriched in compatible ones (which remain in the solid phase). The mantle of the moon is reckoned to be very similar to the Earth's upper mantle, that is, nonchondritic in both siderophiles and lithophiles. Therefore, the moon may be derived from mantle from which the crust had been extracted.

5.2.1.6 Changing Model Parameters Can Achieve Similar Isotopic Compositions for the Earth and Moon

Recently, moon-forming impact models have been proposed that relax the requirement of the "standard" Theia collision for the angular momentum to be fixed, by allowing mass to be lost. Mass loss will carry away angular momentum. Under some circumstances these models result in identical compositions for the Earth and the moon. For example, in 2012 Andreas Reufer of the Center for Space and Habitability in Bern, Switzerland, and colleagues, have run simulations in which more massive (0.15–$0.2M_\oplus$) and higher velocity (1.2–$1.3v_{esc}$) impactors make shallower angle (30–$40°$) glancing blows with the proto-Earth. These are termed "hit and run" impacts because an appreciable amount of the impactor continues after the collision. Crucially, these events result in proto-lunar discs in which most of the silicate material is derived from the proto-Earth. This immediately helps to match isotope ratios of Earth and moon. However, these events have higher kinetic energy, so temperatures are much higher than in the canonical model, with some parts of the Earth's mantle reaching 10,000 K.

Other ways to achieve similar compositions of Earth and moon are to have two bodies of comparable masses (with combined total mass of $1.04M_\oplus$), colliding with modest velocities about or just greater than the escape velocity, or alternately to have high velocity (20 km s^{-1}) impacts between a much less massive impactor and a very rapidly rotating proto-Earth (with rotation period 2–3 h).

These other scenarios require that the impactor's large metal core remains largely intact as it falls through the Earth's mantle. This is needed to explain the identical silicon and tungsten isotope ratios since these elements are strongly siderophilic and would otherwise equilibrate with a metal phase, thereby altering the ratios from those in the silicate disc that formed the moon. These more violent scenarios also have, after the impact, an angular momentum that is too high. A resonance that allows the angular momentum to be transferred from the lunar orbit to the Earth's orbit around the sun has been proposed as a "get out clause" from this difficulty. It would allow the Earth–moon system's angular momentum to achieve its observed value soon after the collision, but it remains unclear whether this would require implausible fine-tuning for it to work.

5.2.1.7 There Are Difficulties with the Giant Impact Theory

A giant impact might be expected to have resulted in a dry moon; water is, after all, very volatile. However, recent measurements of melts trapped in olivine crystals (melt inclusions) appear to refute this long-held assumption. The melts preserve their original volatile content because their crystal prison has prevented degassing. The melt inclusions are found in orange regolith of the Taurus–Littrow valley sampled by Apollo 17 astronauts in 1972. Orange regolith consists almost entirely of volcanic glass beads, 20 to 45 μm across, that erupted in a fire fountain about 3.6 Gyr ago. One study found the melt inclusions contained 615–1410 ppm of water. This is similar to terrestrial mid-ocean ridge basalts (MORBs) and shows that at least some parts of the lunar mantle contain water concentrations comparable to the MORB mantle source, estimated to be 160 ± 40 ppm.

Remarkably, the deuterium/hydrogen (D/H) ratio of lunar magmatic water, sampled by the melt inclusions, is indistinguishable from that of water in carbonaceous chondrites and close to that of terrestrial water. The implication is that there is a common origin for water on the Earth and the moon. The simplest explanation for this is that the Earth was wet at the time of the impact and that widespread exchange of volatiles occurred between the molten Earth and the proto-lunar disc, some of which must have been cool enough to hold onto the volatiles. This requirement imposes constraints on dynamical models for the impact.

5.2.2 The Timing of the Moon-Forming Impact Is Poorly Constrained

The moon-forming impact essentially completed the Earth's accretion. Hence, the age of the moon dates both the impact and the time for the final assembly of the Earth.

Radiometric dating with ^{87}Rb–^{87}Sr gives an age for the moon as 4.48 ± 0.02 Gyr. In other words, the moon formed 70–110 Myr after the start of the solar system. This is consistent with the U–Pb age of the Earth, which probably dates the late stage of core formation.

5.2.3 Tidal Forces Drove Evolution of the Earth–Moon System

Immediately after the impact the moon would have been very close to the Earth, perhaps just $3R_{\oplus}$ away. The high angular momentum of the Earth–moon system implies

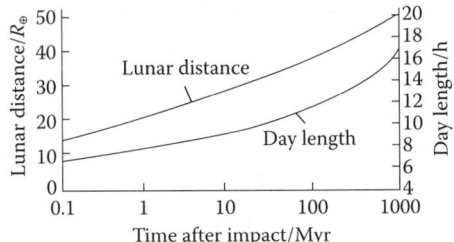

FIGURE 5.3 Evolution of the Earth–moon system. The moon is now $60R_\oplus$ away.

that immediately after the impact the Earth rotated rapidly, so the day length would have been much shorter than now, perhaps 5–6 h initially.

Tidal forces would have been enormous at this time and this had three consequences.

1. High deceleration of the Earth's rotation rate (Figure 5.3).
2. Rapid slowing of the moon's rotation rate until it became tidally locked. This meant that the rotation period and orbital period of the moon became the same (synchronous rotation) so that from the Earth we always see the same face of the moon.
3. The moon rapidly receded from the Earth (Figure 5.3).

The moon continues to move away from us but at a much more sedate pace. Lunar recession has been measured by laser ranging experiments that fire laser pulses at retroreflectors left on the lunar surface by Apollo astronauts and measure the 2-way travel time for their return. The current rate of the lunar recession is 3.82 cm yr^{-1}.

EXERCISE 5.4

A lunar laser ranging measurement records a time of 2.567 s. What is the distance of the moon on this occasion?

5.3 The Early Hadean Was Hot

For a few million years after the moon-forming impact geothermal heat flow was probably about 100 W m^{-2}, 1000-fold higher than now, so the early Hadean eon lived up to its name. Here we look at this hellish time.

5.3.1 Earth's Postimpact Atmosphere Was Largely Rock Vapor

The impact is thought to have vaporized 20% of the mantle. Therefore, immediately afterward the Earth's atmosphere would have consisted of silicate rock vapor at a temperature of about 2300 K and the concentrations of other gases, most probably CO, CO_2, H_2O, and H_2, would have been determined by their solubility in the molten magma that covered the Earth, given its temperature, pressure, and oxidation state. Hydrogen would be lost by thermal escape at the high temperature of a rock vapor atmosphere, but the other gases would be held. At high temperatures, any water would tend to be held in solution in the molten magma, though as the temperature fell, the atmosphere would come to be dominated by outgassed water.

5.3.2 A Magma Ocean Remained After the Moon-Forming Impact

"Hit and run" impact models of the moon's formation predict that most of the mantle would have been heated to temperatures high enough that it was completely melted so that the Earth was covered with a magma ocean. Additional heat energy would have come from core formation as impactor metal sank into the Earth's core, and from huge tidal dissipation due to the proximity of the moon. These would have helped keep the magma ocean in a molten, low viscosity state in which turbulent convection ensured thorough mixing so that it was chemically homogeneous.

The magma ocean may have crystallized from the base up (Figure 5.4). This is because we expect the adiabat (the curve that represents how temperature increases with depth) to be steeper than the solidus (the curve that demarcates the change from solid to partial melting in T–P space).

Convection currents transported heat from the depths to the surface where it would be radiated away. Crystallization at the base would generate latent heat. This could have slowed cooling of the core and so delayed the start-up of the geodynamo to 4.0–3.4 Gyr.

5.3.3 The Hadean Mantle, Atmosphere, and Oceans Could Have Coevolved

We do not know when the Earth acquired its water. It may have accreted wet and retained some of its volatiles after the moon-forming impact or it may have accreted

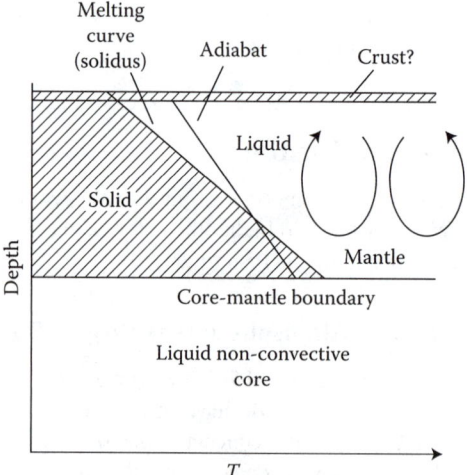

FIGURE 5.4 Structure of the magma ocean. Temperature in the mantle follows the adiabat. With time the temperature falls so the adiabat shifts to the left.

dry and had all of its water delivered after the moon-forming impact. Here we follow through the consequences of Earth accreting wet.

If this were the case, the start of the Hadean would have seen the Earth with a water-dominated dense atmosphere overlying a deep magma ocean.

The density of the atmosphere in the early Hadean is not known. Most of the total inventory of the Earth's CO_2 would have been in the atmosphere, given its low solubility in molten silicate, and this would have amounted to ~100 bar.

Modeling by Linda Elkins-Tanton, Director of the Carnegie Institute for Terrestrial Magnetism, Washington, DC, shows that if early Earth's magma ocean started out with 0.1 wt% (1000 ppm) water this would give a water atmosphere of a few hundred bars since most (between 60% and 99%) of the mass of water would be degassed. Based on the estimates of the Earth's current water inventory, though this is uncertain by a factor of more than 5, this could amount to as much as 500 bar.

The atmosphere would start out supercritical because it would be above the critical point for water ($T = 674$ K, $P = 220$ bar), and no liquid water could condense. The rate of cooling would have depended on the flux of absorbed sunlight plus the geothermal heat flow.

EXERCISE 5.5

    i. On what does the flux of absorbed sunlight depend?

    ii. What factors will contribute to the geothermal heat flow in the early Hadean? What will happen to the geothermal heat flow over time and why?

5.3.3.1 Early Hadean Earth Experienced a Runaway Greenhouse

At high temperature, the atmosphere is transparent to thermal radiation from the surface making it hard for water to condense and allowing the Earth to cool rapidly. As the temperature fell water would eventually condense. This marks the runaway greenhouse limit, the temperature—and hence the total heat flow—at which the phase change between water and steam occurs. The geothermal heat flow for the runaway greenhouse limit, assuming 500 bars of water, is ~140 W m^{-2}.

At this stage, the atmosphere acted as a highly effective thermal blanket, slowing the cooling of the magma ocean. Indeed a negative feedback mechanism operated to keep the surface of the magma ocean molten. As the mantle cooled more water would have degassed into the atmosphere. This raised the temperature, thereby melting some of the crust that had formed, thus increasing the amount of surface covered with liquid magma. Hence, more water dissolved in the magma, removing it from the atmosphere.

Once a significant crust formed geothermal heat flow falls because solid rock is a good thermal insulator. Now the temperature would have fallen enough for water to condense out in appreciable amounts to form oceans and consequently the atmospheric pressure dropped.

Our limited knowledge of heat flux at the surface (from the sun and the interior of the Earth) and atmospheric pressure means that we do not know how long Earth would have been in a runaway greenhouse state and hence how long the magma ocean would take to cool. However, models show that tidal heating could sustain geothermal heat flow above 140 W m^{-2} for a couple of million years. Tidal heating arises when tidal motions are resisted by viscosity: the higher the magma viscosity the greater the heating. This mechanism is most effective where the magma ocean is on the cusp of freezing, usually assumed to be at the base.

EXERCISE 5.6

    i. Why might geothermal heat flow drop below the runaway greenhouse limit after a couple of million years?

    ii. If tidal heating exceeds what the atmosphere can radiate away what happens to the magma temperature and hence viscosity? What are the consequences of this for tidal heating and magma ocean temperature?

Models show it is likely to have taken a few tens of millions of years for the magma ocean to freeze. The earliest time is before 4.4 Gyr if we accept the evidence that by then the Earth had oceans. By this time geothermal heat flow is thought to have fallen to about 0.5 W m^{-2}: a value only five-fold higher than through ocean crust now. The surface temperature would have fallen below the critical temperature allowing steam to form and condense as scalding rain to fill the oceans.

5.3.3.2 CO_2 Must Have Been Removed from the Atmosphere

After the deluge the state of the atmosphere depended on how much CO_2 there was. Assuming that the CO_2 inventory was comparable to that today and the capacity of the oceans to sequester it as carbonate was relatively small (although the solubility of CO_2 in water increases with pressure, it falls with temperature), the atmosphere would have held ~100 bar of CO_2. The greenhouse effect of CO_2 would have resulted in the surface temperature of ~500 K—still more than 100 K higher than the maximum at which we think biochemistry can operate. How soon the surface temperature fell to biochemically plausible values would have depended on how fast the CO_2 could be removed from the atmosphere into the mantle and/or crust as carbonates. This process is called carbonatization, and this requires water. Calcium and magnesium carbonates would have been stable only in the coolest top 500 m of oceanic crust and this could have held only about 10 bar of CO_2 at any time. Repeated carbonatization of the crust and some sort of subduction would have had to operate to remove all the CO_2 into the Earth's interior. This process would have got faster as the temperature fell, and is thought to have taken between 10 and 100 Myr.

5.3.4 A Dry Accreting Earth Would Be Hot During the Hadean

The account above presupposes that the Earth acquired substantial water in the Chaotian eon and held onto an amount comparable to its current water inventory through the moon-forming impact (Figure 5.5).

If the Earth accreted dry, its atmosphere in the early Hadean would have been dominated by ~100 bar of CO_2. In the absence of water or hydrous minerals no carbonates would have formed or any plate tectonics would have operated. Until water was delivered, presumably by the late heavy bombardment at the end of the Hadean, allowing carbonatization and subduction of the crust, Earth would keep its dense greenhouse CO_2 atmosphere, and presumably a surface temperature ~500 K.

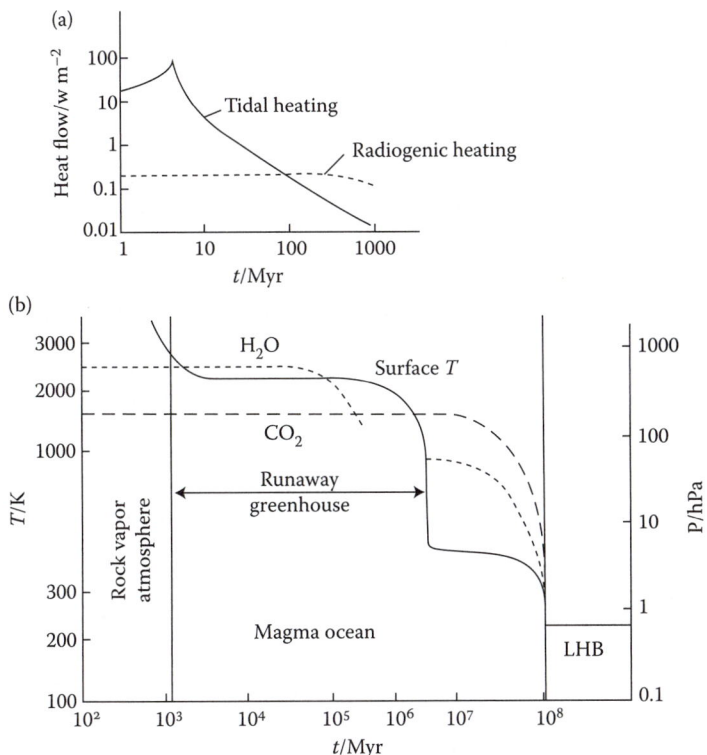

FIGURE 5.5 A history of the Earth during the Hadean. hPa, the hectopascal, is the SI unit of pressure equivalent to one millibar. (a) Major geological heat sources. (b) Atmospheric T, P, and composition. (Adapted from Zahnle et al. *Space Sci Rev* 2007;129:35–78.)

5.4 Late Events Modified the Composition of the Earth

5.4.1 Earth's Mantle and Crust Has an Excess of Siderophile Elements

Highly siderophile elements (HSEs; e.g., Re, Os, Ir, Ru, Rh, Pt, Pd, and Au) in the mantle have abundances that are only 0.5% of chondritic. This seems reasonable since we would expect HSEs to have been taken up by the core during differentiation. However, silicate–metal partition coefficients show that this depletion is not as much as expected. This is termed the siderophile excess problem. In other words, the abundance of highly siderophile elements in the mantle is higher than predicted. There are two frontrunner solutions.

5.4.2 Core Separation Happened at High Pressure and Temperature

One idea is that core separation happened under high pressure and high temperature in which HSEs and some moderately siderophile elements (e.g., Ni) become less siderophilic than at 1 bar pressure. This has been shown by measurements of silicate–metal partition coefficients that mimic the high pressure/temperature conditions at the bottom of a deep magma ocean.

EXERCISE 5.7

Would siderophile element abundances in the mantle be greater or smaller if core differentiation had happened at high P/T conditions rather than low P/T conditions?

We see that high pressure/temperature differentiation of the core might help to explain the excess siderophile problem. There is a difficulty with this idea though. Each element would be removed from the mantle according to its silicate–metal partition coefficient, and this would alter the element ratios from their original chondritic ratios. And this is not what is seen. Broadly speaking, HSE ratios *are* chondritic.

5.4.3 Siderophile Elements Were Likely Delivered by a Late Veneer

Another idea is that after the Earth had differentiated into core and mantle, impacts by asteroids with bulk chondritic composition delivered the last 0.4–1% of the Earth's mass, along with their complement of HSEs. These siderophiles would have remained in the crust and mantle since core formation had finished by this time. This late veneer hypothesis is supported by the fact that siderophiles, although depleted in the mantle with respect to chondritic, are present in chondritic proportions. The late veneer is assumed to have delivered volatiles, such as water, to the Earth. Clearly, the late veneer must postdate completion of the Earth's core formation, so must have happened at least 110 Myr after the formation of the solar system.

Late accretion of chondritic material appears to have been widespread in the inner solar system. Highly siderophile element ratios in lunar basalts, Martian meteorites, and HED meteorites from Vesta are all chondritic, but each element is depleted with respect to CI chondrites.

It is worth pointing out that some researchers argue that there is as yet no quantitative model that convincingly accounts for HSE and volatile budgets and isotope compositions for the Earth and the moon by a late veneer.

5.4.4 The Mantle Has Become More Oxidized with Time

The oxidation state (Box 5.1) of the mantle is important because it determines the nature of the atmosphere that is degassed and the sort of chemistry that can happen when water and rock interact. All this has implications for habitability. Consequently, astrobiologists take an interest in the oxidation state of the Earth and other terrestrial planets.

This is not made easier by the fact that the oxidation state of the modern mantle is not homogeneous. It is lower in the lower mantle. Also it varies with tectonic setting; mantle from active subduction zones is more oxidized than from mid-ocean ridges, probably because subducting mantle has a much higher water content which renders Fe^{3+} more soluble. But on a global scale, theory suggests that the Earth's mantle oxidized soon after the planet accreted.

BOX 5.1   OXIDATION AND REDUCTION

An atom is oxidized if it loses an electron to another atom. The atom that gains the electron has been reduced. In other words, oxidation is the loss of electrons and reduction is the gain of electrons. This can be recalled by the mnemonic OIL RIG (Oxidation Is Loss Reduction Is Gain). Oxidation and reduction are two sides of the same coin, they cannot occur in isolation: oxidation of an atom, ion, or molecule must be accompanied by a corresponding reduction. The electron has to go somewhere!

Iron in the form of native metal is in its most reduced form, and has an oxidation state of zero (Fe^0). It can lose two electrons to some electron acceptor, for example, oxygen, acquiring an oxidation state of two, Fe(II). This is ferrous iron and can also be written as the ion Fe^{2+}. This can be further oxidized, losing another electron to give ferric iron, Fe(III) or Fe^{3+}.

5.4.4.1 Degassing Oxidized the Upper Mantle

Outgassing of reduced gases such as H_2, H_2S, and CO would result in the upper mantle becoming more oxidized. One source for the hydrogen would have been water in hydrated basalts which dissociated to produce oxygen and hydrogen. Oxygen oxidized iron in the mantle. Even today, most iron in the mantle is present as Fe^{2+} ($Fe^{3+}/Fe^{2+} = 0.03$) so the mantle is a huge potential oxygen "sink." Hydrogen was outgassed and lost to space, hence oxidizing the entire Earth system. Once the upper mantle reached its current state, volcanic gases would have been correspondingly oxidized. Since this had happened by 4.4 Gyr, the atmosphere of the early Earth must have been a relatively oxidized one.

5.4.4.2 Large Terrestrial Planets Self-Oxidize

Currently, at depths lower than 660 km the pressure is high enough ($P > 23$ GPa) for magnesium silicate perovskite (Pv) to be stable. It makes up four-fifths by volume of the lower mantle. The formation of perovskite during accretion drives a disproportionation reaction:

$$3Fe^{2+} \rightarrow 2Fe^{3+} + Fe^0. \tag{5.2}$$

More fully, this can be written as

$$3FeO + Al_2O_3 \rightarrow 2Fe\underset{Pv}{Al}O_3 + Fe^0. \tag{5.3}$$

Fe^0 sinks to the core leaving the mantle more oxidized. The Fe^{3+} is locked up in the newly crystallized perovskite (see Equation 5.2), while the Fe^{2+} ends up in the upper mantle where it can undergo further disproportionation as accretion continues and the pressure builds up to levels where perovskite can form.

The disproportionation reaction occurs because perovskite formation at the base of a deep magma ocean requires $Mg^{2+}Si^{4+}$ in the original low pressure crystal structure to be substituted by $Fe^{3+}Al^{3+}$. So stable is perovskite at high pressure, its formation drives the disproportionation reaction to provide the ferric iron.

The fact that the pressure is not great enough at the base of the mantle of Mars to form perovskite may explain why the Martian mantle is reduced compared to Earth's, as revealed by meteorites.

5.5 How Did the Terrestrial Planets Acquire Water?

Venus is a hellish arid planet, Mars has a dribble of water locked up as ice, while Earth has so much water it looks blue from billions of kilometers away. The different water inventories of these terrestrial planets must arise either because they acquired different amounts of water to start with or lost different amounts of water during their subsequent evolution.

Here, we examine how the terrestrial planets may have acquired water. We concentrate on Earth.

5.5.1 The Water Inventory of the Earth Is Not Well Known

Estimates of the water inventory on Earth are compromised by the difficulty in knowing how much is locked up in the mantle. The oceans make up about 2×10^{-4} the mass of the Earth and estimates for mantle water content range from 1.5 to 10 times the mass in the oceans, depending to a large extent on whether the transition zone between 410 and 660 km actually contains as much water as it could theoretically hold: experimental data on this is ambiguous. That said, all the terrestrial planets, Earth included, are depleted in volatiles such as water by comparison with primitive chondritic meteorites, so all appear to have experienced temperatures hot enough to produce some devolatilization, if we accept the argument that the terrestrial planets were made from chondritic material.

> **EXERCISE 5.8**
>
> i. Calculate the mass of the oceans.
> ii. Given the lowest estimate for the amount of water in the mantle, what is the mantle water concentration in parts per million (ppm) by mass? (Take $M_\oplus = 6 \times 10^{24}$ kg, and the mass of the mantle as 4×10^{24} kg.)

It has been argued that Venus might have a larger water budget than Earth on the grounds that it did not experience volatile losses from a moon-forming impact. However, just because Venus lacks a moon it cannot be assumed that it has not suffered a massive impact; its retrograde spin probably came about as a result of an impact that overturned its spin axis.

5.5.2 Did Terrestrial Planets Accrete Dry?

Given the high temperatures in the early solar nebula at 1 AU, and that the snow line was probably somewhere between 2.5 and 3.5 AU it seems sensible to suppose that the Earth accreted dry. If so, Earth must have acquired its water after it had largely

TABLE 5.1

K/U Ratios in the Inner Solar System

Body	K/U	Volatile Depletion (%) Chondritic
Carbonaceous chondrite	60,000	
Earth	10,000	83
Moon	3000	95
Martian meteorites	20,000	67

accreted, perhaps delivered from beyond the snowline by the late veneer/late heavy bombardment.

There is evidence for dry accretion. It comes from studying Earth's inventory of volatile elements. Earth rocks, lunar rocks, and Martian meteorites are depleted in volatile elements that would have behaved like water. The T_{50} values of uranium (U) and potassium (K) are 1610 K and 1006 K, respectively, K/U ratios are used as a measure of the relative depletion of volatile elements with respect to refractory ones. It turns out that the Earth, lunar rocks, and Martian meteorites have low K/U ratios (Table 5.1).

Similar depletions have been determined for other volatile elements (Zn, Ag, As, Sb, Sn, Pb, and S). Of course volatile depletion would also be seen if the Earth had accreted wet and lost water by volatilization due to accretional and impact heating. Remarkably, it may be possible to test this.

When an element is volatilized there is a preferential loss of its light isotope and so rocks become enriched in the heavy isotope. In fact, with respect to several classes of carbonaceous chondrites, Earth is depleted in ^{66}Zn compared with ^{64}Zn, quite the reverse of what is expected if it had lost its volatiles by volatilization. (By contrast, the lunar regolith, generated by four and a half billion years of impacts, is enriched in heavy zinc isotopes.) Thus, this finding suggests that terrestrial planets are depleted in volatiles because they never accreted them in the first place.

If the Earth did accrete dry then we have to be able to show that the late veneer could have delivered at least the minimum estimate of the Earth's water.

5.5.3 Did Terrestrial Planets Accrete Wet?

One possibility is that terrestrial planets accreted wet from the start.

5.5.3.1 Formation of Hydrated Minerals In Situ Is Unlikely

In modeling the solar nebula, Lewis assumed Earth got its water by accreting hydrated (OH-containing) silicate minerals such as serpentine and chlorite. These minerals are stable at temperatures expected in the solar nebula at 1 AU (~600 K), but they could never have formed there by reaction of gas phase water and anhydrous silicate minerals, because given the low pressure in the solar nebula, there would not have been anything like enough time.

Another mechanism is the adsorption of H_2 onto dust grains in the nebula followed by its oxidation to water by iron oxide in the grains:

$$H_2 + Fe(II)O \rightarrow H_2O + Fe^0. \tag{5.4}$$

If this had occurred to an appreciable extent, the Earth's mantle would have been much more highly reduced than it actually is, so this scenario is unlikely.

More recent computational modeling of how mineral surfaces interact with water molecules at the sort of partial pressures expected in the solar nebula (~10^{-3} Pa) suggests that water can remain adsorbed onto forsterite (magnesium olivine, Mg_2SiO_4) up to temperatures of 900 K in the nebula.

Although hydrated minerals may not form in the high temperatures of the inner solar nebula they might survive there if delivered from beyond the snow line.

5.5.3.2 Influx of Wet Planetesimals Could Have Been Responsible for Wet Accretion

Insights into water delivery early in the accretion process come from simulations that model gravitational interactions between bodies in the early solar system by a collaboration between Nice group scientists Alessandro Morbidelli and Kevin Walsh with Sean Raymond (at the Laboratoire d'Astrophysique de Bordeaux, France), and David O'Brien and Avi Mandell in the USA. These simulations show that wet planetesimals that form beyond the snowline can end up in the inner solar system.

Early simulations started out with 1885 planetesimals spanning a range of masses around an average of half the moon's mass, randomly assigned to be between 0.5 and 5 AU, and with different amounts of water (mass fraction between 10^{-5} and $10^{-1.3}$) depending on the distance from the sun. On running the simulation for 200 Myr, planetesimals drift inward and outward, are pumped up to higher eccentricities and inclinations, and accrete to form planetary embryos and eventually a few planets, some within the habitable zone. Running many simulations shows that rocky planets with water are very common. Indeed some have water mass fractions at the high end of estimates for Earth (10^{-3}) and are termed ocean planets.

But these and other simulations of the accretion of terrestrial planets generally produce Mars analogs that are 10-fold too massive. When the simulations are forced to come up with a Mars analog that has the right mass, it is by making the orbits of the gas giants highly eccentric (e ~ 0.1) and results in a dry Earth!

To get the correct mass for Mars, low eccentricity orbits for Jupiter and Saturn and a wet Earth needs rapid inward then outward migration of the gas giants; what researchers call the "Grand Tack" that we introduced in Section 4.5.3.5 (Plate 5.1).

The simulation populates the inner solar system (0.3–3.0 AU) with volatile-poor S-type planetesimals and the outer solar system with volatile-rich C-type planetesimals. The gas giants undergo type II migration inward, become locked into a 3:2 mean motion resonance, and continue until Jupiter reaches 1.5 AU. The effect is to shepherd most of the S-type planetesimals inward. The planets then reverse direction and migrate outward ploughing into the C-type planetesimals pumping them into high eccentricity orbits that take them into the inner solar system. These may have been a source of water for the terrestrial planets.

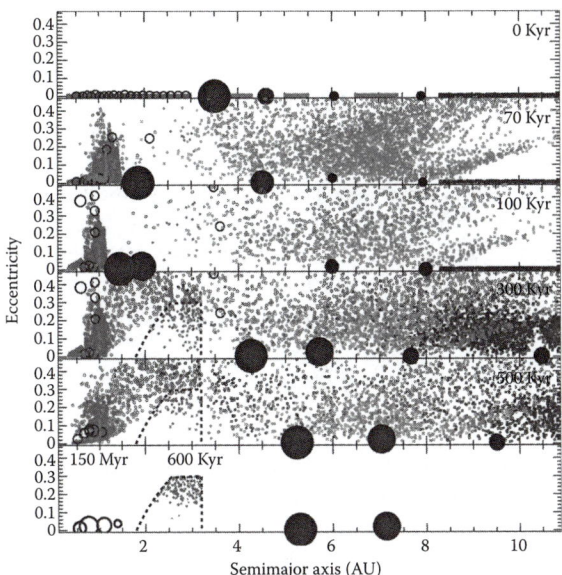

PLATE 5.1 **(See color insert.)** Grand tack model accounts for the distribution of asteroids in the main belt and the small size of Mars. (Photo credit: Walsh et al. *Nature*, 2011;475:206–9. With permission.)

By the time the disc dissipates and migration ceases, Jupiter and Saturn are at 5.4 and 7.2 AU, respectively, in low eccentricity orbits. This all happens in a timescale of 0.6 Myr. Accretion of terrestrial planets in the inner solar system now occurs over a much more leisurely pace (30–100 Myr) and the simulations can produce realistic analogs of the terrestrial planets, including a low mass Mars and the expected architecture for the asteroid belt.

There is nothing special about 1.5 AU for the end-point of Jupiter's inward migration. It is chosen so as to produce our solar system at the end of the simulation. Other choices would result in alternative "solar systems."

This is exciting work but comes with some "health warnings."

i. The simulations do not start from the beginning because the early stages of accretion cannot yet be simulated. Modelers have to put in the mass distribution of planetesimals "by hand" and this is currently just our best guess.

ii. The outcome of simulations is quite sensitive to initial conditions such as the starting positions and velocities of the planetesimals.

iii. The water content of the planetesimals is put in "by hand" and the effects of radiogenic or impact heating on this are not included.

5.5.4 Volatile Delivery Could Have Occurred Late

Regardless of whether the terrestrial planets initially accreted wet or dry, late delivery of water and other volatiles has probably occurred. The main suspects are comets and asteroids. These would have been wet because they formed beyond the snow line. It is assumed that water was delivered by the late heavy bombardment.

5.5.4.1 Comets Are a Potential Source of Water

Comets have been regarded as a problematic source for the Earth's water. The proportion of heavy hydrogen (deuterium, D) to normal hydrogen (H) in water from six Oort cloud comets is $\sim 3 \times 10^{-4}$. This is twice the D/H ratio of seawater (1.56×10^{-4}) and implies that comets cannot have delivered more than about 10% of the Earth's water. However, ESAs Herschel space observatory has recently observed the coma of Comet 103P/Hartley 2 in the far IR and shown that the D/H ratio of its water is comparable to that of the Earth's water, which brings comets back into the frame. Actually the D/H ratio of Hartley 2 is hard to understand in light of current theoretical ideas about how deuterium is thought to be distributed throughout the solar system (see Box 5.2). Clearly there is much still to learn.

5.5.4.2 Asteroids Are a Potential Source of Water

Asteroids have been regarded as promising sources of water. Carbonaceous chondrites, derived from C-class asteroids, are water-rich and can contain up to 20 wt% of water. The D/H ratio of terrestrial water overlaps with that of carbonaceous chondrites. P- and D-type asteroids in the outer regions of the main asteroid belt—low

---

BOX 5.2   D/H DISTRIBUTION IN THE SOLAR NEBULA

A theoretical study suggests that ice grains that crystallize in dense molecular clouds at 10 K should be highly enriched in deuterium, with D/H ratios of $\sim 10^{-3}$. This would have been the starting material from which the solar nebula formed. However, as the density and temperature in the nebula rose, isotope exchange reactions between this presolar water and hydrogen molecules decreased the D/H ratio of the water, for example:

$$H_2 + HDO \rightarrow HD + H_2O. \tag{5.5}$$

Because the temperature and density are greatest at the center, a D/H ratio gradient was established in the solar nebula, highest at the outer edge and lowest close in. Accreting planetesimals should inherit the D/H ratio of their surroundings. It is no surprise then that Oort cloud comets have high D/H ratios (3×10^{-4}). Calculations give D/H at 1 AU of 0.8×10^{-4}, but terrestrial water is higher than this (1.56×10^{-4}), so presumably originated from further out. In fact, the Earth's water lies close to the peak D/H ratio for carbonaceous chondrites, which supports asteroids as a source for terrestrial water.

Comet 103P/Hartley 2 has D/H = 1.6×10^{-4}, so Jupiter family comets (JFCs) are also candidate sources of terrestrial water. Actually, since JFCs are thought to come from the Kuiper belt, also the source of Oort cloud comets according to the Nice model, we might have expected Hartley 2 to have a D/H ratio comparable with Oort cloud comets. Maybe Hartley 2 was born in the vicinity of the asteroid belt and ended up in the Kuiper belt as a result of the extensive radial mixing of planetesimals.

albedo, very red objects, high in carbon, and volatiles—may also be candidates. Intriguingly, the Nice model indicates that D-type asteroids originally hail from the Kuiper belt, so may in fact be cometary nuclei. (Comets, complete with comas and tails, have been seen in the asteroid belt.) One problem with asteroidal delivery of water is that oxygen isotope ratios of asteroids do not match that of the Earth's water.

5.5.4.3 Water Delivery May Have Been Stochastic

Simulations of the orbits of planetary embryos done by the Nice group show that the bulk of the Earth's volatiles could have been delivered by just a few comets or asteroids slamming into the Earth. If few impactors are required to deliver the entire volatile inventory, then it is just a matter of chance what those few objects are and hence what the D/H makeup of the water happens to be. If so, this means that differences in lunar and terrestrial water are accidental. In support of this idea is the fact that the D/H ratios of water of the terrestrial planets and the moon have dissimilar profiles.

5.6 The Temperature of the Late Hadean and Archaean Are Not Well Constrained

Astrobiologists are keen to know when the Earth's surface was cool enough for oceans to condense and what the temperature of the water was. This would provide a handle on when Earth first became habitable. Yet the question about whether the late Hadean and Archaean were hot or cool has been controversial. Here we rehearse some of the arguments. It is worth bearing in mind two facts about this time over which there is no dispute: the sun was appreciably fainter, and geothermal heat flow was higher than now.

5.6.1 Geological Clues Suggest Early Earth Was Warm Rather Than Hot

There are good theoretical reasons for thinking that the Hadean and Archaean eons were hot despite the fainter sun; heat flows in the mantle and crust at this time would have been higher than now for several reasons:

1. Shortly after the moon-forming impact the thermal energy of the impact would be supplemented by heat generated by final assembly of the core. Infalling metal is converting gravitational potential to kinetic energy and this is subsequently converted to the disordered motion of thermal energy.
2. Tidal heating would have been higher due to the proximity of the moon. Indeed, until about 4.4 Gyr, tidal heating probably dominated, but by the start of the Archaean it might have contributed only about one-fifth of the heat flow from the Earth's interior.
3. There would have been higher radiogenic heating. We have seen how important short-lived isotopes were for differentiation in the first few million years

but after the moon-forming impact it is the long half-life isotopes, particularly ^{40}K and ^{235}U, that become the key radiogenic heat sources, generating a steady heat flow of 0.2 to 0.3 W m^{-2}. By the beginning of the Archaean this would have been the major geothermal heat source.

4. There would have been greater impact heating, until at least about 3.8 Gyr ago, and possibly for appreciably longer periods, the magnitude of which would have mirrored the late heavy bombardment.

Direct evidence for higher mantle temperatures in the Archaean are komatiites, magnesium-rich rocks that arise from more extensive partial melting of the mantle than occurs now (see Section 5.8.4.7).

Paul Knauth, a geologist at Arizona State University, is convinced by oxygen isotopic evidence from 3.2 to 3.5 Gyr old cherts in Barberton greenstone belt, South Africa, that the Archaean was warm (70 ± 15°C). The cherts are depleted in ^{18}O relative to their Phanerozoic counterparts (i.e., they have low $\delta^{18}O$; see Box 5.3). Regional metamorphic, hydrothermal, or long-term resetting of original $\delta^{18}O$ values is precluded by the data.

BOX 5.3   OXYGEN ISOTOPES AND PALAEOCLIMATE

About 99% of oxygen is the ^{16}O isotope, most of the remainder is ^{18}O. Water containing the light isotope evaporates slightly more readily than water with the heavy isotope. Conversely, condensation favors precipitation of $H_2^{18}O$ over $H_2^{16}O$. These differences form the basis for using oxygen isotope ratios as a proxy for temperature.

Water evaporating at low latitudes is carried poleward and cools, condensing $H_2^{18}O$ preferentially. By the time it reaches the polar regions, the precipitation is rich in $H_2^{16}O$. If the climate is cold, this falls as snow and will be trapped in ice sheets so that the oceans become enriched in $H_2^{18}O$. By contrast, if the climate is warm the $H_2^{16}O$-rich water falls as rain at high latitudes and will mix with the ocean so that seawater is not rich in $H_2^{18}O$.

Because the abundance of ^{17}O and ^{18}O is so low, the ratios of these heavy isotopes to the abundant light isotope ^{16}O are expressed as deviations from a terrestrial standard ratio (in parts per thousand) as the measure of the degree of fractionation. For $^{18}O/^{16}O$ this is:

$$\delta^{18}O(\text{‰}) = [(^{18}O/^{16}O)_{\text{sample}}/(^{18}O/^{16}O)_{\text{standard}}] \times 1000. \qquad (5.6)$$

The standard is usually the Vienna Standard Mean Ocean Water (VSMOW) which has $^{18}O/^{16}O = 1/498.7$.

A sample which is identical to VSMOW would have $\delta^{18}O = 0.0$‰. If a sample is depleted in ^{18}O in relation to the standard, then its $\delta^{18}O$ will be negative whereas if the sample has a higher $^{18}O/^{16}O$ ratio compared to the standard, then its $\delta^{18}O$ will be positive. (There is a comparable expression for $\delta^{17}O$ (‰).)

Thus, in a warmer climate, marine sediments (e.g., cherts) will incorporate lower $^{18}O/^{16}O$ than when it is colder, and so will have a lower $\delta^{18}O$.

However, oxygen isotopes are only faithful palaeothermometers if the $^{18}O/^{16}O$ ratio of the oceans has remained constant over geological time. A study of oxygen isotopes in Archaean cherts suggests that the Archaean ocean had less ^{18}O relative to today, and this meant that the low chert $\delta^{18}O$ actually indicated a much more equitable 40°C than a scalding hot Archaean. Another study showed that the $\delta^{18}O$ of Barberton phosphates were comparable to those of modern marine phosphates and would have been in equilibrium with water between 26°C and 35°C—not much warmer than now.

For the temperature of early Earth to be as hot as Knauth claims implies requiring large amounts of greenhouse gases in the atmosphere, particularly CO_2, and this is contraindicated by Archaean geology. It would also mean that silicate weathering rates would have been extraordinarily high but several studies find no evidence for this in early Archaean rocks.

In summary, a consensus for a warm rather than a hot Archaean climate is emerging.

5.6.2 When Did the First Oceans Form?

Another approach to discern the surface temperature of the early Earth is to ascertain when oceans first appeared. Of course liquid water on the surface may still have been hot if the atmospheric pressure was high. But we will show later that, at least for the late Archaean, the atmospheric pressure was not too different from now.

5.6.2.1 Do Zircons Provide Evidence for Oceans by 4.4 Gyr?

Many geologists argue that the mineral zircon ($ZrSiO_4$) can provide evidence for early oceans. Conveniently, zircon crystals contain uranium isotopes and are good at hanging on to the lead isotopes that arise from the radioactive decay of the uranium, so they lend themselves to U–Pb dating. Zircon grains from 3 Gyr-old Archaean sandstones (Jack Hills quartzite of Murchison, Western Australia) have been dated to as old as 4.404 Gyr. These are the oldest bits of the Earth we have. Indeed, apart from two remote exposures on the ancient continental crust of the Canadian Shield (see Section 5.2) they are the only traces of the geological record we have from the Hadean. The reason we have zircon crystals this old is because zircon is virtually indestructible, surviving many cycles of weathering, erosion, and deposition as sediments, yet preserving information about the conditions under which they formed.

The key point about zircons is that they crystallize in granites that form when wet sediments are melted. A three-stage argument is as follows: (1) to get wet sediments we need oceans, (2) water lowers the temperature at which rocks melt, and (3) we can determine the temperature at which a zircon crystal grew by its $^{18}O/^{16}O$ ratio. In slightly more detail, zircons that crystallize from mantle-derived magmas have a narrow range of $^{18}O/^{16}O$ values, expressed as $\delta^{18}O = 5.3 \pm 0.6‰$. If their parent magmas contain sedimentary material (e.g., granitic magmas) or source rocks altered by low- (or high-) temperature hydrothermal activity, the $\delta^{18}O$ values are generally higher (or lower), because oxygen isotopes get fractionated by water–rock interactions. Hence, by looking at Hadean zircon oxygen isotope ratios, it may be possible to deduce whether water was around.

The standard story of the Jack Hills zircons is that they crystallized at about 700°C, too low to have frozen from an anhydrous melt but typical for a granite magma that forms from wet melts. What is more, zircons have high $\delta^{18}O$, also evidence for wet sediments in the melt. The implication is that continental crust and water oceans were present 4.4 Gyr ago, and it follows that the Hadean was at least cool enough for surface water. At some time the Hadean crust was completely weathered, releasing the hardy zircons to be transported and eventually incorporated into the Archaean sandstones.

But we need to look more closely at what the zircon crystals are actually telling us. Most Archaean sediments lack a negative Europium anomaly, the depletion in Europium abundance compared with other rare Earth elements (REEs) that is a marker for sediments derived from granites. The obvious inference is that Archaean sediments do not come from granites. Moreover, the number of zircon grains we have from the Hadean is very small. This implies that there was not much granite in the Hadean. In other words, the few zircon grains we have found are not evidence for a *significant* amount of continental crust. Yet, we would expect to see substantial granite crust being formed if there had been oceans of water around.

To be fair, despite being very durable, zircons do not always stay "on message." They contain high levels of radioactive elements and these damage their structure, causing changes that alter U–Pb and oxygen isotopes. Also, many zircon grains have an onion-like structure with an ancient core but successively younger layers. This zoning is caused by pulses of growth under different conditions, making the oxygen isotope story hard to figure out. A study of the few Jack Hills zircons that are undamaged primary magmatic grains and hence have a reliable age showed them to have a high temperature mantle $\delta^{18}O$ signature (lower than 6.5‰) and similar to lunar zircons! The bottom line is that claims for oceans 4.4 Gyr ago may be premature.

5.6.2.2 Water-Lain Sediments Are Found at 3.85 Gyr

There are water-deposited sedimentary rocks by 3.85 Gyr showing there must have been surface water by this time, though the extent of that water is hard to judge.

5.6.2.3 Evidence for Hydrothermal Activity Is Not Evidence for Oceans

The Nuvvuagittuq greenstone belt rocks of Canada are a type of metamorphic rock (one produced by pressure and/or heating) termed amphibolite, but their composition shows that their protolith (the original rock from which they metamorphosed) was a volcanic rock altered by hydrothermal activity. This might imply the existence of deep oceans at 4.28 Gyr since there are hydrothermal vents on the ocean floor. However, much hydrothermal activity occurs in volcanic areas near the surface. Here the surface water (e.g., rain, streams, and freshwater pools) can percolate into rocks, is heated by magma in a magma chamber, and then re-emerge as hot springs, geysers, and fumaroles. Clearly, this does not require an ocean. And some hydrothermal activity is not due to surface water at all but water exsolved from the magma itself. So, hydrothermal activity need not be associated with significant amounts of water or with habitable environments.

5.7 Plate Tectonics on Early Earth

Plate tectonics requires surface water and operates the carbonate–silicate thermostat that regulates the surface temperature of the Earth. It may have allowed Earth to be habitable. This is the motivation for trying to find out when and how plate tectonics got going on Earth.

The amount of continental crust has grown over time. Hence, early on, Earth's surface was largely water covered. Under these circumstances, silicate weathering to sequester CO_2 from the atmosphere would be minimal and would not compensate for CO_2 being vented by volcanoes.

5.7.1 When Did Plate Tectonics Start on Earth?

There is unambiguous evidence for plate tectonics at 3.2 Gyr, but some research-ers argue that plate tectonics was operating in the earliest Archaean (the eon dated between 4.0 and 2.5 Gyr). What is the evidence? Firstly, remnants of an ophiolite have been identified in the 3.8 Gyr rocks in Isua, Greenland. Ophiolites are sequences where ocean lithosphere has been thrust up onto a continent, where they are pre-served, rather than subducted. The ancient ophiolites include basaltic pillow lavas and sheeted dykes just like modern spreading centers, and the basalts have oxygen isotope ratios and textures compatible with hydrothermal metamorphism at a spreading ridge. Moreover, their trace element geochemistry implies that their composition has been influenced by nearby subduction. This shows that they formed at a back-arc spreading center, much like the Oman ophiolite that was generated just 90 Myr ago.

Secondly, there is a geochemical signature of subducted slab material being added back to the mantle from 3.6 Gyr onward. This subducted material would have been generated at mid-ocean ridges at about 3.9 Gyr.

5.7.2 Plate Tectonics May Not Have Operated in the Hadean

Theory says that plate tectonics cannot operate when geothermal heat flows are much greater than 0.1–0.2 W m^{-2} because stresses in a hot Earth are too low for the surface to rupture into plates, or if it did, plates would be too buoyant to subduct. This implies that there was no plate tectonics in the Hadean. It has also been suggested that mantle convection did not operate during the Hadean because a process called mantle over-turning prevented mantle convection. We would normally expect a magma ocean to be hotter at the base and cooler at the top so that thermally buoyant material in the lower mantle would convectively rise. Mantle overturning reverses the normal density and temperature gradients. It happens because the magma ocean freezes from the bottom since that is where the pressure is the highest. The first olivine crystals to form are rich in magnesium, and these have a lower density than those rich in iron. (The densi-ties of Mg_2SiO_4 and Fe_2SiO_4 are 3200 kg m^{-3} and 4300 kg m^{-3}, respectively.) Having low density material overlain by a denser one is an unstable situation. Over time, the deeper hotter low density Mg-rich olivine crystals rise while the cooler Fe-rich olivine sinks. In the absence of thermal convection, calculations show this could have been stable for 400 Myr. If this idea is correct then during the Hadean, Earth was a single plate planet. This state is often referred to as a stagnant lid regimen. During that time the hot magma in the upper part of the magma ocean would have fractionated to form a primordial basaltic mush (mostly crystals with a little melt) topped by a thin frozen basalt crust. Eventually radiogenic heat would have built up sufficiently to restart mantle convection. This rifted the thin crust and heralded the start of some sort of plate tectonics.

5.7.2.1 High Resurfacing in the Hadean May Have Led to Efficient Cooling

Despite the expectations of a hot Hadean Earth, the first rocks come from the Archaean and appear to have formed under conditions not much warmer than now. The Earth must have lost an appreciable heat load, without the help of plate tectonics, to allow it to make the transition between hot Hadean and cooler Archaean. William More and Alexander Webb, at Hampton University, USA and Louisiana State University, USA, respectively, have used numerical simulations of mantle convection in the Hadean to show that hot magma could have erupted rapidly up volcanic conduits, heat pipes, in such a way that the heat largely bypassed the crust. The surface would have been repeatedly covered by mafic lava flows that radiated heat away. The lava flows, exposed to water at the surface would become hydrated. Under the weight of the newer flows the lithosphere would subside *en masse* (advect), and as it descended the hydrated mafic rock would partially melt to form plutons of tonalite–trondhjemite–granodiorite (TTG) rock. The advection would gradually thicken and cool the lithosphere.

This scenario explains many of the features seen at Pilbara (Australia) and Barberton (South Africa), both regions where rocks as old as 3.5 Gyr are represented. They share a common ancient geological history. TTG plutons comprise half of the rock >3.2 Gyr at both locations. Mafic and ultramafic volcanic rocks were erupted in alternating subaerial and marine settings forming sequences 10-km thick at Barberton and 20 km thick at Pilbara.

The model shows that a shift from heat pipe Earth to plate tectonics could have occurred fast, although the two processes could have operated at different sites at the same time depending on local variations in heat flow. Moore and Webb argue that the transition to plate tectonics happened at about 3.2 Gyr when there seems to have been a dramatic change in the rate and style of crust formation (see Section 5.7.3.2) and because neither Barberton nor Pilbara rocks show any trace of horizontal deformation from 3.5 to 3.2 Gyr, nor did they experience further resurfacing after this time. This model of course brings into question the evidence for plate tectonics in the early Archean we saw above (see Section 5.7.1).

5.7.3 What Was the Nature of Early Plate Tectonics?

Plate tectonics would have been faster in the Archaean because the higher heat flow would mean greater partial melting at mid-ocean ridges, faster production of new ocean crust, and so faster seafloor spreading. This would mean plates being subducted when they were younger and warmer than today, making them smaller. Being warmer the plates would be less dense than subducting plates today, and this higher buoyancy would mean subduction was shallower. Today, dense plates bend, so are subducted at quite a steep angle. This carries them to depth where the basalt is metamorphosed to a higher density rock, eclogite, by the high pressure. This increase in density of the subducting slab is an important determinant of the slab pull force, the major driver of subduction. Shallow angle subduction in the Archaean may not have allowed eclogite to form, so slab pull would have been weak.

5.7.3.1 Most Continental Crust Formed in the Archaean

If plate tectonics is needed to make continental crust, estimating the rate at which continental crust has formed over geological time should provide insights into the operation of plate tectonics. There have been a number of attempts to do this, and it is only fair to point out that there are disparities in the models.

Several groups have used zircons. These are good markers for continental crust since they rarely form in the basaltic composition magmas, which generate oceanic crust. But a key problem is to sort out zircons that crystallize in juvenile crust ("geospeak" for newly formed crust) from detrital zircons. These are "recycled" zircons that crystallize in granite formed from sediments derived from older crust that has been reworked (i.e., weathered and eroded). Periods of juvenile crust formation should be associated with zircons which have low $\delta^{18}O$, like the mantle, because new crust is formed from the mantle. These mantle-like zircons should all have similar radiometric ages. (Recall that zircons can be dated.) By contrast, detrital zircons will have elevated $\delta^{18}O$ and will span a wide range of radiometric ages reflecting their diverse origins. These will not provide times of juvenile crust formation.

Distinguishing mantle from detrital zircons in this way reveals that 65% of the continental crust was formed by 3 Gyr. Since very little continental crust older than 2.5 Gyr now remains most of that early crust has been reworked and its trace now turns up as detrital zircons in younger rocks. Mantle-like zircons show there were two periods in which there were substantial peaks in the new crust, one centered on 3 Gyr and the other a steep rise over the past 500 Myr.

A completely different technique shows that a volume of continental crust equivalent to 80% of that present today was generated between 3.8 and 2.5 Gyr. Crust formation after this is thought to be mostly reworking of the older material. The method works as follows: The Earth's mantle contains ^{40}K which decays with a half-life of 1.25 Gyr to ^{40}Ar. This is degassed into the atmosphere by volcanoes. The extent of this degassing can be determined by measuring the $^{40}Ar/^{36}Ar$ ratio in fluid inclusions within quartz crystals that preserve the noble gas signatures of the atmosphere. The ^{36}Ar is primordial and was all degassed from the mantle early in the Earth's history, so its concentration has been fixed for most of geological time. Now for the clever bit. As continental crust is formed it extracts ^{40}K from the mantle, because potassium is an incompatible element. As soon as the mantle begins to melt, incompatible elements enter the melt and remain there as it freezes to form crust. What is more, any ^{40}Ar derived from decaying crustal ^{40}K also remains locked up in the crust rather than degassing. Thus, formation of continental crust leaves a signature in the $^{40}Ar/^{36}Ar$ ratio from which it is possible to model the rate at which it is formed.

In summary, independent methods show that a huge episode of continental crust building happened during the Archaean. This had several far-reaching consequences but arguably the main one was an increase in silicate weathering over several hundred million years that would gradually have lowered the partial pressure of CO_2 in the atmosphere. The Earth appears to have suffered its first glaciation episodes in the late Archaean, about 2.9 Gyr ago, and has experienced several others since. We know that the sun was 25% fainter 3.8 Gyr than now. Given this, lack of glaciation throughout most of the Archaean was probably due to higher carbon dioxide levels. Could the formation of so much continental crust, by sequestering most of this CO_2, have made the Earth system more vulnerable to glaciation?

5.7.3.2 The Style of Plate Tectonics Changed 3 Gyr Ago

Studies imply that the transition to steep subduction, that is, modern-style plate tectonics, occurred 3 Gyr ago. One looked at minerals trapped inside diamonds that have been erupted from deep (125–175 km) in the lithosphere mantle beneath ancient continents. These mineral inclusions can be dated and reflect the composition of the subcontinental lithospheric mantle at that time. Three billion years ago the mineral make-up changed: before 3.2 Gyr the minerals were those found in peridotite, but after 3.0 Gyr they were minerals typical of eclogite, a rock formed only under the high pressures experienced in deeply subducted slabs.

Further evidence for this transition comes from the zircon study above. This found that the cumulative continental crust growth rate (the balance between what was produced and what was destroyed) was high (3 km³ yr⁻¹) for the first 1.5 Gyr and then decreased suddenly to 0.8 km³ yr⁻¹. The researchers attributed the sharp inflection to the onset of subduction-driven plate tectonics. Before 3 Gyr shallow subduction with weak slab pull would have dominated.

5.7.3.3 Vertical Tectonics May Have Generated the First Continental Crust

Before 3 Gyr it is thought that some sort of vertical tectonics operated and this would have formed the first continental crust. Two processes could have contributed.

1. Upwelling mantle caused thickening of overlying basaltic crust. The thicker crust heated up so that it partially melted to form magmas of continental crust composition. This thicker crust would have been underplated by highly depleted and buoyant lithospheric mantle which would have elevated the newly created continent.

2. Modern continental crust has two layers. The upper crust has a high SiO_2 (quartz) content and this makes it relatively weak. The lower crust is rich in Mg and Fe, has a lower SiO_2 content, and is relatively strong. By contrast Archaean TTG continental crust starts out as a single homogeneous layer with quartz distributed throughout its entire thickness. This would make it weaker than modern crust, an effect exacerbated by the higher temperature. This weaker crust subsided to form basins (Figure 5.6). The subsidence produced extension and rifting causing komatiites and basalts—partial melts of the mantle below—to erupt into the basins. This produced further subsidence, so the basins deepened as they filled episodically with lavas and sediments, eventually becoming so thick that extensive metamorphism

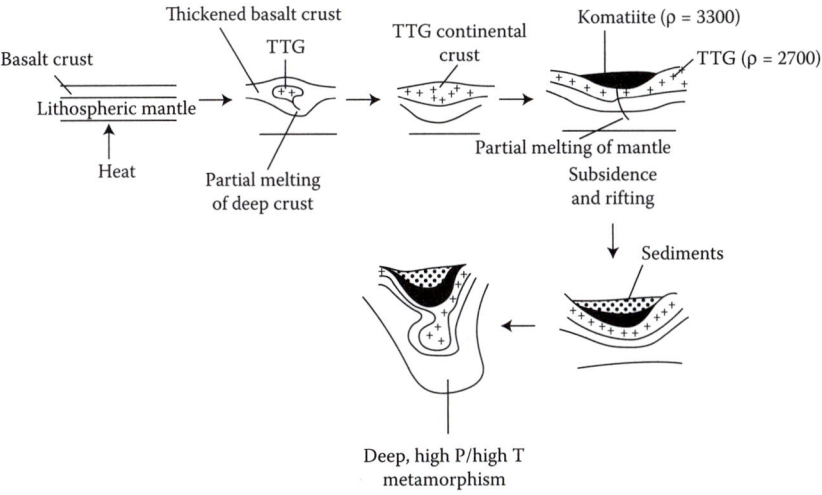

FIGURE 5.6 Early continental crust was formed by sagduction. Densities are in kg m^{-3}.

occurred and the high pressure and temperature produced partial melting of the sediments to form granites. The mélange of very different rock types that resulted are called greenstone belts, and because subsidence features so prominently in their formation, this type of vertical tectonics is sometimes described as sagduction. By the end of the Archaean many of these belts had accreted as plate tectonics brought them together, and they form some of the oldest cores (cratons) of continents. Greenstone belts have formed in three periods, all corresponding to times when supercontinents have been assembling. However, this vertical tectonics did not occur in the last cycle of continent assembly.

5.8 Earth's Atmosphere Has Changed Over Time

5.8.1 Earth's Atmosphere May Come from Two Sources

The origin and early evolution of the Earth's atmosphere is a long-standing puzzle. The relative abundances of noble gases in the atmosphere and mantle are thought to have been controlled by mantle degassing of the atmosphere, and because the noble gases are chemically inert they should have preserved this history. However, the story is far from clear in part because compared to the sun and chondritic meteorites, xenon is highly depleted, relative to krypton and argon, in the atmosphere of the Earth (and Mars). Several solutions have been advanced for this "missing xenon" paradox. It may be that Earth derived its atmosphere from two sources, mantle degassing and comets.

Xenon isotope data is consistent with the mantle degassing of the atmosphere either within 100 Myr of the start of the solar system (i.e., hard on the heels of the moon-forming impact) or not much before 600 Myr.

5.8.2 The Oxidation State of The Atmosphere Has Altered

In the past it was thought that the early atmosphere might have been a highly reducing one of H_2–CH_4–NH_3. There would have been virtually no free oxygen. Molecular hydrogen may have been more abundant than now because:

1. Photodissociation of water by the higher UV flux of the young sun would have been appreciably greater than now and it has been suggested that outgassing of H_2 would have been higher.

2. Sinks for H_2 would have been lower. Currently the exobase (the region from which hydrothermal escape of gases from atmosphere occurs) temperature is 1000 K because of the efficient absorption of solar UV by atomic oxygen. In the Archaean the exobase T would have been less (500–600 K) because the atmosphere was anoxic. Modeling implies that under these conditions the rate of Jean's escape of hydrogen was 100-fold lower than now, so it has been argued that the Earth's early atmosphere could have had as much as 0.1 bar of H_2. However, some researchers reckon that this underestimates the rate of hydrogen escape.

The counterargument to this highly reducing atmosphere is the geochemistry of Archaean komatiites (high magnesium basalt-like rocks) which show the mantle had the same oxidation state then as now. This makes volcanic outgassing of a highly reducing atmosphere unlikely.

In summary, given what we now know about the oxidation state of the early mantle it seems likely that the atmosphere would have been only mildly reducing: largely CO_2, CO, and H_2O plus some N_2 and CH_4. If there had been an appreciable amount of H_2 as well then the combination of H_2 and N_2 would have added significantly to the greenhouse effect. We explore this in Section 5.8.4.2 below.

5.8.3 Nitrogen May Be Derived from Ammonia

Making up 79% by volume, nitrogen is the most abundant gas in the present atmosphere. It may have been delivered to Earth by comets and asteroids as ammonia (NH_3) or amino compounds. The amount of nitrogen delivered by comets is constrained by the $^{15}N/^{14}N$ ratio in cometary CN and HCN, which is about double that of the Earth. However, NH_3 is not stable above 800°C, so at the high temperatures experienced during accretion, NH_3 would have been oxidized to N_2. This would have been outgassed from the mantle. Indeed the main nitrogen species spewed out by volcanoes now is N_2.

The reduction of N_2 to NH_3 may have been possible at deep hydrothermal vents where the high temperature and pressure (300°C; 50–500 MPa) and the presence of metal oxide and sulfide catalysts would allow the very strong N_2 triple bond to be broken. NH_3 produced in this way could experience two abiotic fates. (At this stage we are ignoring any effects of life.)

1. It could be buried in sediments. The ammonium ion $\left(NH_4^+\right)$ can substitute for K^+ in a variety of minerals (e.g., clays). These would be subducted and eventually the nitrogen would by vented by volcanoes as N_2. The timescale for this would be hundreds of millions of years.

2. In the absence of O_2 and O_3 to absorb solar UV, NH_3 in the early atmosphere would be photodissociated to N_2 in a series of reactions by UV radiation between 160 and 230 nm:

$$NH_3 \xrightarrow{hv} H_2 + H. \tag{5.7}$$

$$NH_2 + NH_2 + M \rightarrow N_2H_4 + M. \tag{5.8}$$

$$N_2H_4 \xrightarrow{hv} N_2H_3 + H. \tag{5.9}$$

$$N_2H_3 + N_2H_3 \rightarrow 2NH_3 + N_2. \tag{5.10}$$

Modeling these reactions shows that almost all the nitrogen in the atmosphere would be in the form of N_2.

However, this is to reckon without life. James Lovelock has argued that in the presence of O_2, solar UV, lightening, and combustion, all nitrogen would be converted into the stable nitrate (NO_3^-) ion dissolved in the oceans. High energy or temperature (>1900 K) is required to break the strong triple bond of the nitrogen molecule:

$$N + O_2 \rightarrow NO + O. \tag{5.11}$$

NO is the nitric oxide radical with one unpaired electron. The oxygen radical has two unpaired electrons. Then in the stratosphere:

$$NO + O_3 \underset{UV}{\rightleftharpoons} NO_2 + O_2. \tag{5.12}$$

NO_2 is the nitrogen dioxide radical. Because the back reaction of Equation 5.12 is promoted by UV radiation, in daylight the NO/NO_2 ratio is determined by the intensity of sunlight and the concentration of ozone (O_3); in fact, NO_2 concentration is very low in the day time. Night time favors high concentrations of NO_2 when the nitrate radical is formed:

$$NO_2 + O_3 \rightarrow NO_3 + O_2. \tag{5.13}$$

$$NO_3 + H_2O \rightarrow HNO_3 + OH. \tag{5.14}$$

HNO_3 is nitric acid. This dissociates to form the nitrate ion:

$$HNO_3 \rightleftharpoons H^+ + NO_3^-. \tag{5.15}$$

Hence the presence of N_2 and O_2 in the atmosphere is evidence for disequilibrium. The N_2 in our atmosphere results from the action of denitrifying bacteria. The biochemistry of denitrification is complicated but in essence it proceeds through a series of intermediates to form molecular nitrogen from nitrate:

$$NO_3^- \rightarrow NO_2^- \rightarrow NO + N_2O \rightarrow N_2(g). \tag{5.16}$$

If nitrogen in the Earth's atmosphere depends on life then why has Venus got a nitrogen inventory similar to that of the Earth? The reason is that the hot, acid, dry, anoxic state of Venus means that N_2 rather than NO_3^- is the stable nitrogen compound. Similarly the low N_2 abundance of the Martian atmosphere (2.7%) cannot be taken as evidence about life on Mars since its redox state favors NO_3^- stability. The take-home message is that the presence or absence of nitrogen in a planetary atmosphere is not a good guide to whether it harbors life unless other factors that determine atmosphere equilibrium, such as the planet's redox state, are known.

5.8.4 The Faint Young Sun Paradox

Any discussion of the atmosphere of the early Earth has to take account of the fact that the luminosity of the zero age main sequence sun was 30% less than what it is now. It gradually increased over the next 1.2 Gyr to reach a luminosity that is comparable to what it is today. Consequently, the sun would have been dimmer throughout the Hadean and much of the Archaean eons. This trend is predicted by standard stellar evolution models. The conversion of hydrogen to helium reduces the volume of the sun's core. As the core shrinks its temperature rises, and with it the rate of the nuclear reactions.

However, the Earth's surface was certainly warm enough for liquid water even 3 Gyr ago as shown by sedimentary deposits with ripple marks and mud cracks, and by pillow lavas which only form on the ocean floor, and maybe as early as 4.4 Gyr ago if we accept the zircon evidence for the formation of continental crust from wet melts at this time. What is more there is no geological evidence for glaciation in the Archaean except perhaps at 2.9 Gyr. That the early Earth could have been so warm despite the low solar luminosity is termed the faint young sun paradox and was recognized as early as 1972 by Carl Sagan and George Mullen.

Can it be solved by appealing to the greater geothermal heat? For a few million years after the moon-forming impact, the geothermal heat flow was about 100 W m^{-2}. We can compare this to the amount of sunlight warming the Earth by recourse to the energy balance equation:

$$\sigma T e^4 = I(1 - A)/4, \tag{5.17}$$

where I is the solar irradiance, A is the planetary albedo and σ is the Stefan–Boltzmann constant. (Note this is Equation 3.4 in a slightly different guise.) $I = 1370$ W m^{-2} now but the faint young sun would reduce this to 0.7×1370 W m^{-2}. Assuming that the early Hadean Earth reflected about 30% of the incoming solar radiation, as it does now, the Earth would have absorbed the amount of sunlight given by

$$[0.7 \times 1370\,\text{Wm}^{-2}(1 - 0.3)]/4 = 168\,\text{Wm}^{-2}.$$

So geothermal heating would have made an appreciable contribution to warming. However, by 4.4 Gyr ago, modeling suggests that the geothermal heat flow had fallen to 0.5 W m^{-2}, and by the end of the Hadean it was 0.2–0.3 W m^{-2}. So, in the long run, geothermal heat flow does not help.

You may recall from Chapter 3 that the effective temperature of Earth today is 255 K (−18°C), well below the freezing point of water, but the actual average temperature at the Earth's surface is 288 K (15°C) and the 33°C difference is due to the greenhouse effect of the Earth's atmosphere.

EXERCISE 5.9

What would be the effective temperature of the Earth in the Hadean with $L_\odot$ being 30% less than now, assuming the same albedo?

Assuming the same greenhouse effect then as it is now would make the temperature at the Earth's surface 233 K (−40°C). Actually, the lower H_2O concentration in this cold atmosphere would mean even less greenhouse warming, so −40°C is too warm.

These simple considerations show that there are two ways to solve the faint young sun paradox.

   i. Lower albedo due to less cloud cover or ice.

  ii. Higher amounts of greenhouse gases found in the atmosphere of the early Earth.

5.8.4.1 Lower Albedo Is a Controversial Solution to the Faint Young Sun Paradox

The lower the albedo, the more of the solar flux would be available to warm the Earth. The most important determinant of the Earth's albedo is the amount of cloud cover. This would need to be reduced to near zero to achieve the necessary offset. Clouds are made of water droplets. These form around cloud condensation nuclei (CCN). In our present atmosphere, the majority of CCN arise from the oxidation of gases released by green plants and algae, which did not exist on the early Earth. Also there are likely to have been fewer windblown dust particles to act as CCN given that the area of continents was less in the Archaean. But this would likely be at least partly offset by the higher flux of cosmic rays from the young sun which generate ions that act as CCNs.

5.8.4.2 Higher Amounts of Greenhouse Gases Could Solve the Faint Young Sun Paradox

Earth emits most of its IR between 5 and 500 μm, the thermal IR spectrum. In the late Hadean/Archaean Earth context, three greenhouse gases attract most attention at present: water, carbon dioxide, and methane, though ammonia was proposed in the past. The contribution each gas makes to greenhouse warming depends on the specific characteristics of the gas, that is, how effectively it absorbs over the thermal IR spectrum, its mixing ratio (Box 5.4), its lifetime in the atmosphere, its possible conversion to other greenhouse gases and the total pressure of the atmosphere. All of these properties except the first depend on the particular atmosphere in which the gas finds itself, which is what we are trying to find out.

On a molar basis, the order of greenhouse gas potency (determined solely by the specific properties of the gases) is $CH_4 > H_2O > CO_2$. In fact methane is reckoned to be 21-fold more potent than carbon dioxide. That said, in our current atmosphere CO_2 makes the bigger contribution to global warming because it is present in a much higher concentration than CH_4. Its contribution is also much longer lasting because the lifetime of CO_2 in the atmosphere is thousands of years compared with just 10 for methane.

Water acts in a way different from the other greenhouse gases because its concentration in the atmosphere *depends* on the temperature. This is because the Earth's

BOX 5.4   QUANTIFYING GASES IN MIXTURES

There are several ways in which the amount of a particular gas in a mixture of gases can be quantified.

The mole fraction of a gas, x_i, is the amount (in moles) of the gas, n_i, compared to the total amount, n, of all the gases making up a mixture:

$$x_i = n_i/n. \tag{5.18}$$

Mole fractions are expressed as mol mol^{-1}.

A mixing ratio is a dimensionless number that describes the proportion of a gas in a mixture. Mixing ratios can be defined in two different ways. One is the proportion by volume of a gas in a mixture to the total volume of the mixture. This is usually given as a percentage, or as ppmv or ppbv. Mixing ratios are useful because they stay constant if the density of a gas changes as happens if its temperature or pressure alters. Think of a parcel of air rising adiabatically. Its pressure and temperature fall, so the number of gas molecules per unit volume drops, but the proportion of each gas in the mixture, the mixing ratio, stays the same. For this reason mixing ratios are frequently used when describing the composition of planetary atmospheres.

Because 1 mol of an ideal gas occupies 22.7 dm^3 (at 273 K and 1 bar) a mixing ratio by volume can translate directly into a mole fraction. In this case, ppmv can be expressed as µmol mol^{-1} while ppbv becomes nmol mol^{-1}.

The alternative version of mixing ratio, r_i, can also be given as the amount of a particular gas, n_i, in a given volume compared to the amount of the *other* constituents $(n - n_i)$ in the mixture in the same volume:

$$r_i = n_i/(n - n_i). \tag{5.19}$$

This definition is used in meteorology to define the amount of water present in *dry* air. If n_i is a trace gas in the atmosphere, then this approximates to the mole fraction.

In a mixture of ideal gases, each gas has a partial pressure equal to the pressure it would exert if it alone occupied the volume at a given temperature. Most planetary atmospheres are quite close to being ideal gases because their molecules are sufficiently far apart that they do not interact with each other. Of course the total pressure of an ideal gas mixture must be the sum of all the individual partial pressures of the constituent gases. This is Dalton's Law. Thus, the total pressure of a N_2–CO_2–CH_4–H_2O atmosphere is given by

$$P = pN_2 + pCO_2 + pCH_4 + pH_2O.$$

The ratio of partial pressure of a particular gas, p_i, to total pressure of a gas mixture, P, is equal to the mole fraction of the gas:

$$p_i/P = n_i/n = x_i, \tag{5.20}$$

from which we get $p_i = x_i P$.

So the partial pressure of an ideal gas can be found from its mole fraction, and the total pressure of a gas mixture P. For example, the partial pressure of nitrogen in the atmosphere is

$$P = 0.79 \, \text{mol} \, \text{mol}^{-1} \times 101.3 \, \text{kPa} = 80 \, \text{kPa}.$$

surface temperature and pressure ensure that water is always close to its triple point where it evaporates or condenses readily: as Earth warms, more H_2O evaporates and this increases the greenhouse effect, producing even more warming. This is a classic positive feedback effect. It operates to amplify the effect of an increase in the concentration of other greenhouse gases. Thus, CO_2 is nowhere close to its triple point on Earth but a rise in atmospheric CO_2 concentration for any reason will force an increase in temperature and this in turn will put more water vapor into the atmosphere, so the total greenhouse effect is that due to the carbon dioxide *plus* the additional water vapor. The rise in temperature could be as much as triple the effect of CO_2 alone: in other words, if the CO_2 itself produces a 1°C increase, a 3°C rise will result because of the amplification by water. The raised temperature generates another feedback: because the solubility of CO_2 in water drops with temperature more CO_2 enters the atmosphere as oceans warm up. Of course if the CO_2 concentration drops for any reason, the *fall* in greenhouse effect is amplified by reversing the direction of the feedbacks.

The take-home message is that once early Earth had cooled sufficiently for the steam atmosphere to rain out, the temperature was primarily controlled by the concentrations of CO_2 and CH_4, and water responded to drive the amplifying feedbacks.

It is possible to warm the Earth sufficiently to offset the effect of the faint young sun by having large amounts (e.g., 10 bar) of CO_2 in early Earth's atmosphere. But several lines of geological evidence suggest that there was not that much CO_2 about. For example, fossil soils (palaeosols) laid down 2.7 billion years ago show that CO_2 could not have been *more than* about 10^{-2} bar (roughly 100 times the present atmospheric level). And recent work shows that high CO_2 levels are not necessary. Climate modeling by Jim Kasting at Penn State University and colleagues shows that by adding modest amounts of methane to a nitrogen–H_2O atmosphere with 10^{-2} bar CO_2, and assuming the Earth reflected less incoming sunlight than it does today because it had less land, it is possible to achieve an average global surface temperature up to 15°C, the same as it is today.

The solution above is all very well, but many researchers would argue that the palaeosol constraint on $[CO_2]$ of 100-fold the present atmospheric level (PAL) is too high. Fortunately, it seems that it is possible to have enough warming with lower amounts of CO_2 if H_2 is also present because this adds another source of greenhouse warming. Robin Wordsworth and Raymond Pierrehumbert reckon that a combination of 10% H_2 in a nitrogen atmosphere of 2–5 bar with just 2–25 times the PAL of CO_2 could keep the surface temperature above the melting point of ice. How does this work? Although neither H_2 nor N_2 can absorb infrared radiation from the Earth's surface directly and so are not conventional greenhouse gases, if they collide with each other they can absorb infrared radiation efficiently at exactly the wavelengths where CO_2 and H_2O are poor greenhouse gases. The process is termed collision-induced absorption (CIA).

This is not the first time that hydrogen has been considered as a component of early Earth's atmosphere. CIA due to 1 bar of H_2 was considered by Carl Sagan and George Mullen as a solution to the faint young sun paradox but they dismissed it, partly because they reckoned it would make the Earth too hot. Hydrogen was proposed again in 2005 by Feng Tian at the University of Boulder, Colorado and colleagues. Their calculations suggested that the escape of hydrogen from the Earth's atmosphere was a 100-fold less than previous estimates and that this, coupled to volcanic outgassing of hydrogen, could have bumped up the H_2 content of the atmosphere to 30%. These researchers were motivated by the fact that a H_2-rich atmosphere might have allowed the production of appreciable quantities of organic molecules from atmospheric gases with the aid of energy from UV and lightening, thereby providing raw ingredients for the synthesis of important biochemicals. However, the work was heavily criticized on the grounds that it did not take account of all the possible hydrogen loss mechanisms. But by a fascinating twist it now seems that although these hydrogen losses are important today, they would have had only a small effect on *early* Earth. This puts hydrogen back on the menu.

5.8.4.3 The Pressure of the Earth's Atmosphere Was Less Than Twice Its Present Value 2.7 Gyr Ago

To get some idea of the magnitude of the greenhouse effect, we need to know what the atmospheric pressure was. We have no idea what this was in the Hadean; however, two studies have come up with an upper limit for the surface density of the atmosphere in the Archaean. Encouragingly, they come up with comparable values.

One estimated the partial pressure of N_2 in the Archean atmosphere by analysis of $N_2/^{36}Ar$ isotope ratios in fluid inclusions in ~3.5 Gyr-old hydrothermal quartz in Pilbara, Western Australia, and found it to be between 1.1 and 0.5 bar; that is, comparable to that today. How was the estimate made? The amount of nitrogen in the atmosphere of the Earth may have varied with time as a result of exchange with the mantle or loss to space. By contrast, the amounts of nonradiogenic noble gas isotopes such as ^{36}Ar seem not to have altered in the atmosphere since Earth formed. Hence, the $N_2/^{36}Ar$ ratio of fluid inclusions that have been isolated since they last equilibrated with the atmosphere should allow the partial pressure of N_2 in the atmosphere at the time to be determined. Evidence that the ^{36}Ar concentration in the atmosphere has not changed is that the $^{38}Ar/^{36}Ar$ ratio of the terrestrial atmosphere and of argon trapped in primitive meteorites is identical. This shows that terrestrial argon was not lost to space. If it had been, the $^{38}Ar/^{36}Ar$ ratio would have been higher than that of the meteorite, as lighter isotopes are lost more readily than heavy ones.

It has been suggested that if the partial pressure of N_2 in the atmosphere had been 2–3 bar, this would have enhanced the greenhouse effect of CO_2 and hence have contributed to the solution of the faint young sun paradox. That N_2 seems not to have been so abundant does not support this idea.

The discovery of fossil raindrop imprints in 2.7-billion-year-old lithified volcanic ash in South Africa provided the other estimate. From the area of the imprints and the fact that the terminal velocity of raindrops varies inversely with the square root of air density, researchers were able to show that the density of the air through which the raindrops fell so long ago was no more than 2.3 kg m^{-3}, and could have been 0.6–1.3 kg m^{-3}, compared to today's 1.2 kg m^{-3}. From air density, ρ, the atmospheric pressure can be found from the ideal gas law:

$$P = \rho R_{sp} T, \tag{5.21}$$

where R_{sp} is the *specific* gas constant appropriate to the composition of the atmosphere. The specific gas constant for any gas can be found from

$$R_{sp} = R/M, \tag{5.22}$$

where R is the Universal gas constant (8.314 J mol^{-1} K^{-1}) and M is the molar mass in kg mol^{-1}. R_{sp} for a mixture is the weighted average by mass of the specific gas constants for the individual gases.

What was the temperature in the Archaean? The lack of evidence for glaciation implies that global average temperature must have been 20°C or more. So, assuming $T = 20$°C, and an atmosphere with $0.6N_2 + 0.4CO_2$ by mass (70% N_2 + 30% CO_2 by volume) we can now calculate the atmospheric pressure 2.7 Gyr ago.

EXERCISE 5.10

i. The ideal gas law is usually written (refer Equation 5.21):

$$PV = nRT,$$

where n is the number of moles of the gas and R is the universal gas constant. Derive Equation 5.36 from $PV = nRT$.

ii. Show that $\rho R_{sp} T$ has units of pressure.

iii. Show that an atmosphere which is 70% N_2 and 30% CO_2 by volume has a specific gas constant, $R_{sp} = 254$ J kg^{-1} K^{-1}.

iv. Hence calculate an upper limit for the pressure of the Archaean atmosphere assuming the temperature was 20°C.

An atmospheric pressure less than twice the present level rules out high Archaean temperatures of 70–85°C. To see this, consider a climate model of Earth at 3.3 Gyr when the sun was 23% dimmer, with a N_2 pressure of 0.8 bar, as in today's atmosphere. This showed that a mean surface temperature of 70°C required 2–6 bars of CO_2, in combination with CH_4 mixing ratios of 0.01–0: the higher CH_4 mixing ratios correspond to the lowest CO_2 pressures (pCO_2).

High carbon dioxide levels are also ruled out by geology. For every 100-fold increase in pCO_2, the pH of rainwater in which the CO_2 dissolves drops by 1 log unit.

$$CO_2 + H_2O \rightleftharpoons \underset{\text{carbonic acid}}{H_2CO_3} \rightleftharpoons H^+ + HCO_3^-. \tag{5.23}$$

For example, a 3 bar CO_2 atmosphere would have rainwater with pH 3.7. In fact, at this pH, rainwater would be pure carbonic acid. Coupled with the high temperature and hence high levels of rainfall, this would produce extremely intense silicate weathering, which is not what is seen in the Archaean rock record. This militates against the hot Archaean scenario.

5.8.4.4 Carbon Dioxide Cannot Have Been Greater Than 0.03 Bar in the Late Archaean Atmosphere

Paleosols are fossil soils that have been buried and that preserve evidence of the conditions under which they formed. A crucial observation is that paleosols from the Archaean lack the mineral siderite ($FeCO_3$). Although there is some wriggle room, because the exact condition under which the paleosols formed is uncertain, the lower limit on CO_2 partial pressure (pCO_2) required to form siderite is ~0.03 bar (30,000 ppmv), about 100-fold higher than the present atmospheric level (PAL). Some researchers have adopted the more conservative view that paleosol data restricts the pCO_2 to <75 PAL.

More controversial is the claim, based on the coexistence of magnetite (Fe_3O_4) and siderite in banded iron formations, in ancient marine sediments (see also Chapter 8), that pCO_2 cannot have been higher than 900 ppm. That is because magnetite is oxidized and only stable if CO_2 concentrations are low while siderite is reduced and only stable if there is enough CO_2. Quantifying the balance, given that both minerals are present, constrains pCO_2. If the Archaean pCO_2 was only 900 ppm, then there is currently no solution to the faint young sun paradox. No plausible reduction in albedo can warm early Earth enough in the face of such a small greenhouse effect.

5.8.4.5 Model Late Archaean Atmospheres Can Keep Earth Warm

By constraining CO_2 to no more than 100 PAL and assuming an atmospheric pressure of 1 bar, models of the late Archaean atmosphere are being developed which keep Earth warm by having additional greenhouse gases. Critically, such models should not just maintain the temperature above the freezing point of water; they should prevent glaciation on the grounds that there is no geological evidence for glaciation in the Archaean except at ~2.9 Gyr.

An obvious greenhouse gas to add is methane. The CH_4 mixing ratio in the atmosphere now is 1.8 ppmv, almost all of which is produced biologically. But a small fraction is made by geological processes. Currently methane lasts only 10 years in the atmosphere since it reacts (rather slowly) with hydroxyl radicals:

$$CH_4 + OH \rightarrow CH_3 + H_2O. \tag{5.24}$$

The hydroxyl radicals are largely derived from the reaction of excited oxygen atoms with water:

$$O_3 \xrightarrow{hv} O + O_2 \tag{5.25}$$

and

$$O + H_2O \rightarrow 2OH. \tag{5.26}$$

In the absence of oxygen in the early atmosphere, it has been estimated that the lifetime of methane could be as much as 10,000 years. Hence, even a small source could build up substantial amounts in the atmosphere.

How much methane do we need? One model took account of the fact that nitrogen can enhance the effect of greenhouse gases. It does this by interacting with greenhouse gas molecules broadening their absorption lines so that they absorb more radiation. This process is called pressure broadening or collision broadening (cf. pressure broadening in stellar atmospheres that underpins luminosity classes). In this situation varying N_2 in a H_2O–CO_2–CH_4 greenhouse atmosphere in which the $pCH_4 = 10^{-4}$ bar (about 50-fold higher than now) and $pCO_2 = 10^{-2}$ bar (about 25-fold higher than PAL) showed that just twice the present partial pressure of N_2 would be enough to raise the surface temperature to 11.8°C. This would resolve the faint young sun paradox only in the sense that this would be enough to prevent runaway glaciation (i.e., an ice age), which occurs at 6°C in this model.

James Kasting and colleagues at Penn State Astrobiology Research Center have been modeling the atmosphere at 2.8 Gyr when solar luminosity was 20% less than now. They recognize that to warm the Earth there is a constraint on the CH_4/CO_2 ratio. If it is >0.1 reactions produce an organic haze which reduces the amount of sunlight that can penetrate the atmosphere. In fact, by $CH_4/CO_2 \sim 1$, the haze has opacity of 1. In other words, models cannot add too much CH_4 without producing dramatic cooling. However, the haze-making chemistry does form ethane, which has a significant greenhouse effect. At short UV wavelengths ($\lambda < 145$ nm)

$$CH_4 \xrightarrow{h\nu} CH_3 + H. \tag{5.27}$$

At longer wavelengths ($\lambda < 240$ nm) photolysis of water occurs:

$$H_2 \xrightarrow{h\nu} OH + OH. \tag{5.28}$$

$$CH_4 + OH \rightarrow CH_3 + H_2O. \tag{5.29}$$

In the absence of oxygen, the methyl radicals engage in a three-body reaction:

$$CH_3 + CH_3 + M \rightarrow C_2H_6 + M. \tag{5.30}$$

The trick is to get enough ethane to add to greenhouse warming without generating too much cooling haze.

It turns out that a combination of 0.03 bar of CO_2 and 3000 ppmv CH_4, which allowed the formation of ~1 ppmv ethane, gave a surface temperature of 290 K. This is a couple of degrees higher than it is today, although to avoid polar ice caps the temperature needs to be about 5°C higher than now (293 K). This could be achieved with a modest increase in pCO_2, though this would go over the paleosol limit.

There remains a huge scope for refining climate models, but it seems that in principle the faint young sun paradox can be solved by relatively modest amounts of greenhouse gases, particularly if the requirement is that only part of the Earth's surface maintains liquid water.

5.8.4.6 There Are Several Abiotic Sources for Atmospheric Methane

There are several geological processes that can generate CH_4.

1. *Serpentinization*. When mantle olivine interacts with water at constructive plate boundaries (mid-ocean ridges) and at hydrothermal vents a series of reactions occurs that produce the hydrated mineral serpentine. One of these serpentinization reactions generates methane.

$$18Mg_2SiO_4 + 6Fe_2SiO_4 + 26H_2O + CO_2 \rightarrow 12\underset{\text{serpentine}}{Mg_3Si_2O_5(OH)_4}$$
$$+ 4\underset{\text{magnetite}}{Fe_3O_4} + CH_4. \qquad (5.31)$$

   The reaction is catalyzed by iron and nickel.

2. *Fischer–Tropsch reactions at hydrothermal vents*. Fischer–Tropsch reactions produce hydrocarbons by reacting hydrogen and carbon monoxide. They are catalyzed by metals such as iron, nickel, and cobalt. The higher the temperature, the more methane (rather than bigger hydrocarbons) is produced. In general

$$(2n + 1)H_2 + nCO \rightarrow C_nH_{(2n+2)} + nH_2O. \qquad (5.32)$$

EXERCISE 5.11

Write out the specific Fischer–Tropsch reaction for the synthesis of methane.

There are also astronomical sources, namely comets, which can deliver methane directly, and perhaps surprisingly, impacts by iron meteorites. The idea is that metal distributed by an iron meteorite impact could catalyze Fischer–Tropsch reactions. Estimates show that an impact by a 10-km iron meteorite with $v = 15$ km s^{-1} could produce $10^{12} – 10^{13}$ kg CH_4. Reactions of H_2O and CO_2 with the iron would produce H_2, which could act as a hydrogen source for the Fischer–Tropsch reactions.

5.8.4.7 Is Abiotic Methane Sufficient to Solve the Faint Young Sun Paradox?

How did the atmosphere get the few thousand ppm methane that seems necessary to keep Archaean Earth warm? We have seen that the lifetime of CH_4 on the anoxic early Earth would have been much longer and this might have allowed even small inputs to build up appreciable concentrations.

Abiotic sources of methane would have been larger in the Archaean. One reason is because until about 2.7 Gyr ago magmatism was dominated by the formation of komatiites (Box 5.5). They are rich in olivine; so komatiitic ocean crust would undergo serpentinization as seawater percolates into it. (This has been verified by laboratory studies.) What is more, komatiites contain high amounts of nickel and

---

BOX 5.5 KOMATIITES

Komatiites are igneous rocks produced by very high levels of partial melting of mantle, giving them a composition higher in magnesium and iron, and lower in silica (SiO_2), than modern mantle-derived magmas, basalts. This makes them closer in composition to the mantle peridotite from which they evolved. This similarity allows geologists to describe both mantle peridotite and komatiites as ultramafic rocks.

Whether they are the product of wet or dry melts is a matter of contention. This matters because water lowers the melting point of magmas. If komatiites came from wet melts, it means the mantle temperature in the Archaean was lower than if they were the products of dry melting. Recent measurements of a 2.7 Gyr komatiite magma from Belingwe, Zimbabwe, by Andrew Berry at University College London and colleagues showed the water content to be 0.2–0.3 wt%, much lower than the 3–5 wt% expected of a wet melt. What is more, the iron in the rock had not been oxidized (rusted) as shown by the low $Fe^{3+}/\Sigma Fe$ of 0.1. These results support a high Archaean mantle temperature.

Komatiite magmatism more or less ceased about 2.7 Gyr ago when the mantle temperatures became too cool to achieve the high degree of partial melting needed to generate komatiites.

---

iron chromite that act as catalysts for serpentinization. Modern ocean crust is basalt which is not serpentinized so much, although the mantle is. A second reason is that because of the higher mantle temperature, the rate of formation of new ocean crust at constructive plate boundaries could have been sixfold higher in the Archaean than now, making much more material available for serpentinization.

A recent (2012) attempt to quantify whether serpentinization could have produced the mixing ratio of CH_4 required in Kasting's model Archaean atmosphere calculated the crustal production rate needed to sustain a given global surface temperature as a function of pCO_2. Even the most optimistic values for the parameters in the modeled equation showed that a crustal production rate 10-fold higher than now would be necessary to achieve a surface temperature of 15°C at a pCO_2 of 100 PAL (pretty much the maximum CO_2 allowed by paleosols). This is getting on for double the estimated Archaean crustal production rate and 15°C would probably not mean an ice-free world. More realistic choices of parameters required an implausibly high 25-fold higher crustal production rate to get 15°C.

Of course this model does not take account the other abiotic Archaean methane sources (e.g., Fischer–Tropsch reactions), and this sort of work is in its infancy, so the model and parameters will be refined. We do not even have a good idea how much methane is produced geologically today, because serpentinization and Fischer–Tropsch reactions happen mostly in inaccessible parts of the planet. Consequently, although presently it looks as if abiotic sources could not have provided enough methane to sustain a temperate Archaean climate, we have not had the last word on this point.

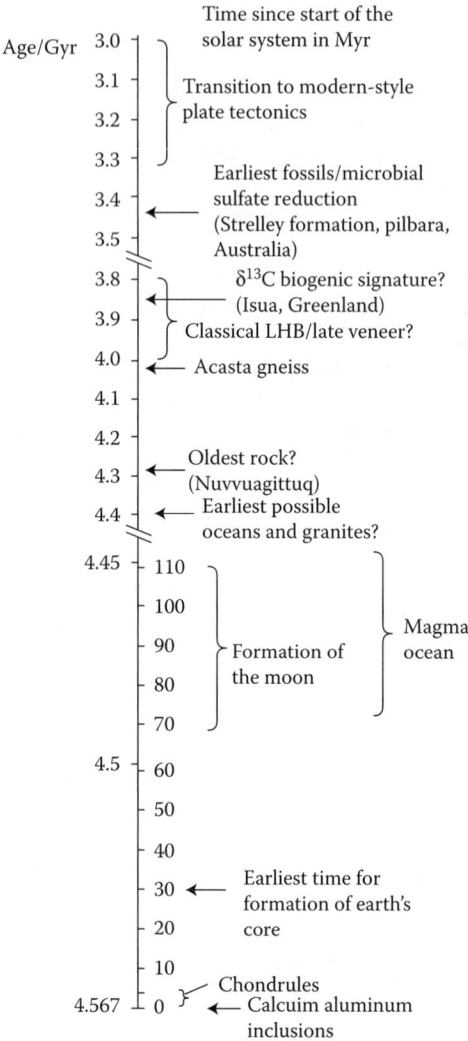

FIGURE 5.7 Putative timeline of key events that shaped early Earth.

That said, a consensus has been emerging over the past few years that methane was supplied to the Archaean atmosphere by what might have been some of the Earth's first organisms, methanogens. If this was the case, then the faint young sun paradox was solved by the emergence of life. To explore this story further and see where it leads us, we must first turn our attention to some properties of biological systems and their ways of making a living. Figure 5.7 summarizes early Earth history.

6

Properties of Life

6.1 Can Life Be Defined?

Biologists have huge difficulty in agreeing on what constitutes life. Maybe, to paraphrase a US Supreme Court judge in 1964, it is like pornography, impossible to define, but everyone knows it when they see it.

Astrobiologists have come up with a few criteria that, taken together, attempt to define what is meant by life. These are based on characteristics shared by all terrestrial organisms and our understanding of physics and chemistry. Of course these criteria may turn out not to be necessary, sufficient, or exclusive. It has been argued that a proper understanding of life requires a general theory of biology. Unfortunately, we do not yet have any such theory, neither will we until we have other examples of life—exobiology—to help us construct it. Until such time, most astrobiologists adopt a pragmatic view of assuming that life elsewhere will resemble Earth life in fundamental ways. The list of attributes below reflects this though it is not without its critics.

6.1.1 Life Is a Complex, Self-Organizing, Adaptive Chemical System

6.1.1.1 Complexity in Living Systems Is Hard to Define

The more complex a system is the more bits of information are required to describe it. It is clear that even a single cell (the basic unit of living systems) would need an extremely long instruction manual to describe it completely. In that sense life is complex. However, the standard mathematical notion of complexity, Kolmogorov complexity, defines a random sequence of digits as having maximum complexity because it can only be described by itself, no data compression can reduce the sequence of bits needed.

However, organisms are not random assemblages, so Kolmogorov complexity is not the best way to define biological complexity. Instead, organisms are said to have a specified complexity, a concept that was first defined by Leslie Orgel in his 1973 book, *The Origin of Life*, but how this is quantified we do not pursue here.

6.1.1.2 Living Systems Have Emergent Properties

The network of interactions and feedbacks that occur between all the components of a living system is extensive, rich, and nonlinear, and from it emerges the remarkable property: life is self-organizing and adaptive.

Self-organization means that patterns at the level of the entire organism emerge from simple rules that operate at a local level. This is perhaps easier to imagine if we think about self-organization seen at the level of groups of organisms. For example, flocking in birds—the coordinated behavior of thousands of animals that creates a large-scale structure that changes over time—is possible because each individual bird operates a set of simple rules according to what its nearest neighbors have just done. No bird has a "master plan" of the dynamic pattern made by the ensemble.

In individual organisms, self-organization emerges from some "simple" rules of chemistry. Macromolecules spontaneously fold or assemble to adopt the lowest energy conformations, and this allows more complex structures to self-assemble. Indeed self-organization is perhaps most impressively seen in the development of multicellular organisms. These arise from a single cell by successive cell divisions, in which each of the newborn cells comes to be in the right place at the right time and with the appropriate set of genes turned on so that it serves its appointed function. This includes ensuring that it acts to guide the next generation of cells to do the right thing. This unfolding of a developmental program that takes a fertilized egg to an adult is self-organization on a grand scale since the original progenitor cell has no master plan that details all the steps.

At the simplest level, adaptation means that organisms are not completely at the mercy of their environment. Thus feedbacks allow organisms to stabilize their internal environment (e.g., pH, ion concentrations, etc.) in the face of external changes. Biologists call this homeostasis.

Complex adaptive systems are those which change their functionality (behavior) in response to their history (i.e., what has happened to them before) and in response to their environment. In other words, adaptive systems have a memory. This allows organisms to regulate their internal processes: metabolism, growth, cell division, etc. in light of external contingencies.

While relatively simple toy models can show self-organization and adaptation, the richness seen in biological systems is much greater.

6.1.2 The Chemistry of Life Is Far from Equilibrium

Living systems do work. This is obvious in the case of complex life forms, such as cats and camels, which move. However, all organisms use energy for a wide variety of basic life processes such as growth, reproduction, and maintaining their integrity. The requirement for living systems to be able to do work shows that the chemistry of life cannot be at equilibrium. That is because it is not possible to get systems at equilibrium to do work.

A thought experiment shows intuitively why this is so. Hydroelectric power stations generate energy that can be used to do work—so-called free energy—at the expense of gravitational potential energy lost by falling water. If the water level either side of the turbine was the same, that is, if the system was at equilibrium, the turbine could not generate free energy because the gravitational potential energy on both sides of the turbine would be the same.

It is the same story with chemical systems. Consider the reversible reaction:

$$A + B \rightleftharpoons C + D.$$

We start with initial amounts of the substrates A and B. These react to produce products C and D. Let us assume that once started we do not add or subtract any reactants and that no energy is lost or gained by the system (i.e., the system is adiabatic). Because the reaction is reversible, after a time it will reach a steady state in which the rate of the forward reaction and back reaction will be the same and all four components will coexist at concentrations determined by the temperature and the thermodynamic characteristics of the system. This system is at equilibrium and it is impossible to get any useful chemical energy from such a system. More technically put, equilibrium systems have zero free energy and their maximum entropy.

Virtually, all biochemical processes occur in conditions that are far from equilibrium, and some are essentially irreversible, so cannot go to equilibrium by definition.

To see how the chemistry of life enables living systems to use free energy, we take a short diversion into thermodynamics.

6.1.2.1 When Will a Reaction Happen Spontaneously?

Thermodynamics tells us how far reactions can go and how much energy is released or consumed in the process. The reactions are the system and the rest of the universe make up the surroundings. An open system can exchange both matter (M) and energy (E) with its surroundings, a closed system can only exchange energy, while an isolated system can exchange neither matter nor energy with its surroundings (Figure 6.1).

The first law of thermodynamics is a conservation law that says that the total energy of a system and its surroundings is constant; in other words, energy cannot be created or destroyed. It follows that energy released/gained by a system must equal energy gained/released by the surroundings. All chemical systems have an internal energy that can change as energy is exchanged with its surroundings. Now, the change in internal energy (ΔU) could come about either because heat is released or absorbed by the system (ΔH) or because work is done by the system to its surroundings, or by the surroundings to the system. Taking the simple case that an ideal gas is a system that does work on its surrounding by expanding at constant pressure then, work

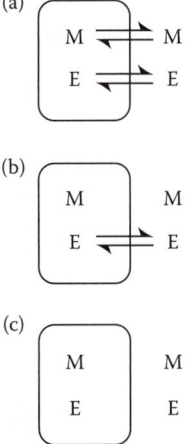

FIGURE 6.1 Open (a), closed (b), and isolated (c) systems.

done = $P\Delta V$, and because work done by the system on the surroundings is defined as negative we can write:

$$\Delta U = \Delta H - P\Delta V. \tag{6.1}$$

ΔH, the enthalpy, is the amount of heat liberated or absorbed by a reaction at a given temperature and constant pressure. By definition, a negative value of ΔH corresponds to heat given out. In these exothermic reactions the heat content of the products is lower than that of the reactants. By contrast, endothermic reactions absorb heat, so the products have higher heat content than the reactants and ΔH is positive. In fact, for biochemical reactions which involve ions and molecules in solution rather than gases, ΔV is generally small, so $\Delta H \sim \Delta U$.

Generally, a low temperature of the surroundings favors exothermic reactions and a high temperature favors endothermic reactions, but unfortunately, whether a reaction is exothermic or endothermic is not a reliable guide to whether a reaction will happen *spontaneously*, that is, whether it is thermodynamically favorable. For this we need to know the entropy, S. This is usually defined as a measure of the degree of randomness or disorder of a system. The higher the entropy the more ways there are of arranging the atoms and molecules in a system, and of distributing energy within the system. This is captured in the Boltzmann equation:

$$S = k \ln W. \tag{6.2}$$

where k is the Boltzmann constant and W the number of ways of arranging the particles in a system (microstates) at constant energy. An increase in entropy can also be viewed as a move toward greater uniformity. Hence, the dissipation of a concentration gradient or a thermal gradient, in other words, a shift toward an equilibrium state, is accompanied by an increase in entropy. The number of ways in which Na^+ and Cl^- ions can be distributed in a dilute aqueous solution at equilibrium is vastly higher than the number of ways these ions can be arranged in salt crystals. Consequently, the probability of the ions assembling within the dilute solution to form salt crystals spontaneously is extremely low. Hence, the dissolution of NaCl is described as thermodynamically favorable. Likewise, a reaction will go spontaneously if the entropy of a system plus its surroundings increases. This is the second law of thermodynamics. More formally, for a spontaneous process we can write it as follows: $\Delta S_{total} > 0$.

Note that it is possible for a reaction to go spontaneously even if its entropy falls, provided that the entropy of the surroundings rises by a greater extent.

It is useful to express the second law of thermodynamics as a change in Gibbs free energy, ΔG. For closed systems:

$$\Delta G = \Delta H - T\Delta S. \tag{6.3}$$

Gibbs free energy is the amount of work that can be obtained from a reaction at a given temperature and constant pressure. The change in free energy of a system, unlike the change in enthalpy, *is* a reliable guide to whether a reaction will go spontaneously. It will if ΔG is negative, but not if it is positive. We can now summarize the conditions under which a reaction is favorable or unfavorable (Table 6.1). A system at equilibrium has $\Delta G = 0$, and no change can occur.

TABLE 6.1

Thermodynamics favoring reactions

Variable	Favorable	Unfavorable
Enthalpy	$\Delta H < 0$ (exothermic)	$\Delta H > 0$ (endothermic)
Entropy	$\Delta S > 0$	$\Delta S < 0$
Free energy	$\Delta G < 0$ (exergonic)	$\Delta G > 0$ (endergonic)

6.1.2.2 Life Obeys the Second Law of Thermodynamics

Living systems are open systems since they exchange both matter and energy with their surroundings, and open systems also obey the Gibbs free energy equation, although in modified form. However, living systems are highly organized; in other words, they have a low entropy. As organisms grow and divide, their level of organization increases, so their entropy falls, in apparent contradiction with the second law of thermodynamics. This is known as the Schrodinger paradox. For the living system itself to maintain a low entropy state it must produce an even larger decrease in enthalpy so that the resultant change in ΔG ($=\Delta H - T\Delta S$) is negative. For this to be the case, life needs a continuous input of energy.

The Schrodinger paradox is resolved by the fact that living systems are open systems. This means that the order they produce, as a result of their chemistry being far from equilibrium, is achieved at the expense of increasing the entropy of their surroundings. Life preserves its internal order by taking free energy from its surroundings (as sunlight or chemical energy) and returning to the surroundings an equal amount of heat and entropy. In fact, detailed thermodynamic calculations show that the net entropy of the living system plus its surroundings increases. Living organisms can be thought of as islands of low entropy—local regions of highly organized matter—created at the expense of increasing the entropy of the surrounding sea. That said, recent calculations have shown that bacterial replication is so efficient that "if it were to produce even half as much heat it would be pushing the limits of what is thermodynamically possible!"

6.1.2.3 Life Messes Up Its Environment

A consequence of living systems being far from equilibrium is that they produce chemical disequilibrium in the environment. The evidence for this on Earth is clear to see. The presence of both oxygen and methane in the atmosphere shows the Earth system to be far from chemical equilibrium. Methane is rapidly oxidized to carbon dioxide and water by oxygen, and has a lifetime of just 10 years in the atmosphere. Its abundance shows that some process must be continually producing it. We know that most methane on Earth is produced by micro-organisms, although there are geological processes that can produce it as well. Far more difficult to explain abiogenically is the extremely high abundance of oxygen. But the real point is it would need extraordinary geological processes to produce methane fast enough to maintain its current steady-state concentration in the face of all that oxygen. In other words, the combination of methane and oxygen is circumstantial evidence for life.

Consequently, one way to detect life beyond Earth is to look for signs of chemical disequilibrium in planetary atmospheres. This was first suggested by James Lovelock in the early 1960s who proposed examining infrared spectra of Mars and Venus, captured by the Pic du Midi Observatory, France, for signs of chemical disequilibrium to test whether these planets had life. This is relatively easy to do but to capture a sufficiently well-resolved spectrum of an exoplanet atmosphere is an altogether harder task although progress in this area is being made.

6.1.3 Life Requires an Energy Source

As we have seen, thermodynamics tells us that all life will require energy. Almost all life on Earth is dependent on energy from the sun. On a warm summer day it is easy to imagine that the sun is a *net* energy source but this is not the case. The Earth radiates back into space the same amount of energy it receives from the sun. If this were not so, the Earth would heat up until it reached an equilibrium in which heat gain did equal heat loss. However, the energy we get from the sun is mostly visible light, while the Earth radiates largely in the infrared. This is because the Earth's atmosphere and surface absorbs 69% of the incoming solar radiation (the remainder is reflected), and consequently the Earth warms up. While the sun radiates approximately like a black body with a temperature of 5770 K, the Earth (including the greenhouse effect of its atmosphere) radiates approximately like a black body at 298 K. Clearly, photons from the sun have a higher energy than those radiated by the Earth. However, as the Earth is in thermal equilibrium there must be far more long wavelength photons emitted by the Earth than short wavelength photons received. The Boltzmann equation, $S = k \ln W$ (Equation 6.3), tells us that there must be more ways of distributing the more numerous outgoing infrared photons than incoming visible light photons. Hence, the sun's energy has much lower entropy than the energy radiated back into space by the Earth. The low entropy of the sun's radiation is crucial.

Much of the free energy available to biological systems comes from solar radiation trapped in photosynthesis. To understand how this works, it is helpful to take an excursion into terrestrial biochemistry.

6.1.3.1 Solar Radiation Can Be Harnessed Directly by Photosynthesis

Capturing solar energy to build organic molecules from carbon dioxide is termed photosynthesis. Organisms that do this include some bacteria, many algae, and all green plants. Overall, for oxygenic (oxygen-producing) photosynthesis:

$$6CO_2 + 6H_2O \xrightarrow{\;h\nu\;} \underset{\text{Glucose}}{C_6H_{12}O_6} + 6O_2. \tag{6.4}$$

EXERCISE 6.1

Is photosynthesis an exergonic or endergonic reaction?

In fact, Equation 6.20 requires an input of free energy, $\Delta G^{\circ\prime} = 2870$ kJ mol^{-1} of glucose. $\Delta G^{\circ\prime}$ is the standard free energy change at pH 7 and signals that the free energy is calculated for standard conditions, in which each of the reactants is present at a concentration of 1.0 M, a pressure of 10^5 Pa (1 bar), and a specified temperature, often 298 K. Biochemists generally work at pH 7 because this is close to the hydrogen ion concentration at which organisms generally operate.

Photosynthesis takes lots of simple molecules to build fewer complex molecules, this is an increase in order, or a decrease in entropy, often referred to as negentropy (negative entropy). In fact, most of the entropy decrease that characterizes living systems comes ultimately from photosynthesis. The photosynthetic machinery only absorbs a small fraction of the incident sunlight; the rest is emitted as heat, which increases the entropy of the surroundings to such a degree as to more than compensate for the negentropy of photosynthesis. In other words, the total entropy of the photosynthetic (open) system and the surroundings increases, as the second law of thermodynamics demands.

Let us explore the consequences of this (Figure 6.2). The total change in entropy during photosynthesis must be greater than or equal to zero; otherwise, it would not be thermodynamically favorable:

$$\Delta S_{TOT} = \Delta S_{PS} + \Delta S_{IR} \geq 0, \tag{6.5}$$

where ΔS_{PS} is the entropy change of the photosynthetic system and ΔS_{IR} is the entropy change of the environment due to the emitted heat. This heat is a loss of enthalpy by the photosynthetic system, that is, $-\Delta H$.

Now for radiation in thermal equilibrium two standard results are as follows:

$$E = \sigma T^4 \quad \text{and} \quad S = (4/3)\sigma T^3,$$

where σ is the Stefan–Boltzman constant. Combining these gives the approximation that $\Delta S \simeq E/T$. Hence, we can write

$$\Delta S_{IR} = -\Delta H/T,$$

and, so Equation 6.0 becomes

$$\Delta S_{PS} - \Delta H/T \geq 0,$$

or

$$\Delta H - T\Delta S_{PS} \leq 0.$$

Hence, $\Delta G \leq 0$.

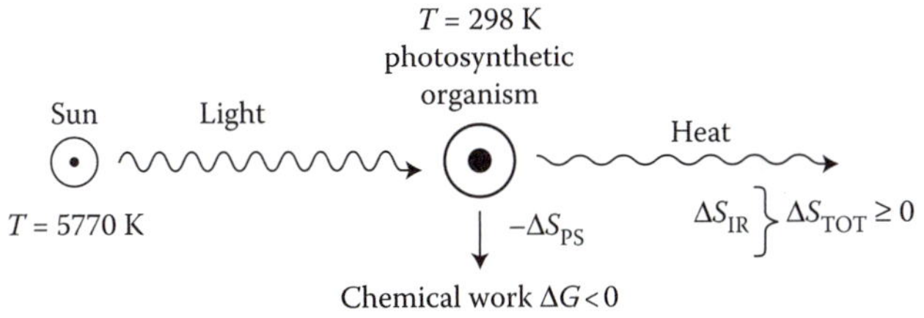

$T = 298$ K
photosynthetic
organism

Sun   Light   Heat

$T = 5770$ K

$-\Delta S_{PS}$ $\Delta S_{IR} \Big\} \Delta S_{TOT} \geq 0$

Chemical work $\Delta G < 0$

FIGURE 6.2 Thermodynamics of photosynthesis. The fraction of energy absorbed must appear as excess ΔG of products over reactants in the photosynthetic organism.

We see that photosynthesis harvests free energy from sunlight, lowering entropy within the organism, but at the expense of an even greater increase in entropy in its environment.

6.1.3.2 Organic Carbon Is Oxidized in Respiration

About half of the organic carbon made by photosynthesizing organisms is subsequently oxidized back to CO_2 to provide chemical energy to fuel their activities. This is termed respiration and is exergonic. Overall,

$$C_6H_{12}O_6 + 6O_2 \rightarrow 6CO_2 + 6H_2O. \tag{6.6}$$

In respiration the organic carbon is oxidized. Here, the oxidant is oxygen and this is aerobic respiration. After the photosynthetic organism has died the remaining half of the organic carbon is ingested by non-photosynthetic organisms (mostly bacteria responsible for decomposition, though a small proportion will be eaten), and its energy is released by the respiration of the ingesting organism. Even a running cheetah is ultimately powered by sunlight since the zebra it has eaten has grazed on grass that grew using photon energy (Figure 6.3).

Comparing Equations 6.20 and 6.21 we see that *overall* respiration is the reverse of photosynthesis.

EXERCISE 6.2

Write down a value for the free energy change of respiration.

The free energy change of a reaction is independent of the pathway. Thus, the free energy change accompanying Equation 6.21 is the same whether the glucose is completely combusted in oxygen at high temperature in a single step or oxidized by living organisms in numerous steps at low temperature. In both cases $\Delta G^{\circ\prime} = 2870$ kJ (as respiration is only about 40% efficient, not all the available energy from glucose oxidation can be harnessed). This seems like a high number (it is about one-quarter of the daily energy requirements of a healthy human) but note that $\Delta G^{\circ\prime}$ is standardized to molar concentrations of reactants; that is, almost 0.2 kg of glucose!

Despite the apparent similarity of photosynthesis and respiration, they are mechanistically very different. What is more, there are many other metabolic strategies. For example, there are organisms that get free energy not from sunlight but from

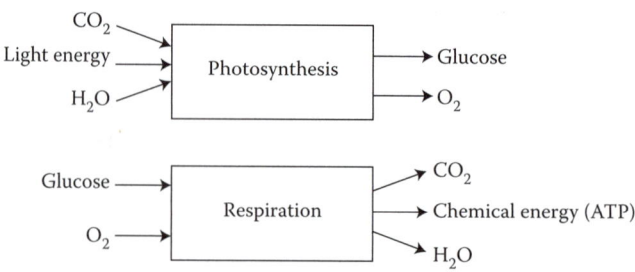

FIGURE 6.3 Highly simplified diagram depicting photosynthesis and respiration.

geochemical sources, and this was the case for the first life on Earth, as we shall see in Chapter 8. There are also many organisms that respire anaerobically, that is, without oxygen. Before 2.4 Gyr, all life on Earth must have been anaerobic because there was no free oxygen in the atmosphere. But the key to metabolism is that ultimately it is tied to sequences of coupled reduction–oxidation (redox) reactions through which electrons are transferred. These redox reactions organize flows of free energy. Remarkably this involves charging and discharging a biochemical battery, as we shall see.

6.1.3.3 Redox Reactions Are Coupled Reduction–Oxidation Reactions

Recall that oxidation is the loss of electrons and reduction is the gain of electrons. However, the oxidation and reduction reactions cannot happen alone, they are always coupled together as redox reactions. For example, the oxidation of a hydrogen molecule can be coupled to the reduction of an oxygen molecule:

$$H_2 \rightarrow 2e^- + 2H^+, \tag{6.7}$$

$$\downarrow$$

$$2e^- + \tfrac{1}{2}O_2 \rightarrow O^{2-}. \tag{6.8}$$

The products of these two half-reactions give

$$O^{2-} + 2H^+ \rightarrow H_2O. \tag{6.9}$$

So, overall,

$$H_2 + \tfrac{1}{2}O_2 \rightarrow H_2O. \tag{6.10}$$

Here H_2 is a reductant, reducing agent, or electron donor, and in giving up its electrons it is itself oxidized. $\tfrac{1}{2}O_2$ is an oxidant, oxidizing agent, or electron acceptor, and in gaining electrons it is reduced.

Half-reactions are often depicted in shorthand as oxidant/reductant, for example, $2H^+/H_2$ or $\tfrac{1}{2}O_2/H_2O$.

6.1.3.4 Redox Reactions Are Quantified by Redox Potentials

The readiness with which an atom or molecule gains an electron and so is reduced is its reduction potential, E_0' (in volts). Reduction potential is measured relative to the tendency of H^+ to gain electrons in the reaction

$$2H^+ + 2e^- \rightleftharpoons H_2, \tag{6.11}$$

which is zero for one molar hydrogen (i.e., pH 0), at 1 bar. However, biochemists quote the reduction potential at the more biologically plausible 10^{-7} M hydrogen (pH 7), which has reduction potential of -0.42 V. The negative of the reduction potential is called the oxidation potential. If we measure reaction (6.24) at pH 0 ($E_0' = 0$), then a half reaction with a negative reduction potential ($E_0' < 0$) has a lower affinity for electrons than H^+, that is, it is more likely to be oxidized. On the other hand, a positive reduction potential ($E_0' > 0$) means that a half-reaction has a higher affinity for electrons than H^+ and so more likely to be reduced.

If two half-reactions are coupled together in a redox reaction, the electrons flow spontaneously from the half-reaction with the lowest reduction potential to the one with the highest; in other words, in the oxidation to reduction direction. This direction of electron transfer has $\Delta E_0' > 0$. The standard free energy change (at pH 7) that is available from redox reactions is given by

$$\Delta G^{o\prime} = -nF\Delta E_0', \quad F = eN_A \qquad (6.12)$$

where n is the number of electrons transferred; F is the Faraday number, the charge on 1 mol of a monovalent ion; e is the charge on an electron, and N_A is the Avogadro number.

EXERCISE 6.3

Calculate the Faraday number. Take e = 1.602×10^{-19} C and $N_A = 6.022 \times 10^{23}$ mol^{-1}.

Thus, $\Delta E_0' > 0$ means that $\Delta G^{o\prime} < 0$, that is, the reaction is exergonic and free energy is available to do chemical work. The greater the difference in the reduction potential between two half-reactions, the higher the tendency of electrons to flow spontaneously between them and the more free energy there will be (Figure 6.4).

From this we see that reduction is endergonic whereas oxidation is exergonic. Organisms make use of this in sequences of redox reactions and electron transport chains, down which electrons cascade spontaneously via half-reactions with successively more positive reduction potentials.

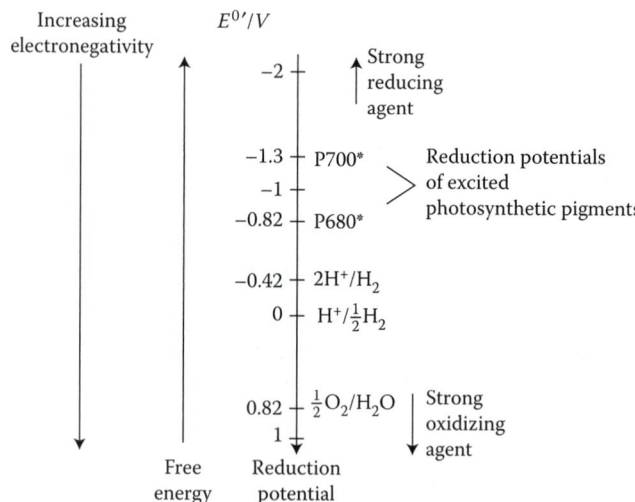

FIGURE 6.4 More negative reduction potentials correspond to higher free energy. Redox reactions with very negative reduction potentials consist of a weak oxidant coupled to a strong reductant, so will be potent electron donors. Redox reactions with very positive reduction potentials consist of strong oxidant and a weak reductant, so will be potent electron acceptors.

6.1.3.5 Are Redox Mechanisms a Universal Attribute for Life?

For life elsewhere the energy need not be that of a star; it could be geothermal, tidal, radiogenic, or whatever. The only crucial thing about it is that it needs to have low enough entropy that the extraterrestrial can somehow turn it into chemical energy. Redox reactions are the way this is done by all life on Earth. Whether a general theory of biology will show that they are a universal feature of life, only time will tell. If it is, it does not mean that alien biochemistry will have photosynthesis or respiration, simply that it will use redox reactions to manipulate free energy to do work. It is worth noting that geological settings on Earth, and presumably on other worlds, naturally provide redox gradients from which free energy could be abstracted. We shall revisit redox reactions again when we consider terrestrial biochemistry and the origin of life.

6.1.4 Living Systems Are Capable of Self-Replication

Organisms make copies of themselves and they can do this autonomously; no other agent is required. For this to be the case, the organism must carry all the information needed to copy itself. The physical substrate for this information is the genome, the DNA molecules that encode all genetic information.

The vast majority of terrestrial organisms reproduce asexually. For example, a bacterium makes a copy of its DNA, and then it divides into two with a single copy of the genome going to each daughter cell. Note that the requirement for autonomous replication excludes viruses. These are the ultimate parasites. They are usually a small genome in the form of DNA (or the closely related RNA) packaged in a protein coat. They cannot reproduce without a host cell. They must hijack the biochemical machinery of their host to make copies of themselves: just as computer viruses postdate the first software, so organic viruses almost certainly arrived after life had been around for a while. Most viruses are probably the degenerate relics of mobile genetic elements, which are a common feature of contemporary genomes.

There are obligate parasites, for example, the unicellular organism that causes malaria, *Plasmodium*, which cannot live and reproduce outside the body of their mosquito or human hosts. Should we regard these as dead on the grounds that they are incapable of autonomous reproduction? No. Although they spend part of their life cycle inside host cells they do not use any of the host cells' biochemistry to reproduce: *Plasmodium* has all the kit needed to reproduce itself, the host is simply providing the right environment. That said, there *are* apparent exceptions to the self-replication edict. For example, mules are the offspring of a male donkey and a female horse, two different species, and they cannot self-replicate; that is, a male and female mule cannot make baby mules.

EXERCISE 6.4

Are mules alive?

6.1.5 Life Exhibits Darwinian Evolution

Arguably the greatest unifying principle in biology is Darwinian evolution. Many astrobiologists argue that any life, however alien its chemistry may be, would exhibit Darwinian evolution.

Evolution theory is based on four central tenets:

1. More individuals are produced than survive.
2. There is a struggle for existence because the environment has limited resources.
3. Individuals show variation—that is, they differ from each other—usually in quite subtle ways. Those with features that confer an advantage, no matter how small, will have a greater chance of survival.
4. Individuals produce similar offspring (the principle of inheritance), and the number of individuals with the advantageous variation will increase in successive generations.

The outcome is the preferential survival to reproductive age of those individuals that are best adapted to the environment—those said to be naturally selected—will leave more offspring than those with less favored variations. The notion that the best adapted individuals are more likely to survive is captured in the pithy phrase, "survival of the fittest."

At the time when Alfred Wallace and Charles Darwin published the idea of evolution by natural selection there was no explanation for how characteristics are passed from one generation to the next (i.e., how inheritance worked), nor how variations arose between individuals. This would have to wait for the molecular biology revolution that was heralded by the discovery of the structure of DNA in 1953.

There are a number of noteworthy points about evolution.

6.1.5.1 Evolution Has No Goal

There is no direction or aim to evolution. It proceeds because randomly thrown up variants are selected by the caprice of nature. For some biologists this means that if the clock were turned back 3.5 Gyr (say) and life were allowed to evolve again, the chance of humans, or any other species, appearing again is virtually zero. However, others argue that similarities would emerge by convergent evolution. There are only so many solutions to the problems posed by nature and this means that organisms that are actually very different may be selected to resemble each other superficially. Thus, sharks and whales are similar in shape because a streamlined body with fins and a tail are successful adaptations to swimming. However, sharks are fishes, and whales are mammals and they have no recent common ancestor. In this view, the replay of life on Earth would throw up very similar organisms.

6.1.5.2 Complexity of Living Systems Tends to Increase with Time

There is, however, a tendency discernible only on geological time scales, for life forms to become more complex with time. Why is this? Random variations should of course throw up simpler forms just as often as more complex ones. Actually this is occasionally seen. Highly specialized parasitic organisms are often simpler than their predecessors. But organisms that evolve to be simpler ones eventually hit a minimum complexity floor below which life cannot operate.

6.1.5.3 Is Darwinian Evolution Universal to Life Everywhere?

Is evolution a necessary ingredient of a general theory of life? Almost certainly. We might imagine that life originated from the abiotic chemical system that happened to produce more copies of itself than other systems around at the time. The temperature, pressure, concentrations of molecules, etc. favored one system over others, perhaps by incorporating a shared reactant faster and so starving its competitors. Note that this is akin to natural selection. Darwinian evolution is a process, it is not dependent on the specifics of the substrate on which it operates; DNA replication is not a necessary precondition.

6.1.6 How Useful Are These Criteria for Detecting Life?

Jerry Joyce of the Salk Institute once defined life as a self-sustaining chemical system capable of Darwinian evolution. The problem with this is that it is not possible to test it by remote sensing. Sustainability and evolution are not susceptible to spectroscopic analysis!

However, chemical disequilibrium generated by the property that life is a far-from-equilibrium system can be tested, at least in principle. Methane and oxygen are out of equilibrium on Earth by 20 orders of magnitude because of the activities of methane-producing microorganisms and photosynthesis, respectively. Both CH_4 and O_2 can be detected in Earth's atmosphere.

However, the disequilibrium argument has to be used with care. Planetary atmospheres can display disequilibrium as a result of straightforward physical processes such as photodissociation (e.g., splitting of water by UV). Moreover, early Earth may not have shown any evidence for disequilibrium since at that time methane was produced in the absence of oxygen-generating photosynthesis. Indeed, if we consider the entire Earth system at this time rather than just the atmosphere it may have tended toward equilibrium as H_2 and CO_2 reacted to produce CH_4.

6.2 Are There Universal Chemical Requirements for All Life?

If there were an astrobiology constitution it would claim two truths to be self-evident; that all life is carbon-based and that liquid water is the universal solvent for the chemistry of life. It is worth briefly examining why this is so, because it seems on the face of it to be an overly proscriptive and unimaginative view. In science, it is wise never to say no, and it might indeed be that life elsewhere is not carbon based or water based, especially if it were adapted to environments very different to those on Earth. That said, no coherent and comprehensive model biochemistries based on other elements and solvents have been proposed. Here, we stick with the orthodox perspective, and leave alien biochemistries for science fiction writers to have fun with.

6.2.1 No Element Is More Versatile in Its Chemistry Than Carbon

Life on Earth is based on the chemistry of carbon (organic chemistry), centered around just six major elements, H, C, N, O, P and S, and a few crucial metals such as Fe and Mg that occur in modest amounts in organisms. What they share in common is that they are all among the most abundant elements in the universe. Carbon is the fourth most abundant element on Earth. It has two properties that, taken together, mean no other element comes close to carbon in the complexity of its chemistry, and that is the major reason for assuming that life elsewhere is carbon-based.

6.2.1.1 Carbon Is Tetravalent

The octet rule in chemistry says that most atoms belonging to the s- or p-blocks in the Periodic Table combine so as to have eight electrons in their valence shell, acquiring an s^2p^6 electronic configuration like that of the noble gases (apart from helium). This allows the atoms to adopt their lowest energy, most stable state. Atoms with incomplete valence shells are more reactive than those with complete shells. Atoms can satisfy the octet rule by forming ionic bonds or covalent bonds with other atoms. Since carbon has a $1s^2 2s^2 2p^2$ configuration, it needs to acquire a share in four valence shell electrons to fulfill the octet rule. It can do this by forming four single bonds (e.g., methane, CH_4), two single bonds and one double bond (e.g., formaldehyde, $O = CH_2$), one single bond and one triple bond (e.g., hydrogen cyanide, $H–C \equiv N$) or two double bonds (e.g., carbon dioxide, $O = C = O$). These bonding options result in just three molecular geometries around a carbon atom: tetrahedral, trigonal, planar, and linear (Figure 6.5). Hydrogen, helium, and lithium obey a duet rule in which their favored electronic configuration is for two electrons in their valence shell ($1s^2$). We see the octet rule for C and the duet rule for H are both satisfied in methane and carbon dioxide. That carbon is tetravalent allows it to form large polymers, a basis for complex chemistry.

6.2.1.2 Carbon Has a Moderate Electronegativity

Carbon has an electronegativity of 2.6 eV which means that it forms bonds of predominantly covalent character with a large number of elements more or less electronegative than itself. Unwilling to give up its electrons entirely or to cling too avidly to electrons from other atoms, carbon delivers an ideal compromise of molecular stability together with the possibility of a rich variety of chemical reactions. In particular, the moderate electronegativity of carbon makes it readily able to act both as an electron donor or an electron acceptor in redox reactions that are central to bioenergetics, as we have seen.

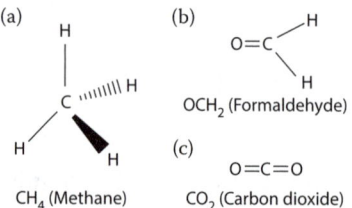

(a) CH_4 (Methane)

(b) OCH_2 (Formaldehyde)

(c) CO_2 (Carbon dioxide)

FIGURE 6.5   Bonding for a single carbon atom. (a) Tetrahedral, (b) trigonal planar, and (c) linear.

6.2.1.3 Can Silicon Substitute for Carbon in Biochemistry?

Is it possible that biochemistry could be based on some element other than carbon? Silicon is the nearest contender. It is extremely abundant. Like carbon it is a Group IV element and so tetravalent. So far, so good.

But the silicon atom is larger than the carbon atom, so the Si–Si and Si–H bonds are weaker than the C–C and C–H bonds; indeed Si–Si bonds are easily hydrolyzed by water. In addition, Si does not form double or triple bonds as easily as carbon, and when it does they are less stable than their carbon equivalents. The outcome is that silanes are less stable and less diverse than their hydrocarbon cousins. Silanes could form stable molecules only at low temperature in anoxic and anhydrous environments.

Silicon has an appreciably lower electronegativity (1.9 eV), so it tends to form positive ions that bond with negative oxygen ions. This paves the way for the formation of silicones and silicates. Silicones are molecular chimeras containing repeating –O–Si–O– units with carbon substituents (usually alkanes) satisfying the tetravalent Si atoms. Silicones can form polymers thousands of units long but they have high melting points and are relatively chemically inert because the Si–O bond, with its significant ionic character, is very strong. Silicones are insoluble in water and organic solvents.

Although silicates (based on the tetrameric SiO_4) are known from IR spectroscopy to be a component of interstellar dust grains, and can form chains, sheets, and three-dimensional arrays, they have high melting points and make up the chemistry of rocks, not biology.

CO, CO_2, and a huge variety of organic molecules have been identified in the interstellar medium (ISM), as we have seen. By contrast, SiO and SO_2 are in low abundance in the ISM and only eight other silicon molecules have been identified, none with more than five atoms. It is clear then that the rich cosmochemistry of carbon is not replicated for silicon.

In summary, silicon is a very unlikely substitute for carbon in biological chemistry.

6.2.2 Water As a Universal Solvent

Although many terrestrial organisms can survive when frozen, growth and reproduction requires that the water inside cells is liquid. The aqueous fluid inside cells has lots of molecules dissolved in it, so its freezing point is lower than that of pure water. This means that seawater temperature down to −15°C (e.g., Antarctic Ocean) is no barrier to many bacteria, with even complex life such as plankton living the good life. The lowest temperature that can sustain active microbial communities seems to be −18°C, though micro-organisms and many cells survive freezing down to −196°C, provided that the freezing is rapid enough that the ice crystals that form are too small to cause damage.

Water can readily donate or accept protons, so is able to act as both an acid or as a base. Liquid water acts as a superb solvent for biochemical reactions because it is a polar molecule (Figure 6.6) and can form hydrogen bonds with other polar molecules and with itself. Hence polar organic molecules such as alcohols, carbohydrates, amino acids and nucleic acids readily dissolve in water. By contrast, carbon forms non-polar molecules when it binds to itself or hydrogen. This means hydrocarbons are not soluble in water, indeed they form hydrophobic ("water-fearing") immiscible

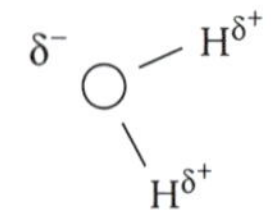

$$\delta^- \;\; O \underset{\diagdown}{\overset{\diagup}{}} \; H^{\delta^+}$$
$$H^{\delta^+}$$

FIGURE 6.6   Polar structure of water. The higher electronegativity of O than H means that the O–H bonds in water have some ionic character, with the oxygen getting more than its fair share of the bonding electrons.

liquids. Perhaps the most remarkable are phospholipid molecules that have distinct polar and non-polar ends. These orient in aqueous environments to form minimum energy conformations—vesicles or bilayers—which keep their hydrocarbon tails away from the water. It is this behavior that allows lipid membranes of cells to assemble spontaneously.

Water has other advantages for life over and above its solvent role. It has a very large heat capacity, so it is able to buffer large swings in temperature. Extraordinarily, ice has a lower density than water, so as a body of water freezes it acquires a lid of ice that thermally insulates the liquid water below.

But water also has disadvantages. Ice has a higher albedo than water, so as water freezes it reflects more sunlight back into space and this can amplify cooling via positive feedback. This has undoubtedly contributed to major near-global glaciations that have occurred in the past. Moreover, as we shall see, a key step toward the origin of life, the abiotic synthesis of macromolecules, involves condensation reactions. To get these to happen in aqueous environments may require special conditions. What is more, the macromolecules once formed will spontaneously hydrolyze in water, albeit slowly in the absence of hydrolytic enzymes.

6.2.2.1 *Alternative Solvents Have Been Postulated*

Substances other than water could act as biochemical solvents though none has anything like the abundance of water in space. The Infrared Space Observatory, an ESA mission that operated from 1995 to 1998, discovered vast quantities of water in molecular clouds of star-forming regions such as the Orion Nebula (M42). Indeed, it is estimated that water is the most common molecule in the universe, after H_2.

The most likely alternative to water is ammonia (NH_3). Its similarity to water in some respects means that a carbon-based biochemistry might be possible with NH_3 as a solvent, although compounds containing $>C = N–H$ will likely be exploited rather than those containing the carbonyl group $>C = O$. Although it forms weaker hydrogen bonds than water, it will dissolve most organic compounds and elemental metals almost as well as water. However, a major problem is that lipid membranes cannot form in ammonia. Another difficulty is that at standard atmospheric pressure, NH_3 is liquid between extremely low temperatures (195–240 K), at which chemical reaction rates are very low. This could be offset by ramping up the pressure. At 6 MPa ammonia is liquid over a very wide range of temperatures, 195–371 K. Alternatively, since ammonia greatly depresses the freezing point of water, ammonia–water mixtures could act as solvents for life in environments colder than traditional habitable zones.

Other putative solvents have been advocated, including methane (CH_4), methanol (CH_3OH), hydrogen fluoride (HF), hydrogen chloride (HCl), hydrogen sulfide (H_2S), and sulfuric acid (H_2SO_4), but all raise serious difficulties including the need for alternative biochemistries, no models for which have been developed.

It is difficult to envisage a solid state biochemistry because the crystal structure of solids would seriously curtail the ways in which macromolecules could interact, and any reactions that did happen would go so slowly that the timescale for any such living systems would be many orders of magnitude longer than for life as we know it.

7

Terrestrial Biochemistry

7.1 Building Blocks for Life

Comets and asteroids probably delivered a large variety of small molecules to early Earth that could react and synthesize key biochemicals. In addition, there are environments on Earth where a wide range of organic molecules will have been synthesized *de novo* as we shall see in Chapter 8. In this section, we look briefly at the broad classes of molecules out of which organisms are constructed: the building blocks for life.

> **EXERCISE 7.1**
>
> List (a) five key small inorganic molecules, (b) five key classes of organic molecules that may have been delivered to Earth from space.

7.1.1 Polymeric Macromolecules

Proteins, nucleic acids, and polysaccharides are high relative molecular mass polymers constructed by linking numerous small chemically similar monomers: amino acids, nucleotides, and sugars. In each case, the covalent bonds between the monomers are formed by condensation (dehydration) reactions and broken by hydrolysis reactions (Figure 7.1).

In all cases, the condensation (forward) reaction is endergonic, and the hydrolysis (backward) reaction is exergonic. Hence, free energy has to be supplied by the cell to build macromolecules.

7.1.1.1 Polypeptides Are Polymers of Amino Acids

Proteins are large molecules that consist of one or more polypeptides. If there are several polypeptides they assemble into a protein by weak bonding.

There are numerous amino acids in nature, but only 20 are incorporated into proteins. They are so called because they have an amino group and a carboxylate group on the same (α) carbon atom. At the pH inside cells the amino and carboxylate groups are charged (Figure 7.2). Because carbon is tetravalent, the α carbon atom can accommodate two further substituents, so all but one of the amino acids in the proteins are chiral and are the L-enantiomer in virtually every case (Box 7.1). When amino acids are made abiotically in nature, we do not expect the chemistry to favor the formation of one enantiomer over the other. Therefore, we expect amino acids in comets and

$$R - H + R' - OH \rightleftharpoons R - R' + H_2O$$

FIGURE 7.1 Condensation and hydrolysis reactions dominate macromolecular chemistry. R and R' are chemical groups.

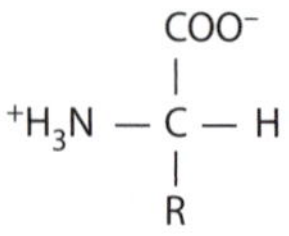

$$
\begin{array}{c}
COO^- \\
| \\
{}^+H_3N - C - H \\
| \\
R
\end{array}
$$

FIGURE 7.2 Structure of a general amino acid.

meteorites to be racemic (50:50) mixtures of the two versions. But this is not the case as we shall see when we investigate the origin of life (Chapter 8).

Each of the amino acids is distinguished by the nature of its R substituent (Figure 7.2). These can be charged (+ve or −ve), uncharged polar or non-polar, aliphatic (linear carbon chains) or aromatic (closed carbon rings), all of which confer distinctive chemistry.

When amino acids polymerize, the carboxylate group of one amino acid undergoes a condensation reaction with the amino group of another to form a peptide bond (Figure 7.4). Once the polymer has assembled, weak bonding between the amino acids, and in some cases covalent disulfide (-S-S-) bonding, causes the linear molecule to fold

---

BOX 7.1   CHIRALITY

If the four substituents on a carbon atom are dissimilar the carbon atom is said to be asymmetric and the molecule can occur in two forms that are mirror images of one another, just as the left hand is a mirror image of the right one (Figure 7.3). In terms of composition they are identical but they are two different versions of the same molecule (called enantiomers or stereoisomers) because no rotation in 3-space can superimpose them. This property is known as chirality ("handedness") and occurs in all organic molecules with asymmetric carbon atoms. Crucially, biological systems can discriminate between the two enantiomers of a chiral molecule

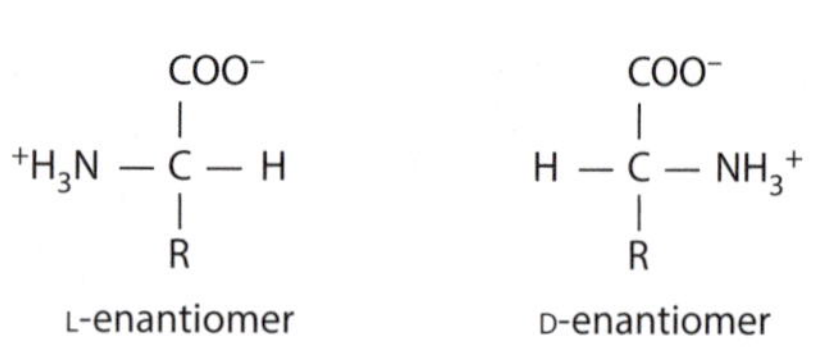

$$
\begin{array}{cc}
\begin{array}{c}
COO^- \\
| \\
{}^+H_3N - C - H \\
| \\
R
\end{array}
&
\begin{array}{c}
COO^- \\
| \\
H - C - NH_3{}^+ \\
| \\
R
\end{array} \\
\text{L-enantiomer} & \text{D-enantiomer}
\end{array}
$$

FIGURE 7.3 Chirality. The two forms of the amino acid above are mirror images and cannot be superimposed other than by reflection. These structures are not actually planar: by convention, the vertical bonds project behind the plane of the page and the horizontal bonds project in front.

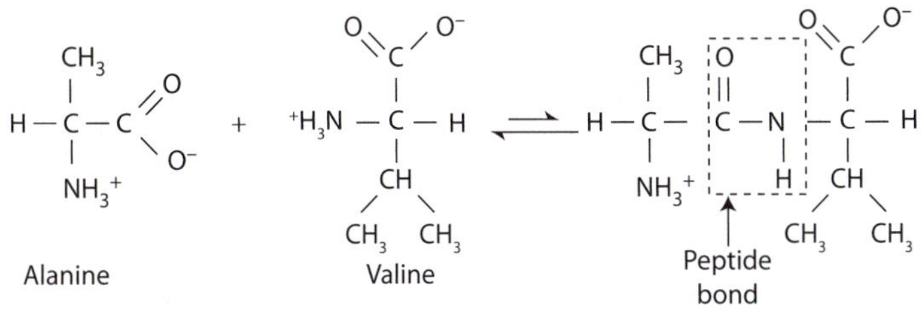

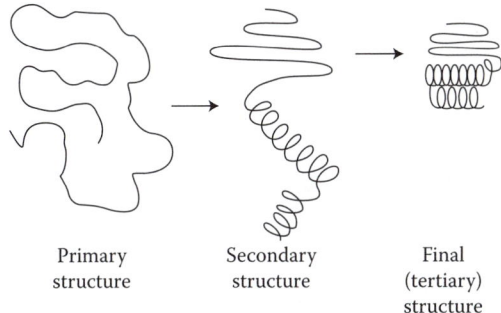

FIGURE 7.4 Polymerization of two amino acids (alanine and valine) forms a peptide bond.

Primary structure → Secondary structure → Final (tertiary) structure

FIGURE 7.5 Folding of the linear amino acid sequence due to covalent and weak bonding results in a 3-D structure.

into a specific shape that is ultimately determined by the primary amino acid sequence (Figure 7.5). The folded protein has a 3-D structure that can dynamically interact with other molecules in very specific ways. This is the secret to how proteins work. But note how the function of the protein is an emergent property of its primary amino acid sequence.

The number of different polypeptides that could be made in theory from 20 amino acids in random order is unfathomable. There are $20^2 = 400$ possible dipeptides and $20^3 = 8000$ tripeptides, but a typical small protein may have 100 amino acid residues—you do the math!

EXERCISE 7.2

How many different proteins could be made from 100 amino acid residues, given 20 amino acids to choose from and no constraint on the order in which they can assemble?

Proteins are broken down to their constituent amino acids by hydrolysis of the peptide bonds, essentially the addition of water.

7.1.1.2 Nucleic Acids Are Informational Macromolecules

Nucleic acids (deoxyribonucleic acid, DNA and ribonucleic acid, RNA) store and process information in cells. The nucleotide monomer consists of a pentose (five carbon atoms) sugar, a phosphate moiety, and a nitrogen-containing base, either a pyrimidine or a purine (Figure 7.6a). The nucleotides are covalently linked via phosphodiester bonds in which the phosphate group attached to C5 on the pentose of one nucleotide is linked to the hydroxyl group on C3 of the pentose on the next nucleotide (Figure 7.6b). This gives nucleic acids a polarity: at the 5′ end is a free phosphate moiety and at the 3′ end is a hydroxyl group. Nucleic acids are always assembled by adding new nucleotides to the 3′ end; that is, they grow in the 5′ → 3′ direction. The precursor molecules for building nucleic acids are not actually nucleotides but nucleoside 5′ triphosphates. When these react with the 3′ hydroxyl groups of a growing nucleic acid, a pyrophosphate is cleaved, providing free energy to drive the formation of the phosphodiester bond, an endergonic reaction.

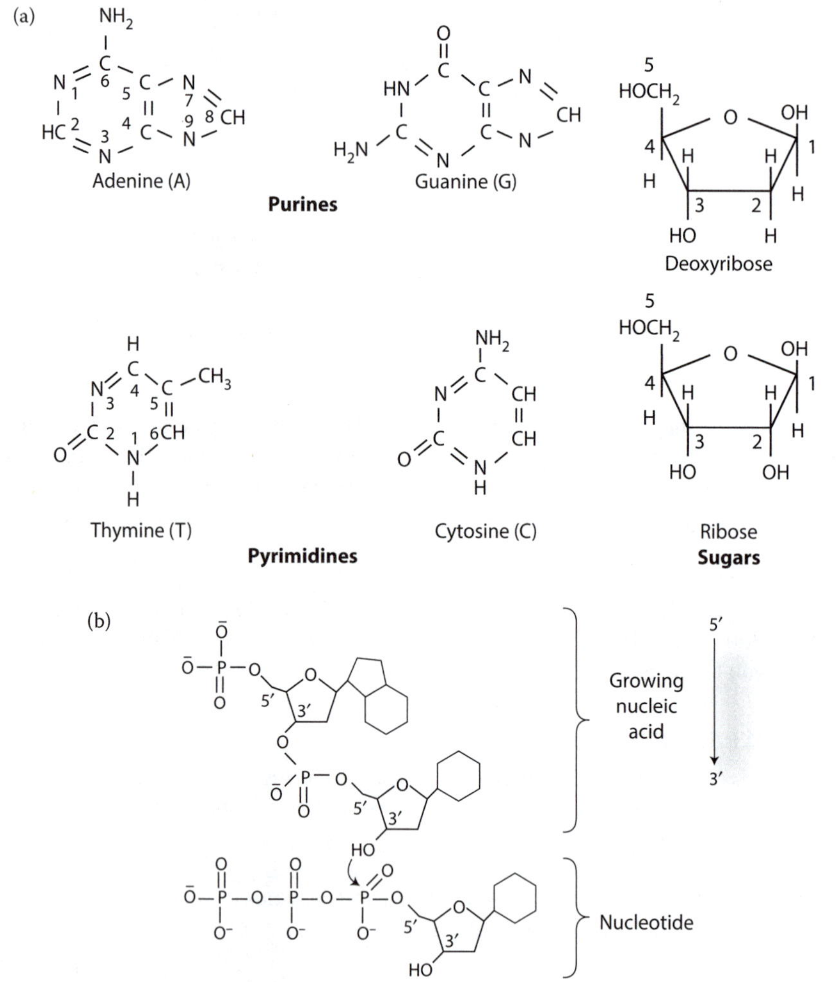

FIGURE 7.6 Nucleic acids. (a) Building blocks for nucleotides. (b) Nucleic acids grow by reacting nucleoside 5′ triphosphate with the 3′ end of a nucleic acid.

Deoxyribonucleic acid, so called because it contains the pentose sugar deoxyribose, is a very stable molecule and is a long-term information store. DNA molecules are double-stranded and replicate. Ribonucleic acid contains ribose, is less stable than DNA, and has several roles in organisms. It is usually single-stranded. Nucleotides contain several chiral carbons (in the sugars and the bases), so nucleic acids themselves are asymmetrical; the strands of the DNA double helix are right-handed.

7.1.1.3 Polysaccharides Are Polymers of Sugars

Polysaccharides such as cellulose, starch, and glycogen are long chains of sugar monomers. Sugar monomers have the general formula $(CH_2O)_n$, and so they and the polysaccharides are called carbohydrates. There are numerous sugars, which typically have 3, 5, or 6 carbon atoms, and in many cases are interconverted in cells. Sugars polymerize by condensation reactions forming glycosidic bonds. Generally, sugars and polysaccharides act as energy substrates, but they also decorate other macromolecules (e.g., proteins) to alter their functionality. Polysaccharides can also act as structural components.

Sugars are chiral molecules, and intriguingly, only the D enantiomers are found in biology (Figure 7.7).

7.1.1.4 Lipids

Lipids are a heterogeneous collection of water-insoluble molecules. They form the major structural component of cell membranes and act as energy substrates, among other roles.

Triglyceride molecules (fats and oils) consist of three fatty acids combined with glycerol. Fatty acids have the general structure of a carboxylate group (COOH) attached to a hydrocarbon residue (R) (Figure 7.8). Each fatty acid reacts by a condensation reaction between its carboxylate group and a hydroxyl group of glycerol to make an ester bond (Figure 7.9).

Fatty acids in which all the bonds between the R carbons are single are said to be saturated, while a double bond between two carbon atoms makes for an unsaturated

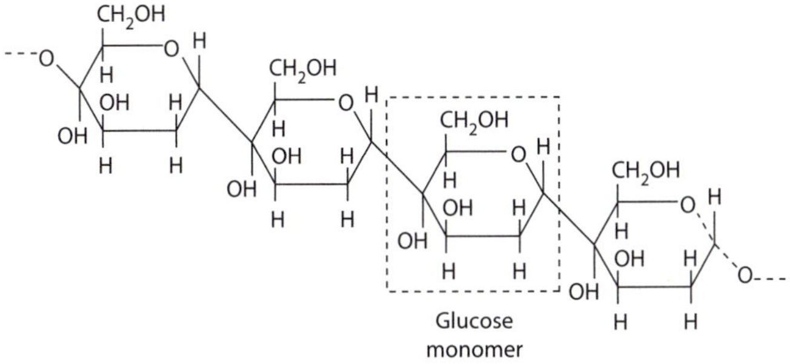

FIGURE 7.7 Synthesis of a polysaccharide (amylose a type of starch) from sugar monomers (glucose).

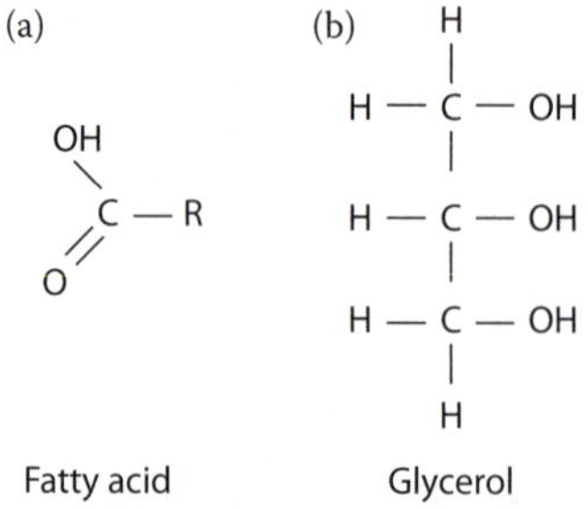

FIGURE 7.8 Triglyceride components. (a) Fatty acid. (b) Glycerol.

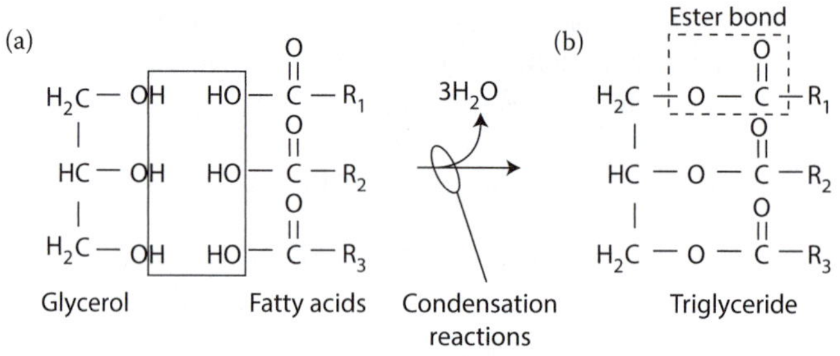

FIGURE 7.9 (a) Condensation reaction between glycerol and fatty acid. (b) General structure of a triglyceride.

fatty acid (Figure 7.10). Saturated fatty acids are more rigid than unsaturated ones and have higher melting points.

If one fatty acid in a triglyceride is replaced with a phosphate-containing group, the resulting molecule is a phospholipid (Figure 7.12a). The charged polar phosphate group forms weak bonds with water molecules, making this part of the phospholipid molecule hydrophilic ("water-loving"), whereas the fatty acids are hydrophobic ("water-fearing"). This polarity means that phospholipids are able to assemble into huge structures held together by weak bonds: the basis for forming membranes.

An entire domain of organism (Archaea, see below) builds their membranes differently by using isoprenoid (terpenoid) residues in place of fatty acids. Isoprenoids

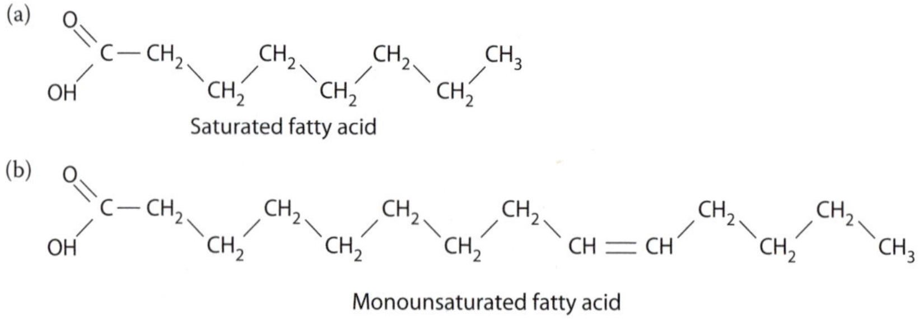

FIGURE 7.10 (a) Saturated fatty acid (caprylic acid, c8:0); (b) monounsaturated fatty acid (myristolec iacid, C14:1ω9).

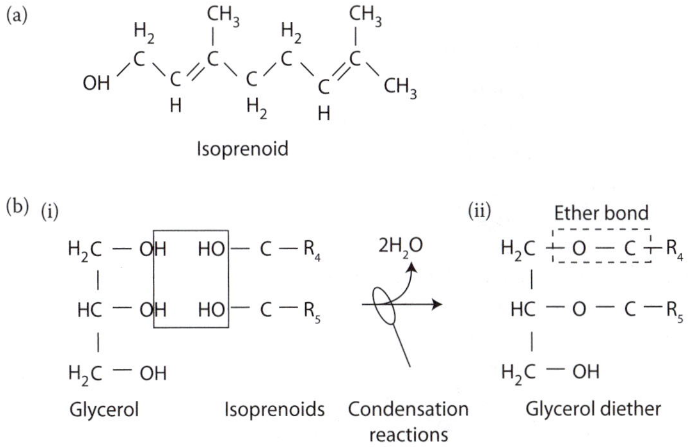

(a)

Isoprenoid

(b) (i)

Glycerol   Isoprenoids   Condensation   Glycerol diether
                                reactions

FIGURE 7.11 (a) Condensation reaction between glycerol and isoprenoids. (b) General structure of a glycerol diether.

are based on repeated 5-carbon isoprene units and are combined with glycerol by condensation to form ether bonds (Figure 7.11).

The simplest resulting lipid is a glycerol diether that has two isoprenoid residues. Isoprenoid phospholipids are synthesized by condensation of a phosphate-containing moiety with the third glycerol hydroxyl group (Figure 7.12b). The membranes produced from isoprenoid-containing lipids are broadly similar in overall architecture if not molecular structure to those produced from fatty acid-containing lipids.

Isoprenoids are also used by all organisms to produce a rich variety of molecules for other functions.

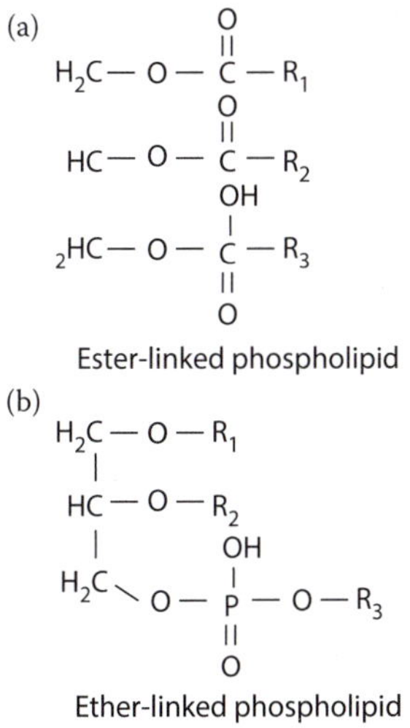

(a)

Ester-linked phospholipid

(b)

Ether-linked phospholipid

FIGURE 7.12 Phospholipid of (a) ester-linked and (b) ether-linked lipids. R_1 and R_2 are fatty acids, R_3's are a range of other chemical groups, often hydrophobic.

7.2 All Life on Earth Consists of Cells

Cells, the fundamental unit of organization in living organisms, are small (typically between 10^{-6} and 10^{-4} m). Most organisms on Earth are just single-cell organisms. More complex, multicellular organisms consist of many cells each specialized for a particular function.

All cells are bounded by a thin membrane that acts as a gatekeeper, maintaining their integrity. Chemically, the cell membrane is composed of lipids in which are embedded a variety of protein molecules. Some of these determine which ions and molecules are allowed to enter or leave the cell; others as receptors allow the cell to sense a number of agents in its surroundings. These include a variety of simple substances, mechanical forces, or electromagnetic radiation. Internal transduction systems convert receptor activation by these signals to biochemical changes that alter the behavior of the cell. For example, bacteria can sense amino acids (e.g., glutamate) in their environment and move in the direction of increasing concentration, maximizing their opportunity to take it up. The amino acid is transported into the cell where it is used to build proteins or oxidized in respiration to provide energy.

Metabolic biochemistry happens inside the cell. Metabolism is the sum total of processes that build large molecules from smaller ones (anabolism) and of those that break larger molecules into small ones (catabolism). Roughly speaking, anabolism is the biochemistry of growth, replication, and repair and usually requires chemical energy, while catabolism is how organisms generate energy for anabolic processes and all their housekeeping needs. Since cell membranes import molecules that are then incorporated into metabolism, and export metabolic waste products (often CO_2 and H_2O) a basic tenet of the cell theory is that every cell is metabolically autonomous. Each one has all the "kit" needed to regulate and utilize the flow of chemical energy through it. For any given type of single-celled organism, each cell/organism has the same chemistry. However, the lifestyles of different unicellular organisms vary widely and hence can have very different biochemistries.

7.2.1 Information Flow in Cells

To first order, the functions of a cell are determined by its proteins. Which proteins a cell makes, and when, is controlled by the flow of information through nucleic acids. At the core of this information system lies the DNA, the main information store within cells. DNA lies at the heart of genetics because when an organism reproduces, a copy of this DNA ends up in the offspring. The sum total of DNA that can be passed to the next generation is termed the genome.

DNA encodes the sequence of amino acids in proteins. The code is the order of the bases along one strand of the DNA molecule. The code is first copied into a transient messenger RNA molecule (mRNA), a process termed transcription. The mRNA is then read out by molecular machines called ribosomes that synthesize the protein encoded in the mRNA. This is called translation. The details are complicated—ribosomes are multi-molecular complexes containing their own type of RNA and many

$$\text{DNA} \xrightarrow[\text{replication}]{\text{transcription}} \text{mRNA} \xrightarrow{\text{translation}} \text{Protein}$$

FIGURE 7.13 The central dogma of information flow in cells.

proteins—but the operating principle is straightforward. Amino acids are covalently bonded to small transfer RNA (tRNA) molecules, with each tRNA responsible for a specific amino acid. The tRNAs recognize the code on the mRNA. This allows incoming tRNAs to be assembled in the correct sequence. Peptide bonds are then forged between adjacent amino acids and on a timescale of seconds a nascent protein is released from the ribosome. The "empty" tRNA molecules are liberated to capture new amino acids.

Ribosomal protein synthesis typically occurs at rates of about 20 amino acids per second for prokaryotes and about half that rate for eukaryotes (see Section 7.2.2). A typical cell may have 20,000 ribosomes, permitting synthesis of order 1000 proteins each second. Proteins are also degraded by regulated mechanisms at comparable rates. The half-lives of *particular* proteins range from minutes to years.

We can now summarize the flow of information through a cell (sometimes described as the central dogma (Figure 7.13)).

DNA also encodes the sequences of a variety of RNA molecules such as ribosomal RNA (rRNA), transfer RNA (tRNA), and others involved in regulating transcription and translation.

In summary, genetic information is the wetware containing instructions for how to make every protein and RNA molecule the cell requires. When a cell divides, a copy of these instructions ends up in both daughter cells.

The section of DNA that codes for a particular polypeptide is termed a gene. Typically, an organism has tens of thousands of genes. Organisms differ because they have different sets of genes; that is, every species of organism has a unique genome. We explore details of the genetic code in Section 7.3.

7.2.1.1 Genes Have Several Components

1. *Coding sequences.* As the name implies, these encode the amino acid sequence of the polypeptide. Some genes can be decoded in more than one way and so are able to instruct the making of more than one polypeptide.
2. *Regulatory sequences.* These constitute the instruction manual for the gene, regulating if and when its product is to be expressed in the cell. It does this in response to internal signals and external signals. Internal signals include proteins made by other genes called transcription factors. Hence, there is a rich network of interactions that occurs *between* genes, even before factoring in the influence on the environment of the genome.

The external signals that link the surroundings of the cell to the genome are a plethora of molecules that interact with cell receptors, thereby switching transduction systems on and off. This changes the biochemistry and behavior of the cell. For example, the sugar lactose is an external signal for many bacteria. Lactose switches on the genes that allow a bacterium to use it as an energy source. Regulatory sequences are not transcribed, they are noncoding regions.

7.2.2 All Life on Earth Has One of Two Basic Cell Architectures

There are two basic types of cells, those with nuclei (eukaryotic cells, or eukaryotes) and those without nuclei (prokaryotic cells, or prokaryotes). Prokaryotic cells are generally small (~1 µm and are rarely >10 µm), while eukaryotic cells are typically 10-fold larger (Figure 7.14).

7.2.2.1 Prokaryotes Have Simpler Genomes than Eukaryotes

Prokaryote genomes come in a variety of architectures, but commonly they are circular double-stranded DNA molecules typically 4×10^6 base pairs (4 Mbp) long. Proteins help them to supercoil into a compact chromosome.

In eukaryotes, the DNA comes as a number of double-stranded linear molecules. Each is tightly packed into a chromosome with the aid of proteins, and all eukaryotes have a minimum of two chromosomes. Eukaryote genomes are about 1000-fold larger than those of prokaryotes.

All prokaryotes and some eukaryotes have plasmids: small double-stranded molecules. These replicate independently of the chromosomal DNA, can be exchanged between cells, and are a vehicle for horizontal gene transfer (see below).

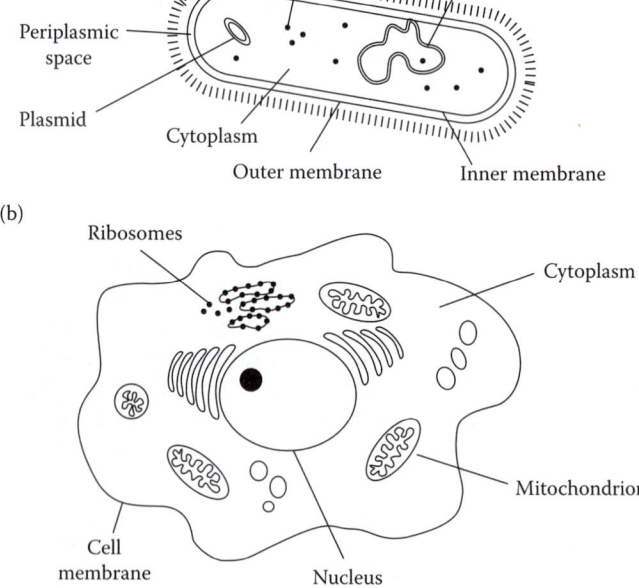

FIGURE 7.14 Diagram of (a) a prokaryote and (b) a eukaryote cell: some eukaryotes have a cell wall.

A key diagnostic feature is that eukaryote chromosomes are enclosed in a double-layered membrane to form a nucleus, a structure absent in all prokaryotes.

7.2.2.2 Eukaryotes Have Mitochondria and Chloroplasts

Almost all eukaryote cells contain membrane-bound organelles termed mitochondria, responsible for generating chemical energy by respiration (Figure 7.15a). Photosynthesizing eukaryotes (e.g., green plants) also have chloroplasts, organelles that harvest the energy of sunlight to generate chemical energy (Figure 7.15b). No prokaryotes have these structures. Both mitochondria and chloroplasts are bounded by two membranes, outer and inner, between which is a space. The inner membrane of mitochondria is thrown into numerous folds, increasing the surface area available to harbor the proteins involved in respiration. Chloroplasts are filled with stacks of membrane discs that contain the photosynthetic machinery.

How eukaryotes came to acquire mitochondria and chloroplasts is truly remarkable.

7.2.2.3 All Cells Have Internal Scaffolding

Eukaryote cells have an extensive cytoskeleton. This is made up of three types of long protein filaments, two of which can assemble and disassemble rapidly and dynamically. The cytoskeleton acts as an internal scaffold affording rigidity, transporting material within the cell, and executing cell division. The cytoskeleton controls how cells change shape, and is needed in particular for phagocytosis (literally, "cell eating"), in which a cell engulfs a small volume of its surroundings forming a vesicle inside the cell. Although prokaryotes have a cytoskeleton, they are not versatile shape changers, and none is able to phagocytose. Prokaryotes rely on transporters in their cell membrane to take up nutrients from their environment. Most prokaryotes are surrounded by a cell wall that confers rigidity.

7.2.2.4 Almost All Organisms on Earth Are Prokaryotes

The number density of prokaryotes ($>10^{30}$) on Earth at any instant is much greater than that of eukaryotes; they have a much wider diversity of metabolic lifestyles, use

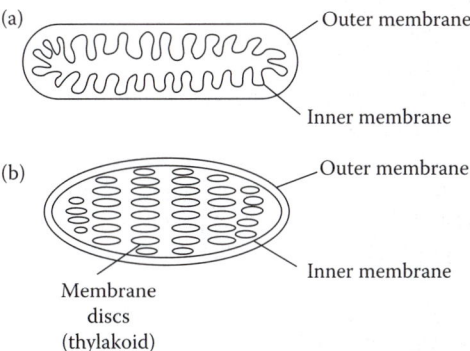

FIGURE 7.15 The structures of (a) mitochondria and (b) chloroplasts share similarities.

a far bigger range of energy sources, and hence populate a greater range of habitats, including extreme environments. The first life on Earth was prokaryotic.

EXERCISE 7.3

Living prokaryotes are estimated to contain 3×10^{40} carbon (C) atoms. What is the prokaryote biomass in kg C? (The atomic mass unit, $u = 1.67 \times 10^{-27}$ kg.)

7.2.3 Gene Transfer Can Occur Vertically or Horizontally

A basic premise of the cell theory is that all cells arise from preexisting cells by division. The division of a prokaryotic cell into two is termed asexual reproduction.

In eukaryotes there are two types of cell divisions: mitosis, which is similar in outcome, if not in detail, to cell division in prokaryotes; namely, two identical cells. The second type of cell division in eukaryotes underlies sexual reproduction and is called meiosis. This produces germ cells (eggs or sperm in animals) each with a half-sized genome. These are brought together and fuse to form a single cell that becomes a new individual (with a complete genome). However, during meiosis, limited shuffling of DNA occurs so that each germ cell is subtly different, making each individual genetically unique. This is thought to confer advantages over asexual reproduction in the evolutionary stakes. Of course, for sexual reproduction it takes two to tango (usually), but this still qualifies as autonomous replication since both contributors are the same type of organisms.

Both asexual and sexual reproduction transfer genetic material from parent(s) to progeny. This is termed vertical gene transfer and has been the focus of most genetics. However, lateral or horizontal gene transfer in which genetic material is exchanged between organisms that do not have a parent–offspring relation is common in prokaryotes and is thought to have shaped much of early evolution.

7.2.4 All Life on Earth Falls into Three Domains

In 1990, Carl Woese argued that there are three fundamental types of life. Each occupies a domain, the largest group (clade) used to classify terrestrial organisms: see Figure 7.16. The Bacteria and Archaea are both prokaryotes. The basis for the tripartite division is the sequences of RNA in one of the subunits that make up ribosomes, the ubiquitous protein "factories" of all cells. This subunit (16S rRNA) changes its sequence of bases (i.e., mutates) over geological time periods as a result of copying errors, or the action of toxins or radiation. The more closely related two organisms are, the more similar their rRNA sequences will be. Thus, it is possible to order rRNA sequences from a wide variety of organisms into a tree that preserves nearest-neighbor relationships. The best fit appeared to be a tree with three branches. Some characteristics that allow the three domains to be distinguished are summarized in Table 7.1.

The implication of Figure 7.16(a) is that a common ancestor called the Last Universal Common Ancestor (LUCA) gave rise to bacteria and to another branch that spawned both Archaea and Eukarya. While Archaea and Eukarya do share features in common (see Table 7.1), this arrangement makes it hard to understand

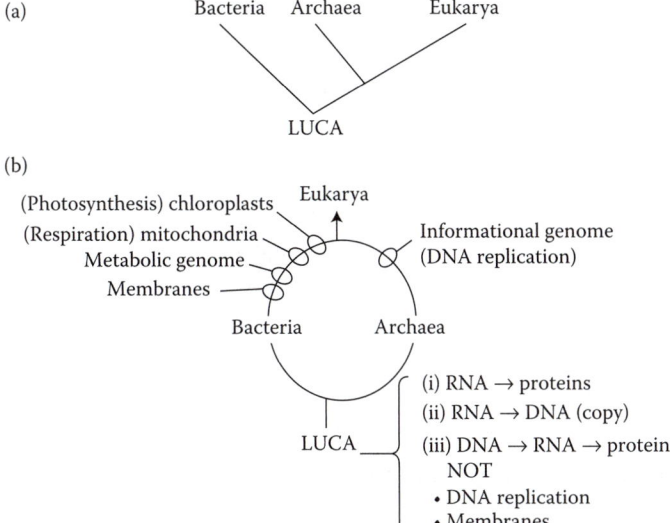

FIGURE 7.16 (a) The three domains of life. This pattern of divergence close to the root of the tree is no longer regarded as the best model. (b) A contemporary "ring of life" model suggests that early life exchanged genetic material as promiscuously as do modern prokaryotes.

TABLE 7.1

Characteristics of Domains

| | Domain | | |
Characteristic	Bacteria	Archaea	Eukarya
Membrane-enclosed nucleus	No	No	Yes
Membrane-enclosed organelles	Yes (photosynthetic membranes)	No	Yes
Introns (noncoding regions within genes)	No	No	Yes
Initiator amino acid	Formylmethionine	Methionine	Methionine
Number of RNA polymerases[a]	One	Several	Several
Ribosomes	Yes	Yes	Yes
Plasmids	Yes	Yes	Unusual
Sexual reproduction	No	No	Usually
Peptidoglycan in cell wall	Yes	No	No
Membrane lipids	Ester-linked, unbranched	Ether-linked, branched	Ester-linked, unbranched
Cytoskeleton	Yes	Yes	Yes
Phagocytosis	No	No	Yes
Some use photosynthesis to fix CO_2	Yes	No	Yes
Some fix nitrogen	Yes	Yes	No
Some are methanogens	No	Yes	No

[a] Although both Archaea and Bacteria have DNA and RNA, the mechanism of DNA replication in the two domains is very different.

how both Bacteria and Eukarya have membranes constructed by ester bonding of fatty acids, while Archaea have membranes formed by ether bonding of isoprenoids.

These difficulties make the Woese tree of life problematic. More recent work—adding data from protein sequences, and using more sophisticated mathematical models to assess the probabilities of alternative relationships—is much more consistent with a *ring* of life (Figure 7.16b). It could be that horizontal gene transfer made an informationally interconnected *network* of early life forms rather than a linearly branching tree; in other words, contemporary life has multiple ancestors. Most recently it has been proposed, on the basis that eukaryote core genes seem to come from Archaea, that there are only two domains: Archaea and Bacteria.

7.3 DNA Is the Universal Replicator

7.3.1 All Life on Earth Uses DNA

As we have seen, a critical feature of life is that it is capable of autonomous replication. For all life on Earth, this ability resides in the DNA. This fact alone attests to the overwhelming probability that all terrestrial organisms are ultimately derived from a single genesis event. If life did get started several times on Earth only one such experiment has left descendents.

There are a couple of caveats to the DNA story. First, it is virtually certain that DNA was not the first replicator. It could have been preceded by RNA for reasons we will explore later, and it may be that earlier still, other replication mechanisms operated. Second, there is no suggestion that extraterrestrial life will use the same chemistry in its replication machinery, though protagonists of convergent evolution might argue that what works well on Earth might work well on similar worlds. That said, in this section, we explore how modern terrestrial organisms reproduce. We shall see whether this can provide any insights into early life.

7.3.2 DNA Replication Was Deduced from Theory

In April 1953, James Watson and Francis Crick published a short paper in *Nature* which, given its huge importance, is the epitome of understatement. It starts, "We wish to suggest a structure for the salt of deoxyribose nucleic acid (DNA)," and finishes with the cliff-hanger, "It has not escaped our notice that the specific pairing we have postulated immediately suggests a possible copying mechanism for the genetic material." It is not an overstatement to say that this work revolutionized biology. It provided an explanation for the mechanism of inheritance, a key to how Darwinian evolution worked at the cellular level, and heralded modern molecular biology and genetic engineering.

A remarkable feature of the Watson and Crick paper is that it was entirely theoretical. They knew the composition of DNA: it contained the sugar deoxyribose, phosphate groups, and four different bases termed adenine (A), thymine (T), guanine (G) and cytosine (C). They made the assumption that the sugars were linked to each other by the phosphate groups, the view held by most chemists. The bases are of two different kinds, purines and pyrimidines, and they were aware of an experimental work in 1951 showing that whenever the base composition of DNA is analyzed, whatever its provenance:

$$\text{amount of A} = \text{amount of T}$$
(purines)   (pyrimidines)

$$\text{amount of G} = \text{amount of C}$$

EXERCISE 7.4

Suggest an explanation for why the amount of A equals the amount of T and the amount of G equals the amount of C in all DNA molecules.

Crucially, Watson had seen an x-ray diffraction photograph of DNA produced by Rosalind Franklin and immediately recognized that the molecule must be a double helix. Watson and Crick proposed that each chain is a right-hand helix and the two chains of the double helix run in opposite directions. The two chains are held together by hydrogen bonds between the bases, in which a purine must pair with a pyrimidine, with the only allowable combinations being A-T and G-C. There appeared to be no chemical constraints to the order of bases along a single strand, but given a particular sequence of bases on one strand, the sequence on the other was fixed by the base-pairing rules. This matching makes the strands complementary to each other. It means that the information needed to make a double-stranded DNA molecule is contained within just one strand: the crucial requirement for self-replication. Watson and Crick correctly assumed that the order of the bases along the DNA molecule encoded genetic information.

7.3.3 DNA Acts as a Template

Prior to cell division, DNA is replicated so that each daughter cell receives a copy of all the genetic information. Although the details are complicated, the basic idea is easy to understand. At specific points along a DNA molecule, called replication origins, the double helix unwinds so that the two strands separate. Each of the strands acts as a template on which is assembled the complementary DNA. The end result is two identical double helices where before there was one (Figure 7.17). This process is described as semi-conservative replication because each of the resulting double helices contains one original strand and one new one.

7.3.3.1 DNA Replication Is Fast

In bacteria, the rate at which DNA can be replicated is about 900 base pairs per second. Although this sounds fast, DNA replication is the rate-limiting step in bacterial cell division. Hence, the rate at which bacteria cells divide is determined by the size of the genome; the smaller it is the faster bacteria can make copies of themselves. Indeed bacteria have small genomes compared to eukaryotes. The implication is that bacteria have experienced selection pressure to maintain small genomes in a way that eukaryotes have not. This has constrained bacteria to remain relatively simple organisms.

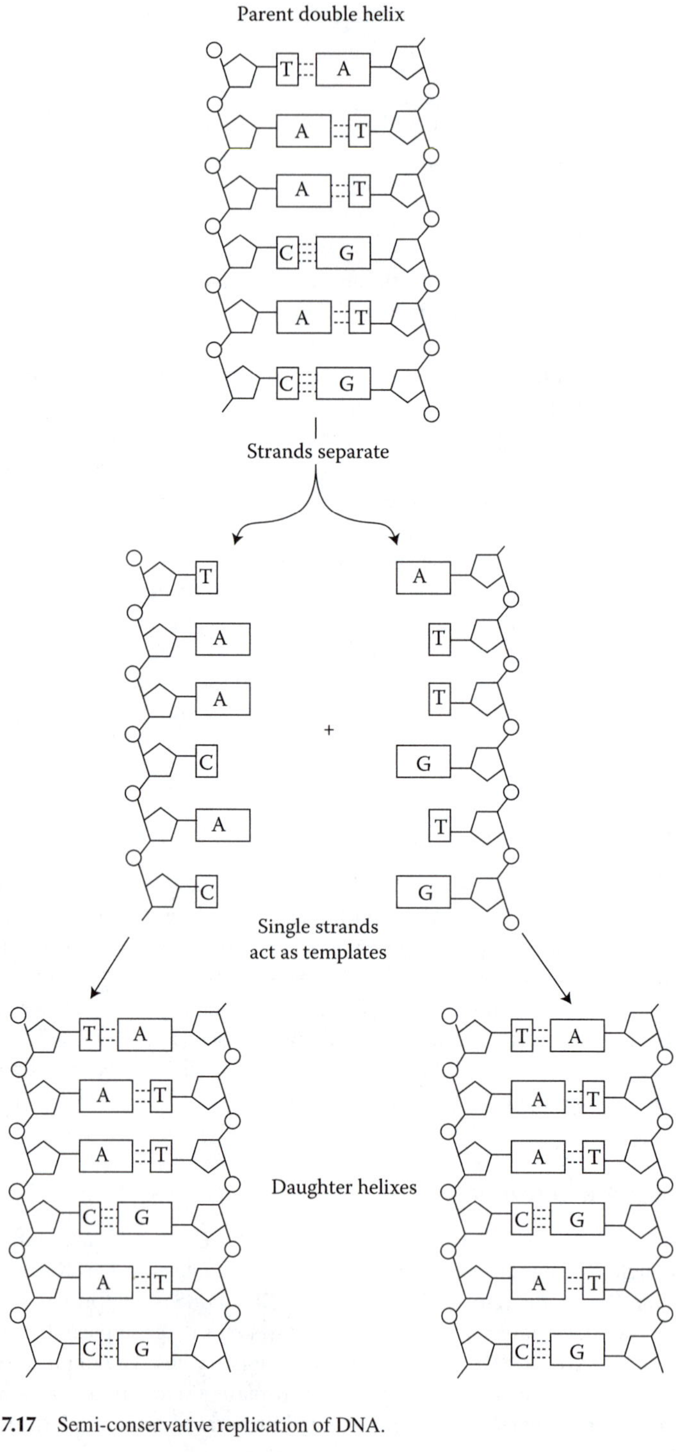

FIGURE 7.17 Semi-conservative replication of DNA.

7.3.3.2 DNA Replication Requires Catalysts

DNA replication requires several enzymes, protein catalysts that speed up the complicated chemistry involved. The details need not concern us except for the fact that DNA replication cannot occur without proteins, and yet DNA encodes the information for making proteins. This is a classic "chicken and egg" conundrum: which came first, proteins or DNA? The solution favored at present is that DNA was not the first replicator.

7.3.3.3 DNA Replication Is High Fidelity

DNA replication is extraordinarily faithful. The error rate is about one for every 10^7–10^8 base-pairs. It translates into one error for every 100 cell divisions of a bacterium. The high fidelity is conferred by proof-reading functions of enzymes that synthesize DNA.

7.3.3.4 DNA Mutations Are the Source of Variation

Changes occur to DNA for reasons other than copying errors. These include ionizing radiation, high energy particles, toxins, and viruses. The rate at which this happens is hard to pin down because it differs with the organism, the region of DNA, and over the life of an organism, but an average of 10^{-8} per base pair per generation is a widely accepted estimate. Without repair enzymes this number would be much higher.

Changes to the sequence of bases in DNA due to any cause are termed mutations and provide a source of the subtle variations in the genome. This translates into variations in the structure and function of the individual, its phenotype. The phenotype can be thought of as the interface between genes and environment which allow particular characteristics to be expressed. It is the phenotype that runs the gauntlet of natural selection. How does this work? Cheetahs kill faster zebras less often than slower ones. Any gene carrying a favorable mutation that enhances zebra running will gradually increase in frequency in the zebra population as cheetahs preferentially pick off the slower ones. Similarly, genes with deleterious mutations will reduce in frequency in the population over time. Neutral gene mutations (those that have no noticeably effect) are retained in the gene pool.

7.3.3.5 The Genetic Code Shows a Single Origin of All Life

As we have seen, there are 20 amino acids found in polypeptides but only four bases in DNA to encode them. This prompted the physicist George Gamow to suggest a three-letter code was likely, because 3 is the smallest integer n such that 4^n is at least 20. Gamow's reasoning was a bit more complicated than this, but he was correct. Hence, 64 triplets of bases, termed codons, represent the 20 amino acids, with appreciable redundancy.

Let us see how this works for a short stretch of one strand of a DNA molecule:

$$5'\text{-TTTGACAGTTAG-}3'$$

The code is read in the 5' to 3' direction. This is an open reading frame, which means that the first base is the start of a codon. Clearly this fragment of DNA could encode $12/3 = 4$ amino acids. However, it actually encodes only 3 because the last triplet (TAG) is a stop codon. In fact, the sequence is: –phenylalanine–aspartate–serine–[stop]. By 1966, the genetic code had been completely deciphered. That is, it was possible to attribute an amino acid to each of 61 codons, with the remaining 3 being stop codons.

It turns out that the genetic code has some remarkable properties. It is unambiguous. Each codon is unique to an amino acid. It is nearly universal in the sense that every organism on the planet uses the same code with only occasional minor variations. The odds at arriving at an identical 64:21 mapping twice by chance are less than 1 in 10^{30}. It is therefore virtually certain that all living things on Earth arise from a single progenitor. The mitochondrial genetic code is significantly different from the nuclear code but this can be explained without postulating a second genesis.

Since 61 codons code for 20 amino acids, at least some amino acids must be encoded by more than one triplet. This is termed degeneracy, which simply means the code has some redundancy. Although three amino acids have no fewer than 6 codons, the outcome of redundancy is that errors in DNA replication are less likely to result in the wrong amino acid being incorporated into a protein than if, for example, each amino acid was represented by a single codon.

7.3.3.6 The Genetic Code Provides Clues to Its Origin

The genetic code is not random. It shows correlations in codon assignments. Amino acids that share the same biosynthetic pathway tend to have the same first base in their codons, and amino acids with similar physical properties tend to have similar codons. This last correlation means that some mutations result in similar amino acids being substituted, with little or no change in the function of the protein. Such mutations are said to be silent. The implication of this is that the code has evolved. Presumably natural selection has favored a code that minimizes the chances of fitness-damaging changes in protein function. Indeed, a detailed statistical analysis has found that only one in a million randomly generated triplet codes is better at reducing errors than the natural code and the way in which the natural genetic codes achieves this efficiency is consistent with natural selection.

The function of transfer RNA (tRNA) molecules is to bring amino acids to the ribosome to be added to a growing polypeptide. They are small RNA molecules, 73–93 nucleotides long and folded into a cloverleaf structure that is stabilized by base-pairing (Figure 7.18). In its activated state, a tRNA molecule is covalently linked to an amino acid. One loop of a tRNA has an anticodon, a triplet sequence that is complementary to the codon in mRNA for the amino acid.

You might reason that there would need to be 61 specific tRNA molecules to achieve a one-to-one correspondence between amino acids and codons, but in fact, just 31 are needed for an unambiguous translation. This is because anticodons can recognize more than one codon. This flexibility is termed wobble base pairing.

Remarkably, experiments show that many amino acids have a selective chemical affinity for the base triplets of their codons. These affinities could have set up a primordial core of amino acid–codon assignments. Maybe, once upon a time, ancient life had a direct template from base sequence to amino acid sequence, and the current complex translation mechanism involving tRNA arose later.

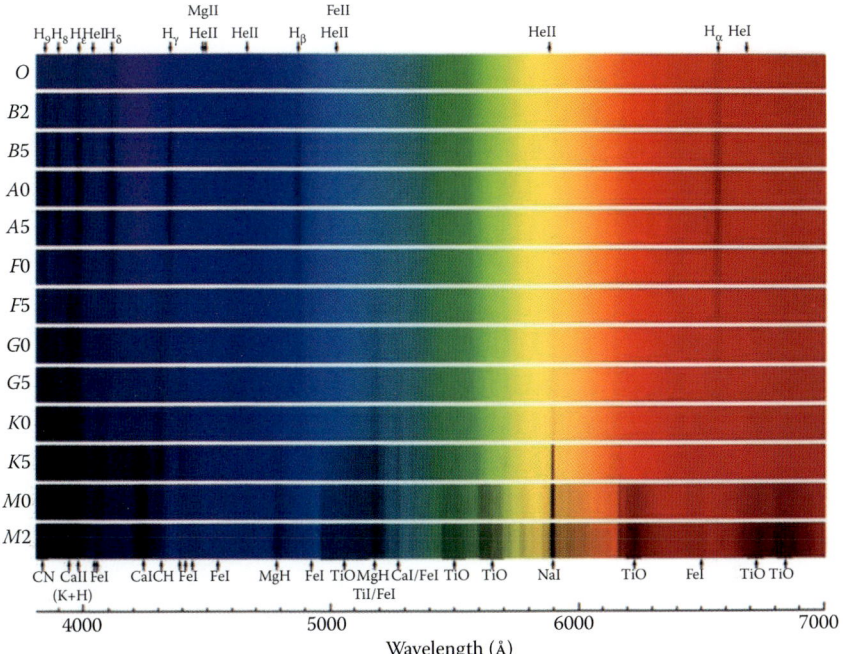

PLATE 1.1 Stellar spectra for spectral classes of the star. (Photo credit: Spectral class/absorption feature diagram © Michael Briley.)

PLATE 2.1 The spiral galaxy M74. The dark dust lanes are giant molecular clouds. Together with diffuse clouds they are the site of much chemistry. The overall blue color of the spiral arms is caused by young, hot stars. HII regions are seen in red H_α light. (Photo credit: NASA, ESA, and the Hubble Heritage (STScI/AURA)-ESA/Hubble Collaboration.)

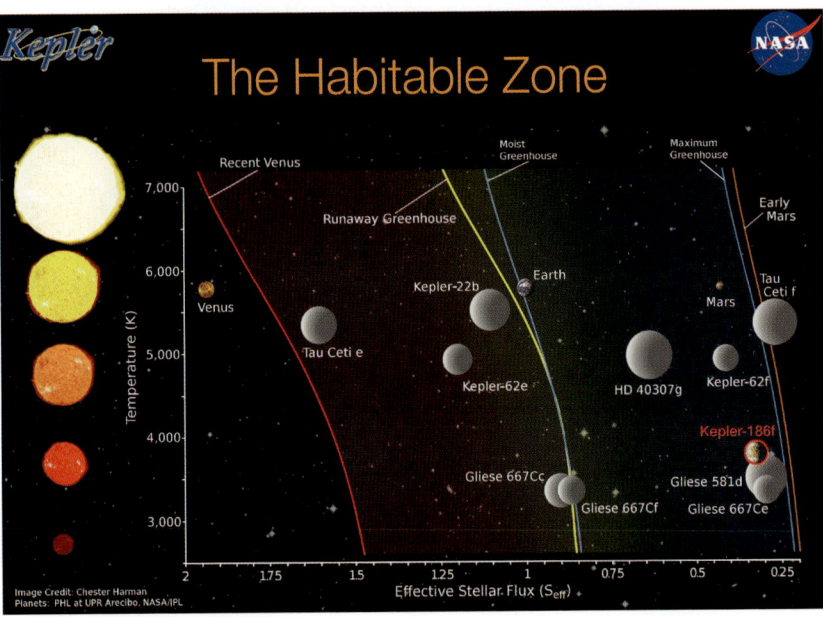

PLATE 3.1 Habitable zones for stars. For definitions of the inner and outer edges of the zone, see text. Kepler 186f is the first Earth-sized planet discovered in a habitable zone. (Photo credit: NASA/Chester Harman.)

PLATE 4.1 Ordinary chondrite. Slice through North West Africa 869. Note the glistening flecks of iron–nickel metal and pale circular section chondrules visible most clearly at the top of the meteorite. (Photo credit: Alan Longstaff.)

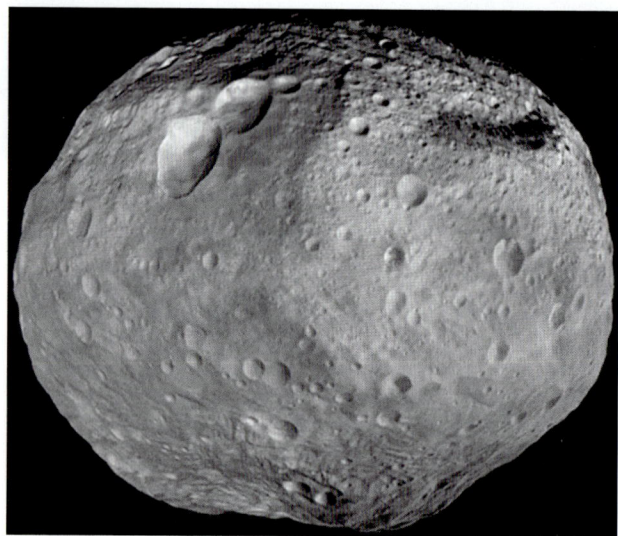

PLATE 4.2 Asteroid 4-Vesta imaged by the Dawn spacecraft. The first million years of the solar system would have been dominated by thousands of objects like Vesta. (Photo credit: NASA/JPL-Caltech/UCAL/MPS/DLR/IDA.)

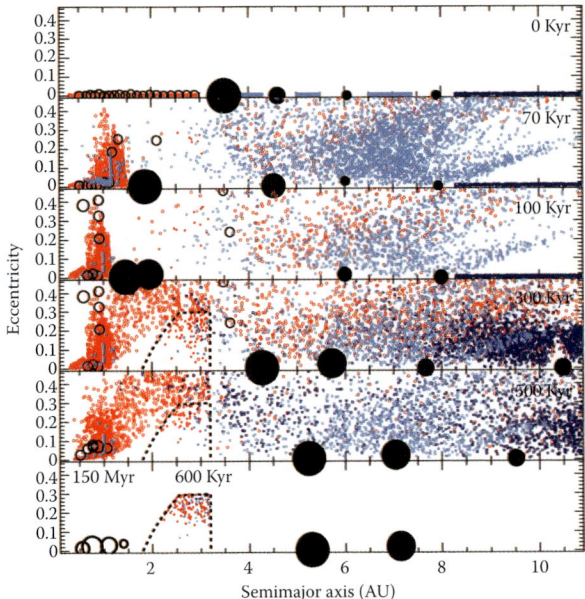

PLATE 5.1 Grand tack model accounts for the distribution of asteroids in the main belt and the small size of Mars. (Photo credit: Walsh et al. *Nature*, 2011;475:206–9. With permission.)

PLATE 8.1 Stromatolites growing in Hamelin Pool Marine Nature Reserve, Shark Bay, in Western Australia. (Photo credit: Paul Harrison, Reading, UK.)

PLATE 8.2 Lost City alkaline hydrothermal vent. This flange has fluids both trapped underneath and flowing through the top of it to form the smaller chimneys on the top. It is about 1 m across. (Photo Credit: NOAA and University of Washington.)

PLATE 9.1 Banded iron formation (BIF), Karijini National Park, Western Australia. Despite their widespread distribution, the origin of BIFs is not yet unambiguously established. (Photo credit: Graeme Churchard, Bristol, UK.)

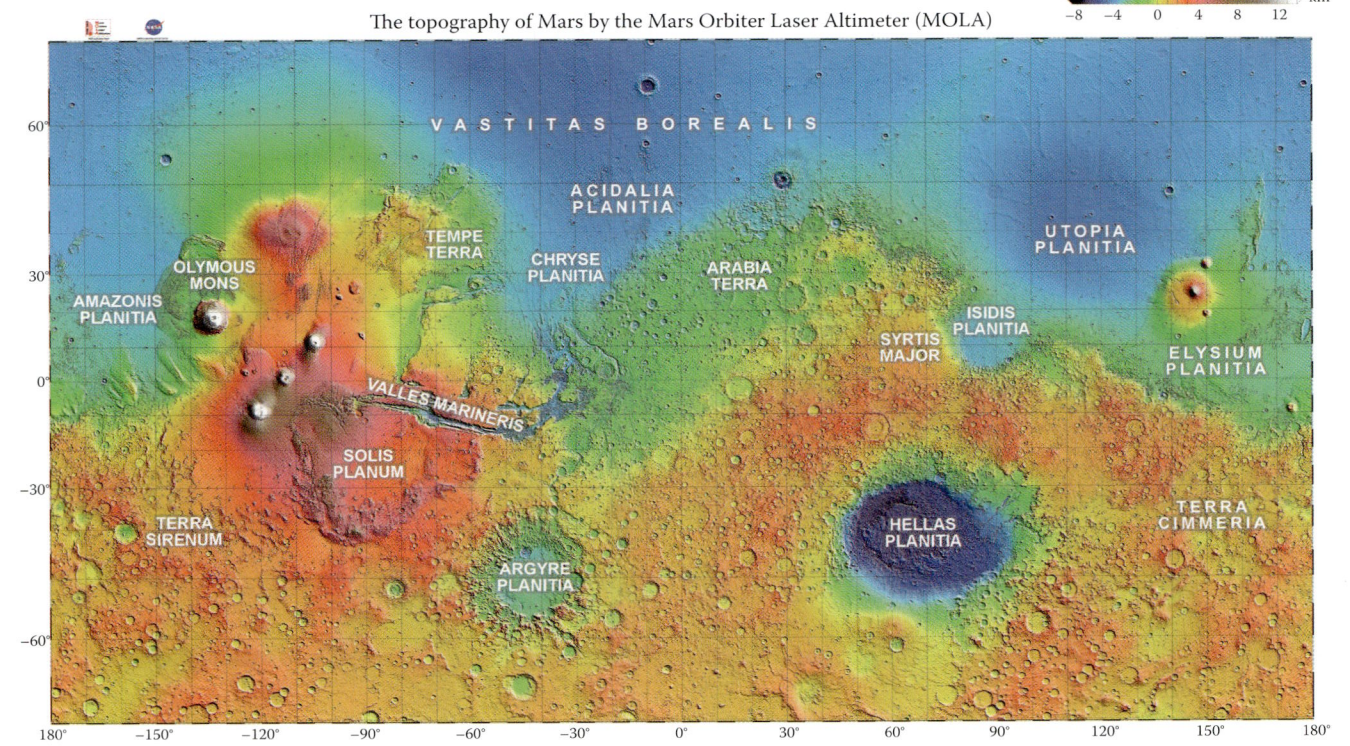

The topography of Mars by the Mars Orbiter Laser Altimeter (MOLA)

PLATE 10.1 High resolution topographic-shaded relief map of Mars constructed from Mars Orbiter Laser Altimeter (MOLA) data. Key features are labeled. Graphic processed by Greg Smye-Rumsby. (Photo credit: NASA/MOLA Science Team.)

PLATE 10.2 Cryoturbation polygons on Mars resemble those in terrestrial tundra that arise from freeze-thaw processes. Imaged by Mars Phoenix Lander's Surface Stereo. (Photo credit: NASA/JPL-Caltech/University of Arizona.)

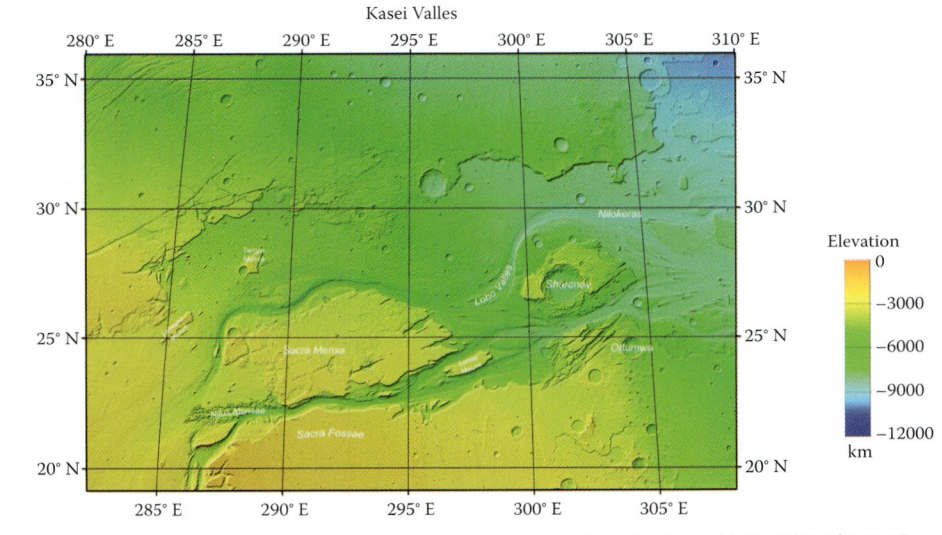

Kasei Valles

Elevation

Projection: Sinusoidal - Central Meridian 286° E
Coordinate System: Mars Sphere - Axis 3396 km

PLATE 10.3   Kasei Valles, an outflow channel which once carried water into the Chryse Planitia. (Photo credit: ESA.)

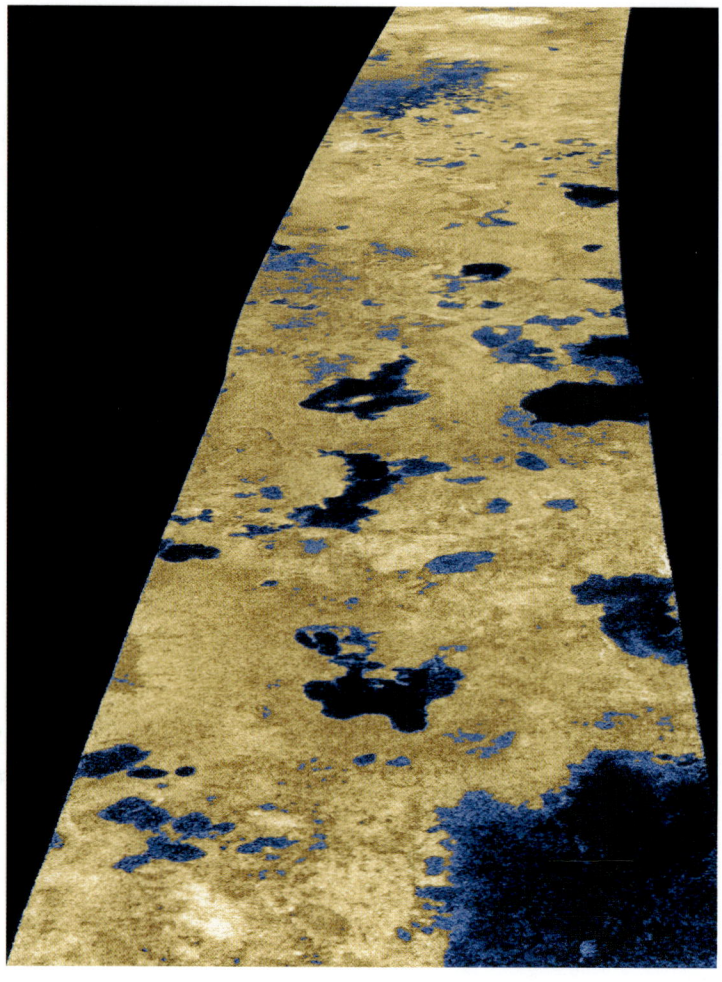

PLATE 11.1 Hydrocarbon lakes in the northern hemisphere of Titan imaged by Cassini radar. Bolsena Lacus can be seen on the lower right. (Photo Credit: NASA/JPL-Caltech/USGS.)

PLATE 11.2 Tiger stripes in the south pole of Enceladus periodically vent plumes of water vapor. Cassini image. (Photo credit: NASA/JPL/Space Science Institute.)

PLATE 11.3 Chaos terrain on the surface of Europa. (Photo credit: NASA/JPL.)

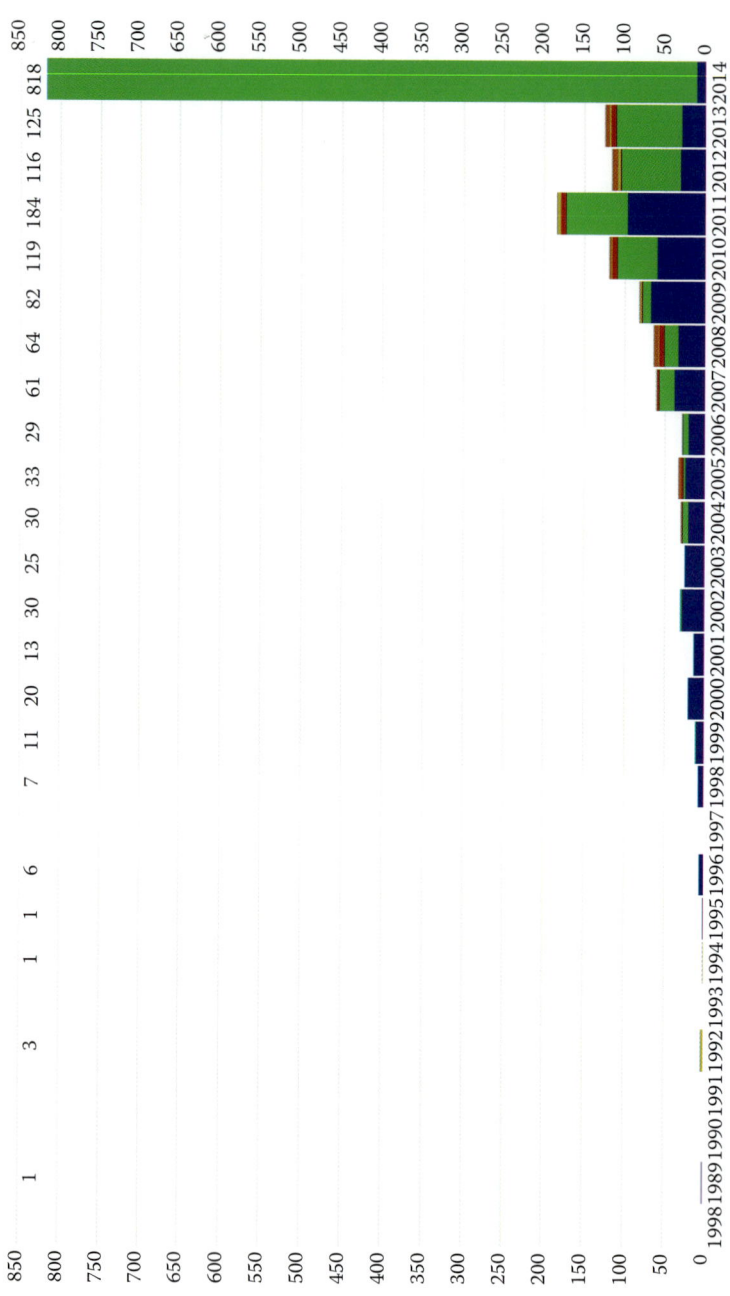

PLATE 13.1 Exoplanet discoveries by year showing the discovery method by colors: radial velocity (blue); transit (green); timing (yellow); direct imaging (red); micro-lensing (orange). (Photo credit: Wikipedia "Exoplanets" May 2014.)

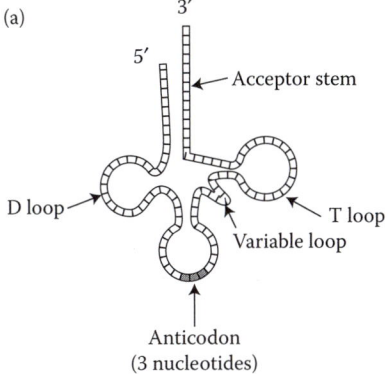

(a)

5′

3′

Acceptor stem

D loop

T loop

Variable loop

Anticodon
(3 nucleotides)

(b)

mRNA codon	AUG	UUC
tRNA anticodon	UAC	AAG
Amino acid	Met	Phe

FIGURE 7.18 (a) The structure of tRNA. (b) Two examples of how tRNA translates the mRNA into amino acid sequence.

It has been suggested that the triplet code was originally derived from codes with longer codons; for example, a quadruple code would provide $4^4 = 256$ codons to play with. This would have higher redundancy and be less error prone than the triplet code. It could have provided greater leeway, allowing reasonably accurate read-out before the evolution of high fidelity translational machinery such as the ribosome or proof-reading by DNA polymerase enzymes.

Another possibility (biosynthetic expansion) is that primordial life used fewer amino acids to build polypeptides, so the modern genetic code grew from a simpler earlier code. The idea is that as life encountered new amino acids—for example, as metabolic pathways evolved—these were later incorporated into the protein synthesis and genetic coding. Indeed there is much circumstantial evidence to suggest that fewer amino acids were used in the past, but there is no consensus about the order in which amino acids were added to life's inventory.

7.4 Metabolism Matches Lifestyle

The metabolism—the sum total of life-sustaining chemical reactions—of an organism is inextricably linked to how it makes a living. Eukaryotes have a limited repertoire of metabolic strategies, dominated by photosynthesis (e.g., plants), using complex organic molecules as energy sources (e.g., animals), and harnessing chemical energy using oxygen in respiration (most eukaryotes). While many prokaryotes also use these strategies, others exploit a wide variety of very different types of metabolism that enables them to thrive in environments that are inhospitable or even lethal to other life forms. In this section, we explore key metabolic strategies used by life on Earth. This will help us to speculate on how life originated and how extraterrestrial life might make a living.

We start by considering a trick used by all life on Earth, namely, catalysis to speed up the rate of chemical reactions.

7.4.1 Living Systems Enhance Reaction Kinetics

Life on Earth can only metabolize and reproduce over a very narrow temperature window. If we include extremophiles (organisms that live in environments that would be hostile to most life on Earth), this is −18°C to +117°C, although most organisms cluster at the cooler end.

Too cold and the fluid inside cells freezes. Even if the water crystals do not damage cell structure, terrestrial biochemistry does not work if molecules are locked in solid phase. Biochemistry relies on reactants colliding with sufficient energy and being orientated in the correct configuration for chemical reactions to happen, and it is liquids not solids that provide the environment for this.

At the other end of the scale, as temperatures increase beyond about 40°C, first weak bonds and then covalent bonds in macromolecules are broken, and this determines the upper temperature limit for life. Disrupting weak bonds changes protein shape, so they can no longer function (the proteins are said to be denatured), Extremophiles that thrive at high temperatures (hyperthermophiles) buck the trend by having ways of curtailing protein denaturation.

A central problem for living systems is that at the temperatures where macromolecules are stable the rates of chemical reactions are extremely slow. Most reactants do not have sufficient kinetic energy on average to overcome the activation energy barrier for most biochemical reactions. The activation energy is the kinetic energy needed for collisions between reactants to be energetic enough to initiate the reaction.

In chemical terms, the activation energy is required to form intermediates with higher free energy than the reactants. Once made, being unstable, these transition state species will either revert back to reactants or form products (Chapter 2; Figure 2.6). The fraction of molecules that have a greater kinetic energy than the activation energy for a reaction can be calculated from the Maxwell–Boltzmann distribution, and it is this that determines the rate of the reaction. Clearly, increasing the temperature increases the reaction rate.

The problem was solved by the evolution of biological catalysts, which speed up the rate of chemical reactions without themselves being consumed. It is important to note that catalysts do not allow reactions to occur that would not take place without them. Catalysts do not change ΔG or equilibrium constants. For reversible reactions they increase the rate of both forward and backward reactions by the same factor, allowing equilibrium to be reached much more quickly in isolated systems. (But recall that most biochemical reactions *in vivo* are far from equilibrium.)

Biological catalysts fall into two broad chemical categories: protein catalysts termed enzymes, and RNA catalysts called ribozymes. Of the 4288 different proteins made by the *Escherichia coli* bacterium (the genome of which was sequenced in 1997), at least a third are enzymes. This attests to how crucial enzymes are.

All catalysts work by lowering the activation energy. Exactly how they achieve this depends on the details of the reaction, but in general the enzyme provides exactly the right chemical surface (3-D shape, charge, polarity) to bring substrates together in the correct orientation and distort them into transition state species. Here, the vagaries of random collisions between molecules are replaced by precisely engineered close approach made possible by the nature of the enzyme's active site.

Enzyme catalyzed reactions can go up to 10^{10}-fold faster than the uncatalyzed rate. A single enzyme molecule will typically convert a few hundred to a few thousand substrate molecules to product each second, with a fast enzyme catalyzing over 500,000 reactions per second. The number of molecules of an enzyme within a cell varies typically from 100 to 10,000. Enzymes are specific for particular reactions. It is the high specificity that distinguishes enzymes from non-biological catalysts. Enzymes can be subject to feedback or feed forward regulation by small molecules and this provides for the self-organizing property of metabolism.

7.4.2 Life on Earth Has Three Metabolic Requirements

To make a living, all organisms need at least three things: a carbon source, a source of energy, and an oxidant. A carbon source is needed because life is based on carbon. The energy source is an electron donor and therefore able to reduce molecules. This reducing power is used to make complex carbon compounds—that is, macromolecules—from simple carbon sources that can be inorganic (e.g., CO_2, HCO_3^- or CO_3^{2-}), or organic (e.g., sugars, organic acids, amino acids, and fatty acids). An oxidant or electron acceptor is required to be able to harness chemical potential energy.

7.4.2.1 Organisms Can Be Categorized by Metabolism

Biologists classify organisms on the basis of general metabolic principles (Figure 7.19). For example, animals, which derive their energy from molecules rather than sunlight, and for which organic molecules act as both electron donor and carbon source, are chemoorganoheterotrophs. The suffix "-troph" is derived from the Greek for food so, for example, heterotrophs are the class of organisms that obtain their nourishment from, or grow using, organic carbon.

The complex organic molecules synthesized by the organism are used either to build cellular components (i.e., growth) or as fuels that can be harnessed to power biological processes.

EXERCISE 7.5

Which terms describe the metabolic lifestyle of plants?

Energy source	Sunlight	Photo-			
	Molecules	Chemo-			
Electron donor	Organic		Organo-		-troph
	Inorganic		Litho-		
Carbon source	Organic			Hetero-	
	Inorganic			Auto-	

FIGURE 7.19 Organisms can be classified according to their metabolism.

7.5 Cells Harness Free Energy

There are two ways in which chemical potential energy can be obtained by the oxidation of organic molecules, cellular respiration and fermentation.

7.5.1 Respiration Requires an Exogenous Electron Acceptor

Cellular respiration requires a final electron acceptor. It is always exogenous; that is, it comes from outside the cell. This distinguishes it from fermentation in which the electron acceptor is endogenous. If the electron acceptor is oxygen the respiration is said to be aerobic, and this is seen in most eukaryotes and many prokaryotes. However, some organisms (mostly prokaryotes) have anaerobic respiration in which the exogenous electron acceptor is something other than oxygen. This alternative oxidant is generally inorganic (e.g., CO_2^- or SO_4^{2-}) but is sometimes a simple organic compound. Usually, the final electron acceptor is the last step in a long sequence of reactions, and if it is not available, all the preceding reactions stop; that is, respiration ceases. The reason hydrogen cyanide (HCN) is so toxic to all organisms that respire aerobically is because it poisons the final step in which electrons are donated to molecular oxygen.

Respiration in prokaryotes occurs in the cytoplasm, but aerobic respiration occurs within mitochondria in eukaryotes.

Respiration involves two processes that are tightly coupled. One is the oxidization of reduced carbon (i.e., organic molecules such as sugars) to CO_2 and H_2O, the other is the spontaneous flow of electrons down a respiratory electron transport chain (ETC) to a final electron acceptor (i.e., O_2) allowing free energy to be abstracted. We take these in turn (Figure 7.20).

7.5.2 Most Carbon Oxidation Happens in the Krebs Cycle

Macromolecules (polysaccharides, fats, and proteins) are broken down into small molecules (sugars, fatty acids, and amino acids, respectively) which in turn are converted into a two-carbon compound, acetate, in the form of acetyl CoA (Figure 7.21). CoA is coenzyme A, a molecule that has a central role in metabolism as a carrier of short fatty acid (acyl) groups like the acetyl group.

CoA is based on adenosine triphosphate (ATP), which is a crucial and universal carrier of chemical potential energy within cells.

In the Krebs cycle (Figure 7.22), acetyl-CoA is combined with a four-carbon compound (oxaloacetate) to yield a six-carbon compound (citrate). A sequence of eight reactions follows that yields $2CO_2$ molecules and regenerates oxaloacetate. The cycle

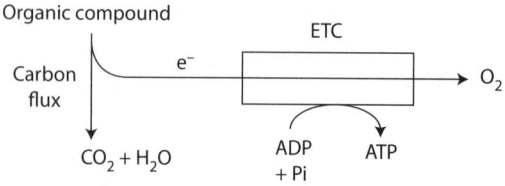

FIGURE 7.20   Respiration couples carbon flow and electron flow.

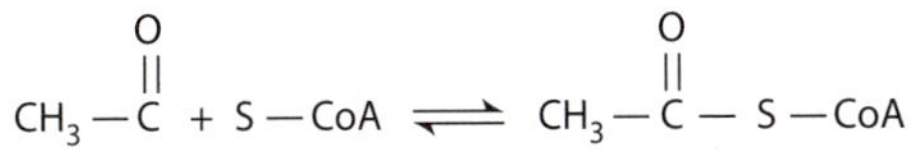

$$CH_3-\overset{\overset{\displaystyle O}{\|}}{C} + S-CoA \rightleftharpoons CH_3-\overset{\overset{\displaystyle O}{\|}}{C}-S-CoA$$

FIGURE 7.21 Acetyl CoA couples the acetyl group to coenzyme A via a sulfur-containing bond.

has four redox reactions. In each, two electrons are released (oxidation) and transferred to a carrier molecule, which is thereby reduced. With the electrons go two protons, which biochemists depict as H^+. One carrier molecule, nicotinamide adenine dinucleotide (NAD^+), is responsible for three of the redox reactions:

$$NAD^+ + 2e^- + 2H^+ \rightarrow NADH + H^+. \tag{7.1}$$

In this reaction, what happens is that $H_2 \rightarrow H^- + H^+$, the hydride ($H^-$) ion reacts with the NAD^+ and the proton goes into solution. The two electrons are bound to a nitrogen atom in the NADH. Biochemists frequently refer to the NADH as reducing power because it is a good electron donor; the reduction potential of the NAD^+/NADH pair is fairly negative at -0.32 V.

One step in the Krebs cycle transfers a pair of electrons to another carrier, flavin adenine dinucleotide (FAD):

$$FAD^+ + 2e^- + 2H^+ \rightarrow FADH_2. \tag{7.2}$$

The ubiquity of the electron carrier molecules, NAD^+ (and its close cousin $NADP^+$) and FAD, which are all based on adenine ribonucleotide, hints at a very early role for RNA in energy metabolism.

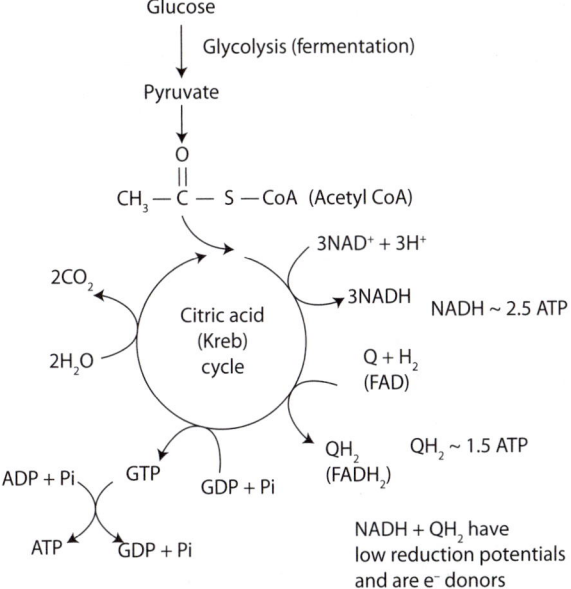

FIGURE 7.22 The Krebs cycle lies at the heart of metabolism in most organisms.

Many Krebs cycle components are not just fodder for oxidation but starting points for biosynthesis. Standard biochemistry textbooks discuss it in the context of aerobic metabolism, but reversed, modified, or incomplete versions of the Krebs cycle operate in organisms that respire anaerobically. The interesting thing about this is that before 2.4 Gyr, there was virtually no free oxygen in the atmosphere, so all life would have respired anaerobically. So, it may be possible to reconstruct the evolution of the classical aerobic Krebs cycle from the anaerobic versions that must predate it.

Other catabolic pathways also generate NADH and $FADH_2$. There are two possible fates for these electron carriers. One is to provide chemical potential energy directly to any number of biosynthetic reactions that are endergonic and so will not go spontaneously. The other is that their electrons are funneled into an electron transport chain. We explore this second option now.

7.5.3 Electron Chains "Quantize" Free Energy Availability

The electron carriers reduced in the Krebs cycle donate their electrons to a final electron acceptor (molecular oxygen in the case of aerobic respiration) not directly but via a respiratory electron transport chain. This consists of a sequence of electron carriers with successively higher reduction potentials, enabling free energy to be liberated in stepwise fashion in a cascade of redox reactions. There are no endergonic reactions with ΔG large enough to accept the free energy were it all to be released in a single step: hence, the "quantization" of chemical potential energy release by electron transport chain redox reactions.

7.5.4 $\Delta G^{\circ\prime}$ of Redox Reactions Can Be Calculated

To understand how metabolic pathways may have got started, astrobiologists need to work out whether they are thermodynamically favorable. Are they likely to "go" and under what conditions? Although we are not going to attempt this here, to get a flavor for what redox reactions in biology are about, we are going to calculate the free energy change associated with a couple of them.

7.5.4.1 Exergonic Reactions Have Positive $\Delta E_0'$

As a first example let us see what we can get from NADH, a ubiquitous electron carrier. The electrons and protons come from the oxidation of reduced carbon in, for example, the Krebs cycle as we have seen:

$$NAD^+ + 2e^- + 2H^+ \rightarrow NADH + H^+. \tag{7.1}$$

The electrons of NADH are relayed to a final electron acceptor, usually O_2. This regenerates the NAD^+. If no oxygen is available, NAD^+ cannot be regenerated. The fact that the electrons flow through a sequence of redox reactions giving up free energy in small jumps is unimportant in working out the total free energy that can be abstracted. We only need to know the reduction potentials of the NAD^+/NADH and $\frac{1}{2}O_2$/H_2O half-reactions—that is, the initial and final states—to do the calculation.

The two redox half-reactions have reduction potentials as follows:

$$NADH \rightarrow NAD^+ + H^+ + 2e^- (E'_0 = -0.32 \text{ V}); \; ½ O_2 + 2H^+ + 2e^-$$
$$\rightarrow H_2O \, (E'_0 = +0.82 \text{ V}).$$

The difference in reduction potential is given by

$$\Delta E'_0 = E'_{0(\text{reduction reaction})} = E'_{0(\text{oxidation reaction})}, \qquad (7.3)$$

or

$$\Delta E'_0 = E'_{0(\text{acceptor})} = E'_{0(\text{donor})}. \qquad (7.4)$$

We work out which of the half-reactions is the reduction and which the oxidation from knowing which way the combined redox reaction goes:

$$NADH^+ + H^+ + \tfrac{1}{2}O_2 \rightarrow NAD^+ + H_2O. \qquad (7.5)$$

In fact, we wrote out the half-reactions in the correct direction to satisfy this above. Clearly the NADH is being oxidized and the oxygen reduced so:

$$\Delta E'_0 = E'_{0(\text{reduction reaction})} = E'_{0(\text{oxidation reaction})} = 0.82 - (-0.32) = 1.14 \text{ V}.$$

Now we use Equation 6.25, and recalling that $V = J \, C^{-1}$,

$$\Delta G^{\circ\prime} = nF\Delta E'_0$$

and two electrons are being transferred, so,

$$= -2 \times 9.649 \times 10^4 \, C\,mol^{-1} \times 1.14 \, J\,C^{-1}$$
$$= -2.21 \times 10^5 \, J\,mol^{-1}$$
$$= -221 \, kJ\,mol^{-1}.$$

We see that $\Delta E'_0 > 0$ and $\Delta G^{\circ\prime} < 0$, so the reaction is exergonic and will yield free energy that can be harnessed to do chemical work.

One of the steps in the Krebs cycle uses the electron carrier FAD rather than NAD^+. This reaction is:

$$\text{succinate} + FAD \rightarrow \text{fumarate} + FADH_2. \qquad (7.6)$$

We do not need to worry about the structures of succinate or fumarate. What counts is that the electrons in the $FADH_2$ are eventually donated to an electron acceptor, usually oxygen, which is thereby reduced.

EXERCISE 7.6

Calculate the free energy change for the redox reaction $FADH_2 + 1/2O_2 \rightarrow FAD + H_2O$. The two half-reactions are $FADH_2 \rightarrow FAD + 2H^+ + 2e^-$ ($\Delta E'_0 = -0.18$ V) and $1/2O_2 + 2H^+ + 2e^- \rightarrow H_2O$ ($\Delta E'_0 = +0.82$ V). State whether the reaction is exergonic or endergonic.

7.5.5 Proton Gradients Are the Core of Terrestrial Metabolism

In both photosynthesis and respiration, the free energy made available by electrons flowing through electron transport chains is used to generate a proton gradient (Figure 7.23a). The energy released by the dissipation of the proton gradient is coupled to the formation of adenosine triphosphate (ATP), a small molecule that stores and transfers chemical potential energy and is universal in terrestrial organisms. This process is termed oxidative phosphorylation (Figure 7.23b). The synthesis of ATP from ADP is endergonic. Thus, ATP drives metabolic reactions that will not go spontaneously, thereby powering any number of energy-requiring processes of organisms: transport of ions and molecules, growth, reproduction, movement, etc. Eventually the energy is degraded to heat that is dumped into the surroundings, the entropy of which increases.

Before 1961 it was unclear how the oxidation of reduced carbon was coupled to the phosphorylation that made ATP. That proton gradients are used to synthesize ATP was first proposed by Peter Mitchell who called it chemiosmosis. There was considerable resistance to the idea for some time, but eventually it was accepted and Mitchell was awarded the Nobel Prize in Chemistry in 1978. The generation of proton gradients lies at the core of biochemistry and is universal to life on Earth. This provides clues to the metabolic activities of the first life on Earth, as we shall see in Chapter 8.

7.5.5.1 Prokaryotes Pump Proteins Across Their Cell Membrane

The respiratory electron transport chains and ATP synthases of prokaryotes and eukaryotes are similar. In prokaryotes, the protons are pumped across the cell membrane into a space (periplasmic space) between two membranes (Figure 7.24).

7.5.5.2 Eukaryotes Make ATP in Mitochondria

In eukaryotes, the respiratory electron chain components are located in the inner mitochondrial membrane. They comprise three large protein complexes and two smaller diffusible carriers.

Several redox reactions occur in each of the protein complexes. These reactions harness free energy from electron flow to pump protons across the inner mitochondrial membrane, from the mitochondrial matrix to the intermembrane space. This creates an H^+ electrochemical gradient (outside positive) across the membrane. Protons subsequently flow down the gradient through ATP synthases in the inner mitochondrial membrane that catalyze the phosphorylation of ADP to ATP. (Under some circumstances, the ATP synthases will run in reverse. Here ATP hydrolysis helps *generate* the proton gradient.)

Electrons flowing through the respiratory electron chain from NADH to molecular oxygen provide $\Delta G^{\circ\prime} = -221$ kJ mol^{-1}. Since ATP hydrolysis to ADP has $\Delta G^{\circ\prime} = -30.5$ kJ mol^{-1}, it might seem as if there is sufficient free energy change for the synthesis of 7 mol of ATP per mole NADH. This is not the case. In fact, the free energy needed to synthesize ATP is variable since it depends on the concentrations of Mg^{2+}, ADP, inorganic phosphate and ATP, and on the size of the proton gradient, quantified by the proton motive force (see Box 7.2). Hence, the synthesis of ATP uses ~50 kJ mol^{-1}, and consequently the actual yield is only 3 mol of ATP per mole of NADH. By a similar argument 2 mol of ATP are synthesized per mole of FADH$_2$.

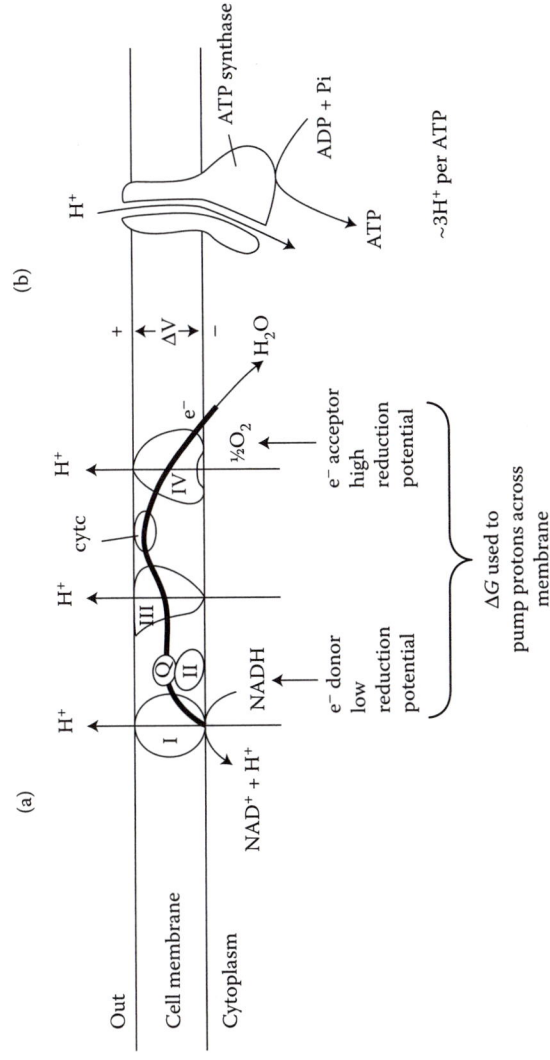

FIGURE 7.23 (a) Respiratory electron chains and (b) oxidative phosphorylation in prokaryotes.

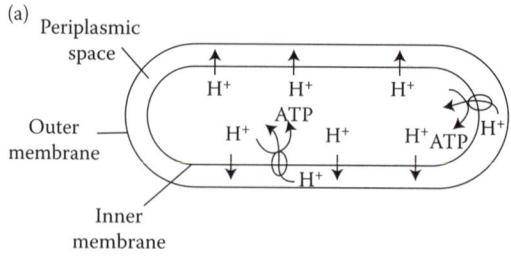

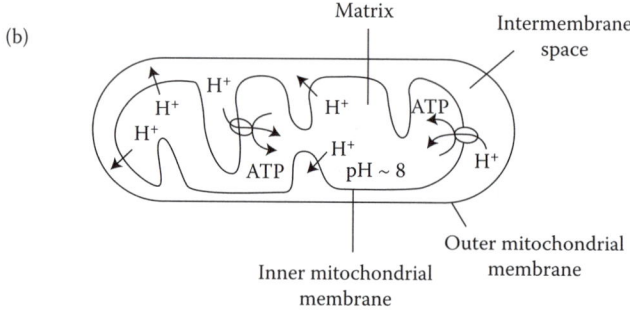

FIGURE 7.24 Location of proton gradients in (a) Prokarya and (b) Eukarya (mitochondria).

7.5.5.3 Electron Transport Uses Metal Ions

Electron transport via electron transport chains happens by change in the redox state of metal ions within the protein carrier molecules. For example, many transport electrons via $Fe^{3+} \rightleftharpoons Fe^{2+}$ transitions. The ferric (Fe^{3+}) iron is reduced to ferrous (Fe^{2+}) by an incoming electron. Subsequently, the Fe^{2+} is reoxidized as the electron spontaneously flows to reduce the next redox couple in the respiratory chain.

Proteins involved in the electron transport show striking structural and functional similarities across widely differing organisms, even across the prokaryote–eukaryote divide. This implies they are very ancient. Many do their redox reactions with iron–sulfur clusters. These could be relics of early life using the mineral pyrite (FeS_2) as an electron carrier. So, why did not life stick with the metal ions alone to power respiration? Partly it is because the rate of electron transfer between donor and acceptor drops off dramatically with their separation. By providing a pathway for electron conduction, proteins increase the rate of redox reactions. For separations of 1.5 nm, protein-catalyzed redox reactions occur on millisecond timescale. The same reactions in the absence of catalyst take hours.

7.5.6 Anerobic Respiration Uses Electron Acceptors Other Than Oxygen

Electron transport systems in anerobes are analogous to those in aerobes but the final electron acceptor is not molecular oxygen but another oxidant, for example, $CO_2^-, SO_4^{2-}, NO_3^-, Fe_3^+$, or organic molecules such as fumarate. Because O_2 is the most oxidizing electron acceptor, anaerobic respiration yields less energy than aerobic

BOX 7.2   CALCULATING THE PROTON MOTIVE FORCE

The pumping of protons across the inner mitochondrial membrane generates a chemical gradient, with higher [H$^+$] outside, and an electrical potential gradient, outside positive.

The free energy of a chemical gradient, where K is the equilibrium constant, is given by

$$\Delta G_C = \Delta G^{\circ\prime} + RT \ln K. \tag{7.7}$$

When there is no net flux of protons the standard free energy of reactants and products are equal, so $\Delta G^{\circ\prime} = 0$ and $K = 1$. (At equilibrium $\Delta G = 0$, so we can write $\Delta G^{\circ\prime} = -RT \ln K$. Clearly for $\Delta G^{\circ\prime} = 0$, $K = 1$.) For protons we can write:

$$\Delta G_C = RT \ln([H^+]_{\text{in}} / [H^+]_{\text{out}}).$$

We can recast this as a difference in pH:

$$\begin{aligned}
\Delta G_C &= 2.3 RT \log_{10}([H^+]_{\text{in}} / [H^+]_{\text{out}}) \\
&= 2.3 RT (\log_{10}[H^+]_{\text{in}} - \log_{10}[H^+]_{\text{out}}) \\
&= -2.3 RT (\text{pH}_{\text{in}} - \text{pH}_{\text{out}}),
\end{aligned}$$

because

$$\text{pH} = -\log_{10}[H^+]$$

so

$$\Delta G_C = -2.3 RT \, \Delta \text{pH}$$

The free energy of an electrical gradient is given by

$$\Delta G_E = zF \Delta \psi,$$

where $z = +1$ and $\Delta\psi = (E_{\text{in}} - E_{\text{out}})$ is the potential difference across the membrane. Since proton pumping maintains the potential outside positive, then $\Delta\psi$ is negative.

The two sources of free energy add: $\Delta G_C + \Delta G_E = \Delta G$,

$$\Delta G = F \Delta \psi - 2.3 RT \Delta \text{pH}. \tag{7.8}$$

The proton motive force, Δp, combines the concentration and electrical potential effects of a proton gradient, and can be defined by $\Delta G = zF\Delta p$, so,

$$\Delta p = \Delta \psi - (2.3 RT / F) \Delta \text{pH}. \tag{7.9}$$

EXERCISE 7.7

For isolated mitochondria at 37°C, $\Delta\psi \sim 0.17$ V, the pH inside the mitochondrial matrix is 7.3, and the pH in the intermembrane space is 6.8. Calculate (a) the proton motive force and (b) the associated free energy change.

The above exercise shows that diffusion of a proton down its electrochemical gradient yields $\Delta G \sim -20$ kJ mol^{-1}. It takes three protons to generate one molecule of ATP. So, provided that the conditions in the mitochondria mean that <60 kJ mol^{-1} of free energy is needed for ATP synthesis; this will proceed spontaneously at the expense of the proton gradient.

---

respiration. In organisms capable of both aerobic and anaerobic respiration in the presence of O_2, aerobic respiration is favored. We will see examples of anerobic respiration as we explore the lifestyles of prokaryotes of interest to astrobiologists.

7.5.7 Fermentation Uses an Endogenous Electron Acceptor

In fermentation, organic molecules form a redox pair in which one acts as an electron donor while the other acts as an electron acceptor. Spontaneous electron transfer generates free energy. The electron-accepting organic molecule is reduced. Only partial oxidation occurs in fermentation, so the energy yield is small. ATP production in fermentation is by substrate level phosphorylation (Figure 7.25):

$$\text{R-OPO}_3^{2-} + \text{ADP} \rightarrow \text{R-OH} + \text{ATP}. \tag{7.10}$$

Fermentation is distinguished from respiration (anaerobic or aerobic) on the grounds that it does not involve oxidative phosphorylation. Fermentation can take place in the complete absence of oxygen. All organisms are capable of fermentation reactions but they get most of their chemical energy from respiration. One of the most important fermentations is glycolysis where glucose is converted into pyruvate.

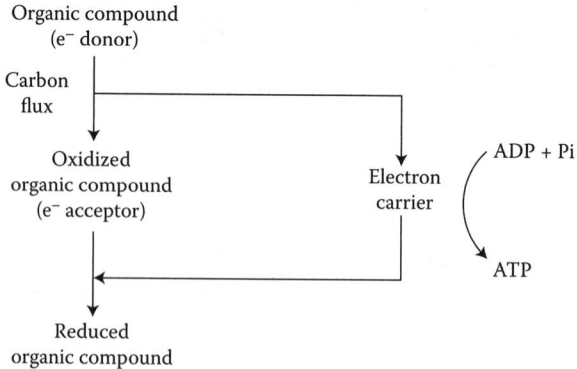

FIGURE 7.25   Fermentation.

7.6 Phototrophs Harvest Sunlight

Photosynthesis uses light energy absorbed by Mg-containing pigments (chlorophylls) to generate a flow of electrons through an electron transport chain. The free energy released is used to synthesize ATP and generate reducing power, NADPH. These constitute the light reactions. Almost all phototrophs are autotrophic, using CO_2 as their carbon source. The CO_2 is reduced to organic compounds using electrons from NADPH and energy supplied by ATP. These are the dark reactions.

7.6.1 Not All Photosynthesis Produces Oxygen

Photosynthesis by eukaryotes (green plants and algae) and cyanobacteria split water to obtain the electrons needed to generate NADPH, producing oxygen as a by-product. This is oxygenic photosynthesis.

Many bacteria use reductants other than water as electron donors. These include molecular hydrogen, reduced sulfur compounds, ferrous iron or organic molecules. Hence, these organisms do not produce oxygen as a by-product. This is anoxygenic photosynthesis.

7.6.2 Oxygenic Photosynthesis Produces ATP and Reducing Power

At this point it is worth noting that in photosynthesis free energy is harnessed via (different) electron transport chains in a way similar to respiration to build reduced organic molecules from CO_2. However, photosynthesis first captures the energy of sunlight to boost electrons to large negative reduction potentials. The electrons come from the splitting of water and yield oxygen:

$$H_2O \rightarrow 2H^+ + \tfrac{1}{2}O_2 + 2e^-. \tag{7.11}$$

This reaction is thermodynamically very unfavorable because it has $E_0' = +0.82$ V. Oxygenic photosynthesis is the only biochemical process that is able to split water and is driven by the extremely low reduction potential of the P680$^+$ molecule, the most powerful biochemical oxidant known. The complex responsible for splitting water contains manganese (Mn) ions.

Electron flow in oxygenic phototrophs uses two linked sequences of light reactions, photosystems I and II (PSI and PSII, Figure 7.26), to generate ATP and NADPH. A set of dark reactions (Calvin cycle) uses this chemical potential energy to reduce CO_2 to organic molecules.

In eukaryotic cells, photosynthesis occurs in organelles termed chloroplasts. These are packed with numerous discs called thylakoids, the membranes of which house the photosystem proteins. Photosynthesis pumps protons across the thylakoid membrane, so the inside of the thylakoid has a lower pH than the stroma of the chloroplast. The proton gradient produces a small transmembrane electrical potential, inside positive. The thylakoid proton motive force is used to synthesize ATP in comparable fashion to respiration. In prokaryotes, the photosystem proteins are in the cell membrane across which protons are pumped.

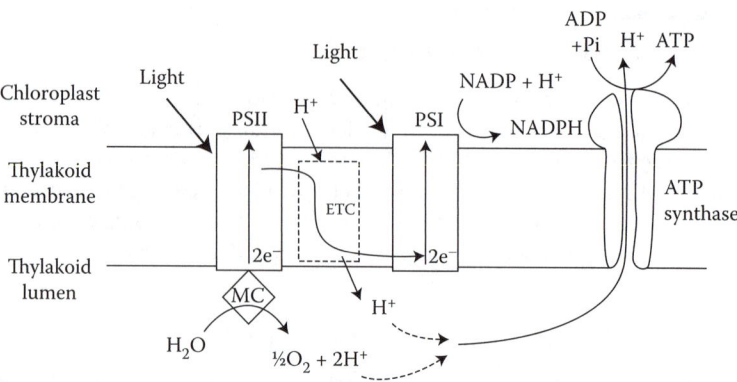

FIGURE 7.26 The Z scheme of oxygenic photosynthesis is so called because of the pattern made by the electron flows. MC; manganese complex.

7.7 Prokaryotes Live in the Crust

Samples recovered from the ocean crust by drilling programs provide biochemical evidence for prokaryotes. Sequencing recovered 16S rRNA has allowed positive identification of species and the discovery of novel organisms. This includes a new species of bacterium, *Bacillus infernus*, found in seafloor sedimentary rocks at a depth of 2.7 km. *B. infernus* is a hyperthermophilic obligate anaerobe; extremely heat tolerant, it can only live in an oxygen-free environment as it is poisoned by oxygen.

Surveys have shown organisms live not just in the seafloor sediments at great depth but also in igneous rocks of both continental and oceanic crust down to depths of 5 km below the seafloor. The limiting factor is probably not depth *per se* but temperature: the maximum temperature at which hyperthermophiles can survive is somewhere between 113°C and 122°C, and temperature increases with depth in the crust.

Exactly how much biomass exists in the crust is not known. Some estimates have suggested it may be comparable to the surface biomass (eukaryotes and prokaryotes) of ~590 × 10^{12} kg C, although more recent work suggests it is not this high.

7.7.1 Crust Provides an Ecologic Niche

Organisms in the crust never see sunlight and there would be little free oxygen, so they have to make a living from simple inorganic compounds provided by the Earth's interior such as H_2, CO_2 and water (this spartan existence is called chemolithotrophy), or from simple organic molecules. For example, a species of proteobacteria identified in igneous rock use hydrocarbons such as methane as carbon sources; that is, they are heterotrophs. These hydrocarbons are likely to have formed abiotically in the crust, in which case these bacteria are independent of the surface biosphere. Organisms living deep in the crust have no choice but to respire anaerobically, using, for example, SO_4^{2-}, NO_3^-, or Fe^{3+} as electron acceptors.

Although the crust may at first glance seem inhospitable it offers opportunities as well as challenges, particularly for heat-loving bacteria and Archaea given that temperature increases with depth. The crust has sufficient pore space to accommodate a large biomass in principle and affords protection from cosmic radiation, solar UV, and provides a relatively constant environment in the face of vagaries of climate change over geological periods of time.

7.7.2 Chemolithotrophs "Eat" Rock

Although most organisms living in the deep crust today are heterotrophs, a minority are chemolithotrophs. These use inorganic molecules as electron donors. Most are able to use CO_2 as sole carbon sources; that is, they are chemolitho*auto*trophs. Many chemolithotrophs use molecular oxygen as an electron acceptor but a number of them—including those in the crust—respire anaerobically. Two types of anaerobic chemolithoautotrophs, acetogens, and methanogens, that get their energy by reducing CO_2 with electrons from H_2, are coming under scrutiny because they might be living fossils, relics of life's first grasp on Earth.

The acetogens are bacteria that reduce CO_2 to acetate:

$$4H_2 + H^+ + 2HCO_3^- \rightarrow CH_3COO^- + 4H_2O, \Delta G^{\circ\prime} = -105 \, \text{kJ} \, \text{mol}^{-1}. \quad (7.12)$$

Acetogens are a diverse collection of organisms, but a recently identified thermophilic species, *Acetothermus autotrophicum*, lies close to the common ancestor of Bacteria and Archaea on the basis of amino acid sequences of four proteins common to all prokaryotes. The implication is that chemolithoautotrophic acetogens arose at the dawn of life.

The methanogens are Archaea, that is, they belong to a domain of life completely different from the acetogens, yet have striking similarities in their metabolism. Methanogens are a large group of extremely heat tolerant (hyperthermophilic) organisms that are so called because they produce methane as a result of their energy metabolism. They reduce CO_2 with electrons donated by hydrogen but using a unique set of electron carriers. Overall:

$$CO_2 + 4H_2 \rightarrow CH_4 + 2H_2O, \Delta G^{\circ\prime} = -131 \, \text{kJ} \, \text{mol}^{-1}. \quad (7.13)$$

Methanogens cluster on the deepest branches of the "tree of life" as defined by sequences of the 16S subunit of ribosomal RNA and so are extremely ancient life forms.

Both acetogens and methanogens fix carbon by means of a reductive acetyl CoA (Wood–Ljungdahl) pathway. Unlike other carbon fixation pathways, such as the Calvin cycle or the reverse Krebs cycle, the reductive acetyl CoA pathway is not cyclical but linear. Overall:

$$4H_2 + 2CO_2 \rightarrow CH_3COOH + 2H_2O. \quad (7.14)$$

Here, in a series of steps, one of the CO_2 molecules is reduced to a methyl group ($-CH_3$) and this is added to a $-CO$ moiety derived from the second CO_2 to yield acetate. In Chapter 8, we will see why acetogens and methanogens are so important.

EXERCISE 7.8

Acetogens and methanogens do not use molecular oxygen for respiration. What do they use?

8

Origin of Life

How life originated on Earth is one of the central questions in science. Our current best guess is that it was conjured by a chemical handshake between warm rock and seawater. In other words, the origin of life was a geochemical event: geology provided the redox and proton gradients to kick-start a primordial, self-sustaining cascade of biochemical reactions.

Elucidating how life may have spawned on Earth should help us identify the necessary and sufficient conditions for its genesis elsewhere. For the moment we neglect the idea that life may have had its origin elsewhere and was delivered to Earth by asteroid or comet. We discuss this notion, panspermia, in Chapter 14.

8.1 When Did Life Originate?

Discovering when life originated on Earth is interesting for astrobiology as it motivates arguments about how easy it is for life to get started. It also gives some insight into the conditions under which life emerged here, and therefore, what to look for elsewhere.

Here we look at attempts to figure out when life originated. They fall into two categories: theoretical models of conditions on early Earth to elucidate when it could have become habitable and evidence for life from the rock record.

8.1.1 When Did Earth Become Cool Enough for Life?

The conventional view of the Late Heavy Bombardment as a 100-million-year fusillade of impactors, including some capable of gouging basins up to 1000 km across, is of a time when the Earth would have been kept sterile. This has been challenged by numerical modeling of impact heating of the crust by Stephen Mojzsis and Oleg Abramov who show that no more than 37% of the planet's surface was sterilized at any given time, and that only 10% reached temperatures above 500°C. Much of Earth's surface would have been cool enough for thermophiles thriving at 50°C to survive. Even if the surface were inhospitable, organisms could live several kilometers into the crust as some do today. If this is realistic, then once there was liquid water on the Earth's surface, which could have been by 4.4 Gyr if we accept zircon grains as evidence for extensive oceans, life that got going this early could have survived the tumult of the LHB to become our ancestor.

8.1.2 Evidence for Early Life

The rock record of Earth's earliest life is sparse. There have been a number of claims for traces of life between 3.85 Gyr and 3.43 Gyr, all of which are controversial, and convincing evidence is elusive until after 3 Gyr. If we cannot be sure about identifying early life on Earth, how much more difficult will it be to recognize the evidence for life on, say, Mars or Europa, where our ability to sample will be far more limited, and we are less sure about what it is we are looking for?

Clearly, reducing our uncertainty about what constitutes evidence for life on Earth is a crucial step in the exploration of life elsewhere (Figure 8.1).

8.1.2.1 Do Light Carbon Isotopes Hint at First Life?

Evidence for earliest life has been based on carbon in the rock record that is enriched in ^{12}C over the rarer, heavier ^{13}C isotope. This might be regarded as a signature of metabolism, because biochemical reactions have a preference for the lighter isotope. This is a matter of kinetics. Reactions involving the light isotope happen faster than those with the heavier one. The result is mass fractionation of carbon isotopes in the rock record. This shows up as alterations in the $\delta^{13}C$ signature (Box 8.1).

BOX 8.1 $\delta^{13}C$

There are two naturally occurring stable isotopes of carbon, ^{12}C and the much rarer ^{13}C. Biochemistry favors the uptake of ^{12}C (for example, $^{12}CO_2$ is taken up preferentially during photosynthesis), so organisms become enriched in ^{12}C and depleted in ^{13}C by comparison with inorganic carbon present as CO_2 in the atmosphere or in bicarbonate and carbonate in water. In other words, organic carbon acquires a lower $^{13}C/^{12}C$ ratio than inorganic carbon. This fractionation of carbon isotopes is given as a $\delta^{13}C$ value expressed as parts per thousand:

$$\delta^{13}C = 1000 \times [(^{13}C/^{12}C)_{sample}/(^{13}C/^{12}C)_{standard} - 1]\%_0$$

The original standard Vienna PeeDee Belemnite (VPDB) was a Cretaceous marine carbonate with a particularly high $^{13}C/^{12}C$ ratio.

If a sample has a lower $^{13}C/^{12}C$ ratio than the standard, which is usually the case, then $\delta^{13}C$ will be negative. The lower the $\delta^{13}C$ value is (i.e., the more negative it is) the smaller the $^{13}C/^{12}C$ ratio. The $\delta^{13}C$ for atmospheric CO_2 is $-8\%_0$, and organic material will always be much more negative, usually less than about $-30\%_0$.

Burying biomass before it has the chance to be decomposed and release its carbon back to the oceans and atmosphere, which can happen if there is a large increase in biomass production, effectively locks up ^{12}C in sediments. This leads to an increase in the $\delta^{13}C$ in oceans and atmosphere. Carbonates precipitated at such a time will therefore have a positive $\delta^{13}C$. On the other hand, a drop in primary productivity or exposure of buried carbon to oxidation will confer a negative $\delta^{13}C$ excursion on carbonates. These changes leave *complementary* signatures in organic-rich sediments.

FIGURE 8.1 Key geographical locations for the study of early Earth. A. Nuvvuagittuq greenstone belt; B. Isua greenstone belt; C. Barberton greenstone belt; D. Acasta gneiss; E. Shark Bay stromatolites; F. Strelley Pool and Tumbiana formations, Pilbara Craton, Western Australia.

EXERCISE 8.1

(a) What is the $\delta^{13}C$ signature of a sample that has the same $^{12}C/^{13}C$ as the Vienna PeeDee Belemnite (VPDB) standard? (b) What is the $\delta^{13}C$ signature of a sample that has a higher $^{12}C/^{13}C$ than the VPDB standard?

However, a major difficulty is that ^{12}C-enrichment is now known to be produced by abiotic chemistry as well as by biology. A further problem is that rocks 3.85 to 3.7 Gyr old, such as the Isua rocks in Greenland, that have been explored for traces of ^{12}C-enrichment are so highly metamorphosed by heat and pressure that it is very hard to determine whether they were sedimentary, or whether the isotope signature is part of the original rock, or was acquired later.

Consequently, we do not have *unequivocal* evidence for life on Earth this long ago.

8.1.2.2 Fossils or Artifacts?

There have been several claims for fossil microorganisms dating back to ~3.5 Gyr. The first to hit the headlines were those found in the ~3465-Myr-old Apex chert of Western Australia. Sections of the cherts 300-μm thick revealed what looked like modern photosynthetic cyanobacteria. But these have long been contentious.

Martin Brasier and colleagues at the University of Western Australia have described more convincing fossils in 3.43-Gyr-old sandstone of the Strelley Pool Formation, Pilbara Craton, Western Australia—once an ancient beach. The fossils are between 5 and 80 μm in diameter, spherical, ellipsoidal or rod shaped, and some are arranged in chains or clusters like modern bacterial colonies. Their boundaries are of uniform thickness, like bacterial cell walls, rather than carbonaceous structures formed by geology. Their carbon has extremely low $\delta^{13}C = -33‰$ to $-46‰$. Intriguingly, micrometer-sized crystals of iron sulphide (pyrite, FeS_2) are associated with the putative cell walls. This pattern of pyrite deposition occurs in modern sulfate-reducing bacteria, and indeed the sulfur isotope make-up of the pyrite suggests that the fossil organisms had sulfate reduction metabolic pathways (Box 8.2). Stromatolites from the same formation also provide sulfur isotope evidence for microbial sulfate reduction.

8.1.3 Precambrian Life Was Dominated by Stromatolites

Modern stromatolites are thin layers of filamentous bacteria, often photosynthetic cyanobacteria, interleaved and bound with carbonate minerals and sediments. Over time they build up into microbial mats about a meter across. The most famous examples are found in Shark Bay, Western Australia (Plate 8.1).

Laminated calcareous structures that resemble modern stromatolites have been found in rocks as old as 3.43 Gyr. In many cases, these have been exquisitely preserved by mineralization in chert. This happens whenever hot SiO_2-rich fluids percolate through the rock and the SiO_2 precipitates out.

The question is whether these earliest examples are actually microbial mats or whether they are abiotic in origin. Several studies have shown that stromatolite-like structures can be sculpted by geological processes without the help of microorganisms.

BOX 8.2   MICROBIAL SULFATE REDUCTION

A substantial number of Bacteria and some hyperthermophilic Archaea use SO_4^{2-} as a terminal electron acceptor (i.e., they respire sulfate) and produce H_2S. In the simplest case, overall,

$$4H_2 + SO_4^{2-} + 2H^+ \rightarrow H_2S + 4H_2O. \qquad (8.1)$$

Sulfate reduction requires an initial input of free energy from ATP. In assimilative sulfate reduction (which the organism uses to grow) the sulfur in H_2S is incorporated into amino acids and other sulfur-containing molecules. In dissimilative sulfate reduction (which is for energy generation) the H_2S is excreted where it can react with ferrous iron in solution to produce insoluble pyrite.

$$Fe^{2+} + H_2S \rightarrow FeS + 2H^+. \qquad (8.2)$$

$$FeS + H_2S \rightarrow FeS_2 + H_2. \qquad (8.3)$$

Most sulfate-reducing organisms are obligate anaerobes, and in the anoxic environments in which they thrive the iron will be reduced.

Modern sulfate-reducing bacteria generally use H_2, lactate and pyruvate as electron donors, but some can oxidize acetate or a variety of other organic acids. Electrons stripped from the donors are transferred through an electron transport chain generating a chemiosmotic membrane potential to synthesize ATP.

The 3.4-Gyr-old Australian sulfate reducers might resemble the modern *Desulfovibrio* because of its simple metabolic requirements. These are chemolithotrophs that use H_2 as an electron donor, sulfate as their electron acceptor, and CO_2 as their sole source of carbon. They use the reductive acetyl-CoA pathway to fix CO_2, which hints at a very early origin as we shall see.

The stromatolites of the 3.43-Gyr-old Strelley Pool cherts have provoked heated debate, with some researchers claiming biogenicity on the basis of morphology and others arguing that the case is unproven. Higher resolution studies of stromatolites using new technology (e.g., see Box 8.3) is helping to clarify this issue. NanoSIMS has allowed microscale measurements of sulfur isotope ratios in organic material (kerogens) in the Strelley Pool stromatolites, showing that they *could* have arisen from the activities of sulfate-reducing organisms.

The most ancient stromatolite for which a biological origin is almost universally accepted is in the 2.72-Gyr-old Tumbiana formation of Western Australia. By the Proterozoic (2.5 Gyr) fossil stromatolites are found in North America and Siberia as well as Western Australia and they peaked in abundance and diversity about 1.5 Gyr ago. These contain fossil microorganisms that resemble cyanobacteria.

PLATE 8.1 (**See color insert.**) Stromatolites growing in Hamelin Pool Marine Nature Reserve, Shark Bay, in Western Australia. (Photo credit: Paul Harrison, Reading, UK.)

8.2 Building the Molecules of Life

Origin of life hypotheses need to account for where and how complex biomolecules came to be synthesized from simple organic molecules, how the first self-sustaining metabolic pathways got started, and how replication was invented.

8.2.1 Where Did Prebiotic Synthesis Happen?

The consensus view is that the origin of life requires first the assembly of macromolecules (nucleic acids, proteins, etc.) from simple organic molecules (bases, amino acids), and these in turn are built from inorganic molecules or atoms (H_2O, CO, H_2, N_2, etc.). The reactions that occurred before the emergence of life constitute prebiotic synthesis. The question is, where did this prebiotic chemistry start? Historically, the first notion was that simple organic molecules were synthesized on the early Earth. Subsequently, as cosmochemists discovered increasing numbers of organic molecules in the interstellar medium, meteorites, and comets, the idea was entertained that prebiotic synthesis could have began in the presolar nebula, and the products were then delivered to Earth via asteroids and comets. In this section, we explore these ideas.

BOX 8.3 SECONDARY ION MASS SPECTROMETRY (SIMS)

Secondary ion mass spectrometry (SIMS) is an ion microprobe technology that allows qualitative mapping of the distribution of elements in a sample. NanoSIMS is a high resolution version. It can be made quantitative by the use of calibration standards. A primary ion beam (Cs^+ or O^-) is scanned across the surface of the sample, and the sputtered ions are extracted to a mass spectrometer which determines their mass/charge ratio. With the Cs^+ primary beam negatively charged secondary ions can be detected, and with the O^- primary beam, positively charged secondary ions can be detected. Images are typically acquired over a 50 µm field of view, and the technique can reveal the elemental composition of the surface to a depth of 1–2 nm. Lateral resolution under 50 nm is achievable for elemental mapping with a Cs^+ primary beam. Several elements or isotopes to be mapped simultaneously with a sensitivity of ppm or ppb depending on the element, and isotope ratios can be determined with uncertainties of a few tenths ‰. Fortuitously, elements commonly found in organic materials—C, N, O, and S—have relatively high negative secondary ion yields when sputtered with a Cs^+ primary ion beam because of their high electronegativity. Although N does not readily ionize during SIMS sputtering, it does have a particularly strong emission when coupled to C as the CN^- ion.

NanoSIMS can be used to map element distributions across rocks, meteorites, or even individual cells. Samples are placed in a high vacuum (10^{-6} mbar) to prevent secondary ions being scattered from gas molecules or gas molecules being adsorbed onto the surface of the sample and contaminating the signal. Rock samples must be highly polished. Biological samples must be fixed and dehydrated. To study isotopes that are mobile, the samples are frozen. If samples are non-conductive, they must be coated with a metal such as gold.

8.2.1.1 Urey–Miller Experiments Have a Fatal Flaw

In some of the most famous scientific experiments ever done, Harold Urey and Stanley Miller in 1952 set out to synthesize organic molecules from gases assumed to be present on early Earth. This was to be the first step in the genesis of life.

The assumption at the time was that life had emerged from a primeval soup, rather as Darwin had surmised in a letter to botanist Joseph Hooker in 1871: "we could conceive in some warm little pond, with all sorts of ammonia and phosphoric salts, light, heat, electricity, etc., present, that a protein compound was chemically formed ready to undergo still more complex changes." Using the spectroscopically determined composition of Jupiter's atmosphere as a proxy for what the early Earth would have captured from the solar nebula, it was reckoned that the major gases in the Earth's early atmosphere were water (H_2O), methane (CH_4), ammonia (NH_3), and hydrogen (H_2). Accordingly, the original Urey–Miller experiment mixed these gases in a flask, passing high voltage sparks through it to simulate lightning that provided the energy for chemical reactions. After a week, more than 10% of the carbon was in the form of organic molecules such as formaldehyde (CH_2O) and hydrogen cyanide (HCN) and, remarkably, these had reacted with ammonia to produce amino acids, the building blocks for proteins.

This was a very exciting result, and inspired Miller and others to do similar experiments. These produced a greater diversity of amino acids, and succeeded in making other biochemicals such as amines, sugars, and the bases needed to build RNA and DNA.

Despite this apparent success the Urey–Miller experiments went out of fashion, partly because ideas about the composition of the Earth's early atmosphere changed. Geological evidence now shows it was not as reducing: CO_2 and nitrogen (N_2) would have been abundant, whilst methane and ammonia would have been only minor components. This changes the chemistry so that the diversity of organic molecules generated is much less, and amino acids are unstable.

Remarkably, Miller did experiments in 1958 which he never analyzed. In them he had attempted to simulate conditions around volcanic vents by including hydrogen sulfide (H_2S) which is outgassed by volcanoes. Vials containing dried residues of these lost studies have come to light since Miller died in 2007 and these have now been analyzed using techniques that are thousands of times more sensitive than the methods available to Miller. These experiments produced 23 amino acids, including 6 sulfur-containing ones and 3 amino acids found in proteins that have never been detected in any other spark discharge experiments. It might be argued that even if most of early Earth's atmosphere was not like those in the Urey–Miller brews, such gas mixtures are bound to have existed locally around volcanoes. A more intriguing possibility is hinted at by the fact that the relative abundances of amino acids in the H_2S experiment are similar to those found in three carbonaceous chondrite meteorites. In other words, the chemical reactions in Miller's H_2S experiment might have happened on the parent body of the carbonaceous chondrites.

Despite this and pioneering as they were, the Urey–Miller experiments have now been superseded for two reasons:

i. As we have seen, a great diversity of organic chemistry happens in the interstellar medium under conditions very different from those in Urey–Miller experiments, and the resulting organic molecules were delivered to early Earth in comets and meteorites. This loosens the requirement for local synthesis.

ii. But the killer flaw is that Urey–Miller experiments were all done under equilibrium conditions. Yet if there is one attribute of life we can be sure of it is that its chemistry is far from equilibrium. It means that these experiments would always stick at the stage of relatively simple organic molecules. We would not expect them to yield complex macromolecules, let alone ordered biochemistry. Moreover, the same stricture would appear to rule out a primeval soup scenario for the origin of life, since the chemistry in a "warm little pond" would rapidly reach equilibrium.

8.2.1.2 Carbonaceous Chondrites Are Rich in Organics

There are eight groups of carbonaceous chondrites, undifferentiated carbon-rich meteorites distinguished by differences in composition and the extent to which they have been heated and/or altered by water over the past 4.5 Gyr. Five of these groups contain amino acids.

The organic chemistry of a CM group rock, the Murchison meteorite, has been particularly well studied. Along with the CI group, the CM meteorites have a high volatile content (water concentration ranges between 3% and 22%), they contain water-altered minerals such as serpentinite, carbonate and clays, and have not been heated above 200°C.

About 2% by mass of the Murchison meteorite is carbon, about three-quarters of which is organic carbon. There is a rich diversity of organic compounds including aromatic hydrocarbons, carboxylic acids, fullerenes, aliphatic hydrocarbons, amines, amino acids, aldehydes, ketones, and alcohols. Many meteorite amino acids are not α-amino acids found in proteins. There are trace amounts (~1 ppm) of purine and pyrimidine nucleobases. However, there are no nucleosides or nucleotides, neither are there any sugars that make up the backbones of nucleic acids, or that are found in polysaccharides.

The amino acids of the Murchison meteorite, present between 17 and 60 ppm, have attracted a deal of attention. A central question is whether they are terrestrial contaminants. This cannot be ruled out but of the 70 amino acids in the meteorite, many have never been seen on Earth and not all of the amino acids found in proteins are present in the meteorite.

However, the waters were muddied somewhat by finding that some of the amino acids in several carbonaceous chondrites, including Murchison, showed a small excess of the L-enantiomer rather than the racemic mix that would have been predicted. Detailed studies have shown that the enrichment in L-amino acids is due to processing of amino acids in the meteorite by chemistry involving water.

Precisely how all the organic compounds in meteorites got synthesized is not known, but there are a number of plausible abiotic reactions.

It has been suggested that the Murchison meteorite L-enantiomeric excess is also consistent with a biotic origin, and hence that life could have emerged in the asteroid belt. This is highly speculative, and unless much more convincing evidence to support such a claim turns up, we should reject it.

8.2.1.3 Most Organic Carbon in Meteorites and Comets Is Unavailable

It is tempting to think of meteorites and comets as vectors for delivery of raw ingredients for more complex prebiotic synthesis on Earth. It is worth noting that most of their organic carbon is effectively locked up in what is termed macromolecular carbon that resembles terrestrial kerogen. This is not a single compound but a mixture of molecules. Each is a highly disorganized arrangement of units of a few coupled aromatic rings linked by different sorts of aliphatic chains. It is thought to have been synthesized at ~1000 K from simple aliphatic hydrocarbons (CH_4 and C_2H_2) in gas phase in the solar nebula. The chemistry is similar to the production of soot particles in combustion and formation of aromatic hydrocarbons in the circumstellar envelopes of C-rich stars. Macromolecular carbon nanoparticles are thought to be responsible for the unidentified infrared emission (UIE) features seen in the interstellar medium (ISM). So, meteorites may have inherited macromolecular carbon from the ISM. But the crucial thing is that macromolecular carbon resembles tar; it is insoluble in water so unavailable to engage in any potential biochemistry, whether in "some small, warm pond" or in deep ocean hydrothermal vents. It is the inventory of *soluble* organic carbon in meteorites and comets that is important for fuelling the chemistry of life.

8.2.2 Did Replication Precede Metabolism?

The first organisms needed to be able to do two things. One was to replicate. Without this life could not be self-sustaining because eventually some catastrophe would be bound to destroy a single entity. The other was to turn disordered prebiotic chemistry into a more ordered sequence of reactions that somehow extracted free energy from its surroundings to do work: metabolism. The formation of an ordered state from a disordered one is not forbidden by the second law of thermodynamics, but it is improbable. Metabolism was needed to control the flow of chemical potential energy, ensuring that order could be assembled while dumping more than the equivalent amount of disorder (entropy) into the surroundings.

These two requirements have led to two schools of thought. The earliest, championed by Aleksandr Oparin, is that metabolism arose first. The alternative is that self-replication was the herald for life. We consider each of these, starting with replication.

8.2.2.1 Did Life Begin with a Self-Copying Molecule?

The molecular biology revolution that started with the Watson and Cricks 1953 paper on the structure of DNA sparked a great deal of interest in replication in the 1960s and the 1970s. Studies on an RNA virus that infects bacteria (Qβ bacteriophage which uses RNA as its genetic material rather than DNA) showed that RNA could replicate exponentially in a test tube provided that ribonucleoside triphosphates and, *crucially*, a protein enzyme Qβ replicase were supplied. By restricting the time that the viral RNA was given to replicate, after 75 generations, the Qβ RNA genome had become much shorter (it had shed genes that served no function in its unnatural test tube environment) and replicated 15-fold faster than the wild type Qβ RNA. A second set of experiments conducted by adding a drug that bound to a specific sequence on the RNA which drastically slowed its replication rate resulted in the emergence of an RNA in which three bases on the binding site had altered. The mutant RNA replicated faster than the normal RNA and came to dominate the nucleic acid population. Both of these studies demonstrated selection at the molecular level. In other words, the replicator-first model satisfies the requirement that life is subject to Darwinian evolution.

However, a key problem is that the RNA replication did not happen by itself. It required a protein catalyst, the replicase. This raised what came to be called the "chicken and egg" paradox. The whole point about replicating nucleic acids is they code for protein, but if proteins are required for nucleic acids to copy themselves, there is an impasse.

8.2.2.2 Ribozymes Support an RNA World

A potential escape from the paradox was offered by the discovery that some RNA molecules had catalytic activities. These RNA enzymes are termed ribozymes. An

RNA molecule that could catalyze its own replication would solve the "chicken and egg" problem. This led to the idea, proposed by Walter Gilbert in 1986, of an RNA world in which RNA was the genetic material and subserved the functions of proteins. The RNA world preceded the "DNA makes RNA makes protein" world in which we live today.

Gilbert's proposal is that RNA molecules that catalyzed their own replication emerged first. Subsequent mutations allowed the evolution of a variety of ribozymes with different catalytic activities, including the synthesis of proteins. Protein enzymes were rapidly selected in favor of ribozymes because 20 side chains in proteins provided far greater versatility than four bases of RNA. Eventually, the emergence of reverse transcriptase led to the ability to store information in the more stable double helix of DNA than the single strand of RNA. At this stage, RNA was left with its contemporary functions as information carrier (mRNA); plug and play hardware (tRNA); and component of the ribosome (rRNA).

8.2.2.3 Evidence for the RNA World

Many cofactors (molecules required in enzyme-catalyzed reactions but not directly responsible for the catalysis) involved in core metabolism contain an adenine ribonucleotide. The classic example is ATP; others are coenzyme A (CoA) nicotinamide adenine dinucleotide (NAD) and flavin adenine dinucleotide (FAD). Their structure is taken to reflect ancient origins in the RNA world, even if their function in modern biochemistry is somewhat different. Thus, NAD acts as an electron carrier in respiration. It can do this as well by interacting with a protein enzyme now as it did, maybe, with a ribozyme 3.8 billion years ago. All NAD has done, according to the RNA world hypothesis, is change its dancing partner.

A number of ribozymes, catalyzing a variety of reactions, have been synthesized in the laboratory. This includes a ribozyme polymerase that can catalyze the synthesis of RNAs up to 95 nucleotides long that was itself able to catalyze the manufacture of another ribozyme. As the researchers pointed out this recapitulates a central aspect of an RNA-based genetic system: the RNA-catalyzed synthesis of an active ribozyme from an RNA template.

Ribozymes exist in nature. Ribosomes contain an RNA component that is a ribozyme; it catalyzes the addition of an amino acid to a growing peptide. A catalytic RNA molecule that cuts and ligates RNA molecules has been isolated from a single-celled eukaryote.

Recently (2013) it has been shown that under early Earth conditions some RNA molecules can catalyze electron transport. The implication is that some sort of bioenergetics, a primordial respiratory electron chain perhaps, could have operated in the RNA world. Today Mg^{2+} is essential for folding of RNA molecules and its catalytic abilities. Loren Williams and colleagues at Georgia Institute of Technology have demonstrated not only that ferrous iron can substitute for magnesium in RNA folding and catalysis but that Fe^{2+} allowed some RNA molecules (specifically 23S ribosomal RNA and tRNA) to catalyze single-electron transfer. The researchers hypothesize that Fe^{2+} was an RNA cofactor when soluble iron was abundant in the anoxic Archaean oceans, and was replaced by Mg^{2+} around 2.4 Gyr ago when Fe^{2+} become much less readily available due to the rise in atmospheric oxygen.

8.2.2.4 Making Ribonucleosides Is a Problem

Advocates of the RNA world hypothesis argue that there must have been abiogenic formation of ribonucleotides from their basic components, a base (A, G, C, and U), the sugar ribose, and phosphate. It has been assumed that this first requires nucleosides, such as adenosine (A linked to ribose). Although nucleobases and ribose are found in meteorites, nucleosides are not, implying that these formed on early Earth. If so, the conditions are not simulated in Urey–Miller type experiments. While these form ribose, the yields of bases are very low and ribonucleosides are not found. In fact, it turns out to difficult to make either adenine or ribose in the laboratory under plausible early Earth conditions.

8.2.2.5 Pyrimidine Ribonucleotide Synthesis Can Happen Under Prebiotic Conditions

A solution to the apparently intractable difficulty of making ribonucleotides from base and ribose is a reaction scheme developed by John Sutherland and colleagues at Manchester University in England. They side-stepped the problem by not trying to make base and ribose but starting with plausible prebiotic molecules (Figure 8.2): the simple sugar glycoaldehyde, cyanamide, glyceraldehyde, cyanoacetylene, and inorganic phosphate. Including the phosphate right at the start, even though it does not get incorporated into the nucleotides until late in the reaction sequence, is crucial to the success of the synthesis. By acting as a pH buffer and a catalyst at earlier steps it short-circuited a network of unwanted reactions, leading to the synthesis of a key intermediary, 2-aminooxazole. Several further steps yield ribocytidine-2′, 3′-cyclic phosphate. Phosphate proved to be important in controlling the chemistry, not just a substrate, in these later steps too. From ribocytidine the activated nucleotide (nucleoside triphosphate) can be synthesized. This is encouraging because there has been some success in polymerizing RNA from activated ribonucleotides abiotically on mineral surfaces. Remarkably, UV irradiation of the reaction mix converted some of the ribocytidine into ribouridine, as well as destroying side products.

Hence, this scheme synthesizes two of the four nucleotides seen in RNA. A mechanism to get to purine nucleotides has not yet been published.

8.2.2.6 Abiotic Synthesis of Self-Replicating RNA Is a Problem

Nucleoside triphosphates (activated nucleotides) are the units from which RNA is synthesized on complementary nucleic acid strands in living organisms. It requires RNA polymerase. Without the enzyme RNA synthesis in this way does not work. Considerable efforts were made in the 1980s to find a way to make RNA nonenzymically as required by the RNA world hypothesis. Eventually, chemically modified mononucleotides that would synthesize RNA on template polynucleotide molecules *without* RNA polymerase were discovered. There was some success getting G-nucleotides to assemble on C-rich templates, although none of the products could self-replicate efficiently. U-, T- and A-based systems proved problematic. The chemically modified mononucleotides do not occur in nature. The abiotic synthesis of an RNA replicator has not yet been achieved.

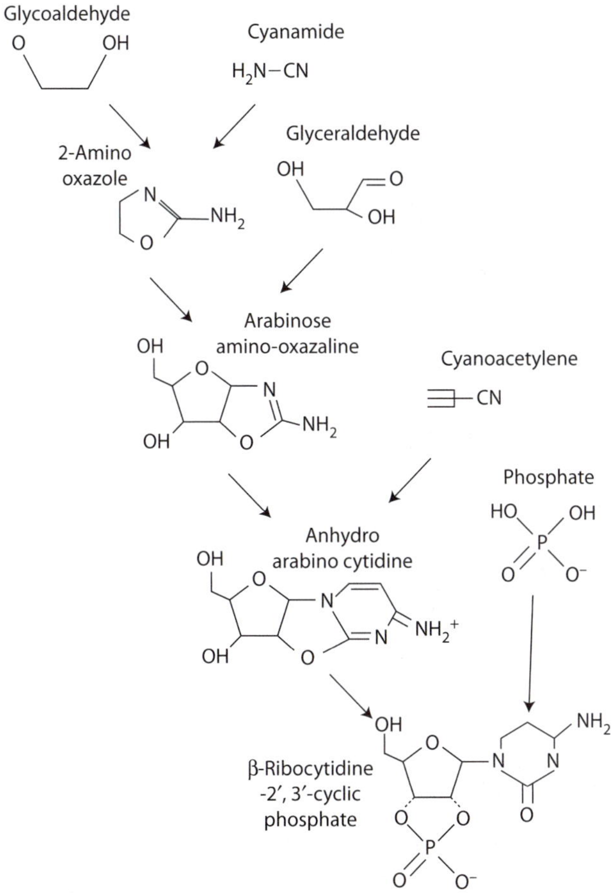

FIGURE 8.2 Prebiotic synthesis of pyrimidine nucleotides.

8.2.2.7 Was There a Pre-RNA Replicator?

The difficulty with abiogenic RNA synthesis has led to the notion of a pre-RNA world in which some other polymer, easier to make on a prebiotic Earth, was the replicator. Several alternatives have been suggested. Although many chemical analogs of nucleic acids have been synthesized, generally only those that are able to form a double helix with RNA are considered candidates, since otherwise it is hard to see how the switch to modern nucleic acids could have happened. There are three front runners (see Figure 8.3), two of which are derived from sugars.

PNA is probably the best bet. It is stable. It makes duplexes with DNA as well as RNA. It could be formed more simply than RNA and DNA under early Earth conditions. It can polymerize spontaneously at 100°C; that is, it does not require enzymes to assemble single-stranded PNAs. Trace amounts of PNA have been detected in Urey–Miller experiments.

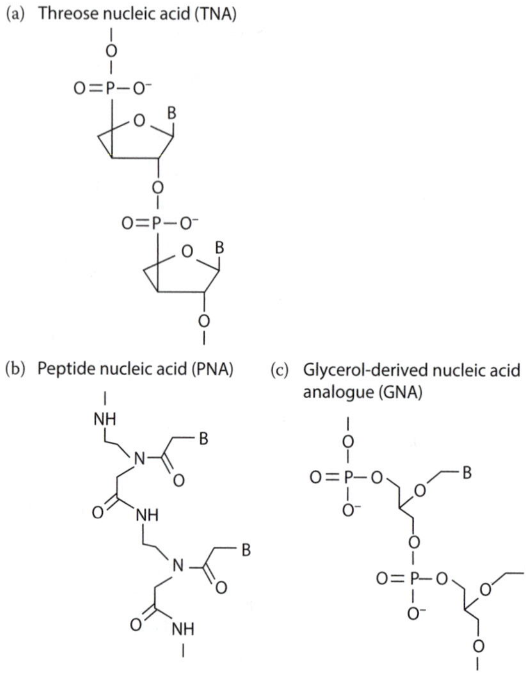

FIGURE 8.3 Structures of alternative replicators: (a) TNA, (b) PNA, and (c) GNA. B, nucleotide base.

8.2.2.8 Why the RNA World Idea Might Be Wrong

A central tenet of the RNA world hypothesis is that RNA appeared before proteins, but works on ribosomes—the "factories" on which proteins are synthesized—contradicts this. Recall that ribosomes are ubiquitous in cells. They are complex, consisting of numerous rRNA molecules and up to 80 proteins. Until 2011 it had been argued that the site in the ribosome where amino acids were actually linked together, the peptidyl transferase center (PTC) active site that is formed by ribozyme rRNAs was the most ancient. However, by analyzing the sequences of the rRNA and ribosomal proteins in detail it has been possible to come up with evolutionary timelines for each of the components. Family trees of rRNA and ribosomal proteins show great congruence with each other and reveal that the proteins surrounding the PTC are as ancient as the rRNAs that form it. What is more, the PTC appeared after the two ribosomal subunits came to be coupled together. The inference is that proteins and RNA of the ribosome have coevolved. Interestingly, not all protein synthesis requires ribosomes. Nonribosomal peptide synthesis (NRPS) is an alternative seen in many bacteria and fungi in which short (5–20 residues) peptides are synthesized by enzymes (synthetase) *without* mRNA. Each synthetase is responsible for producing a specific peptide. Could it be that NRPS predates ribosomal protein synthesis? The discovery of an aminoacyl tRNA synthetase that transfers amino acids not to tRNA, but to an amino acid carrier *protein* is a tantalizing hint of an evolutionary link between NRPS and conventional translation on ribosomes.

8.2.3 Did Metabolism Emerge Before Replication?

Although life now is dominated by DNA, RNA, and proteins, the metabolism-first school as first formulated by Oparin in 1953 suggested that peptides evolved first and were selected for catalytic activity that drove metabolism that favored the survival of the system, for example, the synthesis of adenosine triphosphate (ATP). Crucially, the theory required an encapsulating membrane to maintain integrity, forming a protocell. Whatever the composition of the membrane presumably its components would have to be the products of enzyme-catalyzed reactions. Reproduction of such systems would be by physical splitting. This would happen when the number of molecules increased to the point where osmotic pressure caused the protocell to rupture. A membrane composed of phospholipids would be expected to self-seal. Provided each daughter protocell got its fair share of all the molecules needed to run the metabolic activities of the parent, both could survive. This version of the metabolism-first hypothesis seems to have hit the buffers. One reason is that it does not account for key differences between the two domains of prokaryote: Bacteria and Archaea.

A more recent version of the metabolism-first hypothesis argues that initially there were no macromolecules. Instead, networks of reactions between relatively simple small molecules evolved which harnessed energy from their surroundings to keep their chemistry going; that is, a self-sustaining chemical system. Possible energy sources include redox reactions involving minerals, volatile inorganic or organic molecules, pH gradients, or even light, but arguably the most complete version with the greatest explanatory power is that the energy was provided by redox gradients provided by the geochemical environment. Such systems would potentially be very vulnerable to external disruption (e.g., dissipation of concentration gradients by currents, wind, etc.), but it is not necessary to postulate the early development of protocells with biological membranes if these systems were protected by the geological setting.

Catalysis before protein enzymes or ribozymes is a central issue. Uncatalyzed, most of the reactions would have proceeded with a timescale slower than the time course in which the system itself could be dissipated. However, simple mixtures of amino acids have been shown to achieve modest catalytic activities (up to one thousandth that of protein enzymes), and mineral surfaces can catalyze a variety of reactions. A critical step is the generation of autocatalytic cycles. An autocatalytic reaction is one in which the reaction product itself is a catalyst for the reaction. If a sufficient number of reactions in a set of chemical reactions are autocatalytic, then the entire set of chemical reactions is self-sustaining given an input of energy and substrate molecules. Autocatalytic cycles are nonequilibrium systems that show feedback and nonlinear dynamics, allowing them to exhibit a wealth of complex behavior.

8.3 How Did Life Originate?

The answer to this question at present is we do not know. Almost certainly we will never know for sure precisely how life got started but eventually we should be able to come up with detailed, statistically most plausible accounts that go from simple molecules to the first independent, free-living prokaryote cells. These models will be based on the following:

1. Paleogenomics—using molecular evolution models to make inferences about the origins of life by working backward from DNA and RNA sequences of contemporary organisms,

2. Detailed thermodynamic and kinetic models of developing biochemical systems that include the influence of Darwinian selection,

3. Laboratory experiments which test models under conditions in which life might have originated,

4. Synthetic biology experiments—bioengineering novel biological systems or even entire new organisms—to test inferences deduced from paleogenomics and to determine the minimum biochemistry required for life. A synthetic biology approach could in theory also explore alternative biochemistries.

8.3.1 What Were the First Organisms?

Urey–Miller experiments were motivated by the notion that the first organisms were heterotrophs. The simplest heterotrophs use fermentation, with the substrates they needed (sugars, organic acids, etc.) provided by prebiotic chemistry in a primordial broth. But as we have seen, the equilibrium condition of a primeval soup takes it off the menu.

It gets worse. It had seemed reasonable to suppose that energy flows from catabolic to anabolic reactions were mediated by substrate level phosphorylation (SLP) on the grounds that there are several SLP steps in fermentation and it is much simpler than oxidative phosphorylation. In SLP, phosphate groups are transferred directly from substrate to ADP in an energetically favorable reaction:

$$RO\text{-}PO_3^{2-} + ADP \rightarrow R\text{-}OH + ATP. \tag{8.4}$$

The ATP can then be used to provide free energy to drive an energetically unfavorable reaction. However, all extant free-living fermenters, even those which are thought to be the most ancient, are tied to chemiosmosis. The implication is that chemiosmosis came first. In addition, the enzymes involved in fermentation are completely unrelated in Archaea and Bacteria, so it looks as if fermentation evolved at least twice. In other words, it was not in the biochemical repertoire of the last universal common ancestor.

The smart money is now on chemoautotrophs. Recall these organisms get their energy from simple inorganic reactions and their carbon from CO_2. As we shall see, there are geological settings which provide far-from-equilibrium conditions, free energy, a supply of CO_2, and warm temperatures which would make the synthesis of some biomolecules energetically favorable.

8.3.2 What Was the Last Universal Common Ancestor?

Perhaps we can work out what the first life was like by deducing what characteristics the last universal common ancestor (LUCA) had. She gave birth to Archaea and Bacteria and was closer to life's origin than either. It follows that if Archaea and

Bacteria share a trait in common it was most likely inherited from LUCA, although some traits have clearly been acquired later by horizontal gene transfer. Consequently, it is assumed that most traits not shared by Archaea and Bacteria evolved independently. In contrast, *shared* traits point back to LUCA.

These show that LUCA had DNA, RNA, proteins, a universal genetic code, ribosomes, transcription, and translation. She had ATP and ATP synthase, so was capable of chemiosmosis. The implication is that she had access to, or could create, a proton gradient.

8.3.2.1 LUCA Lacked Several Traits

This all seems very familiar yet there are at least three striking differences between Archaea and Bacteria. First, as we have already seen, they use totally different enzymes for fermentation. Second, they have distinct mechanisms for DNA replication. Third, they have very different membrane lipids. From this it seems that LUCA was:

1. Incapable of fermentation,
2. Did not replicate DNA by any method that exists today,
3. Did not have cell membranes that resemble those in extant organisms.

8.3.2.2 LUCA Was a Hyperthermophile

Ribosomal RNA (rRNA) has a sequence of bases that is unique to each species. However, the more closely related two species are the more similar their rRNA sequences. Hence by comparing the rRNA sequences of several organisms it is possible to work out a family tree on the basis of how closely related they are to each other.

When this is done for the sequences of the small-subunit RNA of extant life, it shows that the hyperthermophiles in the Bacteria and Archaea domains lie at the root of the tree of life: they are the closest living relatives of the hypothetical last common ancestor.

Hence life either started at high temperatures (the hyperthermophilic Eden hypothesis), or it started at low temperature, hyperthermophiles subsequently evolved and were the only traces of early life to survive, the others having been obliterated during the late heavy bombardment (hyperthermophilic Noah hypothesis). The consensus now is that in the Archaean Eon, the ocean temperatures were generally tepid, so the hyperthermophilic Noah hypothesis is plausible. However, hydrothermal vents in the ocean crust do provide temperatures favored by hyperthermophiles (70–110°C) and could have been a hyperthermophilic Eden.

8.3.3 Hydrothermal Vents Are Prime Candidates for Genesis

Hydrothermal vents form wherever water percolates through rock that is heated by underlying magma. Consequently, hydrothermal systems are situated wherever magma gets close to the surface such as on the slopes of volcanoes (subaerial systems) or on ocean crust close to mid-ocean ridges (MORs). Hydrothermal vents may have been much more common on the early Earth than now because of the higher geothermal heat flow.

EXERCISE 8.2

Why might submarine hydrothermal vents be a better bet than subaerial ones for the origin of life?

8.3.3.1 Black Smokers Are Acidic Hydrothermal Vents

Acidic hydrothermal vents have been known since 1979, and were made famous as "black smokers" with the publication of images of roiling black fluids erupting from tall chimneys rising from the seafloor. That they harbor ecosystems that do not depend on photosynthesis, including hyperthermophilic Archaea, coupled with evidence that early organisms were hyperthermophiles, led to the idea that "black smokers" could have been where life originated.

Acidic vents lie close to MORs. Seawater percolating through cracks in the ocean crust is heated at depth and reacts with basalt to produce high temperature (~400°C), low pH (3 to 5), anoxic fluids, typically rich in CO_2, H_2S (and sometimes hydrogen and methane), and laden with metals that dissolve in hot, acidic fluids; for example, Fe, Cu, Zn. When this erupts into cold, oxic, sulfate-rich seawater, tiny particles of metal sulfides and oxides are precipitated forming the characteristic black "smoke." At temperatures above about 150–200°C the mixture precipitates the mineral anhydrite ($CaSO_4$) forming chimneys. The pore spaces in the chimney get lined by chalcopyrite (Fe–Cu sulfide).

8.3.3.2 Acidic Vents Were Initially Proposed for Life's Origin

Acidic vents provide a location where metal sulfides can catalyze basic biochemistry without the requirement for proteins. This was the central idea of Günter Wächtershäuser. Primordial anaerobic metabolism would have derived its energy from the spontaneous combination of hydrogen sulfide with iron to form the mineral iron pyrites:

$$Fe(II)S + H_2S \rightarrow FeS_2 + H_2. \qquad (8.5)$$

This provides an estimated $\Delta G^{\circ\prime} = -38.4$ kJ mol^{-1}, enough energy to reduce CO or CO_2, which are energetically unfavorable reactions:

$$CO + H_2 \rightarrow \underset{\text{formaldehyde}}{HCHO}, \qquad (8.6)$$

$$CO_2 + H_2 \rightarrow \underset{\text{formic acid}}{HCOOH}. \qquad (8.7)$$

Importantly, FeS_2 forms a two-dimensional positively charged mineral surface. The charge would constrain metabolites, most of which would be negatively charged, preventing them from leaking away. This is essentially serving one of the main functions of a biological membrane. The positive charges would also allow reactants to be brought together and oriented, increasing reaction rates. In other words, FeS_2 would act as a catalyst.

Subsequently, Wächtershäuser and colleagues demonstrated a variety of reactions that could have occurred in anoxic acid vents in the presence of iron sulfide, nickel

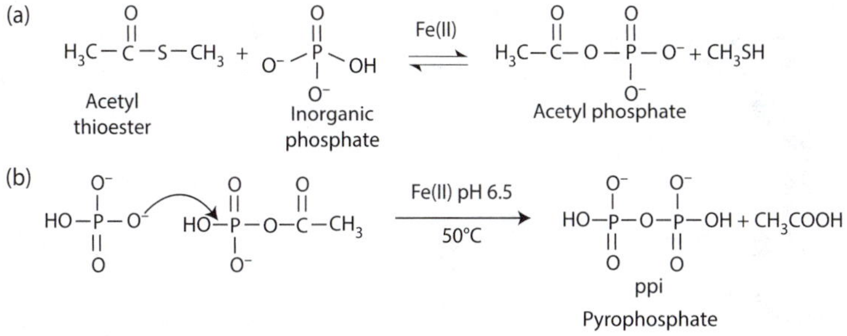

FIGURE 8.4 (a) Acetyl phosphate and (b) pyrophosphate can be synthesized under hydrothermal vent conditions.

sulfide, and other metal catalysts. An important result is the synthesis of the methyl thioester of acetic acid (CH_3–CO–SCH_3) an analogue of acetyl-CoA. Other molecules important in metabolism were formed. These included pyruvate, lactate, and glycerate, and several amino acids, with the nitrogen coming from either hydrogen cyanide (HCN) or ammonia. The energy currency of modern cells is ATP, but much simpler molecules could perform the same function early on. These include acetyl phosphate and pyrophosphate. They are readily synthesized from methyl thioester under vent conditions (Figure 8.4).

The key role of iron sulfide in all this and the fact that there are numerous enzymes involved in core biochemistry that have iron–sulfur clusters as part of their structure has given rise to idea of an iron–sulfur world.

Exciting as all this is, acidic vents pose problems. Although they allow far-from-equilibrium conditions, their internal structure has large spaces in which reactants are not going to be concentrated enough for the chemistry to go very fast. This problem does not appear to be solved by having the reactions occur on the surface of the mineral sulfides. Although this concentrates reactants, reactions do not go to completion unless products are released from the surface, and when that happens, they dissipate. In addition, if acidic vents were as short-lived on the early Earth as they are now it means that if they were to be where life originated, it would all have to happen very fast.

The iron–sulfur world has not vanished. However, the discovery of hydrothermal vents with different internal structure and chemistry points the way forward.

8.3.3.3 Alkaline Vents Are More Favorable Cradles Than Acidic Vents

Alkaline vents are hydrothermal systems that lie several kilometers from mid-ocean ridges. Being off-axis gives these systems different properties from their "black smoker" cousins, including lower temperatures, greater longevity, and different chemistry. Many astrobiologists now regard alkaline vents as the best candidates as nurseries for life.

The most studied alkaline vent system, Lost City, was discovered by accident in 2000 (Plate 8.2). Lying 15 km west of the mid-Atlantic ridge (30°N, 42°W) on ocean crust just 1.5 Myr old, the alkaline vent system consists of numerous spires of calcium carbonate ($CaCO_3$) and brucite ($Mg(OH)_2$) up to 60 m in height. In contrast to

PLATE 8.2 (**See color insert.**) Lost City alkaline hydrothermal vent. This flange has fluids both trapped underneath and flowing through the top of it to form the smaller chimneys on the top. It is about 1 m across. (Photo Credit: NOAA and University of Washington.)

acidic vents, Lost City fluids have a temperature between 40°C and 90°C, a high pH (9–11), and contain H_2 (1–2 mmol kg^{-1}) and CH_4 (1–15 mmol kg^{-1}), and rather low H_2S and CO_2.

Carbon-14 dating of Lost City carbonates shows the vent system has been active for more than 30,000 years and these systems could have life spans 10-fold greater. This longevity makes alkaline vents better candidates than acidic ones as cradles of life. In addition, and importantly, the inside of an alkaline vent consists of a friable network of pores from fractions of a millimeter down to a few micrometers across through which vent fluids can percolate. It is intriguing that some of the pores are comparable in size to cells.

EXERCISE 8.3

What geological process would limit the lifespan of alkaline vent systems such as Lost City?

8.3.3.4 Serpentinization Makes Hydrogen and Methane

At Lost City, H_2 and CH_4 are generated *abiotically* by serpentinization at low temperatures (<150°C).

In the absence of carbon dioxide, the serpentinization of the magnesium end-member olivine (forsterite) produces the mineral brucite, which helps build the chimneys:

$$2Mg_2SiO_4 + 3H_2O \rightarrow Mg_3Si_2O_5(OH)_4 + Mg(OH)_2. \qquad (8.8)$$
$$\text{forsterite} \qquad\qquad\qquad \text{serpentine} \qquad\quad \text{brucite}$$

If the olivine has some iron, then hydrogen is evolved:

$$6(Mg_{1.5}Fe_{0.5})SiO_4 + 7H_2O \rightarrow 3Mg_3Si_2O_5(OH)_4 + Fe_3O_4 + H_2. \qquad (8.9)$$
$$\quad \text{olivine} \qquad\qquad\qquad\qquad \text{serpentine} \qquad \text{magnetite}$$

In the presence of CO_2, serpentinization of olivine produces methane:

$$24(Mg_{1.5}, Fe_{0.5})SiO_4 + 26H_2O + CO_2 \rightarrow 12Mg_3Si_2O_5(OH)_4 + 4Fe_3O_4 + CH_4. \quad (8.10)$$
$$\quad \text{olivine} \qquad\qquad\qquad\qquad\qquad \text{serpentine} \qquad\qquad \text{magnetite}$$

The CO_2 could come from seawater bicarbonate or be outgassed from the mantle.

Serpentinization is exothermic and this heats the vent fluids. The reaction will occur spontaneously but happens much faster in the presence of Fe–Ni catalyst. Serpentinization causes a large increase in volume, expanding the rock by roughly one-third. This expansion shatters the peridotite, which gives seawater access to fresh mantle and so more serpentinization.

Abiotic vent chemistry does not stop with hydrogen and methane. Hydrogen can reduce CO_2 to yield larger hydrocarbons than methane in what are termed Fisher–Tropsch type (FTT) reactions. In general,

$$CO_{2(aq)} + [2 + (m/2n)]H_2 \rightarrow (1/n)C_nH_m + 2H_2O. \qquad (8.11)$$

8.3.3.5 Alkaline Vents Can Provide Energy for Life

A model developed by Bill Martin (University of Düsseldorf) and Mike Russell (at NASAs Jet Propulsion Lab, Pasadena) postulates that life started with the evolution of a self-sustaining metabolism that generated energy by reacting CO_2 with H_2 produced by serpentinization at alkaline vents. The early oceans, unlike now, were slightly acidic (pH ~5.5–6) because of the higher amount of CO_2 in the atmosphere then. In the absence of oxygen the oceans had abundant soluble Fe(II). This would react with H_2S in upwelling vent fluids, precipitating iron monosulfide minerals such as mackinawite onto the walls of the vent pores, essentially creating mineral bubbles. Mike Russell has found these minerals in ancient fossil vents and produced them in the laboratory under simulated vent conditions.

Recall that iron–sulfur minerals can catalyze a variety of abiotic organic reactions. The Fe–S clusters that act as coenzymes in many modern iron–sulfur proteins, for example, in the electron transport chain have identical structures to iron monosulfide minerals. The inescapable implication is that Fe–S clusters in iron–sulfur enzymes are derived from the FeS minerals; relics of a pre-protein biochemistry.

8.3.3.6 Did the First Metabolic Pathway Reduce CO_2 with H_2 to Produce Acetate?

As we have seen (Chapter 7), Archaea and Bacteria both originated from the last universal common ancestor (LUCA). Characteristics shared by the two domains must have been inherited from LUCA. One metabolic pathway that could have come from LUCA is the reductive acetyl–CoA (Wood–Ljungdahl) pathway by which CO_2 is reduced to acetate by H_2 (Chapter 7) since it operates in both Archaea (methanogens) and Bacteria (acetogens).

The hydrogen is an electron donor and the carbon dioxide is both an electron acceptor as well as a building block for biosynthesis. The electrons are transferred via Fe ions in iron–sulfur proteins. A primordial acetyl-CoA pathway would presumably have been catalyzed by iron monosulfide minerals (Figure 8.5).

A feature which makes the acetyl-CoA pathway, or something like it, a prime candidate for the first metabolic pathway is that there is no net energy cost associated with it. Indeed of the five biochemical schemes for fixing CO_2 that have evolved, the acetyl-CoA pathway is the only one that has zero energy cost; it requires no net energy in the form of ATP. This favorable thermodynamics has led to it being described by biogeochemist Everett Shock as "a free lunch you're paid to eat." However, the reaction between CO_2 and H_2 does not go spontaneously because there is an energy barrier at one step. Reacting CO_2 and H_2 requires 1 mol of ATP to make it go but produces only enough energy to make 1 mol of ATP with a little bit to spare. This excess is wasted as heat. In other words, the pathway itself does not provide any net energy.

Extant methanogens and acetogens have an electron transport chain to circumvent the problem. Protons can be pumped using small amounts of energy until there are enough to synthesize one molecule of ATP. In this way, the spare energy can be saved

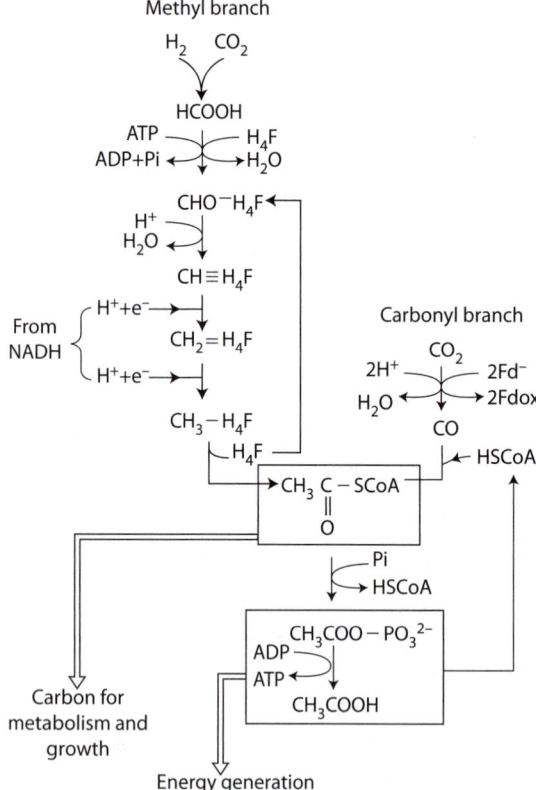

FIGURE 8.5 The acetyl-CoA (Wood–Ljungdahl) in modern acetogens. The H4F (tetrahydrofolate) is a carrier; ferrodoxin (Fd) is an FeS molecule.

rather than wasted. The outcome is that chemiosmotic coupling—the process that links electron transport to ATP synthesis—means that all the energy of the acetyl-CoA pathway can be captured and the system is in energy currency credit.

But this was not an option for life before electron transport chains and ATP synthase had evolved. However, alkaline vents provide the solution.

8.3.3.7 Alkaline Vents Have Proton Gradients

Alkaline vents naturally have proton gradients across their walls which make the reduction of CO_2 by H_2 more thermodynamically favorable, and hence overcome the energy barrier.

The thin (100 nm–5 μm) iron–sulfur mineral walls act as barrier between two fluid compartments. One contains hydrothermal fluids, seawater that has percolated through the crust where it is warmed (70–90°C) and chemically altered so that it erupts as reduced, and alkaline (pH ~9) fluid containing 10–20 mM concentrations of H_2 produced by serpentinization. The other compartment is cooler, more oxidized, mildly acidic (pH ~5.5–6) seawater. Hence there is a proton gradient which is matched by a small potential difference across the walls (Figure 8.6).

The proton gradient makes it easier for the FeS minerals in the walls to transfer electrons from H_2 to reduce CO_2. This is because the reduction potential of FeS minerals falls with increasing pH. So, they are at their most reducing (i.e., most willing to donate electrons) under alkaline conditions and most easily reduced (i.e., gain electrons) under acid conditions. Consequently, electrons are transferred from H_2 to CO_2 across the vent walls.

8.3.3.8 Did the First Pathway Make Acetyl Phosphate?

Before DNA, RNA and proteins, what would the first biochemical pathway look like? On the basis of vent chemistry, Nick Lane and Bill Martin propose something akin to Figure 8.7. It is known that FeS catalyzes the synthesis of methyl sulfide (CH_3SH) from CO_2 and H_2S, and from this acetyl thioester can be formed. This could be the

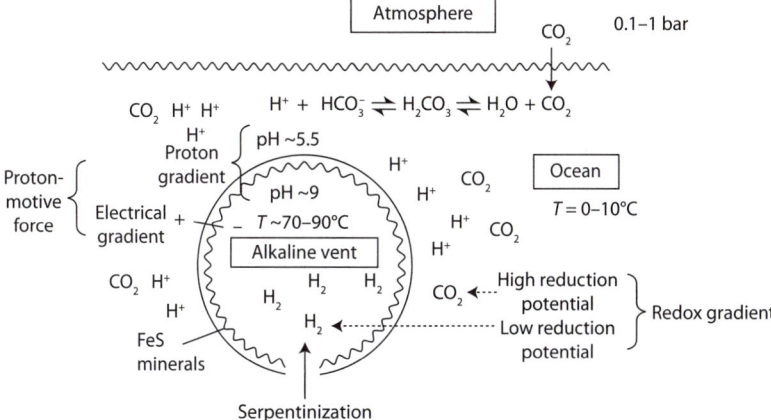

FIGURE 8.6 Hydrothermal vents have a natural proton motive force.

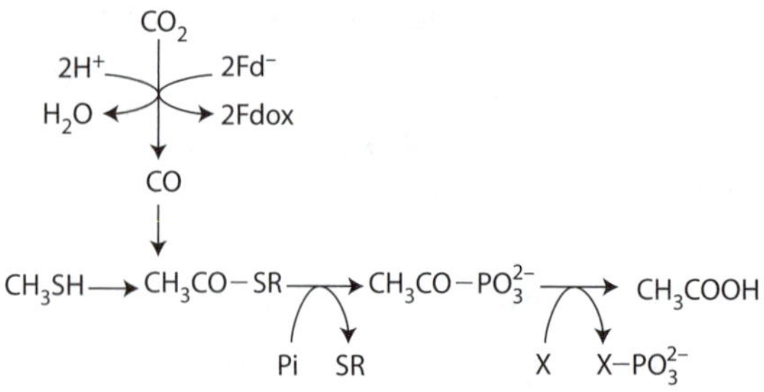

FIGURE 8.7 A possible early biochemical pathway in an alkaline vent. R and X are unknown groups.

starting point for the synthesis of biologically important molecules. Methane and higher hydrocarbons created in the vent could have provided raw materials for this. In addition, acetyl thioester could be phosphorylated to produce acetyl phosphate. This can act like ATP as a store of energy to drive endergonic processes (see Figure 8.4).

8.3.3.9 Temperature Gradients Encourage Polymerization

One of the difficulties in abiotic synthesis of nucleic acids is the requirement for high concentrations of precursor molecules. Temperature gradients exist within the porous labyrinth of the vent and experiments show that these gradients can concentrate organic molecules such as nucleic acids by a process termed thermophoresis. For example, at 3°C DNA molecules diffuse to a region with a higher temperature whereas at 20°C they diffuse to cooler regions. Gradients that arise naturally in vent systems could concentrate nucleotides up to 10,000-fold, enough to favor rapid polymerization.

At this stage, natural selection would begin to operate. Those reactions that went fastest or scavenged substrates from other pathways would come to dominate the chemistry of their part of the vent. A critical stage would be reached when a molecule arose that could catalyze its own replication. Most likely this would have been an RNA molecule.

Furthermore, fluctuations in temperature could allow nucleic acid replication. In the case of double-stranded DNA, as the temperature rises, the two strands unwind as the hydrogen bonds between the bases break. This is termed DNA denaturation. As the temperature falls base pairing can occur between nucleotides and bases in the two strands. This annealing is the first stage in forming two double-stranded DNA molecules where before there was one, though some sort of DNA polymerase would be needed to complete the assembly of the new strands by catalyzing the formation of bonds between adjacent nucleotides.

8.3.3.10 DNA, RNA, and Proteins Evolved in the Vent

We saw earlier that if features shared by Archaea and Bacteria are derived from LUCA, then she had DNA, RNA, proteins, and ribosomes, the universal genetic code, and made ATP by using the energy stored in a proton gradient. How all this happened we do not know.

It is thought that one of the earliest enzymes to evolve, probably predating ATP synthase was pyrophosphatase. This catalyzes the production of pyrophosphate, which acted like ATP, allowing proto-cells to get more energy from the proton gradient between the alkaline vent fluid and the acidic ocean. Pyrophosphatase is one of the most ancient enzymes, found in both Archaea and Bacteria.

What we can surmise is that for life to escape from the vent it must have been capable of coupled chemiosmosis (i.e., be able to create a proton gradient across a cell membrane and use it to synthesize ATP) so that it could free itself from the vent's proton gradient. The selection pressure to do that would have become intense with the advent of lipid membranes, as this would make the boundary of the proto-cells less permeable to protons.

Moreover, Archaea and Bacteria have completely different makeup of membrane lipids. Since these evolved independently it must mean that LUCA had a boundary distinct from either. In the Martin and Russell model, the boundary was provided by the iron–sulfur minerals of the micro-compartments of alkaline vents. This means LUCA cannot have been free-living but must have been encapsulated within a rocky nest.

Membrane lipid evolution must have diverged so that eventually two waves of free-living organisms emerged from a vent system, one giving rise to Archaea, the other to Bacteria. It is remarkable to think that, most probably, both domains would have been spawned from the same vent system.

9

Early Life

After its genesis, life went on to terraform the planet that gave it birth. We shall see that for perhaps 90% of its history Earth has been starved of oxygen and yet whenever oxygen levels did rise, Earth froze. These challenges may account for the late blossoming of complex life.

Life has always been inextricably linked with the rest of the Earth system. It has transformed the mineralogy of rocks, the composition of the oceans and atmosphere, brought about huge climate change, and has in turn evolved to adapt to the modifications it has brought about. We explore that relationship here on the ground that extraterrestrial life presumably messes with its own world in similar ways, and this may give us leverage to detect it.

9.1 A Methane Greenhouse

9.1.1 A Shift from CO_2 to CH_4 Greenhouse Happened in the Late Archaean

The Hadean atmosphere, when the solar luminosity was only 75% of its present value, was kept warm by CO_2 and water. Yet by 2.8 Gyr, evidence such as the absence of siderite ($FeCO_3$) from palaeosols shows there was not enough CO_2 in the atmosphere, possibly by a 10-fold margin, to keep the Earth's surface above 0°C given the fainter young sun. However, Earth was not in a perpetual deep freeze during the Archaean.

Consequently, there must have been other greenhouse gases at work during this time. The most likely candidate is methane. It is a more potent greenhouse gas than CO_2. The major source would have been methanogens.

9.1.2 An Organic Haze Would Form as CH_4 Levels Rose

An organic haze would likely be formed as the atmosphere switched between its prebiotic CO_2-rich state to a CH_4-rich state. Modeling and laboratory experiments show that when $CH_4/CO_2 \geq 0.1$ reactions driven by UV begin to form particles of organic molecules. As methanogens began to pump out methane, the surface temperature of the Earth would rise. This would increase methanogenesis (since methanogens grow better in warm conditions) and drive up temperature still more. The warming would increase silicate weathering and hence the drawdown of CO_2, so that there would be a concomitant drop in the partial pressure of CO_2. As CH_4 levels rose and CO_2 levels fell, there would be a time when the $CH_4/CO_2 \sim 0.1$, and a haze would form.

The haze particles would reflect sunlight, cooling the surface. Consistent with this idea is that the Earth was in the grip of the Pongola glaciation 2.9 to 2.8 billion years ago.

EXERCISE 9.1

Suggest a negative feedback that could have brought the Pongola glaciation to an end.

9.2 The Great Oxidation Event

The Proterozoic Eon (2.5–0.54 Gyr ago) marks the transition from the largely anoxic world of the Archean (>2.5 Gyr) to the mostly oxic world of the Phanerozoic (<0.54 Gyr). It is widely accepted that between 2.45 and 2.22 Gyr ago the abundance of oxygen in Earth's atmosphere rose from extremely low levels (<10^{-5} present atmospheric level, PAL) to between 10^{-2} and 0.1 PAL. This life-changing transition was termed the Great Oxidation Event (GOE) by Heinrich Holland in 2002, and the name has stuck.

9.2.1 The Oxygen Source Was Photosynthesis

The rise in oxygen concentration is attributed to cyanobacteria photosynthesis. But we would expect a time lag between the appearance of cyanobacteria and the increase in atmospheric oxygen content because initially oxygen would have been used up in reducing sinks for oxygen. For example, at first, oxygen would be snapped up to oxidize new continental and ocean crust, which would have afforded huge sinks of reduced metals such as ferrous iron.

EXERCISE 9.2

List other sinks for oxygen that could have existed in the Archaean.

Eventually, the balance between cyanobacteria oxygen production and the oxygen sinks tipped in favor of O_2 accumulation (Figure 9.1).

As oxygen breathers, we naturally think of the life-giving properties of oxygen. Indeed, it may be that complex life would not have emerged without it, so in the long-term the Great Oxidation Event was of vital importance. But the GOE was bad news for Palaeoproterozoic Earth: it produced global cooling and O_2 is lethal for most anerobic organisms. The GOE would have brought about the mother of all mass extinctions.

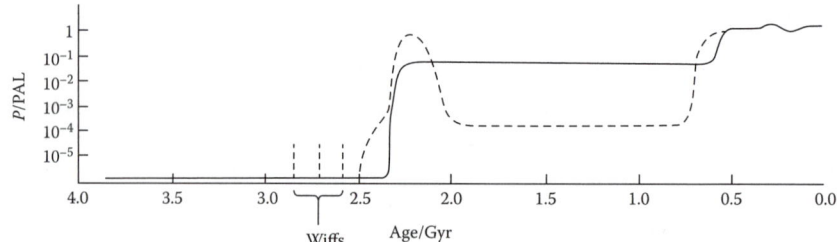

FIGURE 9.1   The Great Oxidation Event (GOE) was an increase in the oxygen content of the Earth's atmosphere about 2.4 Gyr ago. Solid line, original model; dotted line a new model in which it seems to have been preceded by whiffs and to have been followed by a billion year period of near anoxia.

9.2.2 Evidence for the GOE is Geochemical

9.2.2.1 Sulfur Isotopes Time the GOE

Between 2.45 and 2.09 Gyr ago non-mass-dependent fractionation (NMD) of S isotopes disappeared, a sure sign that oxygen levels were on the rise. NMD divvies up isotopes in a way that is *not* proportional to their mass ratios, and its signature is seen mostly in $BaSO_4$ (barite) and FeS_2 (pyrite) in hydrothermal sediments. Sulfur dioxide (SO_2) is one of the main sulfur compounds outgassed by volcanoes. Only one natural process is known to produce mass-independent sulfur isotope fractionation, the photodissociation of SO_2 by UV with wavelengths of 190–220 nm. This radiation is absorbed by ozone (O_3) and O_2 in the modern atmosphere. Therefore, for sulfur NMD signals to be seen in the rock record, oxygen levels in the atmosphere must be very low, probably <0.001% PAL.

Before 2.45 Gyr, a large range of $\Delta^{33}S$ (see Box 9.1) in sulfides and sulfates is seen. This variability falls between 2.45 and 2.09 Gyr, and after this period, $\Delta^{33}S$ is more or less zero up to modern times (Figure 9.3).

BOX 9.1   SULFUR ISOTOPES

There are four stable isotopes of sulfur, ^{32}S, ^{33}S, ^{34}S, and ^{36}S. Of these, ^{32}S and ^{34}S between them make up more than 99% of all the sulfur on Earth.

Microbial sulfate reduction (MSR) produces a mass-dependent fractionation (MDF) of S isotopes. In biochemical reactions, the lighter isotope reacts faster than the heavier, so biogenic material becomes enriched in ^{32}S. MSR produces pyrite, FeS_2 which will be richer in ^{32}S than a sulfur-containing mineral (e.g., gypsum: $CaSO_4 \cdot 2H_2O$) formed *abiotically* in the same water.

The ^{32}S enrichment is commonly expressed in relation to ^{34}S in which the $^{34}S/^{32}S$ ratio of a sample is compared to the $^{34}S/^{32}S$ ratio of a standard in the δ notation:

$$\delta^{34}S = 1000 \times [(^{34}S/^{32}S)_{sample}/(^{34}S/^{32}S)_{standard} - 1]\%o$$

The usual standard is the Vienna Canyon Diablo Troilite (VCDT). Troilite (FeS) is often found in iron meteorites, and the Canyon Diablo meteorite was responsible for creating the Barringer ("Meteor") crater in Arizona about 50,000 years ago. Sometimes the VCDT standard is written as $(\delta^{34}S)_{CDT}$.

MSR is indicated by negative $\delta^{34}S$ and can be as much as $\delta^{34}S = -46\%o$ when there is plenty of sulfate in the seawater. On the other hand, when sulfate concentration becomes limiting (<200 μm) minimal fractionation is seen, so even if some MSR had occurred we would not spot it.

Mass-dependent fractionation can be quantified for $^{33}S/^{32}S$ in the same way, also normalized to the VCDT standard.

If sulfur isotopes are subject only to mass-dependent fractionation, a plot of $\delta^{33}S$ versus $\delta^{34}S$ gives a mass-dependent fractionation line with slope about 0.5. This is because $^{33}S/^{32}S$ always deviates from the terrestrial standard by one-half the amount that $^{34}S/^{32}S$ deviates from the standard. This in turn arises because the difference in mass between ^{33}S and ^{32}S is half the difference in mass between ^{34}S and ^{32}S (Figure 9.2).

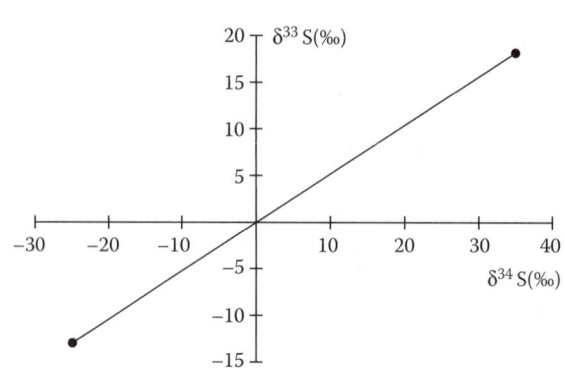

FIGURE 9.2 Mass-dependent fractionation of sulfur isotopes causes them to be plotted along the MDF line with slope ~0.5.

From this we can define

$$\Delta^{33}S = \delta^{33}S - 0.515\delta^{34}S. \tag{9.1}$$

For mass-dependent fractionation, $\Delta^{33}S = 0$.

Photochemical reactions in the atmosphere involving UV can fractionate S isotopes in a way that does not depend on the masses of the isotopes. This is termed non-mass-dependent fractionation (NMD) or mass-independent fractionation (MIF). Crucially, the photochemistry that produces sulfur isotope MIF does not occur if there is free oxygen in the atmosphere. Hence, the detection of MIF forms the basis for a method of estimating when oxygen was present in Earth's atmosphere.

NMD fractionation shifts sulfur isotopes away from the MDF line and gives $\Delta^{33}S \neq 0$. The vagaries of the processes involved means that MIF generates variations, both positive and negative, in $\Delta^{33}S$ values.

Sulfur NMD is reckoned to give the most precise timing of the GOE.

9.2.2.2 Banded Iron Formations: A Red Herring?

The most spectacular rocks attributed to the transition from reduced to oxidized atmosphere are the vivid red and brown-layered sediments termed banded iron formations (BIFs, Plate 9.1).

Generally deposited in shallow continental shelf seas, BIFs consist of alternating iron-rich and iron-poor layers of chert (SiO_2). How they formed is not certain. It is likely that several processes, both abiotic and biotic, may have operated. What they do show is that the ocean was ferruginous; because it was anoxic, it was rich in soluble reduced ferrous (Fe^{2+}) iron derived from weathering of seafloor basalt and erupted from hydrothermal vents.

The standard story is that BIFs formed when deep-water ferrous iron hit shallow, oxygenated water where it was oxidized to insoluble ferric (Fe^{3+}) iron. Most BIFs were

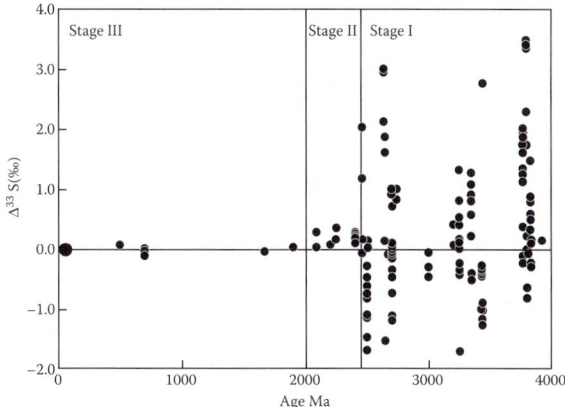

FIGURE 9.3   The $\Delta^{33}S$ signal of mass-independent fractionation of sulfur seen in anoxic conditions disappears at the GOE (Farquhar and Wing. *Earth Planet Sci Lett* 2003;213:1–13).

deposited episodically between 2.5 and 1.8 Gyr. Oxygen is presumed to come from cyanobacteria photosynthesis, which would produce local increases in oxygen in the upper part of the water column (Figure 9.4). However, some BIFs formed throughout the Archaean back as far as 3.7 Gyr ago. If we stick with the conventional story of BIF formation, this would necessitate a very early origin for oxygenic photosynthesis and an extremely long lag time of 1.3 Gyr between biogenic oxygen production and the GOE.

PLATE 9.1   (See color insert.) Banded iron formation (BIF), Karijini National Park, Western Australia. Despite their widespread distribution, the origin of BIFs is not yet unambiguously established. (Photo credit: Graeme Churchard, Bristol, UK.)

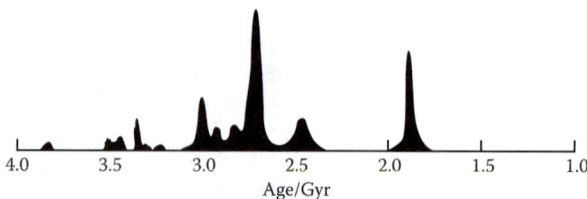

FIGURE 9.4   Banded iron formations appeared episodically.

But BIFs might be formed by the action of bacteria that are able to get energy by oxidizing iron with little or no free oxygen. These include the chemolithoautotroph proteobacterium *Gallionella ferruginea*, which can operate under very low oxygen concentrations (60 µmol L^{-1}):

$$6Fe^{2+} + 0.5O_2 + CO_2 + 16H_2O \rightarrow [CH_2O] + 6Fe(OH)_3 + 12H^+. \qquad (9.2)$$

In addition, there are anoxygenic photosynthesizing bacteria that can couple oxidation of iron to the reduction of CO_2 in the absence of oxygen:

$$4Fe^{2+} + CO_2 + 11H_2O \rightarrow [CH_2O] + 4Fe(OH)_3 + 8H^+. \qquad (9.3)$$

Laboratory experiments have shown that the activities of these organisms could account quantitatively for the bulk of the ferric iron seen in BIFs.

9.2.3 Whiffs of Oxygen Preceded the GOE

During the past decade, studies of trace elements that record the redox state of their surroundings, the weathering of sulfide minerals such as pyrite, and transient episodes in which non-mass-dependent sulfur isotope fractionation disappeared all imply that O_2 appeared in the atmosphere in a number of pulses from about 2.9 Gyr; that is, before the GOE. These transient spikes of oxygen are termed whiffs.

9.2.4 Glaciations Coincided with the GOE

Between 2.45 and 2.3 Gyr, three Huronian glaciations occurred. At 2.22 Gyr, these were followed by the first of the great "Snowball Earth" glaciations, the Makganyene glaciation, so called because it is thought that the entire surface of the world's oceans froze over. That these glaciations spanned the period of the GOE is no coincidence. An obvious explanation for this is that the oxygen dramatically reduced the abundance of methane in the atmosphere by rapidly oxidizing it:

$$2O_2 + CH_4 \rightarrow CO_2 + 2H_2O. \qquad (9.4)$$

This would reduce the greenhouse effect of the atmosphere because methane is a more potent greenhouse gas than CO_2.

9.3 The "Boring Billion"

A great billion year stretch of time in the middle of the Proterozoic is marked by a virtually constant $\delta^{13}C$ ~0 in the rock record. In other words, there was no change in the balance between the rate at which organic carbon was being buried or oxidized. This has earned the interval between 1.8 and 0.8 Gyr the epithet "boring billion."

It saw oceans revert to near anoxic conditions, which in all likelihood retarded the diversification of complex life.

9.3.1 O_2 Levels Plummeted After the GOE

Between 2.3 and 2.1 Gyr ago the largest and longest lived positive carbon excursion in Earth's history—the Lomagundi excursion—played out. During this time, $\delta^{13}C$ values often exceeded +10‰. This is interpreted as evidence for massive amounts of carbon burial (see Box 8.1). When organic material is buried instead of decomposing, less O_2 is taken from the atmosphere by respiration. Hence, carbon burial is associated with rise of atmospheric O_2 levels.

This model is supported by a simultaneous rise in sulfate concentration in the same rocks as indicated by $\delta^{34}S$. This sulfate concentration is thought to reflect the composition of the seawater at the time because coeval sulfate evaporites have similar $\delta^{34}S$ values.

> **EXERCISE 9.3**
> What would a high ocean sulfate concentration indicate and why?

At the end of the Lomagundy event, there is sulfur isotope evidence for a rise in pyrite burial brought about by ocean deoxygenation. About 200 Myr after the GOE, the oxygen level of the atmosphere fell dramatically, perhaps as low as 10^{-4} PAL and perhaps for as long as a billion years. The drop probably happened because carbon buried at the peak of the Lomagundy event was now being uncovered and exposed to oxidation.

9.3.2 The Canfield Ocean Is Anoxic and Sulfidic

The last major episode of BIF deposition was 1.85 Gyr, which implies a return to oceans rich in Fe(II), most likely due to a fall in O_2 levels. In 1998, Don Canfield modeled the outcome of low O_2 levels persisting throughout the mid-Proterozoic (1.8–1.0 Gyr) and suggested that the deep ocean would switch from ferruginous to euxinic. Euxinia is the state in which water is anoxic and sulfidic (rich in H_2S), and is seen in a few places today such as the Black Sea. Euxinia arose because oxygen levels were low enough that the deep water was anoxic, but high enough that there would still be weathering of sulfide minerals on land and hence a flux of sulfate (SO_4^{2-}) into the oceans. This stimulated the growth of sulfate-reducing bacteria that produced H_2S. Hydrogen sulfide reacted with Fe(II) to form the insoluble mineral pyrite, a mineral signature of the sulfidic water left in the rock record. Because H_2S removes Fe(II), BIF deposition shut down (Figure 9.5).

The original model, termed the Canfield Ocean, was taken to mean that the euxinia was global, and this view is not supported by subsequent studies of marine sediments of the mid-Proterozoic, most recently using Mo and Cr as tracers of the state of the oceans (Box 9.2) that do not show widespread deep ocean euxinia. Instead, they indicate deposition mostly under anoxic and Fe(II)-rich (ferruginous) conditions in deep oceans, but with episodic euxinic conditions occurring particularly along continental

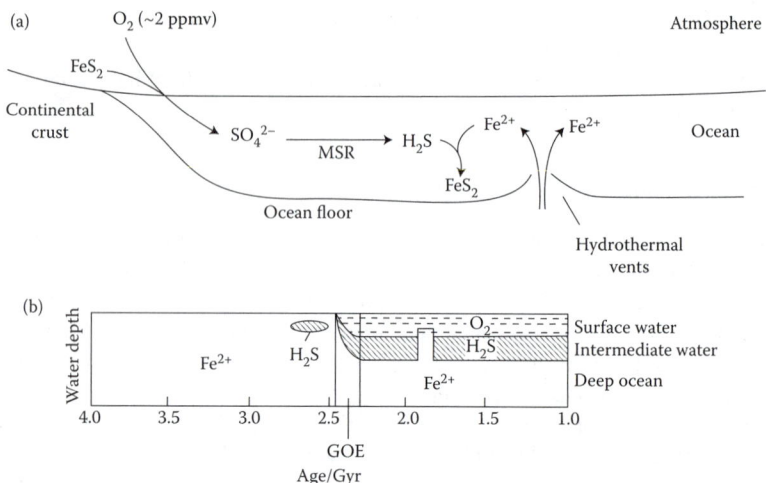

FIGURE 9.5 (a) A combination of low oxygen and microbial sulfate reduction (MSR) produces a euxinic ocean. (b) Deep ocean chemistry during the Precambrian.

margins, presumably where marine organisms flourished. It is estimated that at least 30–40% of the ocean floor, and probably much more, was anoxic for much of the mid-Proterozoic. By contrast, euxinia was limited to 1–10% of the sea floor area (compared to ≪1% today).

BOX 9.2   Mo AND Cr AS REDOX TRACERS

Molybdenum and chromium differ in their behavior to H_2S and O_2. Hence, together they can reveal the extent of ocean euxinia and anoxia.

Molybdenum is insoluble in sulfidic conditions and so is removed from the water and deposited in sediments. This is because sulfur can replace oxygen in the molybdate ion:

$$MoO_4^{2-}{}_{(aq)} + 4H_2S \rightarrow MoS_4^{2-}{}_{(aq)} + 4H_2O. \tag{9.5}$$

The resulting thiomolybdate ion is reduced to the highly insoluble Mo(IV) sulfide:

$$MoS_4^{2-}{}_{(aq)} + 2e^- \rightarrow MoS_{2(s)} + 2S^{2-}{}_{(aq)}. \tag{9.6}$$

Hence, Mo concentration in sedimentary rocks provides a measure of the degree of euxinia in the water at the time the sediments formed.

Cr is present as Cr(VI) in oxic conditions, whereas under anoxic conditions reduction of Cr (VI) to Cr(III) results in isotopic fractionation. The heavy chromium isotope, ^{53}Cr, is more likely to be oxidized and washed into the ocean than the lighter ^{52}Cr. Therefore, the relative amounts of the two Cr isotopes in sedimentary rocks as well as Cr concentrations indicate the abundance of oxygen in the ocean and atmosphere.

9.3.2.1 Euxinia Locks Up Essential Trace Metals

Even though euxinic conditions were not very widespread, it might have been enough to pull concentrations of some trace elements essential to life, such as Mo and Cu, below the comfort zone of many organisms. To some extent, this would have been self-limiting. Thus, trace element deficiency would afflict sulfate-reducing bacteria and thereby curtail the development of euxinia. That $\delta^{13}C$ virtually "flat-lined" during the "boring billion" presumably reflects the operation of negative feedbacks on metal availability that kept a check on biomass.

> **EXERCISE 9.4**
>
> The modified Canfield Ocean is ferruginous with episodic euxinia confined to intermediate depths in the water column. What are the implications of this for BIF deposition?

9.3.3 The Neoproterozoic Oxidation Event

A second phase of oxygenation of atmosphere and oceans occurred late in the Proterozoic and was due to the activities of marine algae as well as cyanobacteria. In other words, eukaryotes were now getting in on the act of terraforming the planet. By the start of the Cambrian period 543 Myr ago, the oxygen concentration of the atmosphere was not much different from today.

9.3.3.1 Oxidation Produced Glaciation

Coincident with the rise in oxygen, the Earth suffered three or four glaciations of which two seem to have been particularly severe: the Sturtian and Marinoan glaciations. This marks the Cryogenian Period from 850 to 630 Myr. In the extreme scenario, Earth was completely covered by ice to a thickness of 1–2 km creating a barrier between atmosphere and oceans that could have lasted for millions of years. This is a full-blown Snowball Earth. Alternatively, a weaker Slushball Earth would have had open water at least at the tropics allowing some gas exchange between atmosphere and oceans.

The Neoproterozoic glaciations may have occurred in part due to oxidation of methane, just as in the Great Oxidation Event, but a particular set of tectonic conditions favored the removal of CO_2 from the atmosphere as well.

9.3.3.2 Continent Configuration Contributed to the Freeze

During the Cryogenian, there was a rare preponderance of continents in the tropics. Perhaps paradoxically this favors global cooling. Most of the sun's energy falls on the tropics and continents have a higher albedo than oceans so tropical continents would produce a large drop in absorption of solar energy by the oceans. The tropics are hot so precipitation rates are high, favoring intense silicate weathering. This would reduce atmospheric CO_2 concentrations and bring about global cooling. The lower temperatures would then curtail the silicate-weathering rate and so the climate system would eventually stabilize at a new colder state.

The high initial silicate-weathering rates were augmented by the fact that the Cryogenian was marked by the breakup of a supercontinent named Rodinia. A supercontinent has all of the continental crust assembled into a single mass. Silicate weathering rates are low when a supercontinent exists, because most land area is far from the ocean and therefore arid. When a supercontinent fragments, formerly dry regions become wetter and weathering rates increase accordingly.

The identification of BIF at 1.0–0.7 Gyr, formed when the atmospheric oxygen concentration was obviously high, seems anomalous.

EXERCISE 9.5

Why might BIFs form at the end of a "snowball Earth" glaciation?

9.4 The Emergence of Life

So far, we have seen that methanogens and acetogens may have been around from the start. The first metabolism we have direct evidence for is sulfate reduction (at ~3.4 Gyr) and the smoking gun for oxygenic photosynthesis is the Great Oxidation Event at ~2.4 Gyr.

Here, in light of our understanding of conditions during the Proterozoic, we explore the evolution of life during this Eon. In particular, we shall look at when cyanobacteria may have first appeared and the emergence of complex life, the eukaryotes.

9.4.1 Methanogenesis and Sulfate Reduction Were Intertwined

On the face of it, methanogenesis and sulfate reduction look like two different respiratory systems so it is hard to see how there could be any evolutionary affinity between them. In fact, generally these two metabolisms are incompatible because sulfite (SO_3^{2-}), an intermediate of sulfate reduction, inhibits methanogenesis. Hence, it seems these two processes could not have developed in the same host.

Actually sulfite is toxic to many cells, and sulfate reducers convert sulfite to H_2S using a sulfite reductase, so do not poison themselves. However, sulfite could have been produced in hydrothermal vents as the oceans became oxygenated, so for ancient methanogens to survive, they would have to be sulfite tolerant. In fact, several deeply rooted thermophilic methanogens have sulfite reductase, which does confer tolerance.

In fact, many methanogens, but not other archaea, have sulfite reductase-like proteins, but they are nonfunctional which is why most methanogens are sensitive to sulfite. Methanogens also have homologs of other enzymes involved in sulfate reduction. Hence, molecular genetics suggests that methanogenesis and sulfate reduction were intimately related early in microbial evolution.

9.4.2 Nitrogen Fixation Probably Evolved Very Early

Nitrogen is essential for biochemistry, yet N_2 in the atmosphere cannot be accessed by eukaryotes or most prokaryotes; it is biologically unavailable.

EXERCISE 9.6

(a) List some classes of biomolecules that contain nitrogen atoms. (b) In what form was nitrogen in the atmosphere of the Earth before life emerged?

The way round this impasse is provided by a critical biochemical pathway we have ignored until now, nitrogen fixation, which can be done by some prokaryotes including methanogens, and many phototrophs (cyanobacteria), and sulfate-reducing bacteria.

Nitrogen fixation requires 16 moles of ATP for every mole of N_2. This high free energy requirement is perhaps surprising because the overall reaction is exothermic. It is needed to overcome the high activation energy caused by the great strength of the N_2 triple bond:

$$N_2 + 8H^+ + 8e^- \xrightarrow{\text{nitrogenase}} 2NH_3 + H_2 \quad (E_0' = -0.314\,\text{V}). \quad (9.7)$$

The reaction consumes eight electrons but why two of the electrons are used to synthesize H_2, an apparently wasteful step, is unclear. The ammonia produced can be taken up by all non-nitrogen fixing organisms and hence is the ultimate nitrogen source for all life on Earth.

The first organisms would presumably have had access to biologically available forms of nitrogen (e.g., NO_3^-) from local chemistry but a reliable supply of nitrogen to sustain a sizable biomass ultimately came from nitrogen fixation which must therefore have evolved very early. It also means that biochemistry that was able to supply high amounts of chemical energy evolved very early. The key enzyme in nitrogen fixation is nitrogenase. At its heart is a FeMoCo complex, so iron, molybdenum and cobalt are essential trace metals for nitrogen fixation.

We do not know whether Archaean sulfate-reducing organisms were capable of nitrogen fixation, but it is highly likely since the deeply rooted sulfate reducer *Desulfovibrio* can also fix nitrogen. Nitrogen fixation in archaea (which is confined to the methanogens) and in bacteria operates very similarly, and so was likely a characteristic of LUCA. Intriguingly, the molybdate ion (MoO_4^{2-}) can be taken up by cells using the same membrane transport system as the sulfate ion, SO_4^{2-}.

The nitrogenase enzymes that catalyze nitrogen fixation are inhibited by oxygen. Consequently, many modern nitrogen-fixing bacteria are anaerobic. Those living in an oxic environment, and which may even respire aerobically, have developed biochemical strategies to prevent O_2 reaching the enzyme.

EXERCISE 9.7

Both bacteria and archaea use the same FeMoCo cofactor in their nitrogenase enzyme. Is this convincing evidence for the FeMoCo complex having been present in the last universal common ancestor (LUCA)? Justify your answer.

9.4.3 Genome Expansion Occurred in the Archaean

Phylogenomics uses multiple gene sequences in extant organisms to establish evolutionary relationships. It attempts to quantify the origin, duplication, horizontal transfer, or loss of specific genes over time—in other words, it reconstructs gene histories. Lawrence David and Eric Alm, both at the Massachusetts Institute of Technology, have used a novel phylogenomics method, together with a model of evolution rate, to map the evolutionary history of nearly 4000 genes from all three domains of life onto a geological timeline. They show that there was a large expansion of the genome size 3.33–2.85 Gyr ago brought about by an increase in the birth rate, duplication rate, and horizontal transfer of genes. A smaller and slightly delayed rise in gene loss was also seen. The expansion featured genes involved in microbial respiration and electron transport. The earliest of these redox genes were likely to have been involved in anaerobic respiration, but the fraction of genes coding for proteins using oxygen steadily increase after the expansion. Thus, the timing of the birth of gene coding for enzymes using oxidized forms of nitrogen, N_2 and NO_3, that an aerobic nitrogen cycle was in place by the late Archaean.

There is also an increase in genes for transition metal-using enzymes (e.g., those using Cu and Mo) during the Proterozoic. Why would such genes evolve if the transition metals were unavailable? However, if euxinia was relatively localized, and the seas were largely ferruginous, making metals available, the evolution of genes for transition metal-using enzymes makes sense.

Hence, DNA seems to have an imprint of oxygenesis hundreds of millions of years before the Proterozoic. An earlier model of evolution rates made the expansion of bacterial genomes later (2.75–2.5 Gyr), but even this supports the idea that oxygen was in the atmosphere before the classical GOE.

9.4.4 When Did Oxygenic Photosynthesis Start?

We do not know when oxygenic photosynthesis evolved. Estimates range from 3.7 Gyr to 2.3 Gyr.

Early protagonists have to show that the inventory of oxygen sinks is big enough to account for the huge lag time between the onset of oxygen production and the GOE; 1.4 Gyr at its greatest. They also have to account for how such a complex metabolic pathway could have evolved so soon in the history of life. Those arguing for a late appearance have to explain the geochemical evidence for oxygenation before the GOE.

9.4.4.1 Anoxygenic Photosynthesis Evolved First

Anoxygenic photosynthesis uses either system I or II, but never both together. Modern oxygenic photosynthesis originated with the evolution of the oxygen-evolving Mn_4 reaction center in organisms coexpressing I and II, since splitting of water allows the two systems to work in tandem.

It is energetically more expensive to use H_2O to reduce CO_2 than it is to use reduced compounds such as H_2, Fe(II), or H_2S. For these reasons, it is thought that anoxygenic photosynthesis evolved first. Oxygenic photosynthesis is more complex but it has the advantage that water is available everywhere, whereas the electron donors for anoxygenic photosynthesis are not widespread in the modern world. This presumably explains the success of oxygenic photosynthesis despite its higher energy cost.

EXERCISE 9.8

At what periods of Earth's history would anoxygenic photosynthesis be favored by widespread availability of electron donors?

9.4.4.2 *¹²C, BIFs, and Cyanobacterial Biomarkers Are Red Herrings*

Three lines of evidence that have been used in the past to time oxygenic photosynthesis are no longer secure.

1. The enrichment of ^{12}C in rocks may not be biotic, and even if it is, it may not be exclusive to oxygenic photosynthesis.
2. Banded iron formations show that oceans were rich in Fe^{2+} (ferruginous) but not necessarily that they were becoming oxygenated.
3. The identification, in 1999, of 2-methyl hopanes (membrane lipids thought at the time to be predominantly of cyanobacterial origin) in 2.7 Gyr-old Archaean rocks, has turned out to be unreliable evidence. Recent work shows that not all cyanobacteria make 2-methyl hopanes, while other bacteria (e.g., acidobacteria, α-proteobacteria) can. In addition, NanoSIMS revealed a discrepancy in $\delta^{13}C$ values between the soluble 2-methyl hopanes and insoluble kerogens in the 2.7 Gyr-old rocks, showing the biomarkers were contaminants that must have entered the rock after 2.2 Gyr.

9.4.4.3 *Did Oxygenic Photosynthesis Start ~2.4 Gyr Ago?*

The appearance of oxidized manganese in the rock record is probably the best marker for the onset of oxygenic photosynthesis. That is because Mn is oxidized in photosynthesis (the Mn_4 reaction center donates electrons to split water), yet Mn does not oxidize in the absence of appreciable amounts of oxygen in the atmosphere. In other words, oxidized manganese will only be seen if oxygenic photoautotrophs are producing it, or there is plenty of oxygen around as a result of their activities.

Massive Mn oxides deposits from before 2.4 Gyr have been found in South Africa. This is the key evidence for a late onset of photosynthesis. It must be weighed against indirect evidence for Mn deposits at the time of the Pongola glaciation at 2.9 Gyr and other geological evidence for oxidation in the Archaean well before the GOE.

9.5 Eukaryotes: Complex Life

9.5.1 When Did Eukaryotes Appear?

The earliest *possible* fossil eukaryote is the 2.1-Gyr-old *Grypania*, which resembles marine algae. Not until 1.8 billion years ago, in China, is there unambiguous evidence for eukaryotes. These are spheroidal acritarchs that measuring up to about 245 μm across show evidence of a complex, multilayered wall (Box 9.3). South African acritarchs that date back to 3.2 Gyr have been described. Their large size (300 μm) is suggestive of a eukaryote but they show no eukaryotic diagnostic features.

The earliest valid claim for the appearance of steranes, chemical biomarkers for eukaryotes, is ~2.45 Gyr.

BOX 9.3   ACRITARCHS

Acritarchs are un-mineralized, organic-walled, large (20–150 µm) microfossils of uncertain biological affinities. They are composed of kerogens such as polycyclic aromatic hydrocarbons that have been so chemically altered as to leave no vestige of their original cellular material. This renders them insoluble in acid, allowing them to be isolated from carbonate and silicate rocks by using HCl and HF. Most are interpreted to be unicellular photosynthetic protists; that is, eukaryotic marine algae, or their cysts (resting stages). Some are decorated with surface ornamentation such as spines. They are found throughout the Proterozoic and Phanerozoic, though they are rare until ~1.6 Gyr, and their diversity reduced during the Cryogenian.

In summary, we do not know with any precision when eukaryotes evolved.

Most modern eukaryotes respire aerobically which implies some O_2 must have been available after the GOE. Although this is hard to square with recent geochemical evidence from the mid-Proterozoic, as we have seen, presumably there were local ecosystems in which eukaryotes lived side-by-side with cyanobacteria.

9.5.2 Eukaryotes Are Archaeon–Bacteria Chimeras

The accepted view for the origin of eukaryotes is that an Archaeon and an α-proteobacterial cell developed a symbiotic relationship in which the bacterium ended up fully enclosed within its host, retained its capacity for division, and gave rise to mitochondria. This process, in which one cell is engulfed by another cell, is termed endosymbiosis. The resulting hybrid organisms diversified into the heterotrophic eukaryotes, such as animals. A second episode of endosymbiosis occurred subsequently in which a mitochondrion-bearing host subsumed a cyanobacterial cell, giving rise to the photosynthetic eukaryotes such as the photoautotrophic protists and plants.

9.5.2.1 *Endosymbiotic Origin for Eukaryotes Is a Nineteenth Century Idea*

The idea of endosymbiosis was originally proposed by the French botanist Andreas Schimper in 1883 who observed that the division of chloroplasts in plants closely resembled the division of free-living cyanobacteria. It was developed by the Russian botanist Konstantin Mereschkowski in the first decade of the twentieth century and resurrected by Lynn Margulis in 1966, but did not gain widespread acceptance until the 1980s when it became obvious from gene sequencing that mitochondrial and chloroplast DNA were different from nuclear DNA.

9.5.2.2 *Eukarya Only Evolved Once*

Modern biochemical and genetic evidence for the endosymbiotic origin for eukaryotes is overwhelming. This is not the place to review it. Suffice it to say that it is supported by studies that find that the best fit to the origin of the three domains of life is a ring rather than a tree. It explains how bacterial cell operational genes were transmitted to eukaryotes, presumably by the endosymbiont ancestor of mitochondria.

Genetic evidence strongly suggests that eukaryotic cells only evolved once. In other words, all extant Eukarya are the result of a single primary endosymbiotic event in which an Archaeon engulfed a bacterial cell that subsequently gave rise to mitochondria. It appears that the endosymbiosis of a cyanobacterium leading to chloroplasts was also a one-off event. (Biologists will be aware that *secondary* endosymbioisis, in which one eukaryote engulfs another eukaryote has happened several times, but this should not be confused with *primary* endosymbiosis in which both of the partners are prokaryotes.)

The evolution of eukaryotes, which are generally equated with complex life, is probably a very hard step, given that it has happened only once in the lifetime of the Earth. By comparison, multicellularity is reckoned to have developed on several occasions and even structures we think of as highly complex, such as eyes, have evolved many times. We might argue that the origin of complex life is hard if it happened relatively late. However, the validity of this argument has been questioned (see Chapter 14).

9.5.3 The Hydrogen Hypothesis Explains Endosymbiosis

The endosymbiotic event that led to the eukaryotes was most likely between a methanogen and an α-proteobacterium. The genes of eukaryotes are most closely related to methanogens and genetics supports an α-proteobacterial origin for mitochondria (Figure 9.6).

But this raises a problem. Methanogens are restricted to anoxic environments while proteobacteria generally respire aerobically; how did they ever meet, let alone get into a cosy symbiosis? The solution—the hydrogen hypothesis—was posed by Bill Martin and Miklós Müller in 1998. The idea came from the identification of hydrogenosomes in primitive eukaryotes. These organelles generate H_2 and CO_2. Gene sequencing shows that both hydrogenosomes and mitochondria are descended from a common α-proteobacterial ancestor. Now α-proteobacteria are a metabolically diverse group; while many respire aerobically, some generate energy by fermentation, releasing H_2 and CO_2, and methanogens use H_2 as an energy source and CO_2 as their carbon source. Hence, the first proteobacterial symbiont provided all that is needed for a methanogen lifestyle: it was hydrogen metabolism, not oxidative phosphorylation that gave the earliest eukaryotes their selective advantage. What is more, α-proteobacteria that ferment do not require O_2. In fact, it is unlikely that O_2 and H_2 could coexist in appreciable quantities in the same ecological niche.

9.5.3.1 Symbiosis Is a Two-Way Exchange

In return for H_2 and CO_2, the methanogen generated glucose that the proteobacterium could ferment. However, while proteobacteria have glucose transporters in their cell membrane that allow them to take up glucose from the environment, methanogens lack glucose transporters. For the proteobacterium to access the glucose, it had to pass its genes for glucose transporters to the methanogen. Moreover, since methanogens use glucose to build more complex molecules (anabolism) but not as an energy source,

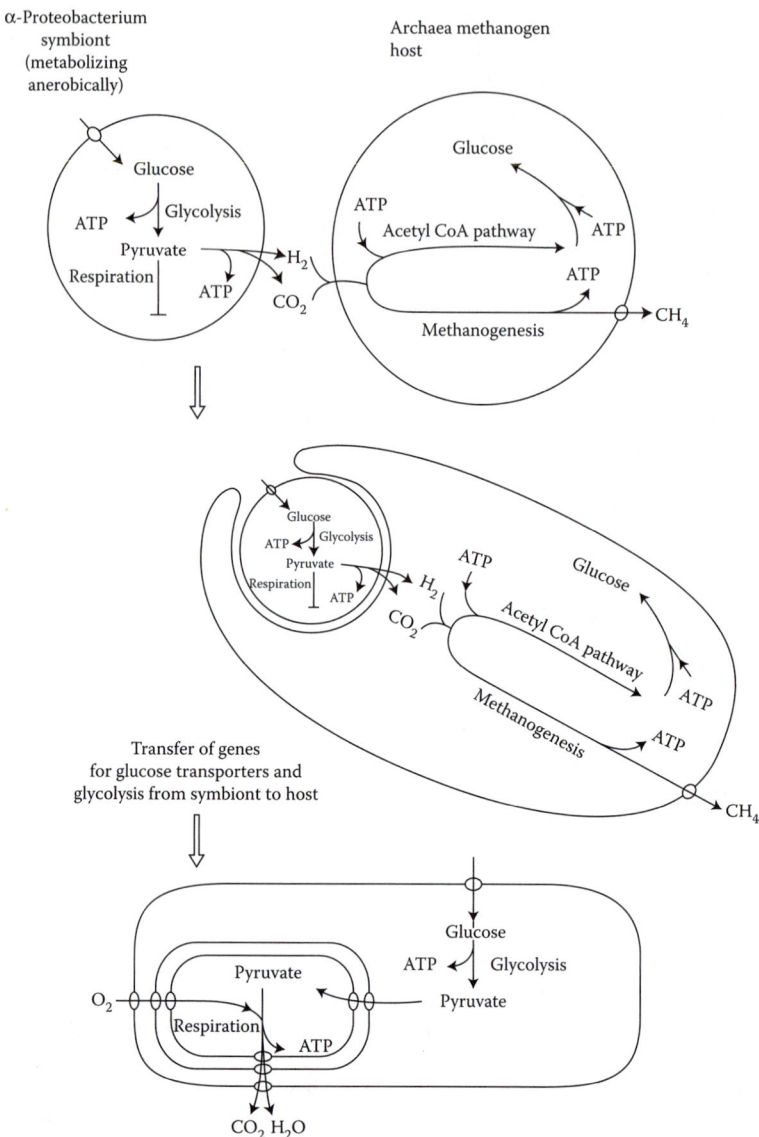

FIGURE 9.6 Endosymbiosis conferred Eukarya with mitochondria.

the proteobacterium risked starvation unless it could prevent the methanogen from using all of its glucose for anabolism. This could be achieved if the proteobacterium passed genes for glucose fermentation to the methanogen. How does this gene transfer happen? In a population, it is likely that some methanogens will have several closely associated α-proteobacteria, and if one of them dies as it is engulfed because it cannot get enough glucose from its environment, then its genes are shed and incorporated into the methanogen.

9.5.3.2 Symbiosis Had to Be Rapid

Once the symbiosis was complete, the host no longer had to act as a methanogen and could therefore survive in an oxic environment. This meant that the α-proteobacterium symbiont could now respire aerobically. This required that the genes for oxidative phosphorylation had not been lost from the α-proteobacterium during the symbiosis. Given how rapidly bacteria discard genes that are no longer required, and given that a symbiont had no way of re-acquiring these genes, the transition must have been fast.

In this model, eukaryotes probably emerged at the time when the oxygen concentration in the atmosphere was increasing because this might make life tough for methanogens. This is because the rise in oxygen concentration led to an increase in sulfate concentration in the ocean and hence the burgeoning of sulfate-reducing bacteria. These invariably outcompete methanogens for hydrogen: hence the powerful selection pressure to drive methanogens to establish symbiosis with the proteobacteria.

9.5.3.3 Mitochondria Lost Genes to the Nucleus

Mitochondria retained the ability to divide from their ancestral proteobacteria. Dead mitochondria would have released their genes into the cytoplasm where they would be transferred to the host genome. Perhaps surprisingly this is a common scenario. The human genome project shows that there have been at least 354 separate independent transfers of mitochondrial DNA to the nucleus.

9.5.4 Eukaryotes Inherited Bacterial Lipids

One of the things that any theory of the origin of the Eukarya domain has to explain is how the cell membranes of eukaryotes are all derived from bacteria. One scenario is that an α-proteobacteria cell transferred genes for making lipids to the host cell but without the signal sequences that would normally target them to the correct location in the cell. Hence, they assembled into flattened vesicles in the cytoplasm, which eventually formed the nuclear membrane and other internal membranes; for example, the smooth endoplasmic reticulum and Golgi apparatus. Presumably, the bacterial lipids had a selective advantage over the Archaeal ones in that bacterial membrane proteins functioned optimally within them.

Intriguingly, eukaryotic cells retain the ability to synthesize and metabolize the isoprenoid building blocks of Archaeal membranes. However, in eukaryotes, isoprenoids are not used to make membranes but a host of small molecules such as steroids.

9.5.5 Eukaryotes Have Huge Advantages Over Prokaryotes

Although bacteria and archaea evolution has resulted in a remarkable diversity of metabolic strategies, all prokaryotes are physical small, have tiny genomes (the limit appears to be 10 million base pairs), and are relatively simple compared with eukaryotes, despite having had 3.5 billion years to evolve. An explanation for this, persuasively argued by Nick Lane of University College London, UK, is that prokaryotes are irrevocably constrained by energetics in a way that eukaryotes are not.

9.5.5.1 *Prokaryotes with Small Genomes Are Favored*

Replicating DNA is the slowest step in bacterial cell division, so the larger the genome the slower it takes for a bacterial cell to divide. In addition, cell division is energetically expensive so organisms that are less efficient in producing metabolic energy produce fewer copies of themselves in a given time than more efficient organisms. This matters because bacterial reproduction is generally curtailed by lack of resources such as nutrients. The faster an organism can reproduce, the more copies of itself it can make before the "goodies" run out. Hence, there is selection pressure on bacteria to keep their genomes small and have energy metabolism that is as efficient as possible.

One strategy bacteria have for maintaining small genomes is by discarding genes they do not require at a particular time, a process that can occur in a few hours. For example, in the absence of a particular substrate, bacteria can lose the genes responsible for making the enzymes that metabolize that substrate without loss of function in their present environment. This is not irrevocable: lateral gene transfer will ensure that at least some cells in a bacterial colony can reacquire genes if and when the need arises. The reappearance of the missing substrate enables those cells that have reloaded the discarded genes to take advantage of the substrate, giving them a selective advantage over the cells that have not.

9.5.5.2 *Small Is Best for Energy Efficiency*

Energy efficiency is optimized by small physical size. That is because the smaller a cell is, the larger its surface area/volume ratio. The rate at which nutrients (electron donors) and oxygen (or any other electron acceptor for respiration) enters the cell, or protons can be pumped out by an electron transport chain, depends on its surface area, and this determines how much ATP can be synthesized. On the other hand, the demand for nutrients, oxygen, and ATP rises with volume, since large cells need more energy for the biosynthetic (anabolic) metabolism that makes their components. Therefore, if the surface area/volume ratio gets too small, a cell risks being starved of electron donors, electron acceptors, and ATP.

> **EXERCISE 9.9**
>
> Show that an increase in the linear size of a cell reduces its surface area/volume ratio.

Hence, we see that larger genomes are likely to be selected against in bacteria. In general, this is borne out, although large genomes can hold their own if they provide the metabolic flexibility that allows an organism to utilize novel substrates in lean times that cannot be harnessed by the small genome competition.

9.5.6 Mitochondria Are Advantageous for Energetics

Mitochondria retained the genes for oxidative phosphorylation and they kept the ability to reproduce so that a eukaryotic cell can have thousands of mitochondria. Given the high energy cost incurred by each mitochondrion to transcribe and translate these genes, why did these organelles retain any genes? John Allen proposed in 1993 that having multiple centers for oxidative phosphorylation within a cell gave local control over respiratory electron chains, which conferred a huge selective advantage for cell energetics.

9.5.6.1 Oxidative Phosphorylation Is Normally Demand Led

The rate of oxidative phosphorylation is normally regulated by demand.

EXERCISE 9.10

If demand for ATP in a cell is low:

1. What will happen to the concentrations of ATP and ADP?
2. What will happen to the rate of the mitochondrial ATPase?
3. Consequently, what will happen to the rate at which protons are pumped against their concentration gradient and hence the rate at which electrons are transported in the electron transport chain?

The above exercise should have convinced you how the rate of oxidative phosphorylation is normally demand led.

9.5.6.2 Oxidative Phosphorylation Can Be Supply Led

The rate of oxidative phosphorylation will also drop if the supply of glucose, ADP, phosphate, or O_2 falls off, or if proteins in the electron chains are damaged, so work less effectively. Normally, each carrier in the electron transport chain (ECT) has a higher affinity for electrons than its predecessor, so is reduced and then oxidized as it gains an electron and then passes it to the next carrier down the line. But what if the next carrier is already reduced because electron transport has stopped?

EXERCISE 9.11

How would a lack of (a) glucose or (b) oxygen slow electron transport?

In the event that electrons cannot be transported down the ECT, they are transferred directly to O_2 forming toxic superoxide radicals. This unpleasant scenario is mitigated if slowing of oxidative phosphorylation in each mitochondrion can be corrected by local protein synthesis.

Local control over respiration conferred by mitochondria means that eukaryotic cells are much more energy efficient. They can afford to have much larger genomes, and being no longer so constrained by surface area/volume ratio, they can grow much larger.

9.5.7 Large Size Is Advantageous for Eukaryotes

The large size of eukaryotes allowed them to phagocytose, to engulf whole particles of foodstuffs, rather than rely on membrane transport of individual molecules to gain nutrients. In phagocytosis, food is encapsulated in a vesicle into which enzymes are secreted which break down macromolecules into their small molecule constituents.

The vesicle membrane contains transporters (just as bacterial cell membranes do) which then take up the small molecules into the cytoplasm. That said, the first eukaryotes may not have been phagocytic but instead might have secreted the digestive enzymes to degrade macromolecules externally and then take up the resulting small molecules.

9.6 The Fate of Life on Earth

9.6.1 Earth Will Be Habitable for Another 1.5 Billion Years

The luminosity of the sun will continue to increase even while it remains a main sequence star. This is because fusion of hydrogen to helium reduces the volume of the sun's core. As the core shrinks its density and temperature rise, boosting the rate of nuclear reactions which powers the increase in luminosity. This will continue to increase so that the average global temperature at the Earth's surface is predicted to rise from its current value of 15°C to 50°C by 1.3 Gyr and to 100°C by 1.5 Gyr in the future.

As the temperature rises so does silicate weathering and drawdown of CO_2. By about one billion years, the CO_2 concentration of the atmosphere will become so low (10 ppmv) as to limit photosynthesis to the 3% of terrestrial plants that use the C4 pathway: most plants do the less efficient C3 photosynthesis. Actually, CO_2 starvation may not be the rate limiting factor for plant growth since as the temperature rises CO_2 will degas from the ocean into the atmosphere, possibly maintaining sufficient atmospheric CO_2 for C4 carbon fixation. Of course a billion years is a long period and undoubtedly organisms will evolve that are adapted to the high temperatures and low CO_2 concentrations, but eventually even the deep ocean will become too hot to support all but hyperthermophiles. Life on Earth looks likely to end as it may have begun.

The inexorable temperature rise will increase the amount of water in the atmosphere, thereby increasing the greenhouse effect by absorbing outgoing long wave radiation, and reducing the Earth's albedo by absorbing incoming solar short wave radiation. Both of these effects will further raise the temperature in a positive feedback manner. Once the surface temperature gets to about 70°C, water will find its way into the stratosphere where it will undergo UV photodissociation and the hydrogen will be lost to space. Modeling suggests that sometime after about 1.5 Gyr the Earth will lose all of its water in this manner. Without water, plate tectonics will cease, and so does silicate weathering. Outgassing of CO_2 by volcanoes will no longer be buffered, so atmospheric CO_2 will rise irrevocably. The Earth will end up, about 2.5 Gyr from now, looking like Venus does today. Life on Earth will have been wiped out by some combination of CO_2 starvation, high temperature, and lack of water. By the time the sun becomes a red giant, about 5.4 Gyr from now, the oceans will have been long gone and Earth's surface will have become a magma ocean.

10

Mars

10.1 Martian Romance

Mars has been the target of over 40 missions. There are good reasons. It is the easiest planet to get to, and has a much less hostile surface than Mercury or Venus. This is not to downplay the many engineering and technical difficulties; only about 40% of Mars missions have been successful. But the major driver has been astrobiology. For a long time the red planet has been thought of as the most likely habitable world in the solar system other than Earth.

10.1.1 Mariner Missions Reveal a Cold Arid World

Mariner 4 was the first spacecraft to flyby Mars, in 1965, followed in 1969 by Mariners 6 and 7 which mapped 20% of the Martian surface. The images returned by these three missions disappointed the astronomical community. They revealed a heavily cratered, barren world; a desert planet with no evidence of life.

10.2 Martian Geology

Despite the depressing Mariner images, subsequent missions have revealed that liquid water has flowed across the Martian surface, for brief periods in large volumes. This has given rise to the notion that early Mars was warm and wet and might therefore have been a cradle for life. The central issue for Mars science is establishing whether this view is right. Perhaps surprisingly the evidence is ambiguous and it is not yet possible to say with any certainty whether Mars was ever warm and wet for any appreciable length of time; that is, whether Mars has ever been in the habitable zone of the solar system. The purpose of this chapter is to explore the validity of the warm, wet Mars hypothesis.

10.2.1 Mars Is a Planetary Embryo

Martian meteorites allow Hf-W dating of the time at which the core of Mars had formed (Box 10.1). Although there is considerable uncertainty it implies that Mars accreted rapid. This is consistent with its small size. With about half the radius of Earth, but just one-tenth the mass and one-third the surface gravity, it may be that accretion of the red planet was truncated by the Grand Tack (5.5.3). So, Mars is more like a planetary embryo than a fully-grown planet.

BOX 10.1   MARTIAN METEORITES

More than 120 Martian meteorites have been identified. They have a chemistry showing that they were formed from a much more volatile-rich oxidizing environment than would be expected for an asteroid. Martian provenience is confirmed by measuring the abundances and isotopic compositions of N_2, CO_2, and noble gases found in glassy inclusions within the meteorites. These are identical to Martian atmosphere sampled by the Viking missions, as are the deuterium/hydrogen ratios.

Most meteorites from Mars are termed Shergotitte–Nakhlite–Chassignite (SNC) meteorites, named for the locations of the type specimens. Shergottites are by far the most common and are basalts containing the unusual pyroxene mineral pigeonite. They have suffered severe shock metamorphism and their radioisotopic age, only 180 Myr, probably dates the impact event that launched them from the Martian surface; the original rock is presumably much older. Shergotittes are presumably derived from a lava flow on the Martian surface. Nakhlites and Chassignites both date to ~1.3 Gyr and have compositions showing that they are cumulates that crystallized in an ultramafic magma chamber. Nakhlites are mostly the pyroxene augite and iron-rich olivine, some of which has been altered by water. Chassignites resemble terrestrial dunites in being >90% olivine. The SNC meteorites testify that Mars was volcanically active just 1.3 billion years ago.

Two Martian meteorites do not fit into the SNC package. One, NWA 7409, is a basaltic breccia that is 2.089 Gyr old and has the highest water content of any Martian meteorite. The other is the infamous 4.09 Gyr-old ALH84001, of which more is discussed later.

10.2.1.1 *The Martian Core Is Totally or Partly Liquid*

The core of Mars has a radius of about 1800 km and moment of inertia and tidal deformation measurements by Mars orbiters such as Mars Global Surveyor show that the core could be entirely liquid or partially solid, but not completely solid. There is reasonable circumstantial evidence to support sulfur as the light element in Mars' core. The crust of Mars is depleted in sulfur-loving (chalcophile) elements that were presumably removed to the core as the planet differentiated. Laboratory experiments on Fe–Ni–S mixtures under conditions in the Martian core ($P \sim 23$ GPa, $T \sim 1800$ K) imply that the core could be entirely liquid. The sulfur lowers the melting point of the metal alloy and could make up as much as 17% of the core.

Despite having at least a partly liquid metallic core, Mars has no global magnetic field today. However, it did.

10.2.1.2 *When Mars Lost Its Magnetic Field Is Significant*

Once upon a time Mars had a global magnetic field. We know this because its remnant magnetism is imprinted in the most ancient Martian crust. The loss of the magnetic field could have set in motion the decay of the Martian atmosphere, so it is of some interest to know when this happened.

The Martian meteorite ALH84001 has a crystallization age of 4.091 Gyr. This bears remnant magnetism showing that it cooled on a Mars that still had a global magnetic field. The magnetic field over the Tharsis volcanic plateau is very weak indicating that the dynamo was decaying by the time this was emplaced about 3.7 Gyr ago.

EXERCISE 10.1

What mechanism might have driven the loss of the Martian atmosphere after the planet lost its magnetic field?

10.2.2 Mars Has Two Very Different Hemispheres

Even a cursory glance shows that the northern and southern hemispheres of Mars are very different. This is termed the Martian dichotomy. The two hemispheres can be distinguished by their elevation, age, and crater density. The southern crust is topographically high, ancient, heavily cratered, and reckoned to be about 60-km thick on average. The northern one-third of the planet is a huge depression 3–6 km lower than the southern highlands. It is younger, much smoother, less cratered, and thinner than the southern crust, averaging 35 km in thickness. The northern hemisphere is not all it seems, however. Impact basins hidden beneath the surface were discovered by radar altimetry that presumably represent impacts into crust every bit as ancient as that of the southern highlands. These so-called quasi-circular depressions (QCDs) have subsequently been covered by younger deposits. A topographic map of Mars is shown in Plate 10.1.

10.2.2.1 The Origin of the Northern Lowlands Is Uncertain

There are two classes of theory to account for the Martian dichotomy. Both acknowledge that it originated within a few tens of millions of years of Mars' accretion. One is that the crust in the north was thinned by mantle convection or overturning or an early episode of plate tectonics. The other is a giant impact. Modeling shows the Boreal basin could have been excavated in a grazing impact by a Pluto-sized body. Two features of this idea are appealing. One is that such an impact could explain how Mars lost most of its early atmosphere. We will return to this later. Another is that it could account for why the remnant magnetic field is stronger in southern hemisphere. The early magnetic field seems to have been asymmetric: stronger in the south. Computer models show that impact heating of the northern hemisphere would shut off convection on which the dynamo depended by reducing the thermal gradient, whereas in the southern hemisphere a large temperature gradient would remain.

10.2.2.2 The Tharsis Bulge Is the Highest Region on Mars

A huge volcanic-tectonic province, the Tharsis plateau, rises on average 7–10 km above the Mars zero elevation datum (Box 10.2) and hosts three huge volcanoes. Olympus Mons, a shield volcano almost three times higher than Mount Everest lies adjacent to the bulge proper. The plateau is thought to result from mantle plume magmatism. In the absence of plate tectonics, this continually acts beneath the same

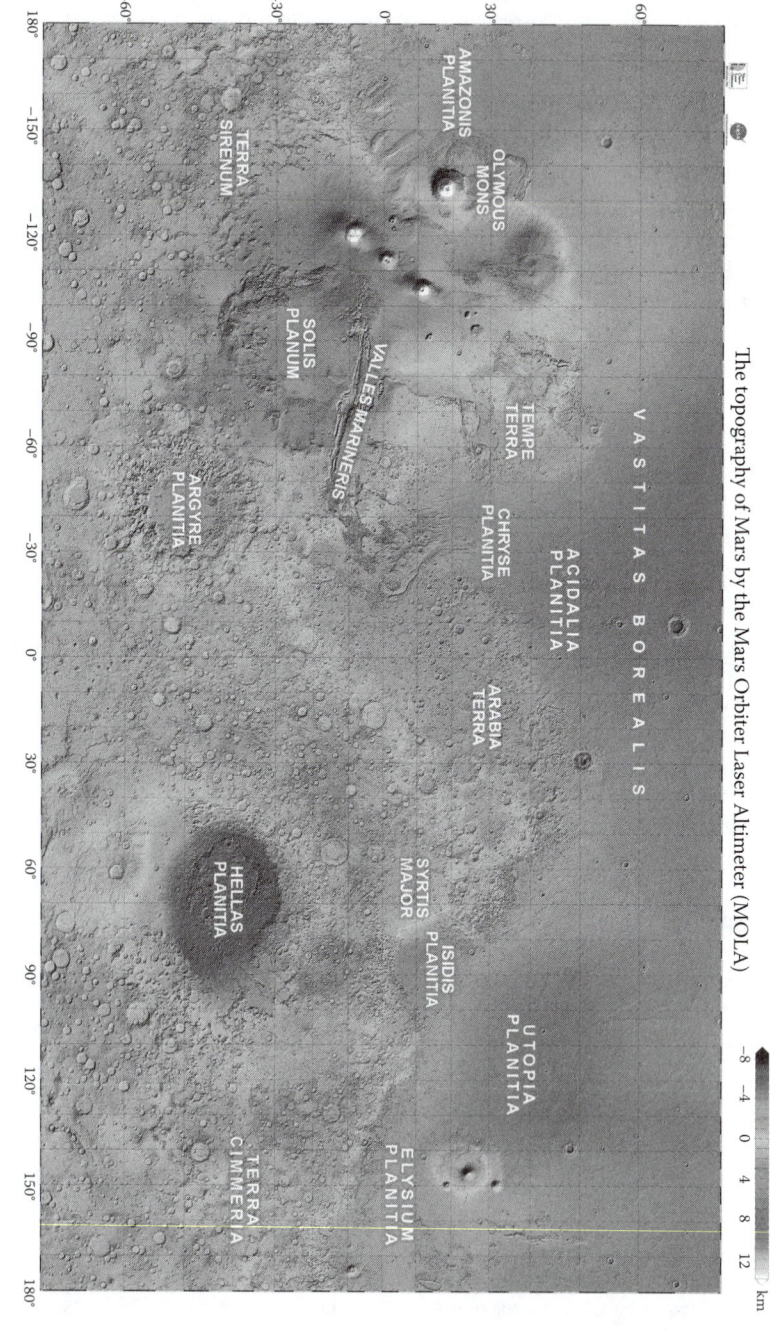

The topography of Mars by the Mars Orbiter Laser Altimeter (MOLA)

PLATE 10.1 (See color insert.) High resolution topographic-shaded relief map of Mars constructed from Mars Orbiter Laser Altimeter (MOLA) data. Key features are labeled. Graphic processed by Greg Smye-Rumsby. (Photo credit: NASA/MOLA Science Team.)

BOX 10.2 MARS ZERO ELEVATION DATUM

On Earth, topography—the height of a mountain or the depth of an ocean trench—is expressed in relation to mean sea level. This datum cannot be defined for a planet with no oceans. The topography of Mars is given in relation to the altitude at which the atmospheric pressure is 610.5 Pa, chosen because it is the triple point pressure of water.

---

region of lithosphere and is therefore able to build much larger structures than hot spot volcanism does on Earth. Subsidence and extension generated by this loading of the Martian lithosphere has produced a huge series of graben and rifts (Figure 10.1) which together comprise the Valles Marineris (VM), a system of canyons near the Martian equator that extends east from Tharsis for 4000 km and which in places is 300 km across and 10 km deep. At its most easterly, it forms an outflow channel which empties into the northern lowlands. Formation of VM had probably begun by 3.7 Gyr but continued to be eroded, possibly in part due to water, until maybe as late as 2 Gyr ago.

EXERCISE 10.2

Assuming that water ice is present, what two conditions must be met for liquid water to be stable on Mars?

10.2.2.3 Cratering Rates Provide Clues to Mars' History

The number of craters on a planetary surface provides an estimate of the *relative* age of the surface. The longer a surface has been around the greater the chance that it has been impacted. The only unambiguous way to find the absolute age of a surface is to radiometrically date a representative rock from the surface. This has been done for the parts of the moon visited by the Apollo missions and allows crater numbers to be calibrated to the absolute ages. This is not possible for Mars since we have had no sample return mission and the provenience of Martian meteorites is unknown. Attempts to get an absolute chronometry for Mars by comparing estimated cratering rates on Mars to those on the moon are fraught with uncertainties. Crater counts need to distinguish between primary and secondary craters (those formed by ejecta thrown up by the primary event), and become dependent on assumptions about the flux rate of impactors for surfaces that are saturated. Translating from one planetary surface to another has to take account of the fact that bodies of different size and surface gravity will experience different flux rates and crater size distributions. Larger

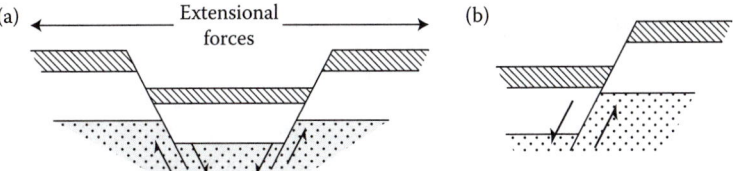

FIGURE 10.1 (a) Graben. (b) Strike-slip fault.

TABLE 10.1

Three Cratering Ages of Mars

Epoch	Time (Gyr)
Noachian	4.1–3.7
Hesperian	3.7–3.0
Amazonian	3.0–0.0

bodies are a bigger target and more massive ones will experience a higher velocity strike than a less massive one for a given mass projectile. What is more, craters get covered, eroded, and exhumed by geological processes that are likely to differ from one world to another. Consequently, the precise age of any part of the Martian surface is not known.

Nonetheless, differences in crater density have defined three broad periods of Martian geological history, each named after an archetypal location on the planet (see Table 10.1). Noachian-aged surfaces (named for Noachis Terra) are the most heavily impacted visible surface. During this period the Tharsis plateau and extensive valley networks were formed. Most of the surface of the southern hemisphere of Mars is thought to be of Noachian age. Cratering rates declined dramatically during the Hesperian and Tharsis volcanism and large outflow channels date to this time. The Amazonian period is characterized by low crater numbers consistent with current cratering rates, though there is evidence for modest levels of volcanic and glacial activity.

It is important to be aware that the age ranges in Table 10.1 are the usually quoted values. Several cratering models have been formulated that make different assumptions and have different age ranges. For example, different models have the Noachian period ending either 3.5 or 3.8 Gyr ago, and estimates for the end of the Hesperian range from 3.2 to 1.8 Gyr.

The Noachian is usually regarded as starting with the Hellas impact and being coeval with the classical timing of the late heavy bombardment. However, laser altimetry has uncovered a large population of hidden pre-Noachian craters (QCDs) eroded by Noachian impacts. This pre-Noachian terrain is saturated with craters and represents the oldest crust on Mars.

10.2.3 Spectrometry Reveals the Nature of Planetary Surfaces

Several methods are used to work out the composition of a planet's surface. A variety of instruments can be accommodated on orbiting spacecraft for remote sensing, or in landers to analyze surfaces directly.

We now briefly look at five techniques, although there are others.

10.2.3.1 Gamma Ray Spectrometry (GRS)

GRS allows elements to be detected in a surface. It relies on the fact that γ-rays with distinctive energies are emitted when Galactic cosmic rays collide with atoms. The collisions excite the atoms, so that they eject neutrons which in turn collide with other nuclei, exciting them. On de-exciting, a nucleus emits a gamma ray with specific energy. Hence, GRS produces an emission spectrum with line energies characteristic of the emitting nuclei and amplitudes proportional to the number of nuclei. For example, neutrons ejected by cosmic rays that collide with H atoms in water excite the emission of γ-rays of 2.223 MeV.

Astronomical γ-ray spectrometers consist of a high purity germanium crystal diode scintillator, the light output of which is proportional to the energy of the incoming gamma ray photon. The light is detected by photomultiplier tubes which generate a voltage proportional to the number of light photons via the photoelectric effect. GRS can penetrate tens of centimeters into the surface, but have poor spatial resolution of many kilometers.

EXERCISE 10.3

Why would it be difficult to use GRS for remote sensing of the surface of a planet with a dense atmosphere such as Venus?

Gamma ray spectrometers can also double as neutron detectors, measuring the energy of the neutrons emitted by the top 1 m or so of a planetary surface as a result of cosmic ray strikes. Neutron detectors have proved useful in detecting hydrogen, which is usually assumed to be present either as hydroxyl groups in minerals or water ice. The hydrogen detection is indirect. It depends on the fact that when a neutron collides multiple times with protons, it is slowed sufficiently that it can be captured by a nucleus to form a nuclide that decays by emitting gamma rays. For example,

$$^{10}_{4}B + n \rightarrow {}^{7}_{3}Li + \alpha + \gamma. \tag{10.1}$$

10.2.3.2 Alpha-Particle-X-Ray Spectrometry (APXS)

APXS uses α-particles from a radioactive source such as curium-244 (half-life 18.6 years) to examine the elemental composition of a surface. Three mechanisms provide data:

1. Alpha particles of a defined energy are backscattered at close to 180° into a detector if they strike an atomic nucleus. By applying the conservation of energy and conservation of linear momentum to the spectrum of the backscattered α-particle energies, the mass of the deflecting nuclei can be found.
2. Atomic nuclei absorb α-particles and emit protons with characteristic energy in response. This (α, p) process can be used to identify light elements such as Si, Mg, Al, Na, and S.
3. Alpha particles can eject electrons from the inner shell of an atom. These vacancies are filled by electrons from outer shells, which results in the emission of an x-ray with characteristic energy. This can be used to identify some heavy elements.

APXS instruments were/are part of the toolkit of the Mars Pathfinder Rover, the Mars Exploration Rovers, and Mars Science Laboratory (MSL) Curiosity Rover.

10.2.3.3 Thermal Emission Spectrometry (TES)

TES measures the spectrum of energy emitted in IR and this allows the spectra of individual minerals to be distinguished. The TES instrument on Mars Global

Surveyor has a spatial resolution of 1 km, but can only penetrate to a depth of 10–100 μm, so may be picking up minerals formed by weathering rather than the native rock beneath.

10.2.3.4 Reflectance Spectrometry

When photons strike a mineral some are absorbed by the grain and some are scattered, that is, either refracted through it or reflected from its surface. Reflected photons can be detected remotely, producing an absorption spectrum which is characteristic of a particular mineral. The eye is a crude reflectance spectrometer. Light reflected from an olivine crystal, detected by photoreceptors, is in the yellow-green part of the spectrum because red light is absorbed by the olivine.

Photons are reflected mostly by irregularities in the surface of a mineral grain, whereas absorption is determined by the path length of the photon through the grain.

> **EXERCISE 10.4**
> a. What is the effect of grain size on reflectance?
> b. A rock is a mixture of a light mineral and a dark mineral. What effect will the dark grains have on the reflectance spectrum of the rock?

Absorption of photons is due to electronic and vibrational transitions in the ions within the minerals. Electronic transitions are seen in UV and visible light. The most common electronic excitation boosts electrons into unfilled d orbitals of transition elements, such as iron, either as Fe(II) or Fe(III). But it is not just the element or its oxidation state that determines the energy of the photon absorbed. Many other factors, including the geometry of the crystal and how distorted it is by the presence of other metal ions, and the transfer of electrons between different ions within a crystal—that is, the environment in which the metal ion finds itself—give the mineral a specific spectral signature.

Vibrational transitions occur in the IR, but will only be seen in molecules that have a dipole moment so that they are IR active, as we saw in Chapter 2. Water and OH produce particularly diagnostic vibrational absorptions in minerals at about 1.9 μm and 1.4 μm, respectively. Carbonates (and other oxides) also have diagnostic vibrational spectra.

10.2.3.5 X-Ray Diffraction (XRD)

A crystal can be regarded as being a set of stacked planes, each plane acting as a semi-transparent mirror. X-rays with a wavelength close to the distances between these planes are reflected at an angle equal to the angle of the incoming ray. This diffraction is described by Bragg's Law:

$$2d\sin\theta = n\lambda$$

where d is the spacing between the planes, θ is the angle of incidence of the x-rays with wavelength λ. When Bragg's Law is satisfied, constructive interference of diffracted x-ray beams occurs and a reflection will be picked up by a detector scanning at the appropriate angle (Figure 10.2).

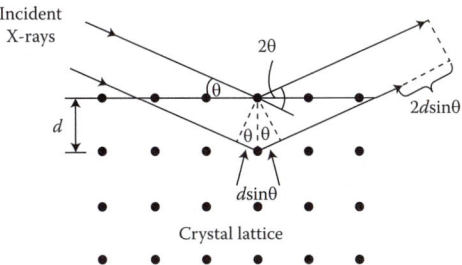

Incident X-rays

2θ

$2d\sin\theta$

d

$\theta\;\theta$

$d\sin\theta$

Crystal lattice

FIGURE 10.2   The refraction of x-rays from a mineral can be used to determine the inter-plane distance.

The positions of the reflections encode the interplane spacing of atoms in the crystal. Peak intensities represent the degree of x-ray scattering in the reflection, providing information on where particular atoms lie in the structure, and how much of a phase is present in a sample. In this way, crystal sizes, structures, and lattice parameters can be worked out.

The Chemistry and Mineralogy (CheMin) instrument on the MSL Curiosity Rover includes the first x-ray diffractometer ever to be used off-Earth. The instrument scoops up regolith or drills out a piece of rock, pulverizes it, and sieves it to particle size <150 μm. The sample is then bombarded with Kα x-rays emitted by a cobalt target in response to electrons above a critical voltage. These have enough energy to kick an electron from the cobalt atom K shell. The gap is filled by an electron dropping down from the L shell to produce the Kα emission line with a wavelength characteristic of cobalt. Cobalt is chosen as a target because its Kα x-rays have a wavelength of order 0.1 nm, close to the interplanar spacing in silicate mineral crystals. These monochromatic x-rays are collimated to make a beam 50 μm wide that is passed through a sample chamber disc with diameter 8 mm and 175-μm thick. The reflections are imaged by a 600×532 pixel CCD cooled to $-60°C$. Each analysis is over 1000 exposures and takes take up to 10 h. The goal is to analyze 74 samples during the nominal mission lifetime, but the instrument is capable of analyzing many more.

EXERCISE 10.5

Show that cobalt Kα x-rays (6.93 keV) are appropriate for determining the interplanar spacing of silicate mineral crystals.

10.2.4 The Martian Crust Is Mostly Basalt

Similar to the Earth's ocean crust, the lunar maria, and the crusts of Mercury and Venus, remote sensing by orbiters and direct analysis by Mars Landers show that most of the Martian crust is basalt, although some continental crust-type rock (e.g., granite) has also been identified.

EXERCISE 10.6

Briefly outline how basalts are produced on Earth.

10.2.4.1 Much of the Surface of Mars Is Covered by Regolith

Mars has had only a tenuous atmosphere for much of its history and so its surface has been bombarded continuously by meteoroids. This has pulverized the surface forming a regolith. It consists of particles of basalt and volcanic glass laced with various amounts of sulfates (e.g., $MgSO_4 \cdot 6H_2O$) and iron oxides.

The regolith is highly oxidizing and chemically degrades organic molecules. In 2008, the Phoenix lander discovered perchlorate (ClO_4^-) in the regolith near the north polar cap. Perchlorate is a weak oxidizing agent that is formed in arid deserts on Earth (e.g., Death Valley) by the action of UV on chloride salts. Intriguingly, some terrestrial bacteria use it for respiration, reducing perchlorate to chloride (Cl^-) and oxygen (O_2).

The finer particles in the regolith can be lifted in the atmosphere by winds generated by the large thermal gradients experienced on Mars due to its thin atmosphere. Even particles as large as 20 μm can remain suspended by wind velocities as low as 0.8 m s^{-1}. Suspended iron oxide particles give the Martian sky a pink color. At perihelion, atmospheric dust particles can be heated sufficiently by sunlight to warm the gas molecules in their vicinity. This sets up convective winds which blow more dust into the atmosphere to be warmed, a positive feedback that launches dust storms that can be global and long-lasting. Since perihelion of Mars corresponds to southern hemisphere summer, the dust storms often start in the south.

10.3 Water Ice Is Abundant on Mars

Water ice exists in the polar ice caps, and in a cryosphere of permafrost in the crust. The depth of the cryosphere is hard to gauge, since only the shallow subsurface can be probed directly, but estimates suggest Mars has more 5×10^6 km^3 of ice.

10.3.1 Mars Has Substantial Frozen Water

Mars has permanent polar ice caps. These are mostly water ice but in winter the caps acquire a veneer of dry ice. CO_2 condenses to form a cloud layer termed the polar hood and then falls as snow thickening and enlarging the cap. This results in ~30% of the atmosphere of Mars freezing out during both northern and southern hemisphere winters. Because the seasons are more extreme in the southern than the northern hemisphere, and because the south pole is at high altitude, the dry ice covering of the south polar cap is 8-m thick compared with 1 m on the north polar cap.

Both polar caps are built up from numerous layers. These polar layered deposits reflect seasonal melting and deposition of ice and dust and should allow changes in the Martian climate in the recent past (for example, as a result of swings in obliquity) to be deduced.

10.3.1.1 Obliquity Swings Redistribute Water Ice

While Earth's obliquity is thought to have been constant over geological time, possibly because it is stabilized by our large moon, Mars' obliquity fluctuates widely. For example, over the past 3 Myr, there has been a periodic shift between 15° and 35°

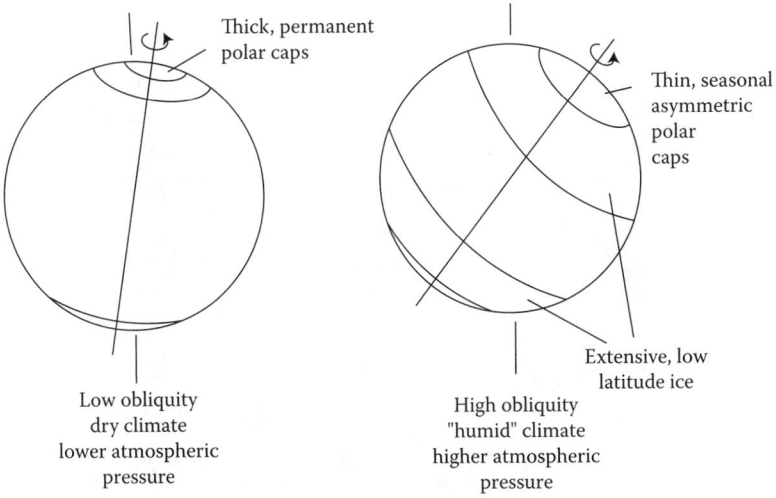

FIGURE 10.3 Changing obliquity of Mars causes water ice to shift.

with a period of 125,000 years. Going further back, the swings in obliquity seem to have been chaotic.

At high obliquity, the poles are warmed, ice is sublimed, and re-deposited at mid-latitudes where the insolation is less. This is a Martian ice age, and it is thought the planet spends more of its time in this state rather than not (Figure 10.3).

At low obliquity the warming of mid-latitudes shifts ice back to the poles.

10.3.1.2 Subsurface Water Ice Is Revealed by Geological Features

Several features on Mars attest to subsurface frozen water by analogy with similar regions on Earth. One is a region 800 km across in southern Elysium imaged by Mars Express that is thought to be an equatorial frozen sea, because its appearance and geological features have a striking resemblance to frozen pack ice in the Antarctic Ocean.

Other features resemble periglacial landscapes on Earth. One, imaged by Phoenix in the far north, is exactly like cryoturbation polygons seen in Arctic tundra where permafrost undergoes episodes of freeze–thawing (Plate 10.2).

10.3.1.3 Subsurface Water Ice Has Been Detected by Geophysics

The gamma ray spectrometer and neutron spectrometer on Mars Odyssey have found subsurface water ice. Generally, there is more ice in regolith the higher the latitude. If all the ice were in the top 1 m it would amount to a water equivalent global layer of about 14 cm.

10.3.1.4 Water Ice Has Been Seen Directly

Phoenix, located at latitude 68°N, dug a shallow trench uncovering bright reflective material that disappeared over 4 days. This fits with the timescale over which water ice would sublime. Dry ice would also sublime but under these conditions would have

PLATE 10.2 **(See color insert.)** Cryoturbation polygons on Mars resemble those in terrestrial tundra that arise from freeze-thaw processes. Imaged by Mars Phoenix Lander's Surface Stereo. (Photo credit: NASA/JPL-Caltech/University of Arizona.)

done so much faster. In addition, the Phoenix mass spectrometer detected water in a sample heated to 0°C. More recently, Curiosity Rover's gas chromatogram shows superficial regolith to contain 1.5–3 wt% water.

10.3.1.5 The Cryosphere Is a Thick Permafrost Layer

The Martian cryosphere is the region of the crust where the temperature remains continuously below the freezing point of water. The cryosphere thickness is estimated by modeling the heat flow in the Martian crust, the thermal properties of an ice-rich crust, its porosity, and topography. These models give 2.3–4.7 km at the equator to 6.5–13 km at the poles. However, recent work indicates that heat flow in the Martian crust may only be half of what we previously thought. The thickness of the cryosphere would also change with time as a result of changes in obliquity and eccentricity which alter the global distribution of sunlight and total annual insolation.

> **EXERCISE 10.7**
>
> What impact would halving heat flow in the Martian crust have on our estimate of cryosphere thickness?

10.3.1.6 Extensive Groundwater Aquifers Are Missing

It has long been assumed that beneath the cryosphere is a groundwater aquifer within pore spaces of the crust. It would lie beneath the cryosphere, and its lower boundary would be basement rock that is so compacted by the mass above as to have no pore space. The argument is that geothermal heat could be enough to liquefy water at depth, where the pressure is higher, especially if the water contained high concentrations of salts that would lower its melting point. This might provide a haven for chemolithotrophs, by analogy with the bacteria that are found several kilometers below in the Earth's crust.

Measurements by ground penetrating radar (GPR) such as Mars Advanced Radar for Subsurface and Ionosphere Sounding (MARSIS) on Mars Express—which can probe down about 4 km if the ground is porous or low density, or a few hundred meters if it is high density rock—have provided evidence for a water ice cryosphere beneath the northern lowlands and the south polar region. MARSIS maps correspond well to abundances of water ice inferred from gamma ray and neutron spectrometry. But no aquifers have been seen beneath the cryosphere.

> **EXERCISE 10.8**
>
> MARSIS shows evidence for high abundance of water ice in equatorial regions as well as at the poles with rather less at mid-latitudes. Suggest what might account for this distribution of water ice in the Martian crust.

10.4 Water Has Flowed on Mars

There is little doubt that liquid water has flowed on Mars at times. The central questions are when and for how long.

10.4.1 Mars Has Numerous Fluvial Features

The two most obvious fluvial features (those caused by the action of water in rivers or streams) on Mars are valley networks and outflow channels.

10.4.1.1 Valley Networks Are Not Like Terrestrial River Beds

Valley networks are U-shaped valleys typically 60 km long and <1 to 10 km wide found mostly in the equatorial regions of the southern hemisphere. Over 90% are Noachian with the remainder split evenly between Hesperian and Amazonian in age. Despite the fact that at first glance they look rather like dry river beds (wadis) on Earth, they have a much sparser drainage pattern—fewer tributaries and much less complex branching—than their terrestrial counterparts, and they are often isolated. Their shape and structure shows water would have flowed in them sporadically and they are likely to be groundwater seepage/runoff channels. This implies that valley networks are not the result of long-term greenhouse warming and a fully fledged water cycle, but processes which allowed Mars to be wet episodically. These episodes may have lasted for hundreds of years. Impacts or melting of ice deposited during high obliquity periods as Mars swung to low obliquity are possible causes.

10.4.1.2 Outflow Channels Were Formed by Episodic Floods

Outflow channels are broad, steep-sided, flat-floored valleys with rectangular cross sections. Most drain into northern lowlands with the greatest number around Chryse Planitia (Plate 10.3).

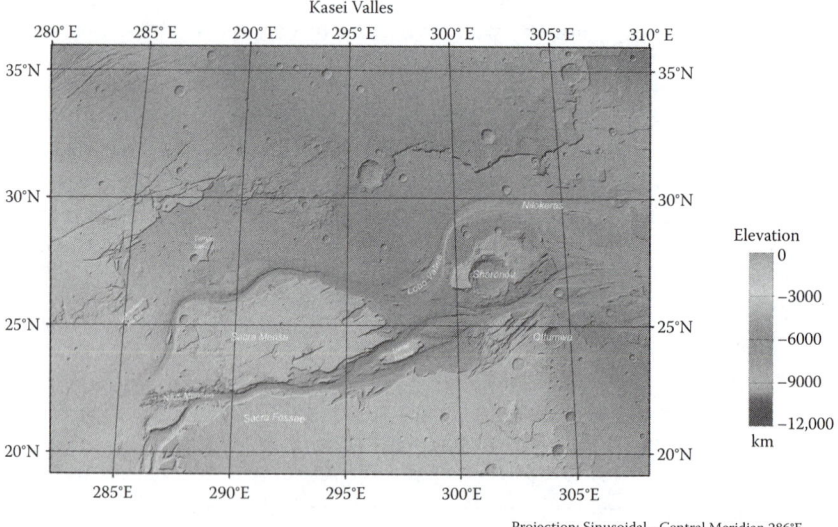

Projection: Sinusoidal - Central Meridian 286°E
Coordinate System: Mars Sphere - Axis 3396 km

PLATE 10.3   **(See color insert.)** Kasei Valles, an outflow channel which once carried water into the Chryse Planitia. (Photo credit: ESA.)

Generally, outflow channels are Hesperian and early Amazonian. Their scale is huge—typically hundreds of kilometers long and more than a kilometer wide—similar to terrestrial flood channels formed by the sudden release of prodigious quantities of water. Although this is the majority view, arguments have been made for outflow channels being carved by glaciers, debris flows driven by CO_2, and lava.

Martian outflow channels usually start at graben (Figure 10.1) or chaos terrain. Chaos terrain typically consists of irregular groups of tilted kilometer-scale blocks a hundred or more meters high, cut through by valleys, and lying in a depression caused by collapse of the underlying cryosphere. This collapse could be due to impacts and volcanism melting the permafrost, or the explosive decompression of carbon dioxide or methane clathrates. Chaos terrain appears to have been shaped by over 100 floods, and calculations show that outflow channels are unlikely to have been carved by a single event.

An alternative model for chaos terrain formation is less violent. Studies of Aram Chaos propose that it started out as an impact crater. This filled with ice-rich sediment that was then overlain by sediment that insulated the ice beneath from sublimation and became cemented into sedimentary rock. Radiogenic heating eventually melted the ice, causing the break-up and subsidence of the rocky cap, while the melt water rose and eroded the channels. If this occurred episodically it would explain the morphology of the chaos terrain and outflow channels. In this model, recharge of the groundwater following outflows comes from melting of snow and ice that accumulated on the large volcanic provinces, Tharsis and Elysium, during high obliquity. Calculations show that this would account for the estimated outflow channel volumes.

10.4.1.3 Did Early Mars Have a Northern Ocean?

One way of explaining how the aquifers that sourced the outflow channels were recharged was if there had been a large ocean in the northern lowlands. On this basis, the existence of an ice-covered Hesperian Boreal Ocean was proposed by Stephen Clifford and Tim Parker. They used Mars Orbiter Laser Altimeter (MOLA) data to define a palaeo-shoreline, the Deuteronilus shoreline, inside which the surface was much smoother, and that would have contained a water equivalent global layer of 100 m, consistent with estimates of the water inventory of early Mars. The shoreline is controversial. It has terraces with ocean-facing ridges that resemble tectonic stress ridges created by early volcanism, not those of a shoreline. It is not flat but ranges in elevation from about −2 to −4 km below the Mars datum, though this could be due to subsequent load changes due to the formation of Elysium volcanoes or the Utopia basin, or discharges from the outflow channels during the Hesperian period that covered any surviving frozen relic of the ocean.

Recent support for the ocean comes from MARSIS which has mapped the dielectric constant across the northern lowlands. This suggests the presence of vast deposits of sedimentary materials, as would be expected on an ocean floor, extending into the three main sites for outflow channels. Intriguingly, this circumscribes an area that corresponds well to the palaeo-shoreline proposed by Clifford and Parker in 2001.

10.4.1.4 Martian Gulleys Are Enigmatic

Martian gulleys, like their terrestrial cousins, are narrow channels cut into steep slopes. Each gulley has an up-slope alcove, a triangular cut-out region etched with a drainage pattern that converges into the channel, and a fan-shaped apron of debris down-slope. Material is removed from the alcove and deposited on the apron by fluid flow. On Earth the fluid is usually liquid water but whether this is the case on Mars is not known. Avalanches caused by small impacts or debris flows driven by carbon dioxide are alternative explanations.

10.4.2 Mars Exploration Rovers Have Searched for Signs of Liquid Water

The Mars Exploration Rovers reached Mars in January 2004. Spirit landed in Gusev crater and operated for over 6 years, Opportunity landed on Meridiani Planum and is still functional (2014). Both landing sites were selected because they looked as if liquid water had been there in the past. How have they got on?

10.4.2.1 Water Interaction at Gusev Has Been Limited

Gusev crater lies at the end of the Ma'adim Vallis network. Although the crater was probably created 4 Gyr ago, its floor dates to 1.8 Gyr, the same age as the floor of Ma'adim Vallis. The thought was that 1.8 Gyr marked the last time sediments were deposited from water supplied by the valley that ponded in the crater. That Gusev crater was a lake bed no longer seems likely. The rocks of Gusev plains are a primitive picrobasalt with a veneer of water-altered minerals at the surface and in veins. A thin film of moisture is all that would be needed to produce these features.

Spirit Rover left the plains for the Columbia hills where it did find evidence for aqueous weathering. The almost ubiquitous presence of olivine, a mineral easily weathered by water, shows that the exposure to water must have been limited, but sulfates (mostly $MgSO_4$ and $CaSO_4$), carbonates (salts of the CO_3^{2-} ion), and the iron mineral goethite, $Fe(III)O(OH)$, testify to moderate levels of water weathering. It is possible that aqueous aerosols of SO_2, HCl, and HBr outgassed from volcanoes were responsible for the weathering. The acidic conditions would promote the oxidation of Fe^{2+} (in olivine) to Fe^{3+} oxides (e.g., goethite) and formation of sulfates.

10.4.2.2 Meridiani Hosts Water-Formed Minerals

It was the identification of the water-formed mineral hematite by the Thermal Emission Spectrograph on Mars Global Surveyor that made Meridiani Planum such an attractive Rover destination. Several lines of evidence point to liquid water at Meridiani. Firstly, in places, sedimentary deposits composed of layers of sand-sized particles of olivine basalt show festoon cross-bedding. Geologists associate this feature in which the layers are nested concave-up, centimeter-scale shapes (rather like little smiles), with deposition in running water. Secondly, the sediments are as much as 40% magnesium and iron sulfates. The Mars Exploration Rover principle investigator Steven Squyres of Cornell University argues that the sulfates are evaporites that were deposited when a large body of briny water dried up. Thirdly, the hematite (Fe $(III)_2O_3$) is important because this mineral usually (though not invariably) requires liquid water for its formation. At Meridiani, hematite is embedded in sulfate evaporite sediments, filling channels and hollows. In places it forms spherules 1–5 mm that came to be known as "blueberries" because the false color imaging NASA used made them look bright blue. When cut open by the rover's grinding tool, blueberries are revealed to be made of concentric layers, rather like an onion. This is taken to mean that they were formed by repeated episodes of wet and dry conditions. Although alternative volcanic or meteoric origins for hematite concretions are possible, the fact that they are randomly distributed through the sediments rather than in discrete layers is more suggestive of a watery genesis. Finally, Opportunity has identified jarosite, $(KFe(III)_2(SO_4)_2(OH)_6)$, a mineral formed in oxidizing acidic sulfur-rich water. Its presence shows us that standing water on Mars would have been essentially dilute sulfuric acid!

10.4.2.3 Was Meridiani Plain a Sulfate Brine Lake?

Given the evidence above, the standard story is that a sulfate-rich acidic water lake existed at least periodically at Meridiani. The water would have been kept liquid by geothermal heat, the high concentration of $MgSO_4$ which lowered the freezing point of water, and a covering of ice which acted as an insulator.

But this interpretation has been questioned. Two papers published in the journal *Nature* in December 2005 completely reinterpret the findings of the Mars rover scientists, removing the need for standing water. One, by Paul Knauth, a geologist at Arizona State University and colleagues, argues that ground-hugging surges of gas and ash, either from a meteorite impact or volcanic explosion, can account for all the Opportunity findings at Meridiani. They point out that surges on Earth mimic a great diversity of depositional environments, including festoon cross-bedding.

The other paper by Thomas McCollom and Brian Hynek of the University of Colorado reckon that if the sulfates had been deposited by evaporation they should have been balanced by cations (e.g., Ca^{2+}, Mg^{2+}, or Fe^{2+}), but this is not what is found at Meridiani. They argue that the most likely source of sulfate without metals is SO_2 outgassed by volcanoes. Their idea is that the sedimentary deposits started out as volcanic ash flows that were subsequently percolated by sulfur dioxide-rich steam vapors erupting from geothermal fumaroles.

At present, then, there is doubt about standing water at Meridiani.

10.4.3 Global Geological Markers Can Reveal How Long Mars Was Wet

Gusev crater and Meridiani were selected for detailed examination because planetary scientists expected to find indications for the past action of liquid water there, but they are unlikely to be representative of Mars as a whole. Yet we need a global assessment of the extent to which water has altered the red planet over its history if we are to understand its climate and how that has changed. This can be done by looking at several geological proxies of water action. These can furnish insights into the timescale that water was present at the Martian surface.

10.4.3.1 Lack of Carbonates Is a Red Herring

Since Mars started out with a substantial inventory of carbon dioxide, if it had plenty of surface water and/or a water cycle, then carbonates should have precipitated as CO_2 dissolved in water:

$$H_2O + CO_2 \rightarrow H_2CO_3 \rightarrow H^+ + HCO_3^-. \tag{10.2}$$

Given this, we need to account for why regolith dust has only 2–5 wt% and meteorites only ~1 wt% by weight of carbonates. This may mean that water was present so briefly there was no time for carbonates to form, but not necessarily. The presence of jarosite shows that any Martian surface water would have been very acidic and that carbonates dissolve at low pH, so the lack of surface carbonates does *not* mean Mars was not warm and wet.

10.4.3.2 The Martian Surface Has Not Suffered Much Alteration by Water

That water alteration of the surface of Mars has not been great can be inferred from three observations. One is the widespread appearance in the southern highlands of olivine, a mineral rapidly weathered by water. Another is that the rind of water-altered palagonite on basalt surfaces is only a few microns thick. A third is that regolith contains pristine magnetite (Fe_3O_4) grains despite the fact that fine particles of magnetite are rapidly weathered by water.

Finally, as we have seen jarosite requires water to form—but there is a twist; jarosite is converted into iron hydroxide as aqueous weathering of basalt raises the pH. Hence jarosite shows that water cannot have been around for long.

10.4.3.3 Clays Provide Insight into Water Action

Clay minerals (phyllosilicates) incorporate OH or H_2O into their structures and are formed when water weathers minerals in igneous rocks, such as basalts. On Earth most commonly this happens at the surface where it produces soils, but it also occurs in hydrothermal systems, or in water-rich magmas at volcanic settings. Each of these processes produces different clay minerals, so clay composition reveals how it was made. Clay formation at the surface requires surface water, implying a clement climate, whereas the other processes do not, so clays can act as a window onto the climate.

10.4.3.4 Martian Clays Are Generally Noachian

Clay minerals were first identified on Mars in 2005 by the OMEGA spectrometer on board ESA's Mars Express. Subsequent mapping has revealed thousands of outcrops of clays on the ancient heavily cratered Noachian (4.1–3.7 Gyr) terrains of the southern highlands. By contrast they are detected on Hesperian- and Amazonian-aged surface units of the northern lowlands only where large craters have punched deep, excavating materials beneath.

Most Martian clays seem to have been formed beneath the surface by hydrothermal groundwater circulation. How do we know? Below ground, only small amounts of water tend to be available and the pH and redox state are determined locally by the rock. Although weathering causes different minerals to form, there is little transport of soluble ions, so elemental abundances do not alter much and the chemical reactions that occur will be close to equilibrium with the precursor rock. Under these conditions, Fe- and Mg-rich smectite clays are formed.

In the case of hydrothermal activity, subsurface temperatures increase to 200–400°C, additional phyllosilicates such as chlorite and serpentine will form, and other hydrated silicates such as silica will be precipitated.

By contrast, at the surface there is lots of water, so soluble ions tend to be removed, changing the bulk elemental composition, and the atmosphere helps to determine the pH and redox state. In other words, the chemistry is not at equilibrium with the parent rock. This leads to acid conditions and the formation of Al-rich clays called kaolinites, but not smectites.

Across all geological settings on Mars, Fe and Mg smectites are the most common clays, although Al-rich clays overlie the smectites, and crater counts show they date to the Late Noachian/Early Hesperian. These Al-rich clays are almost certainly markers of water action, and in which case, there must have been episodes, between 3.9 and 3.7 Gyr ago, in which there was open water on the Martian surface. This coincides with the time when the valley networks were being carved. If these relative dates are right, then it could be that surface water came about by impact melting of the cryosphere during the LHB. Extensive volcanism at this time may also have been important as outgassing would have pumped additional greenhouse gases into the atmosphere boosting the surface temperature. After the Noachian, clay deposition ceased implying much colder, highly arid conditions, and was replaced by the precipitation of sulfates and silica under much more acid conditions, presumably as a consequence of copious volcanic outgassing of SO_2.

In summary, throughout most of the Noachian, clays were made in hydrothermal subsurface environments; while in the Late Noachian/Early Hesperian when valley networks were active, clays were formed at the surface.

10.4.3.5 Much Hydrothermal Activity Was Produced by Impacts on Early Mars

On Earth, most hydrothermal activity occurs at conservative plate boundary spreading centers, with some generated as groundwater is heated by magma which can happen anywhere there is volcanism. On Mars hydrothermal activity could have been volcanic or as a result of impacts. Modeling impacts on Mars suggests that impacts producing craters greater than 5–10 km across are energetic enough to generate hydrothermal systems and there is evidence that this has happened on Mars. The best example is Toro crater located on the edge of Syrtis Major. Importantly, studies of this and other craters provide a case for an impact-generated global subsurface hydrothermal environment that extended throughout the Noachian and into the Hesperian. Indeed opaline silica, a hallmark mineral of hydrothermal systems has been detected by the Spirit Rover at Gusev crater and by Opportunity Rover in Victoria crater at Meridiani.

The interest for astrobiology is that hydrothermal deposits can provide nutrients and habitats for life long after hydrothermal activity has ceased. Those hydrothermal systems associated with impact craters provide the most numerous, and possibly the best preserved environments for finding fossil evidence for life on Mars.

10.4.4 Mars Has Experienced Three Climates

By seeing which minerals are present on Martian surfaces of different inferred ages it has been possible to put together a history of major climate changes that have happened on the red planet. This defines three climatic eras (Table 10.2, Figure 10.4).

10.4.4.1 Clays Formed in the Phyllocian Era

Until about the mid-Noachian, subsurface phyllosilicates were formed as a result of water alteration under generally alkaline or neutral conditions in what is termed the Phyllocian era. While there is an absolute requirement for water in the Phyllocian, it did not have to be warm or wet at the surface if the clays formed from wet magma or hydrothermally, as seems likely. Hence, early Mars could have been warm and wet, cold and wet, or cold and dry. But the fact that surface clays were only produced at the end of the Noachian and start of the Hesperian shows that groundwater availability decreased at this time. In other words, there was a transition phase after the Phyllocian when the planet was becoming more arid.

TABLE 10.2

Three Climate Ages of Mars

Epoch	Time/Gyr	Geology/Climate
Phyllocian	4.5–3.9	Clays formed in alkaline (subsurface) aqueous environment. Surface conditions poorly constrained
Theiikian	3.8–3.5	Volcanic SO_2 and H_2O react to form sulfates. Episodic surface water
Siderikan	3.5–0	$Fe^{2+} \rightarrow Fe^{3+}$ by atmospheric peroxides to form red iron oxides. Cold and dry

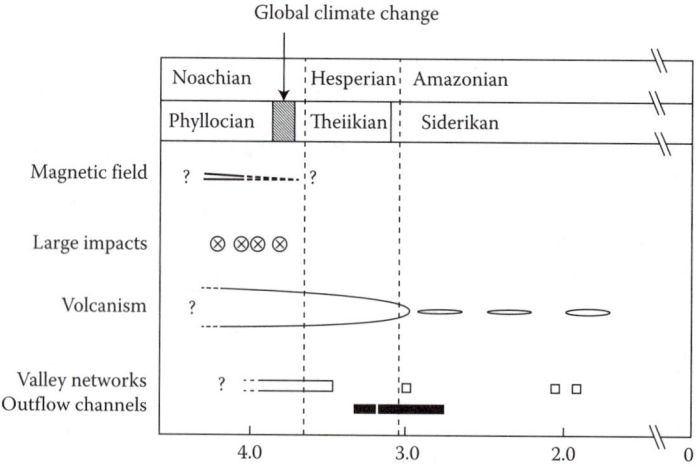

FIGURE 10.4 Martian timeline.

10.4.4.2 Sulfates Were Produced in the Theiikian Era

Sulfates formed in the Theiikian era from the Late Noachian into the Hesperian, as a result of the degassing of water and sulfur dioxide by the extensive volcanism that built the Tharsis Plateau and flooded the northern lowlands:

$$SO_2 + \tfrac{1}{3}O_3 + H_2O \xrightarrow{h\nu} H_2SO_4. \tag{10.3}$$

The sulfuric acid rained out and reacted with surface rocks to produce sulfates. This involves substantial amounts of water and hence surface water was likely to have been around for some of the time. This presumably corresponds to the period when valley networks were being carved.

10.4.4.3 The Siderikan Era Made Mars Red

From about 3.5 Gyr, Mars slid into its last era, the Siderikan era, dominated by iron oxidation and the formation of red ferric oxides. These oxides are anhydrous, so the climate must have been arid.

The most likely oxidizing agent is hydrogen peroxide, H_2O_2. This is formed in a series of reactions starting with the production of OH via photolysis of water by solar UV. Peroxide may also be produced by electrostatic fields generated during dust storms since this too yields the precursor OH:

$$
\begin{aligned}
CO_2 + e^- &\rightarrow CO + O^- \\
H_2O + e^- &\rightarrow OH + H^-.
\end{aligned}
\tag{10.4}
$$

Hydrogen peroxide concentrations in the Martian atmosphere now are very low. If this was the case throughout the Siderikan, then the "rusting" of Mars has been a slow process and explains why most surface rocks have not experienced much weathering.

10.5 Atmosphere

Curiosity Rover has confirmed that the composition of the Martian atmosphere is 96% carbon dioxide. Most of the rest is argon (1.93%), N_2 (1.89%), O_2 (0.14%), and CO (0.1%).

EXERCISE 10.9

What proportion of Earth's flux of solar energy does Mars, at 1.5 AU from the sun, receive?

With a pressure at the zero elevation datum defined to be 610.5 Pa, or about 6 mbar, the Martian atmosphere contributes little greenhouse forcing. This and Mars' lower insolation in comparison to Earth gives a mean surface temperature of −53°C. The tenuous atmosphere and solid surface have little thermal inertia, so the surface stays in equilibrium with absorbed solar energy. This means it warms and cools rapidly; at night the temperature plummets to a low value that is relatively independent of latitude or season. Hence there are large daily excursions in surface temperature as great as 100°C in mid- and low latitudes. In the daytime, summer air temperatures near the equator recorded by the Spirit Rover regularly exceeded 0°C with a maximum of 35°C. By contrast, the temperature at the poles in winter is typically −125°C although −153°C has been recorded.

EXERCISE 10.10

The subsolar point on a planet's surface is where the sun is directly overhead. (a) At what latitude is the subsolar point on Earth at noon on the June solstice? Hazard a guess at the temperature at this time assuming a cloudless sky. (b) The temperature at the subsolar point on the Moon is 110°C. Explain this temperature. (c) The temperature at the subsolar point of Mars is 24°C. Explain this temperature.

10.5.1 Liquid Water Cannot Exist at the Martian Surface Today

The daytime temperature on Mars can rise above 0°C which implies that ice can melt to form liquid water at the surface at times. However, the saturation vapor pressure of water is 611 Pa at 0°C and rises exponentially with temperature, so is almost always above the Martian atmospheric pressure. Since the boiling point of a liquid is defined by when its saturation vapor pressure is equal to the atmospheric pressure, any liquid water on the surface of Mars would rapidly boil away. This contrasts with hot deserts on Earth, such as the Sahara, where water evaporates but does not boil. In other words, desiccation on Mars is more thorough than in arid terrestrial environments. In general, water ice at the Martian surface sublimes straight to a gas.

Because rocks are very poor conductors of heat, the temperature 20–30 cm below the surface is always below the freezing point of water. Radiogenic heating and higher pressure might be sufficient for liquid water to exist at a depth of several kilometers in the crust, but we have not found it.

10.5.2 Detection of Methane in the Martian Atmosphere Is Dubious

Both ground-based and Mars Orbiter detections of methane between 2003 and 2006 at 10–250 ppbv (cf. 1800 ppbv in the Earth's atmosphere) have been reported. Methane detection was at the limits of the sensitivity of the instruments, and ground-based observations were made at wavelengths where interference from methane in Earth's atmosphere is very hard to remove, making them unreliable. Curiosity Rover, which has far greater sensitivity than previous experiments, failed to detect methane in six samples analyzed between October 2012 and June 2013. This shows that methane could not be present at greater than 1.3 ppbv. Hence, speculation about whether methane on Mars comes from serpentinization or methanogens was premature.

10.5.3 Atmosphere of Early Mars

Assuming that Mars accreted in the same way as Venus and Earth, it would have started out with a CO_2 atmosphere at a pressure of 6–10 bar. Was there enough greenhouse warming by the early atmosphere to put Mars in the habitable zone, given the faint young sun? To answer this we need to know how long the atmosphere lasted, its pressure and composition. If Mars had a dense greenhouse atmosphere for an appreciable time, say several hundred million years, then life may well have got started and flourished, and may even now exist in some haven deep in the crust. On the other hand, if Mars was never warm and wet for long, then the probability that life originated on Mars is low.

10.5.3.1 *Mars Probably Lost Most of Its Water Early On*

Ground-based spectroscopy in 1988 showed that the D/H ratio of Martian water to be sixfold greater than terrestrial. This enrichment in the heavy isotope, or equivalently the preferential loss of the light isotope indicates that Mars must have lost a substantial quantity (60–74%) of its original water inventory.

> **EXERCISE 10.11**
> a. Suggest a plausible mechanism for water loss on early Mars.
> b. What would be the effect of loss of water on Martian temperatures?

10.5.3.2 *Mars Probably Lost Most of Its Atmosphere Early On*

A number of isotope systems attest to the loss of atmosphere. For example, both ^{40}Ar and ^{36}Ar are stable isotopes but ^{40}Ar comes from the decay of ^{40}K in the mantle while ^{36}Ar is primordial. Curiosity Rover in Gale crater has measured $^{40}Ar/^{36}Ar$ as 1900, a large enrichment in the heavy isotope compared to terrestrial. This shows that Mars must have lost most of its primordial ^{36}Ar. The high $^{40}Ar/^{36}Ar$ now is the consequence of continuing decay of ^{40}K.

However, the long half-life of ^{40}K, 1.28 Gyr, does not provide much of a handle on when the loss happened. For this we can turn to earlier results obtained with a similar isotope system, $^{129}Xe/^{132}Xe$, in which ^{129}Xe comes from the decay of ^{129}I while ^{132}Xe is primordial. The Martian $^{129}Xe/^{132}Xe$ ratio is high. But the half-life of ^{129}I is only 16 Myr, so the enrichment in ^{129}Xe must mean that Mars lost most of its ^{132}Xe within a few half-lives of ^{129}I, perhaps 200 Myr.

10.5.3.3 *Several Atmosphere Loss Mechanisms Likely Operated on Mars*

There are a number of ways in which a planet can lose an atmosphere. All probably played a role on Mars though may have dominated at different times.

10.5.3.4 *Impacts Can Remove Substantial Amounts of Atmosphere*

Impact erosion would act early on. Relatively small impactors can remove a mass of atmosphere equal to that which it encounters as it travels through the atmosphere. Large impactors, such as a dichotomy-making Boreal impactor, are capable of removing the entire atmosphere above the plane tangent to the impact point (Figure 10.5). But impacts do not preferentially erode light isotopes of gases, so cannot have been responsible for the mass fractionation discussed above.

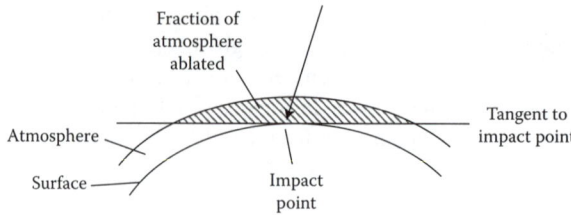

FIGURE 10.5 Large impacts can remove a substantial amount of a planet's atmosphere.

10.5.3.5 Sputtering Mass Fractionates Isotopes

Once Mars lost its magnetic field its atmosphere became increasingly prey to sputtering (Box 10.3). Sputtering mass fractionates isotopes. The degree of isotope mass fractionation suggests that 90% or more of the atmosphere was lost in this way.

> **EXERCISE 10.12**
>
> Briefly say why sputtering mass fractionates isotopes.

A process associated with sputtering, photochemical escape, is important for oxygen. This starts with dissociative recombination of a substantial number of O_2^+ ions to produce neutral oxygen:

$$O_2^+ + e^- \rightarrow O + O. \tag{10.5}$$

BOX 10.3   SPUTTERING

Sputtering is the erosion of the top of a atmosphere by collisions with high energy particles. A planetary magnetic field affords some protection from sputtering.

On the dayside molecules and atoms high in the atmosphere are ionized by solar extreme ultraviolet (EUV).

$$CO_2 \xrightarrow{\; h\nu \;} CO_2^+ + e^-, \tag{10.6}$$

$$CO_2^+ + O \rightarrow O_2^+ + CO. \tag{10.7}$$

These ions are picked up by the solar magnetic field and swept in the anti-sun direction. Some of these pick-up ions collide with neutral atoms and molecules (C, O, CO, CO_2, N, and N_2) at the top of the atmosphere.

Ions in the solar wind (H^+, He^+, etc.) can collide with neutral species.

As a result of collisions some of the neutral species and pick-up ions will be ejected into the exobase of the atmosphere on ballistic collisions with sufficient energy to escape the gravitational field.

These oxygen atoms are hot ($T = 200$–1000 K) and can directly escape from the atmosphere. In doing so, some will collide with other atoms adding to the sputtering.

Several studies have attempted to estimate the loss of O and CO_2 from the Martian atmosphere from measurements of the loss of O^+, O_2^+, and CO_2^+ ions by, for example, the Ion Mass Spectrometer of the ASPERA-3 (Analyzer of Space Plasma and Energetic Atoms) instrument on Mars Express, and by modeling the complex interactions between solar wind and Martian ionosphere. These suggest that oxygen is lost at a rate of $\sim 10^{25}$ s^{-1} by sputtering and $(0.3$–$4) \times 10^{26}$ s^{-1} through direct photochemical escape. Surprisingly, the absolute escape from the Earth, with its magnetic field, is *higher* than that for Mars.

Several studies have attempted to estimate the total loss of atmosphere over the past 3.5 Gyr from present day measurements by modeling how the loss rate changed over time, but the uncertainties are very large.

MAVEN (Mars Atmosphere and Volatile EvolutioN), a Mars orbiter launched in November 2013 should provide better measurements of current loss rates.

10.5.3.6 What Was the Atmospheric Pressure of Early Mars?

Another way to track the history of the Martian atmosphere is to determine what its surface pressure was like in the past. There are a few ways to do this. One approach is to compare the sizes of craters embedded within ancient-layered sediments to models of the how the thickness of an atmosphere filters impactors of different sizes: the smaller the minimum crater size the thinner the atmosphere must have been. When this is done for craters in 3.7 Gyr-old Aeolis river deposits, close to Gale crater, the best fit of a model to crater size gives an *upper* limit of 1640 mbar, or of 760 mbar if small craters at the surface are also included. If the Aeolis deposits are rocks harder than assumed in this model, then the upper limit on pressure could be 1.8 bar. At these pressures, it is hard to see how water could be stable at the surface for the 10^4–10^5 years necessary for water to cycle between deep aquifers and the surface even if the atmosphere was all CO_2. This makes it hard to understand how rivers could flow at this time. General circulation models show that the Martian atmosphere could periodically collapse by condensing out of solid CO_2 at the poles as the obliquity swung to a low value. It might be that the small craters formed during this time, while rivers flowed during high obliquity intervals when the atmosphere was thicker. But for this to happen, the atmospheric pressure must have been appreciably lower than the 1640 mbar upper limit.

A second technique looks at the size and depth of imprints made by volcanic bombs when they penetrate the ground: a bomb sag. One such discovered by Spirit Rover at Gusev Crater, caused by a rock fragment erupted 3.5 Gyr ago, was compared with bomb sags made in the lab by firing projectiles into sand with the same particle size as those observed by Spirit. It gave a minimum pressure of 120 mbar and, in addition, showed that the impact had been into water-saturated sand.

Thirdly, we can appeal to isotope ratios of gases trapped at the time to see how much of the atmosphere had been lost. Modeling based on $^{40}Ar/^{36}Ar$ in gases entombed in the meteorite ALH84001 implies that the atmospheric pressure did not exceed 1.5 bar in the Pre-Noachian and was less than 400 mbar by 4.16 Gyr. Such pressures of CO_2 could only stabilize liquid water at low latitudes during a summer day.

In summary, such experimental evidence as we have shows that the pressure of Mars' atmosphere by the Hesperian could have been comparable to that of Earth today or considerably less.

10.5.3.7 Early Mars Probably Had a Reduced Atmosphere

Mars, being further from the sun than Earth, might have more volatile-rich material, and this is supported by geochemical evidence. Consequently, Mars would have started out with plenty of N_2 and would have outgassed other volatiles from volcanoes, but the exact make up of this would depend on how reduced or oxidized the mantle was. The Martian meteorite ALH84001, with a crystallization age of 4.1 Gyr, is reduced, so it is reasonable to assume the mantle of early Mars was too. A reduced mantle would result in outgassing of H_2O, CO, CH_4, and even H_2 from volcanoes. CO and CH_4 would have reacted to produce CO_2. Hence the primordial Martian atmosphere would have been predominantly CO_2 and N_2, laced with H_2O, CH_4, and possibly H_2. After the catastrophic early loss of atmosphere there is no source from which to regenerate N_2, but outgassing could replenish the other components. The significance of this is that for almost its entire history Mars has had an atmosphere extremely depleted in nitrogen, yet this element is a component in amino acids and nucleobases: no nitrogen—no proteins, RNA, or DNA.

10.5.3.8 Climate Models Struggle to Warm Early Mars

To get the surface of Mars above 0°C at the end of the Noachian requires serious greenhouse forcing. With the sun 25% fainter the solar flux at Mars would have been just $0.75 \times 0.43 = 0.32$ that of the present Earth. Climate modelers have looked at greenhouse gases like CO_2, CH_4, SO_2, and water vapor, but so far have not found a convincing way to warm ancient Mars using 1-D radiative–convective models which model radiative transfers (including absorption and emission by greenhouse gases) and convection in a vertical column of atmosphere.

The case of CO_2 is instructive. We might think that simply increasing the pressure of CO_2 would rack up the temperature, but this is not the case. A model CO_2–H_2O atmosphere in which water is allowed to be present at saturating concentrations so as to maximize the greenhouse effect fails to get the temperature of early Mars above −48°C. One reason is that CO_2 is a good Rayleigh scatterer of light, so as the CO_2 concentration rises; so the planet reflects more sunlight. Another reason is that as CO_2 pressure increases it forms CO_2-ice clouds. This condensation releases latent heat, reducing the tropospheric lapse rate, the rate at which the temperature drops with altitude (for Earth this is ~10 K km^{-1}). This means that the stratosphere is slightly warmer, and this is bad news because it means more infrared radiation is lost to space. This effect is made worse by the low mass of Mars. Its lower surface gravity gives it a lower tropospheric lapse rate, about half that of the Earth.

Now it is the case that CO_2-ice clouds themselves can exert a warming effect, just as high altitude cirrus clouds do on Earth. This is because they are quite transparent to incoming short wavelength sunlight, but scatter back down some of IR emitted by the surface that has absorbed and been warmed by that sunlight. However, the warming effect only becomes globally significant for Mars if cloud cover is greater than 50%.

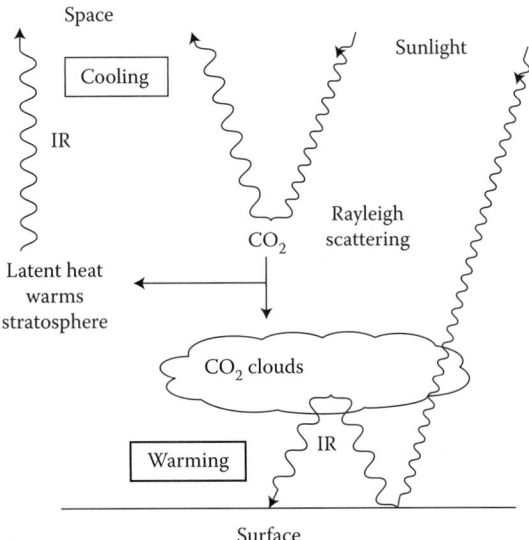

FIGURE 10.6 Several processes conspire to prevent CO_2 warming early Mars above the freezing point of water.

The combined effect of all these processes is that there is a maximum greenhouse effect that can be achieved by CO_2 (Figure 10.6).

EXERCISE 10.13

Given the maximum greenhouse effect that CO_2 can generate suggest why Venus suffered runaway CO_2 greenhouse warming. (Hint: Check out the composition of the Venusian atmosphere.)

If CO_2 cannot warm early Mars what about other greenhouse gases? Unfortunately, models which lace a CO_2–H_2O atmosphere with plausible amounts of CH_4 do not boost the greenhouse enough to ensure above 0°C temperatures. Sulfur dioxide (SO_2) must have been erupted by Martian volcanoes in the past since sulfates are abundant on surfaces younger than Noachian, and SO_2 is a good greenhouse gas in tens to hundreds of ppm. But calculations show that the sulfate aerosols formed by SO_2 photolysis would more than overcome the greenhouse effect of the SO_2 gas.

As computing power has increased researchers are increasingly using more complex, computationally expensive, 3-D climate models which expand the 1-D radiative–convective approach by adding the global transfer of fluids and heat, surface topography, atmosphere–surface interactions, CO_2-ice clouds, and representations of the water cycle. This allows climate response to changes in a variety of factors to be explored.

When this was done for early Mars with a CO_2 atmosphere with pressure between 0.1 and 7 bar it was found that although CO_2-ice clouds covered much of the planet their greenhouse effect did not exceed 15 K and it was not possible to get the annual mean temperature above 0°C anywhere on the planet. Summertime diurnal mean surface temperatures above 0°C did occur at high latitudes when the obliquity was made larger than 40°. This would have permitted rivers to run at least seasonally, but not in locations where most valley networks are actually seen. For pressures greater than a fraction of a bar, adiabatic cooling causes temperatures in the southern highland valley network regions to fall significantly below the global average. Long-term climate evolution simulations indicate that in these circumstances, water ice is transported to the highlands from low-lying regions for a wide range of orbital obliquities, regardless of the extent of the Tharsis bulge, and an extended water ice cap forms on the southern pole. Even a multiple-bar CO_2 atmosphere did not allow liquid water to exist at the surface long term. Episodic melting of the highlands ice as a result of volcanism or impacts could occur because ice migration to higher altitudes is a robust mechanism for recharging highland water sources after such events. In this way, the fluvial features seen in the late Noachian might be explained without the need to invoke a warm, wet early Mars.

10.5.3.9 Hydrogen Might "Rescue" a Warm Climate

A 1-D climate model in which 5–20% H_2 was added to a CO_2–H_2O atmosphere does raise the mean surface temperature of early Mars above the freezing point of water. H_2, CH_4, and CO would have been outgassed from volcanoes if the mantle of early Mars was highly reduced. CO_2 could then have been produced by photochemical oxidation of outgassed CH_4 and CO to form a CO_2 and H_2 greenhouse. A difficulty with this model is the 1.3–4 bar pressure that is required. As we have seen, this is higher than most estimates of the surface pressure on early Mars.

10.5.4 Was Mars Episodically Warm and Wet?

If climate models cannot provide clement conditions for early Mars, then how can we explain its many fluvial features? One model developed over the past decade by Teresa Segura at Space Systems and colleagues is that large impacts would produce instantaneous climate change resulting in global warming and wet episodes, and this would have carved rivers and recharged aquifers. The timing of these impacts (3.9–3.8 Gyr) roughly coincides with when many valley networks apparently formed.

The researchers developed a 1-D radiative–convective model that tracked the evolution of atmospheric and ground temperature after an impact. It computed the evaporation, condensation, and precipitation of injected and surface-evaporated water, taking account of greenhouse warming by CO_2, water vapor, and water clouds, and the latent heat changes as clouds condense and evaporate. Model runs changed impactor size and atmospheric pCO_2.

The model showed that impacts filled the atmosphere with vaporized rock and ice at 1000–2000 K, some of which escaped to space. Hot rock ejecta condensed first to produce global blankets of hot ejecta, ranging in thickness from meters to hundreds of meters. This kept the global surface temperature above 0°C for 95 days to 6 years depending on impactor size and CO_2 level, even without the warming effect of clouds. Water ice from the impactor, the Martian cryosphere, and polar caps injected into the atmosphere as super-heated gas in amounts equivalent to tens of meters of precipitable water eventually rained out at a rate of order two meters per year. The temperature of the rain exceeded 350 K, comparable to the

373 K at which hematite crystals form on Earth. Hence, impact rain could have made the hematite spherules at Meridiani Planum. Intriguingly, when the radiative effect of water clouds is included, a greenhouse climate resulted from impactors 50 km across that could be centuries long, although the equilibrium temperature was only 250 K.

A different radiative 1-D model that did not include clouds found that sufficiently large impacts with a kinetic energy of order 10^{26} J, such as that which punched out the Argyle basin, could flip Mars into a runaway greenhouse that was stable until the atmosphere lost water. The model showed that there were two temperatures for one radiative flux, one cold like present day Mars, and one a hot runaway greenhouse, that depended largely on the partial pressure of water in the atmosphere (Figure 10.7).

For example, for the solar flux at Mars during the Noachian (80 W m^{-2}), the hot runaway state could exist for the pressure range of 0.006–1 bar of CO_2, and for water inventories 6.5 bar or greater. The temperature was 420 K, so liquid water would not exist at the surface because it would evaporate, but liquid water could exist in the crust for a time that depended on how fast it was transported to the surface. The greenhouse would collapse as Mars lost water (e.g., by photodissociation and hydrodynamic escape of H_2) and as Mars cooled to the lower branch, precipitation of rain then snow could have formed fluvial features. Alternatively they may have been formed during the greenhouse state as subsurface ice melted and erupted onto the surface to evaporating into the greenhouse atmosphere. A fractal analysis of valley networks suggests that Mars experienced enough rainfall to begin carving them but not to produce the complex drainage patterns brought about by Earth's water cycle. Hence, groundwater sapping was important on Mars.

EXERCISE 10.14

By a bizarre twist a runaway greenhouse triggered by an Argyle basin-sized impact could also be terminated by a similar size impact. How would this happen?

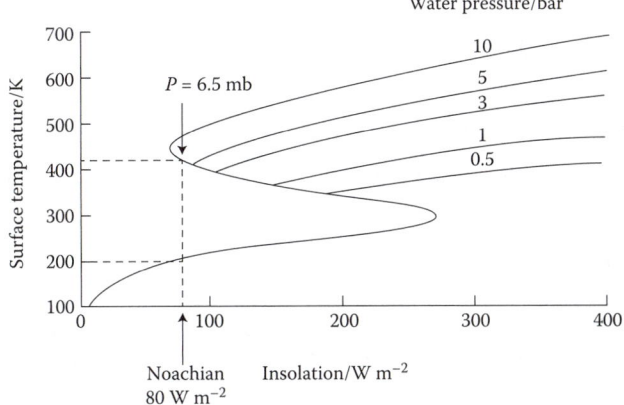

FIGURE 10.7 Dual solutions to the climate of early Mars might exist, one a runaway greenhouse. For a flux of 80 W m^{-2}, 6.5 bar of water is required to flip the planet into the greenhouse; $pCO_2 = 6$ mbar. (From Segura et al. *Icarus* 2012;220:144–8. With permission.)

10.5.4.1 Can Impact Precipitation Account for the Fluvial Features?

For the impact hypothesis to be credible it has to account for the total erosion estimated to have occurred on early Mars. At present there is no agreement on this point. One of the reasons is that a robust understanding of how much rainfall, or water flow beneath a glacier, it takes to carve fluvial features is lacking. One estimate is that it would take 5000 km of rainfall (over 17 Myr) to produce Nanedi Vallis on Mars, which is comparable in size to the Grand Canyon; another estimate is just 0.5 km. The first estimate is orders of magnitude more than can be expected from impacts. Hence not all planetary scientists subscribe to the impact idea.

There is also little appreciation of how long fluvial features take to form. Although we might expect Martian valley networks to take a long time to form, the Salton Sea flood in California carved a river similar in depth and length to some of the channels on Mars in approximately 9 months. Deltas/alluvial fans, of which there are a number on Mars, were thought to need water flow between 10^3 and 10^6 years until an analog simulation suggested it could happen in just tens of years. To determine how long it took to form fluvial features on Mars it will be necessary to have absolute dates for rock samples from the features. At present this would require sample-return missions.

10.6 Mars and Life

The strongly oxidizing regolith, high fluxes of solar UV and cosmic radiation, extremely low temperature and pressure, and lack of water make the surface of Mars very hostile now and this has probably been true for the past ~3.7 Gyr. If life did get started on early Mars we might assume that it could have survived in an aquifer of liquid water in pore spaces several kilometers deep in the crust. The failure to detect these aquifers is therefore disappointing.

10.6.1 The Viking Experiments Were Designed to Detect Life

There were three biological experiments on both of the Viking missions, which landed in July and September of 1976.

10.6.1.1 Pyrolytic Release Tests for Carbon Fixation

Samples of Martian regolith were put into a medium which would supply nutrients to any putative carbon-fixing (i.e., photosynthetic) organisms. ^{14}C-labeled CO and CO_2 were added and the samples incubated for several hours in sunlight. Unreacted gases were flushed out. Any carbon-fixing organisms would be expected to take up the radiolabeled gases. This could be measured by subsequently heating the samples to 580°C to liberate any radiolabel that had been assimilated and counting it. Assimilation was seen and was prevented by preheating the sample for 3 h at 175°C, but not for 2 h at 90°C. While this might seem that the high temperature had sterilized the soil while the lower temperature had not, it is so improbable that thermophiles capable of surviving 90°C exist on the frigid surface of Mars that the assimilation was taken to be chemical, particularly given that it also occurred in control regolith that had not been fed growth medium.

10.6.1.2 Labeled Release (LR) Tests for Metabolism

Martian regolith was incubated with dilute aqueous solutions of ^{14}C-labeled small organic molecules to act as nutrients. The idea is that growing organisms would take up the nutrients, metabolized them, and then release ^{14}C-labeled gases, such as $^{14}CO_2$, just as we exhale CO_2 as a result of our metabolism (though not so radioactive!). Hence, labeled release could be interpreted as evidence of microbial metabolism. Release was seen which was prevented by preheating to sterilizing temperatures. However, when the second and third doses of nutrients were added to unsterilized samples no additional release was forthcoming. This implied chemistry rather than biology at work; organism should have responded to subsequent meals.

These findings could be explained by superoxide radicals (O_2^-) generated by the action of UV light on the regolith. These would oxidize the organic molecules to CO_2. Superoxide radicals are destroyed by heating, which is why sterilizing temperatures stopped the release. Adding additional doses of organic molecules would not work if all the superoxide radicals had already reacted.

Interestingly a recent reanalysis of the LR results shows that the release followed a 24.66 h cycle. This is the length of the Martian day, and is reminiscent of circadian rhythms seen in terrestrial biological processes.

10.6.1.3 GCMS Detects Organic Compounds

A gas chromatograph–mass spectrometer (GCMS) to detect organic compounds given off by heated samples of regolith failed to find any at a level of 0.1–1.0 ppb, except for chloromethane and dichloromethane. At the time these were assumed to be cleaning fluid contaminants. The failure to find organic compounds was unexpected—given the organic content of comets and many meteorites—leading most researchers to conclude that the labeled release experiment was due to oxidation of the labeled compounds by the regolith rather than by metabolic activity.

The nature of the regolith oxidants were not known then, but after the discovery of perchlorate by Phoenix, experiments were undertaken to elucidate the Viking results, in which perchlorate was added to Atacama Desert soils. (The Atacama Desert is the driest hot desert on Earth.) Perchlorate is not able to oxidize organic molecules at Martian temperatures. But when perchlorate was added to soil and heated to >350°C, simulating the Viking GCMS protocol, chloromethane and dichloromethane were produced. Modeling suggests this is because heating releases chlorine from perchlorate which then reacts with methane released from the organic compounds. If this is correct it would allow the Martian regolith to contain up to 0.1 wt% of organic compounds yet not be detected by the Viking GCMS.

10.6.1.4 What Was the Legacy of Viking?

Most astrobiologists reckon the Viking biology experiments were negative or inconclusive.

EXERCISE 10.15

Speculate on why the strategy of Mars exploration changed from biosignature detection to assessing habitability after Viking.

The apparently negative Viking result poisoned the political will for further Mars missions for over a decade. The first post-1976 mission was the NASAs Mars Observer spacecraft launched in 1992.

10.6.2 Does ALH84001 Harbor Evidence of Life?

Found in December 1984 in Allen Hills, Antarctica, the Martian meteorite ALH84001 has been dated to 4.091 Gyr. It is an igneous rock that cooled slowly in the Martian crust. Four features in the rock were taken as *prima facie* evidence for past life on Mars.

10.6.2.1 ALH84001 Has Had a Shocking History

Sometime between 3.9 and 3.6 Gyr the rock was fractured by an impact event and between 3.6 and 1.3 Gyr it was percolated by a water-rich fluid which deposited 20–250 μm carbonate globules at a temperature ~18°C, according to recent carbon and oxygen isotope studies. Presumably during this time the putative biomarkers were deposited. The meteorite was ejected into space by an impact 16 Myr ago, which fractured the carbonate globules. It spent almost 16 Myr in space; a period established by measuring the amount of spallation products formed by cosmic rays colliding with atomic nuclei. The length of time the meteorite spent on the surface of the Earth, just 13,000 yr was deduced by ^{14}C dating, since cosmic radiation produces ^{14}C by spallation reactions on ^{16}O in silicates.

10.6.2.2 Four Features Hinted at Relic Life

In 1996, David McKay and colleagues published a paper in *Science* claiming ALH84001 contained four possible traces of past Martian life.

10.6.2.3 Complex Organic Molecules Were Detected

ALH84001 contains polycyclic aromatic hydrocarbons (PAH). These could result from processing (diagenesis) of biochemicals in dead organisms. For example, they are found in coals that result from diagenesis of plants that have been buried in sedimentary rocks. A major problem is that PAHs are common on the surface of the Earth. (You make some every time you accidentally char meat on a barbeque!) Evidence cited that the ALH84001 PAHs are not Antarctic contaminants was that they are concentrated toward the center of the meteorite and neither were they found in Antarctic ice.

However, PAHs can arise nonbiologically: they can result when CO_2 or CO reacts with hydrogen in the presence of metal catalysts. They have been found in some carbonaceous chondrites.

EXERCISE 10.16

What suggests that PAHs found in carbonaceous chondrites are unlikely to be biogenic?

10.6.2.4 Iron Sulfide (Fe(II)S) and Iron Oxides Appeared Together

The conjunction of iron sulfide and iron oxides was thought to be suggestive of microbial sulfate reduction, but this view was not sustainable after it was shown that the sulfur isotope signature resembles that of minerals rather than of biogenic materials.

10.6.2.5 Magnetite Crystals Are Enigmatic

One of the iron oxides present in ALH84001 is magnetite $(Fe(III)_2O_3)$. It occurs as high purity nanocrystals, sometimes arranged in linear chains, with structural features that resemble magnetite crystals in magnetotactic bacteria. These use the magnetic mineral to sense which direction is up in the ocean by detecting the local magnetic inclination. The magnetite might originate from thermal changes to the carbonate (an abiotic origin) or be independent of the carbonate which allows for (but does not prove) a biological origin. There is now considerable work that the magnetite is indeed independent of the carbonate.

10.6.2.6 "Fossil Bacteria" Seem Too Small

Structures were identified that were interpreted to be fossil bacteria. At 20–100 nm long, they are smaller than the tiniest terrestrial bacteria, the Mycoplasmas, which are 200–300 nm.

An argument biologists have used to dismiss ALH84001 fossils is that the minimum size for terrestrial life is about 200 nm, determined by the need to accommodate ribosomes that are 40 nm in diameter. The counter argument is, of course, that extraterrestrial life might operate in ways that allows it to be smaller.

10.6.2.7 ALH84001 Does Not Provide Evidence for Life

Of the four features only the magnetite crystals remain hard to explain abiotically. That said, the general consensus is that the case for past Martian life is not made by ALH84001.

10.6.3 Martian Life Could Be Based on Iron and Sulfur Metabolism

Despite ALH84001 providing no support for metabolism based on iron or sulfur, the widespread global distribution of Fe and S has led a number of researchers, including Charles Cockell of the Centre for Astrobiology, University of Edinburgh, Scotland, to explore how Martian life may use these in thermodynamically favorable *anaerobic* redox reactions. Basalt contains much iron as Fe^{2+} and as we have seen there is no shortage of sulfur as sulfate (SO_4^{2-}) on Mars.

10.6.3.1 Terrestrial Iron-Oxidizing Bacteria "Breathe" Nitrate

One possibility is the oxidation of iron coupled to the reduction of nitrate. In terrestrial heterotrophs with this lifestyle, the nitrate is reduced all the way to nitrogen gas and the ferric iron deposited as ferrihydrite.

$$10Fe(II)CO_3 + 2NO_3^- + 10H_2O \rightarrow Fe(III)_{10}O_{14}(OH)_2 + 10HCO_3^- + N_2 + 8H^+. \quad (10.8)$$

TABLE 10.3

Postulated Martian Life Energetics

Electron Donors	Electron Acceptors
Fe^{2+} (Fe-rich silicates)	Fe^{3+} (water-altered/oxidized minerals)
H_2 (serpentinization?)	SO_4^{2-} (as Ca and Mg salts)
Organic molecules (delivered by carbonaceous chondrites)	NO_3^- (not found on Mars so far)
Organic molecules	ClO_4^- (identified on surface)

These organisms use simple organic molecules such as acetate as carbon sources. However, many terrestrial iron oxidizers can use CO_2 as their sole C source, which is more plausible on the red planet. Unfortunately, nitrate has not been detected on Mars but perchlorate has been proposed as an electron acceptor for iron oxidation, though whether this could support microbial growth is not currently known. In summary, it is not clear that electron acceptors for iron oxidation are available on Mars.

10.6.3.2 Mars Is Rich in Electron Acceptors for Fe- and S-Reduction

Iron oxides (e.g., hematite) and oxyhydroxides such as ferrihydrite are viable terminal electron acceptors for the reduction of Fe^{3+}, and sulfate (SO_4^{2-}) to sulfite (SO_3^{2-}). On Earth iron- and sulfate-reducing bacteria use a big range of organic molecules as electron donors (e.g., acetate and lactate) which appear to be lacking on Mars. However, estimates suggest carbonaceous chondrites could deliver 8.6×10^6 kg of organics to the surface of Mars each year, and reduced carbon has been found in Martian meteorites that could act as electron donors. Even more simply, some terrestrial Fe- and S-reducing organisms use H_2 as electron donors and CO_2 as a carbon source (Table 10.3).

EXERCISE 10.17

List two potential sources of H_2 on Mars.

However, it has to be said that proof of available electron donors for iron or sulfate reduction is presently lacking.

11

Icy Worlds

11.1 Life Might Exist Beyond the Conventional Habitable Zone

Three icy moons—Titan and Enceladus (in the Saturnian system), and Europa (a moon of Jupiter)—have intrigued astrobiologists: Titan because it has a dense nitrogen atmosphere rich in organic compounds, and all three because they probably harbor subsurface liquid water oceans. These moons lie well outside the conventionally defined habitable zone of the solar system.

Any habitable environments on these worlds could be cold and the lack of sunlight would rule out photoautotrophs. Heterotrophs might make a living if there is abundant organic carbon, but the relative simplicity of the metabolism of chemoautotrophs would make these a better bet.

To assess the astrobiological potential of these worlds we need to know whether they can supply, in biologically available forms, the essential elements for life (CHONPS): the composition of the oceans; the possibility of redox gradients; and quantitative estimates of the energy available both now and in the past, mostly from radiogenic and tidal heating.

11.2 Titan

Titan, discovered in 1655 by the Dutch astronomer Christiaan Huygens, is the largest of Saturn's moons (Box 11.1). Its interest for astrobiologists stems partly from its substantial atmosphere, the only moon in the solar system so endowed. Composed mostly of nitrogen laced with methane, this atmosphere hosts a rich prebiotic chemistry cooked up with the help of solar UV radiation despite the frigid temperatures. It is a world where water is frozen as solid as rock but methane rain falls from methane clouds into methane lakes: here a methane cycle operates rather than a water cycle. Titan is sometimes described as early Earth in "deep freeze" but it is important to recognize that this does not make Titan an analog for early Earth. Titan's volatile inventory is very different from that of the Earth, as illustrated by its low bulk density, and it is far too cold for water to contribute to the chemistry of Titan's surface or atmosphere in any significant way. Nonetheless, considerable prebiotic chemistry is happening on Titan in the absence of water, making it an important astrobiological laboratory.

Most of what we know about Titan comes from the Cassini–Huygens mission which landed the Huygens probe on the surface of the moon on January 14, 2005. It sent back data for 2.5 h during its descent through the atmosphere and for a further 90 min on the surface before its batteries failed.

BOX 11.1   TITAN IS A LARGE MOON

Titan is the second biggest moon, after Ganymede, in the solar system, large enough to be seen easily with small aperture amateur telescopes. With a radius of 2575 km it is larger than Mercury and about three-quarters the size of Mars. Its mass (1.346×0^{23} kg) and surface gravity (1.35 m s^{-2}) allow it to retain a substantial atmosphere given the low temperature (150 K) at its exobase, which is at an altitude of 1500 km.

EXERCISE 11.1

(a) Calculate the escape velocity for Titan. (b) Show that N_2 is likely to be retained in Titan's atmosphere.

Despite the extreme conditions on Titan, some astrobiologists speculate that life might have originated there.

11.2.1 Titan Has a Dense Atmosphere

Voyager and Huygens have provided data on Titan's dense atmosphere. It is composed mostly of nitrogen (98.4%), with 1.4% CH_4, 0.2% H_2, and a large number of trace components, many of them hydrocarbons and other organic molecules. It has a surface pressure of 1.5 bar, and is much more extensive than Earth's atmosphere, though it is shielded somewhat from the solar wind by Saturn's magnetosphere (Figure 11.1).

EXERCISE 11.2

Why is the atmosphere of Titan more extensive than that of Earth?

11.2.1.1 A Methane Cycle Operates on Titan

Titan has a surface temperature of 94 K which is virtually the triple point temperature of methane. Hence, methane can exist as a gas, liquid, or solid at the surface of Titan, depending on the latitude and season. About 10% of incoming solar radiation reaches the surface of Titan, heating it and the atmosphere above so that the troposphere convects. Adiabatic cooling in the troposphere causes the temperature to fall to the point where methane reaches its saturated vapor pressure, so it could condense and precipitate. It seems then that a methane cycle could operate on Titan. There is considerable evidence that this is the case.

EXERCISE 11.3

Determine the adiabatic lapse rate in the lowest part of Titan's atmosphere.

During its descent, the Huygens probe imaged networks like river systems running from bright highland regions to darker, flatter lowlands that resemble lakebeds,

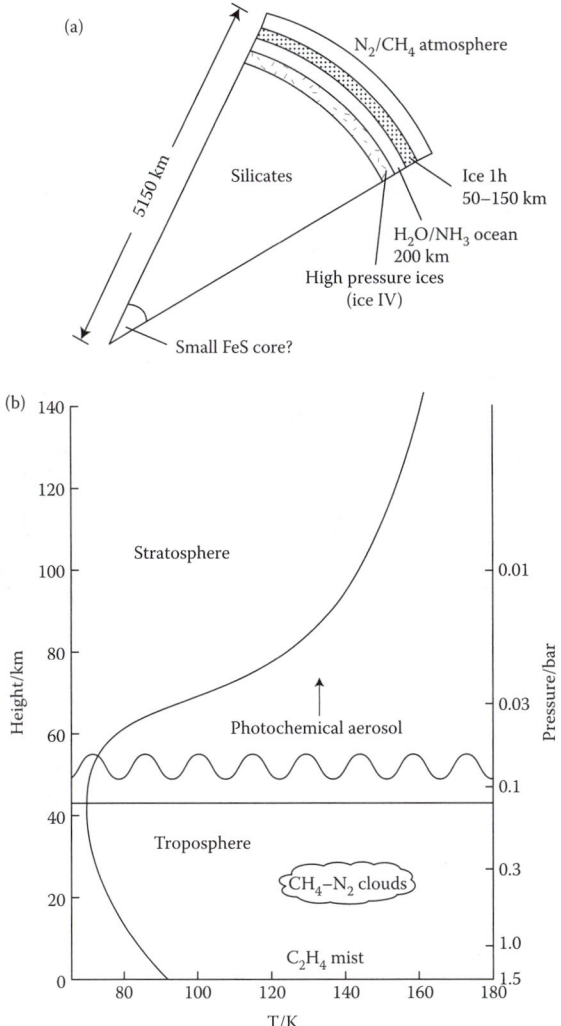

FIGURE 11.1 (a) Model of the internal structure of Titan. (b) Vertical profile of Titan's atmosphere.

complete with what looked like coastlines and offshore islands. Pebbles close to the lander are rounded in shape. Their distribution—strewn across a valley floor—suggests deposition from a powerful current that weakened as the fluid broadened out into shallows. The most likely fluid is methane.

Cassini images show lakes, most at high northern latitudes (Plate 11.1). The lakes are filled with hydrocarbons, predominantly ethane with some methane, although they may vary in composition. Ligeia Mare, the second largest lake, appears to be pure methane. Ethane and methane have a lower viscosity than water, so even wind speeds of 1 m s^{-1} should whip up detectable waves. These have been seen.

Clouds, thought to be made of droplets of methane with dissolved nitrogen, are located at high latitudes in winter and summer and over the tropics in spring and

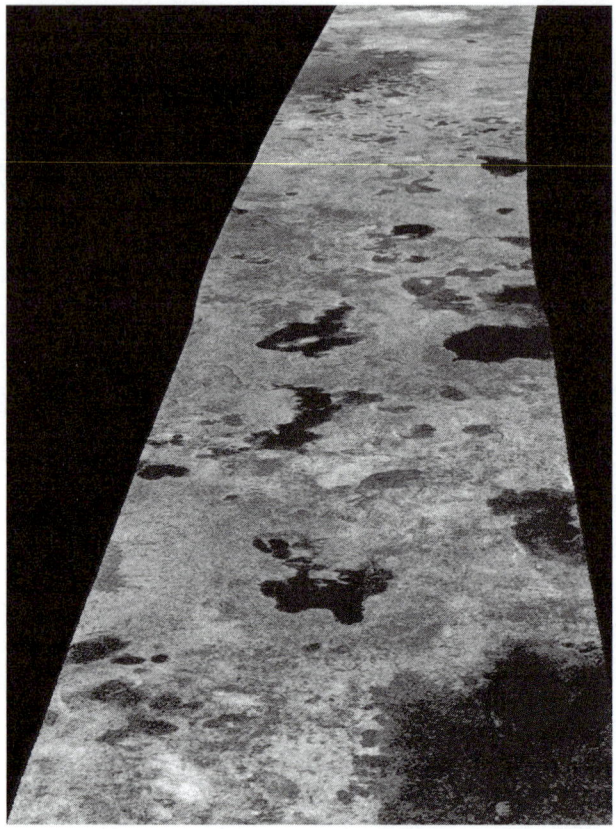

PLATE 11.1 (**See color insert.**) Hydrocarbon lakes in the northern hemisphere of Titan imaged by Cassini radar. Bolsena Lacus can be seen on the lower right. (Photo Credit: NASA/JPL-Caltech/USGS.)

autumn. There is evidence for rainfall at high latitudes in summer. It seems that methane evaporates from the warm tropics over summer and is transported toward the poles where it falls as rain. This recharges the lakes but leaves the tropics a dry desert with large areas covered by dunes. The eccentricity of Saturn's orbit makes northern summer longer than the southern summer and, so the rainy season is longer in the north. This is one reason why most of the lakes are in the north.

In summary, a methane cycle operates on Titan that is analogous to the water cycle on Earth (Figure 11.2).

11.2.1.2 Titan Is an Ice–Silicate World

The bulk density of Titan (1.88×10^3 kg m^{-3}) suggests it is composed of ice and silicates in roughly 0.4–0.6 proportions. Gravity measurements obtained during the numerous Cassini flybys of Titan that show how the satellite responds to tidal and rotational forces allow its moment of inertia to be calculated (see Chapter 2, Box 2.1). This calculation only works if the satellite is close to hydrostatic equilibrium and this is supported by the data. It gives Titan an inertial constant between 0.335 and 0.342,

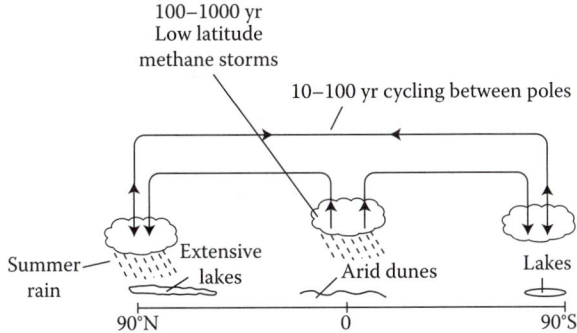

FIGURE 11.2 The methane cycle on Titan evaporates methane in equatorial regions in summer and transports it to polar regions where it rains.

compared with 0.4 for a uniform sphere and 0.33 for Earth. This shows that Titan has its mass only moderately concentrated toward the center. In other words, Titan is a partly differentiated satellite and current models give it a core of hydrated silicates (serpentinite) with a radius of ~2000 km, a mantle of high pressure ices IV and V and a crust of ice I_h. Although the moment of inertia measurement cannot rule out a small inner metal core (<500 km), the low internal temperature (<1000 K) is not hot enough for such a core to segregate.

EXERCISE 11.4

(a) How does the moment of inertia for a sphere in which the mass is more concentrated at the center differ from the moment of inertia of a sphere in which mass is distributed uniformly? (b) Earth's moon has $k = 0.394$. Comment on the moon's metal core.

11.2.1.3 Titan Harbors a Subsurface Water Ocean

Beneath Titan's icy crust there is a global water–ammonia ocean. This seems not to be an unusual feature. There is evidence for subsurface water oceans on three of the four Galilean moons of Jupiter: Europa, Ganymede, and Callisto.

Evidence for a subsurface ocean is good. Several independent observations provide evidence for a subsurface ocean on Titan:

1. Radar mapping by Cassini shows that surface landmarks shifted 30 km between 2004 and 2007. Titan is so close to Saturn (its orbital period is just under 16 days) that it is tidally locked into synchronous rotation (i.e., it orbits in the same time that it takes to rotate once on its axis). The shifting landmarks are most straightforwardly explained if the crust is disconnected from the interior of the satellite by an ocean.

2. Extremely low frequency radio (ELF) waves were detected by the Huygens lander during its descent. On Earth ELF waves with particular (resonant) frequencies are reflected back and forth between the surface and the ionosphere about 100 km above, and are reinforced. Titan's low conductivity icy surface

means that ELF waves readily penetrate the crust rather than being reflected. So, for them to resonate they must be reflected from some boundary in Titan's interior. This is most likely from the high conductivity water–ammonia liquid ocean beneath the crust. By modeling how ELF waves resonate on Titan, it will be possible to work out how far down the ocean is.

3. Heat flow from the interior of the moon to the surface provides insight into the thickness of the crust. The higher the heat flow the thinner the crust and the deeper the ocean will be. Heat flow estimates for Titan are ~4–7 mW m^{-2} mostly due to tidal heating and suggest that the crust is between 50 and 150 km thick, with an ocean ~300 km deep, although more recent data suggests the ice shell is thicker than these thermal models suggest (see below).

EXERCISE 11.5

Explain how tidal flexing comes about for a moon in synchronous rotation in an eccentric orbit. Assume a constant rotation speed.

11.2.1.4 Ammonia Is an Anti-Freeze

To keep Titan's putative ocean liquid requires the addition of an anti-freeze. Ammonia is especially effective in this role and our understanding of how Titan formed supports a NH_3–H_2O ocean. Adding enough ammonia can lower the melting point of water by ~100 K. How does this work?

If an ammonia–water mixture containing 1–2 mol% NH_3 is cooled below 0°C, water crystals begin to freeze out and remain in equilibrium with NH_3–H_2O liquid. As the temperature continues to fall more and more water ice is formed and the NH_3 concentration of the remaining liquid rises (Figure 11.3). At −97°C (176 K) ammonia

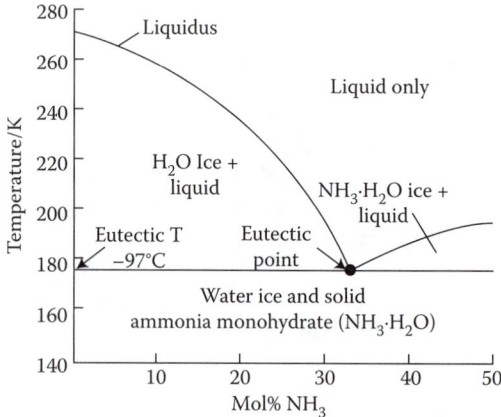

FIGURE 11.3   As an ammonia–water mixture with low NH_3 concentration cools, it drops vertically onto the liquidus. As cooling continues, the mixture moves down the liquidus getting richer in NH_3 as water ice freezes out. At −97°C, the NH_3–H_2O liquid reaches the eutectic point and freezes instantly and completely. This phase diagram is for 1 bar, but the higher pressures on Titan do not change the figures very much.

concentration reaches 35.4 mol% and the entire mixture freezes as crystals of water ice and ammonia monohydrate ($NH_3 \cdot H_2O$). This point in the temperature-composition phase diagram of the NH_3–H_2O binary system is termed the eutectic point. If the frozen mixture is rewarmed, on reaching the eutectic temperature it partially melts giving a melt with eutectic composition. The point is that if Titan's ocean started out as a pure NH_3–H_2O mixture, it could remain liquid down to 176 K but no colder.

11.2.2 Atmospheric Methane Must Be Continually Regenerated

Methane is irreversibly destroyed by UV in Titan's atmosphere. Without replenishment atmospheric methane would vanish within 10–100 Myr. The loss of the greenhouse effect may result in condensation of N_2 and a collapse of the atmosphere to a few millibars. Hence, there must be some reservoir which regenerates it. Radar measurements of the depth of the methane lakes give estimates for the total volume of methane at the surface. This proves to be too small to maintain methane in the atmosphere over geological timescales. Neither is cometary delivery of methane over the past 3 Gyr enough.

There is a "get out" clause which currently cannot be excluded. Titan might now be in an unusual state in that most of the time it has lacked methane in its atmosphere. In this case, a much more modest methane reservoir is all that is needed.

11.2.2.1 Subsurface Methane Reservoirs Must Exist

Methane reservoirs in Titan's interior are needed to supply atmospheric methane. To estimate how much there might be we need some sense of how Titan acquired its methane. One possibility is that methane may have formed internally, possibly from CO_2 by reaction with H_2 produced by the serpentinization of olivine. This would have happened early on in Titan's history when accretionary and radiogenic heating meant water could interact with the rocky core.

EXERCISE 11.6

What would terminate the production of methane by serpentinization in Titan?

The serpentinization would have occurred at low temperatures (40–90°C) even on early Titan, and this is seen at the Lost City hydrothermal vents so is feasible. Calculations show this could have produced sufficient methane to supply the atmosphere. CO_2 condensates could have been accreted by Titan from the Saturn subnebula and would presumably be present as CO_2 clathrates (Box 11.2). These have a high density and, so would exist at depth. However, the D/H ratio of methane in Titan's atmosphere is threefold higher than theory predicts it should be if it were derived by serpentinization.

It now seems more likely that Titan accreted primordial methane from the nebula around Saturn.

However acquired, methane would have been locked up in Titan as methane clathrates. Because they have a much lower density than the high pressure ices in Titan's interior they would buoyantly rise until they got to the ice crust which has a similar density. Mass balance arguments indicate that little of Titan's carbon inventory has

---

BOX 11.2   CLATHRATES

Clathrates are water molecule crystalline lattices, icy cages, in which "guest" molecules are locked. Most low molecular mass gases can form clathrates; for example, O_2, H_2, N_2, CO_2, H_2S, Ar, Kr, and Xe, as do small hydrophobic molecules. Hydrocarbons such as methane, ethane, and propane, but not higher members of the series, can be accommodated into the lattice. Ammonia clathrates have been made in the laboratory but have never been seen in nature. The water molecules in the lattice are hydrogen bonded but the structure cannot exist without the guest molecule. Clathrates are stable at low temperature and high pressure and dissociate if warmed or decompressed.

On Earth, methane clathrates are abundant in tundra permafrost and in ocean floor and lake bed sediments, and methane comes largely from methanogens. On Titan, methane, ethane, N_2, and CO_2 clathrates have been proposed, and clathrates are thought to play a role on other icy worlds also.

---

been recycled through its atmosphere, so there could be kilometers of methane clathrates underplating or within the icy crust. These are surmised to be the source of atmospheric methane.

11.2.2.2 *Why the Ocean Cannot Be the Source of Atmospheric Methane*

Methane is poorly soluble in water even at pressure. This means that methane cannot be replenished from the subsurface ocean. Calculations show that 3×10^8 km³ of cryolava (enough to cover the surface of Titan to a depth of 3.8 km) over the last 10–100 Myr would be needed to replenish Titan's methane once from an ocean reservoir, but crater counts show that Titan's surface is 0.1–1 Gyr old, so it cannot have experienced such extensive recent resurfacing.

11.2.2.3 *Cryovolcanism*

We know that appreciable outgassing from the interior of Titan has happened because of the high ^{40}Ar content of Titan's atmosphere. Hence, there is presumably a route for methane in the interior of Titan to make it into the atmosphere.

EXERCISE 11.7

Why does high ^{40}Ar indicate significant amount of outgassing?

An obvious process to allow this is cryovolcanism. This has been seen on Neptune's moon Triton, Mars, Enceladus, and Europa. The Visual and Infrared Mapping Spectrometer and Radar Mapper instruments on Cassini have imaged features that could be cryovolcanic in origin. These include a dome 180 km across, and what appear to be calderas and lava flows. Could partial melting of the crust be responsible?

Pure water cryovolcanism on Titan is impossible. The temperature never gets as high as 273 K and pure water could not buoyantly rise through the ice crust because water has a higher density than ice. Under the relevant pressures, a binary H_2O–NH_3 system could produce a cryomagma that was almost neutrally buoyant provided the ammonia concentration was close to the eutectic. Hydrostatic pressure gradients might then be able to drive cryomagma through deep crevasses.

A key question is how cryovolcanism leads to methane release. High-pressure anvil cell experiments on H_2O–CH_4 and H_2O–CH_4–NH_3 mixtures show that dissociation of methane clathrates and subsequent outgassing can only occur in Titan's icy crust where it is relatively warm ($T > 240$ K) and in the presence of large amounts of ammonia. The idea is that local convective warming by a rising ice plume generates a cryomagma chamber of liquid H_2O–NH_3 at shallow depth in the crust warm enough to dissociate surrounding methane clathrates thereby releasing methane.

Note that this model requires the ice crust to be convective. But the cold temperature predicted for the ocean and low degree of tidal dissipation (transformation of tidal energy to heat suggested by modeling) suggests the crust is conductive. More data is needed to (a) confirm cryovolcanism on Titan and (b) better understand exactly how it happens.

11.2.3 Organic Chemistry Occurs in Titan's Atmosphere

Spectra of Titan's atmosphere show numerous features of hydrocarbons (e.g., C_2H_6, C_2H_2), nitrogen-containing molecules (e.g., HCN, CH_3N), and oxygen-containing molecules (e.g., CO, CO_2, and H_2O) and it is clear that a rich chemistry occurs. It starts from the action of UV radiation and cosmic rays on CH_4 and N_2.

Several research groups are involved in deciphering chemistry on Titan. Three broad strategies are employed:

1. Spectroscopy of Titan's atmosphere to refine our knowledge of the chemical species and how they are distributed vertically and horizontally.
2. Laboratory studies of individual reactions under Titan conditions to measure equilibrium constants and rate constants.
3. Building models of the chemical network using observational and laboratory data as inputs.

11.2.3.1 Low Temperatures Make for Slow Kinetics

From the Arrhenius equation that quantifies the temperature dependence of chemical reactions it is easy to show that reactions requiring activation energy of ~100 kJ mol^{-1} or greater are so slow at 94 K as to be insignificant: that is, not much chemistry happens at the surface. High in the stratosphere where the temperature gets above 130 K, and the extended atmosphere means the pressure is a few tens of millibars, rate constant are >20 orders of magnitude higher, although reactions are still sluggish by comparison with those at the Earth's surface.

Building thermodynamic models of chemistry at low temperatures is complicated. Normally the idea is to calculate the equilibrium composition of a mixture at various temperatures so as to work out which reactions will happen and to what extent. It is not difficult to measure how equilibrium constants vary with temperature in the laboratory, but at low temperatures phase changes occur; gases condense into liquids and these may freeze. This is certainly so for hydrocarbons at the temperatures in Titan's atmosphere. Phase changes that remove a species will alter the equilibrium of a reaction, sometimes considerably. Hence phase changes have to be included in the calculations, and for mixtures of several compounds the thermodynamics becomes hard to model and difficult to analyze in the laboratory.

11.2.3.2 Free Radical and Ion-Neutral Chemistry Dominates

Absorption of UV photons and cosmic rays by chemical species happens high in the atmosphere. It starts with the dissociation of methane and nitrogen but there are two distinct chemical routes. One proceeds via free radical production in the thermosphere some 600–800 km above the surface and makes hydrocarbons. The other occurs by ion-neutral reactions much higher up (1000 km) in the ionosphere and results in high molecular mass ions. Let us look at these in turn (Figure 11.4).

Free radicals are highly energetic species because they have unpaired electrons. They are therefore usually short-lived, with notable exceptions (e.g., O_2 which has two unpaired electrons in its ground state). Starting with photolysis of CH_4 and N_2, larger molecules are built, mostly hydrocarbons and nitriles (compounds containing the $-C \equiv N$ functional group), mostly by free radical reactions. These gases are transported down, cool (see the temperature profile in Figure 11.1b) and condense to form an orange aerosol haze in the stratosphere. Few hydrocarbons can remain in the gas phase down to the tropopause which is a chilly 74 K. Eventually some of the hydrocarbons precipitate.

It was the detection of benzene by the Ion-Neutral Mass Spectrometer on Cassini at concentrations 1000-fold higher in the ionosphere than the stratosphere which revealed that ion-neutral reactions were happening at ~1000 km to generate PAHs and tholins: positively and negatively charged high molecular mass heteropolymers. These aggregate to form an aerosol of particles that are small enough to remain

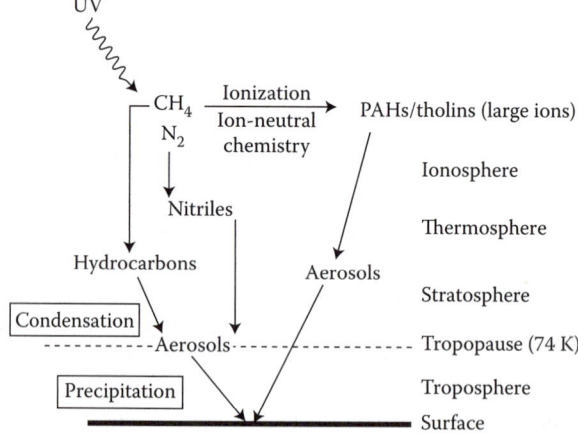

FIGURE 11.4 Ion-neutral chemistry that produces PAHs and tholins occurs at a higher altitude than the free radical chemistry that forms the hydrocarbons.

suspended for long periods, and scatter blue light to give Titan's upper atmosphere its distinctive color. Gradually the particles fall, eventually coating the surface with a dark organic layer. At least that is the theory. However, it is important to note that the make-up of the aerosol particles has not been directly determined.

The photochemical haze provides shielding from UV radiation, thereby protecting some of the more fragile chemical species from destruction.

11.2.3.3 Methane Is the Precursor Hydrocarbon

UV radiation splits methane in several ways:

$$CH_4 \rightarrow \xrightarrow{h\nu} CH_3 + H,$$

$$\rightarrow CH_2 + H_2,$$

$$\rightarrow CH_2 + 2H,$$

$$\rightarrow CH + H + H_2. \tag{11.1}$$

Each of the carbon–hydrogen moieties is a radical which now reacts in a variety of ways to produce a number of hydrocarbons. Key among them is ethane (C_2H_6) and acetylene (C_2H_2, ethyne).

Two methyl radicals generate ethane:

$$CH_3 + CH_3 + M \rightarrow H_3C\text{-}CH_3 + M. \tag{11.2}$$

Here M is a catalyst which carries away excess kinetic energy in the collision so that the ethane is not so vibrationally excited as to immediately dissociate. Ethane is not only a major component of many lakes, but it may also be present as aquifers in equatorial regions.

Acetylene is generated by the photolysis of ethylene (ethene)

$$CH + CH_4 \rightarrow \underset{\text{ethylene}}{H_2C\text{=}CH_2}. \tag{11.3}$$

$$\underset{}{H_2C\text{=}CH_2} \xrightarrow{h\nu} \underset{\text{acetylene}}{HC\text{≡}CH} + H + H. \tag{11.4}$$

As we shall see there is currently a big mystery surrounding acetylene on Titan, making it doubly important to understand in detail how it is formed and destroyed.

Two molecules of acetylene can react to produce diacetylene:

$$HC\text{≡}CH + HC\text{≡}CH \rightarrow \underset{\text{diacetylene}}{HC\text{≡}C - C\text{≡}CH} + H. \tag{11.5}$$

Photolysis of diacetylene yields the HC≡C radical:

$$HC\text{≡}C - C\text{≡}CH \xrightarrow{h\nu} 2HC\text{≡}C. \tag{11.6}$$

All the ingredients are then in place for lengthening the polyyne species by successive addition of the HC≡C radical. The first step is:

$$HC\text{≡}CH + HC\text{≡}C \rightarrow HC\text{≡}C\text{—}C\text{≡}CH + H. \tag{11.7}$$

In general,

$$C_{2n}H_{2n} + C_2H \rightarrow C_{2n+2} H_2 + H, \quad \text{where } n = 1, 2, 3, \ldots \tag{11.8}$$

This generates a large number of hydrocarbons, including cyclic compounds, which give rise to the orange photochemical haze.

11.2.3.4 Nitrogen Leads to Nitriles

Nitrogen is dissociated by high energy photons at high altitude and cosmic rays lower down.

$$N_2 \xrightarrow[\text{cr}]{h\nu} N + N. \tag{11.9}$$

It is also ionized by cosmic rays:

$$N_2 \rightarrow N^+ + N. \tag{11.10}$$

The nitrogen atom reacts with a methyl radical to produce hydrogen cyanide, a deadly poison for oxygen-respiring organisms, yet central to prebiotic chemistry:

$$N + CH_3 \rightarrow H_2C\equiv N \rightarrow \underset{\text{hydrogen cyanide}}{HC\equiv N} + H + H. \tag{11.11}$$

Photolysis of HCN gives rise to CN, a key intermediary in the formation of nitrile polymers:

$$HC\equiv N \rightarrow CN + H. \tag{11.12}$$

One such polymer is cyanoacetylene (prop-2-ynenitrile):

$$CN + HC\equiv CH \rightarrow HC\equiv C—C\equiv N + H. \tag{11.13}$$

The stage is now set for nitrile polymerization which generates a component of tholins.

11.2.3.5 Tholins Are Formed by Ion-Neutral Reactions

At altitudes of ~1000 km hydrocarbons are ionized by solar radiation allowing ion-neutral reactions to produce benzene and thus polycyclic aromatic hydrocarbons (PAHs) and tholins. The term "tholin" was first used by Carl Sagan and Bishun Khare in a 1979 *Nature* paper to describe the red-brown viscous residue formed in their Urey–Miller-type experiments on gas mixtures that simulate Titan's atmosphere. Tholins are complex hydrocarbon-nitrile negative ions with relative molecular mass up to 8000 Da. Such a molecule would form an aerosol particle 260 nm across. A flavor of the ion-neutral chemistry involved can be seen in the reactions that lead to benzene.

There are several reactions that lead to $C_4H_5^+$; one is

$$C_2H_2 + C_2H_5^+ \rightarrow C_4H_5^+ + H_2. \tag{11.14}$$

From this the benzyl cation ($C_6H_7^+$) can be synthesized in several ways; one is

$$C_4H_5^+ + C_2H_4 \rightarrow C_6H_7^+ + H_2 \tag{11.15}$$

and

$$C_6H_7^+ + e^- \rightarrow \underset{\text{benzene}}{C_6H_6} + H.$$

11.2.3.6 Titan Chemistry Is Being Explored in the Laboratory

Several studies are attempting to simulate the chemical environment in Titan's atmosphere to see what gets cooked up.

An international team at three French universities, lead by Robert Yelle, University of Arizona, undertook simulations of Titan atmosphere chemistry motivated by the discovery that oxygen ions flow into Titan's atmosphere. Aerosol particles comparable in size to those on Titan were formed when N_2–CH_4–CO gas mixtures were irradiated with microwave radiation. CO was added to provide a source of oxygen. There was no water. The aerosols were found to contain 18 molecules with molecular formulae that correspond to amino acids and nucleotide bases. Very high-resolution mass spectrometry confirmed that glycine and alanine and all five of the nucleobases found in RNA and DNA were produced in the simulation chamber. These experiments are fascinating since it has shown that water is not necessary to make amino acids and bases, although *low* levels of oxygen apparently are. On early Earth oxygen would come from photodissociation of carbon dioxide, carbon monoxide, and water from volcanoes and comets. On Titan now, oxygen appears to be coming from the icy geysers of Enceladus, molecules from which can spread long distances through the Saturnian system.

How much does this work tell us about the origin of life on Earth? Firstly, it need not change our thoughts about where life on Earth was spawned. Even if the organic molecules were made high in the atmosphere they would have rained out into pools and oceans, although as we have seen the redox chemistry of hydrothermal vents might be a richer source of biochemical precursors than Earth's atmosphere. Secondly, in reproducing conditions in Titan's atmosphere there was no water present: this is not a plausible scenario for early Earth. Thirdly, the experimental Titan atmosphere does not seem to have solved one of the biggest barriers to understanding how life originated on Earth, namely, the chemical synthesis of proteins from amino acids, and of nucleotides (let alone DNA and RNA) from the raw bases, under plausible prebiotic conditions.

11.2.4 Could There Be Life on Titan?

11.2.4.1 Is the Subsurface Ocean Habitable?

The subsurface ocean of Titan provides an environment that is at least familiar to astrobiologists in the sense that it is mostly water. Not surprisingly, there has been speculation about the metabolic possibilities for life in Titan's ocean—for example, nitrate reduction, sulfate reduction, or methanogenesis—so it is worth seeing how well this might stand up to scrutiny. The essential question is whether the ocean is habitable. Its temperature could be $-13°C$ now; cold adapted terrestrial extremophiles—including methanogens—can just about grow and reproduce at these temperatures, and it would have been warmer on early Titan. The pH of a 15 wt% NH_3–H_2O ocean would be ~11.5, at the limit of what is tolerable to terrestrial alkaliphiles. The pressure could range between 100 and 500 MPa and some terrestrial barophiles would be happy at the lower end. A potentially fatal difficulty is that sulfur and phosphorus might not be available on Titan.

EXERCISE 11.8

It has been calculated that radiogenic heating alone (5×10^{11} W) would provide an energy flux into the ocean of 5×10^8 W, enough for the production of 4×10^{11} moles of ATP per year, equivalent to 2×10^{13} g of biomass per year. Assuming an average turnover of 1 year this gives a biomass density of 1 g m^{-2}, some three to four orders of magnitude less than Earth. Assuming that the numbers are correct why is this calculation unrealistic?

11.2.4.2   Terrestrial-Style Methanogenesis Is Dubious

For methanogens to live in the ocean they would need sources of CO_2 and H_2. The CO_2 abundance in the atmosphere is extremely low (14 ppbv) though dry ice has been detected on the surface by Cassini's Visible and Infrared Mapping Spectrometer. In some models of Titan's evolution it accretes large quantities of CO_2 from the Saturn subnebula, and this got locked up as high density clathrates deep inside the moon.

Serpentinization could supply hydrogen, although whether this process could operate inside Titan now is not known.

EXERCISE 11.9

Terrestrial methanogens are inhibited by acetylene. How much weight would you attribute to this in assessing whether methanogens existed on Titan?

11.2.4.3   Titan's Atmosphere Poses Paradoxes

In 2005, Chris McKay and Heather Smith at NASA posed the possibility of methane-based life on Titan. This would use methane as a biochemical solvent, as the metabolism is based on the reduction of hydrocarbons which are largely insoluble in water. Being methane-based it would be expected to be widespread across the surface of Titan, given the ubiquitous distribution of methane, and the tendency of terrestrial life to fill any available ecological niche. They would get energy by reacting acetylene or ethane with hydrogen (H_2) and produce methane as a waste product. (So it is a methanogen, but not like any we know.) Here, hydrogen is acting as an electron acceptor, just as oxygen does in aerobic metabolism. Theoretical work in the same year by McKay and Smith and by Dirk Schulze-Makuch, and David Grinspoon show that the key reactions are thermodynamically favorable. For example acetylene, which provides the most free energy of any hydrocarbon, reacts to give

$$C_2H_2 + 3H_2 \rightarrow 2CH_4. \tag{11.16}$$

which has a ΔG of approximately -100 kJ mol^{-1} under the surface conditions on Titan. The reaction has an energy barrier, so a catalyst would be required. (There are terrestrial bacteria that use acetylene as an electron donor but they convert it to ethanol and acetate, not methane, and of course they use water as a solvent.)

Intriguingly observations have thrown up three currently inexplicable observations that are consistent with these ideas.

First, it has long been recognized that there should be far more ethane on the surface of Titan than is observed. Photochemical models predict Titan's surface should be covered to a depth of many meters by ethane: the hydrocarbon lakes are far too small to accommodate this.

Second, in 2010 Roger Clark showed that Cassini's Visual and Infrared Mapping Spectrometer has not detected as much acetylene on the surface of Titan as predicted by how rapidly we think photochemistry should convert methane to acetylene and ethane. Although the Huygens probe detected methane shortly after landing there was no evidence of acetylene.

Third, H_2 is generated by UV splitting of acetylene and methane in the upper atmosphere. Hydrogen produced in this way should be distributed fairly evenly throughout the atmosphere. However, Cassini has discovered an inexplicable disparity in the hydrogen densities between upper and lower atmosphere: higher in the upper. Detailed calculations by Darrell Strobel of John Hopkins University show this could result from a flow of hydrogen down to the surface, where it disappears. The low density of H_2 near the surface implies that some surface process is absorbing hydrogen at 10^{28} molecules per second: equivalent to hydrodynamic escape to space.

There are of course potential nonbiological explanations for these observations. An obvious one is that there is something wrong with our models of how photochemistry happens in Titan's atmosphere and this leads us to overestimate the production of ethane, acetylene, and hydrogen. In particular, the flux of hydrogen into the surface is not an observation. Instead, it comes from a computer simulation designed to fit measurements of the hydrogen concentration in the lower and upper atmosphere in a self-consistent way. How closely this simulation matches Titan chemistry is not clear.

There may be a process that transports hydrogen from the upper atmosphere into the lower atmosphere. Adsorption of hydrogen onto aerosol particles which eventually fall to the ground would fit the bill. However, this would be a flux of H_2, not a net loss.

There is also the possibility that surface catalysis by an unidentified mineral is destroying hydrogen in some way, perhaps by converting it to methane.

Finally, we have to acknowledge the two serious impediments to methane-based surface life. One is the frigid temperature (94 K) that would make biochemistry crawl at a snail's pace. To offset this would require heroic enzymes the like of which terrestrial life has not evolved. The other is that methane is a nonpolar molecule; it barely dissolves the interesting organic molecules needed for metabolism. To circumvent this, organisms could develop ways to maximize their transport of nutrients.

EXERCISE 11.10

Suggest how an organism could maximize its transport so as to be able to take up adequate amounts of a nutrient from a low concentration in its environment.

11.3 Enceladus

Enceladus is a tiny moon of Saturn with a mean diameter of just 505 km. It was discovered by William Herschel in 1789 using what was then, with an aperture of 1.26 m, the largest telescope in the world. Its bulk density (1609 kg m^{-3}) implies a silicate–ice composition, and modeling its likely thermal evolution shows that short-lived radio-isotopes would have produced sufficient heating for Enceladus to differentiate a rocky core and icy mantle. But its small size means it should have cooled long ago into a frigid, geologically inactive world, yet it is not. Cryovolcanism from fissures in its icy surface shows Enceladus to be anomalously warm. It appears to have liquid water beneath its surface, energy, and sources of carbon and nitrogen. Not surprisingly this has attracted the attention of astrobiologists.

11.3.1 Enceladus Has Been Resurfaced Multiple Times

Images of Enceladus taken by the Voyager spacecraft revealed a geologically active world. It has several distinct geological units. Going north to south there is a 500-fold variation in crater counts from heavily cratered terrain dominating the northern polar region to south polar terrain which has very few craters. This trend in crater density shows that, to a first approximation, the surface gets younger in a north–south direction. This makes the heavily cratered unit the primary crust.

However, the ages implied by the crater counts depend on how the impact rate is assumed to have fallen off after the late heavy bombardment. Two scenarios were examined by Caroline Porco of the Space Science Institute, Boulder, Colorado, and colleagues. One was an exponential drop in impact rate, the other a linear drop.

> **EXERCISE 11.11**
>
> How would the relative ages deduced from crater counts differ between the assumption of exponential or of linear drop-off in impact rate?

In these two different models the oldest regions of Enceladus are either 4.2 or 1.7 Gyr old while the youngest surface, the south polar terrain is either 100 or 1 Myr old.

Craters in the younger, moderately cratered terrain show evidence of much more viscous relaxation than those in the older heavily cratered terrain. Viscous relaxation refers to the plastic flow of ice over geological timescales in response to warming. It shows itself in a smoothing and flattening of crater rims and doming of their floors. Calculations show that a thermal gradient of tens of degrees per kilometer must have existed beneath the moderately cratered terrain to account for the extent of topographic relaxation seen. This implies that warm convecting ice or ammonia–water magmas were only a few kilometers below the surface. Craters on nearby heavily cratered regions are unaffected, so the warming was localized.

11.3.2 Enceladus Cryovolcanic Plumes Contain Water Vapor

In January 2005, Cassini imaged plumes of gas which we now know are mostly ionized water vapor and ice crystals erupting up to 450 km above its southern polar regions. The plumes are vented more or less continually from "hot spots" along four linear fissures or sulci (130-km long, 2-km wide, and 0.5-km deep) bounded on either side by ridges; structures that have come to known informally as "tiger stripes" (Plate 11.2).

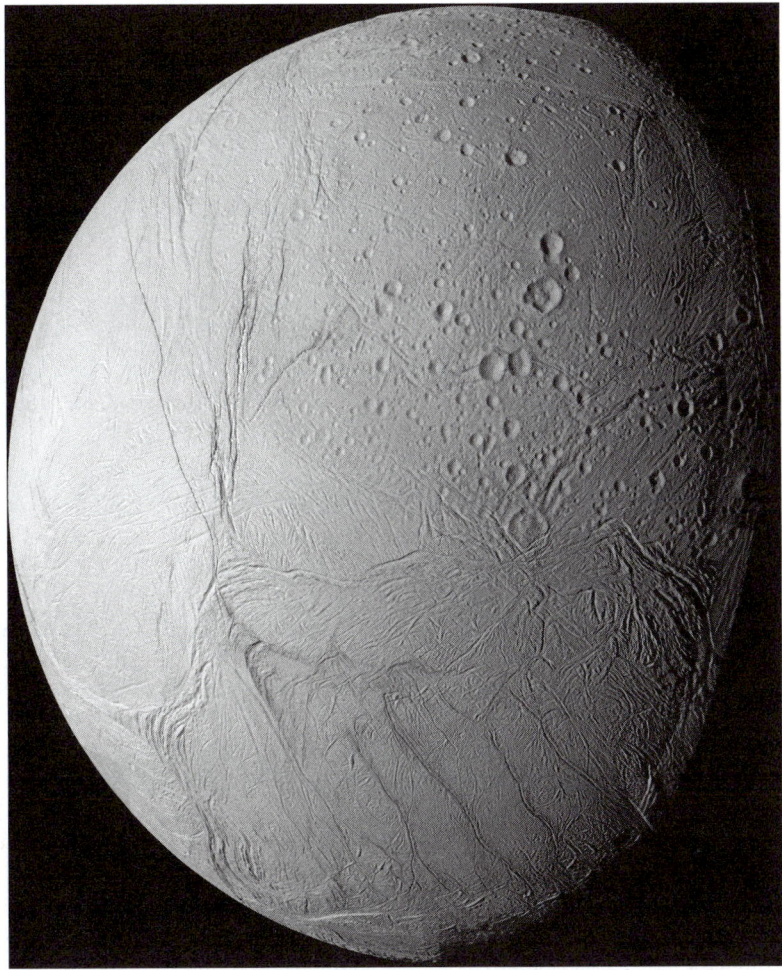

PLATE 11.2 (**See color insert.**) Tiger stripes in the south pole of Enceladus periodically vent plumes of water vapor. Cassini image. (Photo credit: NASA/JPL/Space Science Institute.)

Most of the erupted ice falls back to the surface. Indeed the reflectance spectrum of Enceladus is dominated by water ice with traces of CO_2 ice, and the moon's very high bond albedo (0.99) is explained by this covering being fresh.

Over a number of flybys Cassini has examined the make-up of the cryovolcanic plumes. It has detected the following (mixing ratio): water vapor (91%), N_2 (4%), CO_2 (3.2%), CH_4 (1.7%), and NH_3 (0.8%), plus trace amounts of hydrocarbons such as propane, ethane, and acetylene. Sodium salts, in the same concentration as in Earth's oceans, were also found.

The eruption of water vapor could imply that there is a magma chamber of liquid water beneath the surface, or even a subsurface global ocean, that is the source of the cryovolcanism.

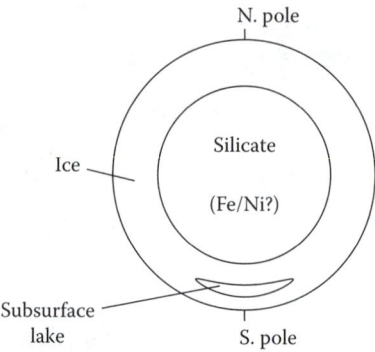

FIGURE 11.5 Enceladus is likely to be differentiated and may have a subsurface ocean at least some of the time.

EXERCISE 11.12

What evidence suggests that subsurface liquid water might exist even if the temperature at the surface where the eruptions are occurring is only 167 K?

The presence of sodium salts in the plumes suggests the source of the cryomagma has had access to silicate minerals from which the salts have leached. This might mean that there is contact between the cryomagma source and the rocky core (Figure 11.5).

11.3.3 What Heats Enceladus Now?

The mean surface temperature on Enceladus is 75 K but over the south polar terrain, where it should be 68 K, it varies generally between 113 and 157 K. Indeed, temperatures of 180–200 K have been measured over the most active of the tiger stripes, Baghdad Sulcus. The tidal deformation needed to account for the tectonic features on Enceladus needs at least 8 GW heating globally and the most recent Cassini infrared spectrometry indicates a heat flow from the south polar terrain of 15.8 GW.

There has been much effort expended to understand what could warm Enceladus so much. Current radiogenic heating is trivial; just 0.32 GW. Enceladus has a slightly eccentric orbit ($\varepsilon = 0.0047$) which allows it to experience tidal heating. The eccentricity is maintained by a 2:1 mean motion resonance with Dione. Every two orbits Enceladus makes, it passes Dione once, when tidal torques excite its eccentricity (Figure 11.6).

Calculations show that this could provide no more than 1.1 GW of tidal heating, assuming the eccentricity is not changing over time. So this will not do. But being in synchronous rotation and having an eccentric orbit, Enceladus should experience a spin–orbit libration with Saturn. Libration is the periodic oscillation ("wobble") in longitude of an orbiting body (see Exercise 11.5). Libration of Enceladus has not been detected, so it can be no bigger than 2°. This constrains heating by this mechanism to only 0.12 GW. On this analysis, calculated and measured heat flows disagree by an order of magnitude (1.54 GW versus 15.8 GW). Clearly something is missing.

The clue is that Enceladus is not a perfect sphere but a triaxial ellipsoid with its rotation axis flattened. This asymmetry could increase the tidal heating produced by

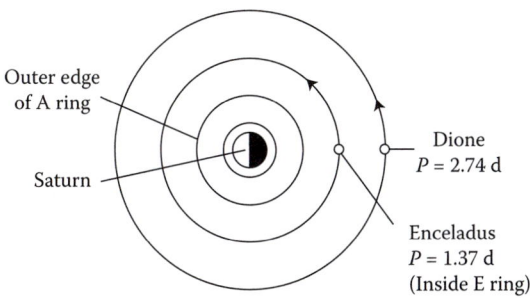

FIGURE 11.6 A 2:1 mean motion resonance provides tidal torques on the inner world once every two orbits at closest approach.

libration dramatically. One computer simulation gives a fivefold increase. Another surmises that sometimes Enceladus gets locked into a 4:1 forced spin–orbit libration with Saturn in which the moon librates four times for each orbit around the planet. This could increase tidal heating perhaps as much as 100-fold. These active periods might be interspersed with more quiescent intervals. This raises the question about whether subsurface liquid water is a permanent feature of Enceladus.

Recent geophysical evidence from tiny perturbations in the trajectory of Cassini as it passes Enceladus suggests the presence of a regional ocean about 50 km beneath the south pole. Whether this is the source of the cryovolcanism is not known.

11.3.3.1 Tidal Stresses Control Eruptions

Cassini's Visual and Infrared Mapping Spectrometer (VIMS) team have shown that the plumes triple in brightness when Enceladus is close to apoapsis (furthest from Saturn) compared to the half of the orbit around periapsis. This is because the tidal stresses on the south polar regions cause the tiger stripes to gape around apoapsis and thereby vent more volatiles. During the time Enceladus is closest to Saturn, the stresses change from extension to compression, closing the fissures and damping down cryovolcanism.

Computer simulations show that shear stress can produce lateral (strike-slip) motion of the tiger stripe fissures by 0.5 m when tensile stress at right angles to the fissure holds it open. Half an orbit later when stresses change from tensile to compressional and the fissure closes, friction prevents the lateral offset from relaxing completely. So over successive orbits a substantial strike-slip displacement builds up. This has now been detected in Cassini images (Figure 11.7).

The importance of this is that tidal deformation would not be able to produce this sort of lateral motion if the crust were thicker than 40 km, which is just about enough for it to be convective; that is, to allow plumes of warm ice to rise to sustain volcanism.

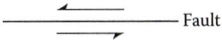

Fault

FIGURE 11.7 Strike-slip faulting has directionality. This is a sinistral fault because the opposite side of the fault slips to the left with respect to an observer facing the fault.

11.3.3.2 Cryovolcanism Is Probably Driven by Ammonia–Water Magma

Several mechanisms have been proposed for the cryovolcanism.

Ice particles at 157 K would sublime to form a plume. Thermodynamic calculations show that only a small amount of vapor would then condense to ice as the plume expanded and cooled adiabatically, not nearly enough to account for the high ice/vapor ratio of the plume. Hence, sublimation has to be ruled out.

Another possibility is that the plumes are generated as clathrate hydrates of CO_2, CH_4, and N_2 degas explosively when exposed to the vacuum of space. This would work at temperatures 80–100 K cooler than liquid water. However, salts do not form clathrates, and it is not clear that ammonia clathrates can form in nature, though they have been made in the laboratory. So this model cannot easily explain the presence of salts and NH_3 in the plumes.

Cryovolcanism from a liquid water magma chamber is not credible given estimates of the heat flux, even with the large uncertainties. However, it could work with water–ammonia magma. The temperature could be as low as 170 K if this was a eutectic composition, and this would make it almost neutrally buoyant with respect to the icy crust. The difficulty with this idea is that nitrogen and methane are barely soluble in $H_2O–NH_3$, so it could not account for N_2 and CH_4 vented by the plumes.

EXERCISE 11.13

Suggest how the problems raised by the clathrate and $H_2O–NH_3$ magma models of cryovolcanism on Enceladus might be overcome.

11.3.3.3 Putative Organisms Are Water-Based

As usual there has been speculation about what sort of life could inhabit any liquid water niches inside Enceladus: methanogens living on CO_2 and H_2, and acetylene-using organisms that produce ethanol or acetate are key suspects. Methane has been detected in the plume, but it is hardly a biomarker. Ethanol and acetate in plume vapor would be more suggestive but they have not been detected so far.

It is worth bearing in mind that neither the orbital dynamics nor geology of Enceladus provides any guarantee that there has been a long-lasting subsurface liquid water habitat.

11.4 Europa

The lure of Europa for astrobiology is not just that it probably hosts a subsurface ocean of liquid water beneath an ice crust but that the ocean floor may have hydrothermal vents. Could they provide the redox gradients needed to "kick start" metabolism just as those on Earth may have done? Radiation generates oxidants including oxygen by dissociating water ice. The surface of Europa is very young and has numerous tectonic features which attest to on-going geological activity and the possibility of communication between the oxidizing surface and ocean below adds to the intrigue.

Most of what we know about Europa comes from the Galileo mission which spent 8 years in the Jovian system before plunging to its death in Jupiter's atmosphere in 2003, a fate designed to prevent the probe from contaminating any of the moons with terrestrial organisms.

Europa was imaged by New Horizons in 2006 but the next mission to visit this world, the European Space Agency's (ESA's) Jupiter Icy Moon Explorer (JUICE), is not due to launch until 2022.

From an astrobiology perspective there are several key questions to ask regarding Europa.

1. What is the thickness of its crust? This is important because, as we shall see, chemistry generates oxidants, including oxygen, at the surface and if the crust is thin then these could get into the ocean below. Crustal thickness depends on where and how much heat is generated in Europa and the properties of the ice.

2. What is the composition of the subsurface? Models predict that the icy crusts should contain salts. Which salt (magnesium sulfate, sodium chloride, etc.) determines whether a subsurface ocean can exist at the temperature predicted by heat flow calculations and has implications for the sort of metabolism any life could indulge in.

3. Is the heat flowing through Europa's silicate mantle enough to produce hydrothermal activity? The presence of hydrothermal vents makes it possible that Europa could have environments like those on early Earth that may have spawned life.

4. What is the oxidation state of the mantle? This will determine whether hydrothermal activity can generate a redox gradient which incipient life could harness for energy.

These are the main issues addressed in this section.

11.4.1 Europa Is a Differentiated World

Europa is the smallest of the Galilean moons of Jupiter that were discovered in 1610 using a refracting telescope with a magnification of just 20. With a radius of 1561 km it is a little smaller than Earth's moon. Its bulk density is 3010 kg m^{-3}. Galileo gravity measurements require that Europa is differentiated with a metal/metal sulfide core, silicate mantle, and an outer crust of water ice which may harbor a subsurface water ocean. Europa's bulk density is comparable to the density of silicates and only 15% less than the density of its nearest neighbor, Io. This means that Europa has a relatively small inventory of water compared to the giant Jovian icy satellites Ganymede and Callisto. Consequently, the thickness of the water layer can be no more than 80–170 km and a layer of high pressure ice between an ocean and the silicate mantle is ruled out. Water must therefore be in direct contact with silicates, and intriguingly, the temperature and pressure at the bottom of a hypothetic Europan ocean 100 km deep is comparable to that on the floor of a terrestrial ocean (Figure 11.8).

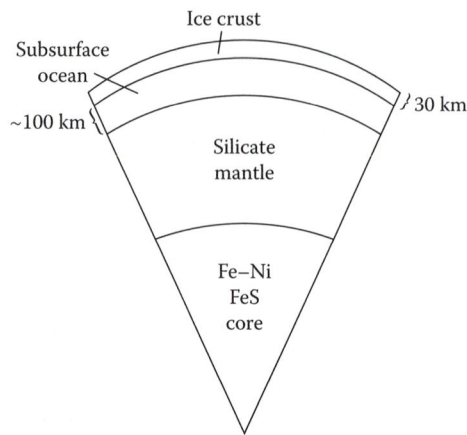

FIGURE 11.8 Internal structure of Europa. Not to scale.

11.4.2 There Is Good Evidence for a Europan Ocean

Two lines of evidence support the notion of a subsurface ocean.

1. The Galileo magnetometer recorded an induced magnetic field. Europa has no magnetic field of its own but it orbits within the strong Jovian field. Jupiter's magnetic field is tilted at 10° with respect to its rotation axis, so as the planet spins, Europa experiences a change in the Jovian magnetic field over a period of ~11 h. This time-varying magnetic field induces a current loop in the interior of Europa. This current in turn induces the magnetic field measured by Galileo. For currents to flow there must be a continuous conducting layer beneath the Europan surface; pockets of conducting fluid would not produce the observed field. The periodic variation rules out production by a metallic core and the field is far too strong to be produced by the ionosphere of Europa. The results imply the existence of a brine ocean several tens of kilometers thick, but allow a range of values for the conductivity of the ocean, its thickness and how far below the surface it is.

2. Surface features show that the icy shell of Europa is mechanically isolated from the rocky mantle beneath.

However, these features are also consistent with the presence of a slush of ice and liquid beneath an ice shell rather than a fully molten ocean. At present, we cannot discriminate between these two alternatives on geological grounds.

11.4.3 How Thick Is the Europan Crust?

The thickness of the crust has serious implications for Europa's habitability. How close Europa's ocean is to the surface is the key to how effectively oxidants created at the surface could be transported to the putative liquid water ocean. The thinner the crust the better will be the coupling between the surface of Europa and its ocean.

While surface tectonic features imply a thin crust (~3 km), crater impacts, thermal, and mechanical models argue for a thick crust (~30 km).

11.4.3.1 Europa Experiences Tidal Heating

If Europa does have a subsurface ocean, then almost all of the heat energy keeping it liquid now is thought to come from tidal dissipation in the ice crust.

Europa orbits Jupiter in 3.551 days, with an orbital radius of about 670,900 km and is currently in synchronous rotation so that one hemisphere always faces Jupiter. Europa's small orbital eccentricity of 0.009 is maintained by a 1:2:4 mean motion Laplace resonance between Ganymede, Europa, and Io. The eccentricity causes two effects:

1. As Europa approaches periapsis, tidal forces by Jupiter elongates Europa along an axis that goes through the planet and satellite, and as Europa moves toward apoapsis it relaxes back into a more spherical shape: tidal flexing.
2. Europa's sub-Jovian point librates due to the periodic mismatch between its orbital speed and its rotational speed.

In combination, these effects produce tidal heating.

> **EXERCISE 11.14**
>
> The semi-major axis of Io's orbit is 421,700 km. Show that Europa and Io are in a 1:2 mean motion resonance.

Calculating the tidal dissipation depends on the exact nature of the orbital dynamics and therefore the tidal forces acting over time. This is uncertain. Some models of Europa's evolution suggest its orbit was more eccentric in the past or that it may have had a larger obliquity, both of which would increase the tidal heating up to 100-fold the present levels in one model. Tidal dissipation also depends on the viscous and elastic properties of the ice crust and the silicate mantle. These have to be inferred from studies of how the Europan surface has deformed in response to loads and from laboratory studies of the viscoelastic properties of ice subjected to cyclical loading with frequencies comparable to that of tidal flexing.

These show that that the cooler the ice crust is, the higher its viscosity and the lower the tidal heating will be.

> **EXERCISE 11.15**
>
> (a) Assuming that tidal heating of Europa is currently enough to sustain a subsurface ocean of pure water, how might the state of Europa differ if the ocean contained a high concentration of ammonia or other antifreeze? (b) What might be the astrobiological consequences of such a difference?

Current models suggest that tidal heating is several times larger than radiogenic heating. Moreover, for plausible values of silicate viscosity, tidal heating of the silicate mantle is very much less than radiogenic heating, which is insufficient to produce partial melting of the mantle. Consequently, there is unlikely to be active volcanism at the mantle/ocean interface now.

Thermal modeling that takes account of the physics above suggests that the crust is 20–30 km thick.

11.4.3.2 Crustal Thickness Can Be Estimated from Mechanical Properties

Crustal thickness can be estimated from its rigidity. Ice has elastic strength only at relatively cold temperatures, and because the temperature gradient across the ice crust depends on its total thickness, the measured elastic thickness can be used to infer the crustal thickness (effective elastic thickness). The elastic thickness is calculated from the deformation of the surface, as shown by topography, produced by estimated loads. This method yields a present day crustal thickness for Europa greater than 15 km, which agrees with impact crater studies.

11.4.3.3 Surface Features Are Evidence for an Ocean

The low crater density gives the surface of Europa an age between 0.5 and 50 Myr if the impactors were mostly comets. The high albedo (0.64) shows that Europa has been resurfaced relatively recently. The moon is criss-crossed by numerous long linear features that resemble mid-ocean ridges. These are thought to be fractures opened in the icy shell by tidal stresses. Given the age of Europa's surface these features must be young. Salts identified on disrupted regions of the surface have been taken to imply that eruptions of warm brines through the fractures coat the surface. If this is upwelling from a subsurface ocean then it was obviously liquid recently and is therefore likely to be so now.

Because Europa is tidally locked to Jupiter it is possible to predict the pattern of stress fractures that occur on Europa. Only the youngest regions of the surface conform to this pattern. Successively older surfaces bears stress patterns that are rotated in a coherent fashion. Many of the ridges are cycloid, showing that as fractures elongated, the direction of extensional stress was changing continuously. All this is evidence that the Europan surface rotates faster than its interior with respect to Jupiter. This would be explained by an ice shell floating on an ocean.

An alternative model is that in the recent past Europa had a small obliquity. The shift in stress pattern matched well with what would have occurred as Europa precessed about this tilted spin axis. This does not rule out the asynchronous rotation of shell and interior of the moon, but a precession pattern would be produced much more quickly than the asynchrony pattern. Hence Europa's fractures may be a lot younger than we had thought. Obliquity would also increase tidal heating.

11.4.3.4 Chaos Terrain Implies a Thin Crust

Thermal and mechanical models suggest the Europan crust is a few tens of km thick. Yet some of the most iconic close-up images of Europa show chaos terrain (Plate 11.3),

PLATE 11.3 (**See color insert.**) Chaos terrain on the surface of Europa. (Photo credit: NASA/JPL.)

which looks like icebergs trapped in a frozen ocean. If this interpretation is correct, then the crust of Europa must be thin: a few kilometers at most.

EXERCISE 11.16

What other aspects of the surface of Europa imply that the crust is thin?

Chaos terrain consists of roughly circular regions of disrupted ice raised on average ~100 m above the surrounding surface. The disrupted ice consists of large polygonal blocks that resemble icebergs surrounded by mounds of ice matrix that rise ~200 m higher. A recent model accounts for how chaos terrain forms over a period of 10^5–10^6 years (Figure 11.9). The model gives the thickness of crust above the subsurface lake as 2.8 km.

The model explains many of the morphological features of chaos terrain. What is more, it suggests a way in which conflicting evidence regarding the thickness of the Europan crust might be resolved. At chaos terrain it will appear thin at least in respect of tectonic features, but elsewhere it will look thick.

11.4.3.5 Water–Ice Cryovolcanism May Occur on Europa

During November and December of 2012 the Hubble Space Telescope (HST) imaged UV emissions from Europa that provide evidence for a 200-km high transient plume of water vapor. This finding is exciting because it provides support for subsurface liquid water, implies that at least some regions of the crust are thin enough to allow cryovolcanism, and shows there is sufficient heat energy to drive it.

11.4.4 The Europan Surface Is Chemically Altered

Spectroscopy reveals regions of the Europan surface that are not water ice. Signatures of salts and oxidants have been discovered. Where these come from we explore now.

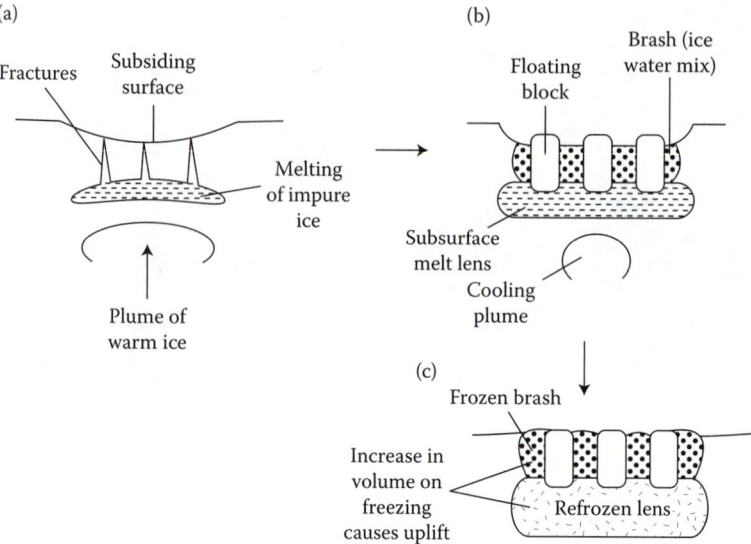

FIGURE 11.9   Stages in chaos terrain formation: (a) A rising plume of pure water ice produces extensive melting of the impure ice. Melt forms a lake and the reduction in volume causes subsidence, pressurization, and fracturing of the ice above. (b) Ice calves into blocks as fractures propagate to the surface, allowing brash to erupt causing depressurization and further subsidence. (c) Refreezing of the ice matrix causes it to expand to form mounds higher than the blocks.

11.4.4.1 Europa's Surface Is Altered by Radiolysis

Some salts and oxidants are generated at the surface of Europa by a process termed radiolysis. This is dissociation of molecules by nuclear radiation (i.e., associated with radioactive decay) or by high energy charged particles generated in other ways. Most important is the splitting of water ice to produce hydrogen and hydroxyl radicals (OH):

$$H_2O \rightarrow H + OH, \tag{11.17}$$

$$H + H \rightarrow H_2. \tag{11.18}$$

The hydroxyl radicals then react to form oxidants such as hydrogen peroxide and oxygen species:

$$OH + OH \rightarrow H_2O_2, \tag{11.19}$$

$$2H_2O_2 \rightarrow H_2O + O_2. \tag{11.20}$$

In the context of Europa, most of the radiolysis is thought to be electrons and ions accelerated to high energies (tens of keV) by Jupiter's magnetosphere.

In combination with the dissociation of water, radiolysis of CO_2 ice produces many organic compounds such as methanol (CH_3OH), formaldehyde (HCHO), and ethenone ($H_2C=C=O$), a ketene.

11.4.4.2 The Europan Ocean May Resemble Earth's Oceans

There has been considerable debate about the salt composition of Europa's putative ocean. Theoretical models have postulated several candidates including magnesium sulfate ($MgSO_4$), sodium sulfate (Na_2SO_4) or sulfuric acid hydrate ($H_2SO_4 \cdot nH_2O$).

In 2013, the spectroscopic signature of the hydrated magnesium sulfate mineral, epsomite ($MgSO_4 \cdot 7H_2O$), was obtained using the 10-m Keck II telescope. Intriguingly, it was confined to the trailing hemisphere of the satellite. We have known for some time that Europa's leading hemisphere is relatively pure water ice with only a modest abundance of material that is nonwater ice. By contrast, the trailing hemisphere, particularly at low latitudes, has a higher proportion of nonwater ice. This hemisphere is bombarded by sulfur erupted by Io's volcanoes, so the nonwater ice is thought to be a sulfur compound such as sulfuric acid hydrate (Figure 11.10).

Mike Brown and Kevin Hand of the California Institute of Technology propose that the magnesium sulfate has been generated from a magnesium salt erupted from the ocean. A strong candidate for this salt is magnesium chloride. Here is why: models of the Europan ocean show that if it is reduced, the dominant salts will be chlorides such as sodium, potassium, and magnesium chlorides: $NaCl$, KCl, and $MgCl_2$. (Atomic Na and K have been identified in the very tenuous—10^{-12} bar—Europan atmosphere.) Sulfates cannot form under reduced conditions but would dominate an oxidized ocean. In other words, any ocean must either be chloride-rich or sulfur-rich depending on how oxidized it is. The fact that $MgSO_4$ is only found on one hemisphere rules out a sulfur-rich ocean. By implication, the ocean must be chloride-rich and the most likely magnesium salt is $MgCl_2$. Unfortunately, none of the chloride salts have distinctive IR spectral features, so their presence on the surface cannot be proved directly.

So where does the magnesium sulfate come from? The idea is that sulfur from Io on Europa's trailing side reacts with oxygen produced by radiolysis of water. The resulting sulfate ion reacts with the endogenous $MgCl_2$ to form $MgSO_4$. The overall reaction is:

$$MgCl_2 + 4H_2O + S \rightarrow MgSO_4 + 4H^+ + 2Cl^-. \qquad (11.21)$$

Perhaps the most important thing to emerge from all this is that the Europan ocean could have a similar composition to Earth's oceans (Figure 11.11).

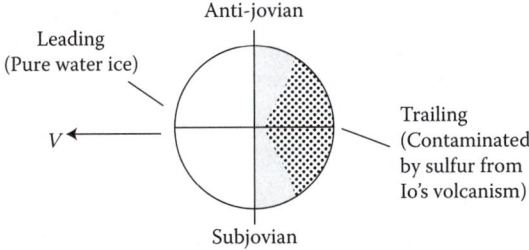

FIGURE 11.10 Europa has distinctive leading and trailing hemispheres.

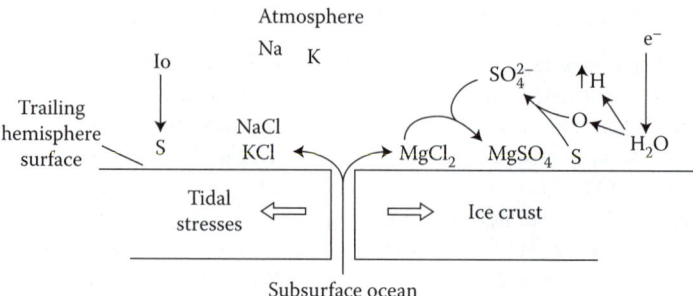

FIGURE 11.11 A model in which $MgSO_4$ is formed from endogenous $MgCl_2$ and sulfur from Io explains Europa's asymmetrical distribution of $MgSO_4$.

11.4.4.3 Low Hydrogen Peroxide Abundance May Limit Oxidation of the Ocean

The atmosphere of Europa contains oxygen, and hydrogen peroxide (H_2O_2) has been identified on the surface. These, and other oxidants, are produced by radiolytic splitting of water.

Because of the synchronous rotation of Europa it is the trailing hemisphere that gets the brunt of the radiation. In 1999, infrared spectroscopy by the Galileo probe detected hydrogen peroxide in the anti-Jovian region of the leading hemisphere at an abundance of 0.13% by number relative to water. Data from other regions was too noisy to be reliable. Much more recent (2013) data from an IR spectrograph on Keck II shows that the trailing hemisphere has virtually no peroxide. This surprising finding is because despite the higher radiation flux on the trailing hemisphere, there is less water ice at its surface from which to generate the peroxide, because it is covered by nonwater ice. The outcome is that globally the peroxide abundance is only 0.044% by number relative to water, lower than previously assumed. This has astrobiological consequences as we shall see.

11.4.4.4 How Effectively Do the Surface and Ocean Communicate?

On Earth, a volume of water equivalent to the entire oceans circulates through hydrothermal systems at mid-ocean ridges in ~10 Myr and would equilibrate with basaltic magmas on this timescale if the ocean were a closed system, perhaps during a major "snowball Earth" glaciation. This has been used to argue that the Europan ocean would equilibrate on a similar timescale if it were a closed system. But equilibrium is anathema to life. To maintain the redox gradients and energy flows, the needs for life, the ocean has to be an open system. Hence there is the desire to know how efficiently the oxidants and organic molecules produced by radiolysis of CO_2 ice formed on the surface find their way into the ocean.

EXERCISE 11.17

What factors suggest that the timescale for equilibration of a closed Europan ocean might not necessarily be the same as the 10 Myr quoted for the Earth's oceans?

The rate at which the oxidants produced at the surface are transferred to the liquid ocean depends on the turnover of oxidant at the surface (i.e., the balance between the yield of oxidant from radiolysis and its destruction by radiolysis and UV), how much of the initial yield is lost by sputtering of the surface, the effectiveness of impact gardening in getting oxidants buried beneath the surface, and finally the rate at which convection and tectonic processes (tidal stresses) overturn the crust.

Impact gardening is a vertical mixing of a planetary surface by micrometeorite and meteorite impacts. It churns up the surface making regolith. Based on the mass flux of meteoroids at Jupiter it is estimated that impact gardening can mix radiolysis products down to a depth of 1.3 m in 10 Myr. Of course, if the sputtering rate is higher than this, then the surface will be removed faster than it can be gardened inward. Estimates for sputtering rates range from 0.02 to 2 μm yr^{-1} with the lowest rate being the current best guess.

EXERCISE 11.18

Comment on the likelihood of oxidants and organic molecules reaching the ocean if the sputtering rate is 2 μm yr^{-1}.

How efficiently radiolysis products buried in the upper few meters of the crust are transported to the ocean depends on whether the crust is conductive or convective. Some models suggest that the solid ice shell of Europa consists of two layers: an upper, cold, conductive layer (a few km thick) in which deformation occurs by brittle processes and a deeper warmer region in which ductile deformation predominates. This lower layer may allow solid-state convection if there is sufficient tidal heating. Hence, if chaos terrain formation got radiolysis products through the upper layer they might then be convected down through the lower layer into ocean beneath. Much work is needed to see whether this is feasible.

11.4.5 What Is the Astrobiological Potential of Europa?

Despite the huge uncertainties, attempts have been made to quantify potential delivery of oxidants and organics to the putative Europan ocean and hence to estimate the biomass that might be sustained. Unfortunately, these estimates vary by six orders of magnitude!

11.4.5.1 A Reducing Ocean Would Be a Sink for Oxidants

Once oxidants reach the ocean their fate would rest on how reducing the ocean is. If Europa has hydrothermal systems erupting reductants—for example, H_2, CH_4, and Fe (II)—these would act as a sink for the oxygen. In this situation whether the ocean has any free oxygen depends on the relative balance between hydrothermal system activity and O_2 delivery. Take, for example, the much lower recent estimate for peroxide abundance (Section 11.4.4.3). This reduces one low-end estimate for peroxide delivered to the ocean per year from ~10^9 moles to ~10^8 moles. Compared to models for seafloor production of reductants, such as methane and hydrogen sulfide pumped into the ocean each year of about 3×10^9 moles, it appears that the new results for peroxide on Europa could imply an ocean that was completely anoxic.

One escape route from this dismal scenario is if the ocean did not mix well. Stratification of the ocean might establish a redox gradient between top and bottom water which could drive bioenergetic processes.

11.4.5.2 Could H_2O_2 Support Biology?

In 2001, Christopher Chyba, now at Princeton University, and Cynthia Phillips, at the SETI Institute, Stanford University, calculated that $\sim 8 \times 10^{10}$ kg of formaldehyde (HCHO) and $\sim 7 \times 10^{14}$ kg of H_2O_2 would enter the Europan ocean every 10 Myr. Once in the ocean, in the absence of a catalyst the peroxide would dissociate with a half-life of 10 years at 273 K:

$$2H_2O_2 \rightarrow 2H_2O + O_2. \tag{11.22}$$

It is worth putting this in perspective. Assuming there are no oxygen sinks it would take up to 200 Myr to oxygenate the entire Europan ocean to a concentration of 20 μm, equivalent to levels found in deep oceans on Earth and which is capable of supporting life with modest oxygen demands.

Chyba and Phillips speculated that a Europan organism may have evolved to use formaldehyde as a carbon source, just as the terrestrial bacterium *Hyphomicrobium* does:

$$HCHO + O_2 \rightarrow H_2O + CO_2. \tag{11.23}$$

Calculations that assume the number of organisms is limited either by available carbon or energy, and based on Galileo data for peroxide abundance, show that the ocean could support 10^{23}–10^{24} prokaryote-like cells. For comparison, one 1998 estimate for the number of bacteria on Earth is 5×10^{30}.

Other calculations, by Eric Gaidos, Kenneth Nealson, and Joseph Kirschvink, lead to the conclusion that the energy supplied by transport of oxidants into the ocean are 1000-fold lower than those needed to support terrestrial ecosystems, leading these researchers to reject this as a mechanism that could sustain life in Europa's ocean.

11.4.5.3 Energy for Hydrothermal Vents

Global heat flow of Europa currently is thought to be in the same ball park as average heat flow for Earth (0.089 W m^{-2}), but on Earth the heat source is largely decay of radioisotopes in the silicate fraction while on Europa it is tidal dissipation, much of which heats the ice shell rather than the silicate mantle. It is not possible to say with any certainty whether there is enough heating of the mantle to produce magmatism or hydrothermal circulation within Europa's ocean floor at present, but several researchers suggest it is plausible. Hydrothermal activity would have been greater for early Europa when radiogenic heating was higher.

The much lower acceleration due to gravity on Europa compared to Earth (1.31 m s^{-2} compared with 9.8 m s^{-2}) means that buoyancy-driven hydrothermal flows would be an order of magnitude less on Europa than Earth, other factors being equal, so heat transport through Europan hydrothermal systems would be less. It has been proposed that in consequence, compared to Earth, hydrothermal activity on Europa would be characterized by more numerous vents, each with lower heat flux.

11.4.5.4 What Is the Oxidation State of the Europa Mantle?

We do not know what the oxidation state of Europa's mantle is. Because it now seems that $MgSO_4$ at the surface of Europa does not reflect a sulfate-rich ocean, the case for an oxidized mantle is not as strong as it once was, and a reduced ocean seems plausible.

In a closed ocean, might it still be possible for local redox gradients to exist at hydrothermal vents that could provide enough free energy for life? On Earth, hydrothermal alteration of the upper crust and mantle oxidizes iron to produce hydrogen. The relatively oxidized state of the mantle means that carbon is released in the form of CO_2. The H_2 and CO_2 are harnessed by methanogens. But this might not be possible on Europa if the mantle is reducing because it has not been oxidized. In this situation, hydrothermal vents would outgas hydrogen, but carbon would be in the form of CH_4. This implies no methanogens. However, there might be a "get-out" clause. Thomas McCollum has shown that if the temperature of the vent fluids were high enough (>425°C) fluid–rock interactions can oxidize methane to CO_2 while reducing ferric to ferrous iron. He has further shown that there is ample potential metabolic energy available from methanogenesis above 6.5°C, although biomass production would be 10^8–10^9-fold less than that made by terrestrial photoautotrophs.

Sulfur would be vented from a reduced mantle in the form of H_2S rather than S^0 or SO_2, which would seem to rule out organisms that live by sulfur reduction.

Given a reduced mantle, Gaidos and colleagues argue that low temperature reduction of an oxidized metal such as Fe^{3+} by hydrogen, methane, or sulfide would be thermodynamically favorable. Iron-reducing bacteria exist on Earth, including some that use hydrogen as a reductant. Perhaps an organism like this could live on Europa.

12

Detecting Exoplanets

12.1 Exoplanet Bonanza

Over the past 2 decades, almost 2000 planets have been found orbiting other stars. The first discoveries were generally large gas giants close to their host stars because these were the easiest to detect. However, the sensitivity and resolution of detection methods has rapidly improved, so we are now discovering worlds that are comparable in size to Neptune, the Earth, and even Mercury. Moreover, exoplanets have been detected to a distance of 2×10^4 light years (ly), and one detection technique offers the potential to discover exoplanets in other galaxies.

The Holy Grail for astrobiologists is to find Earth-mass rocky planets orbiting in the habitable zone of a sun-like star.

12.1.1 The First Exoplanets Were Found by Timing Pulsars

The first two confirmed exoplanets were reported in 1992 by Aleksander Wolszczan and Dale Frail. The exoplanets were in orbit around a millisecond pulsar, 2000 light years away in Virgo. Wolszczan had discovered the pulsar, PSR B1257 + 12, in 1990 using the Arecibo radio telescope, but there were puzzling irregularities in the timing of its radio pulses. Analysis revealed the presence of two planetary mass objects orbiting 0.36 and 0.46 AU from the pulsar.

12.1.2 The First Exoplanet Around a Main Sequence Star Was Discovered in 1995

The star, 51 Pegasi, is a naked eye sun-like (G5V) star 50 ly away. It shot to fame in 1995 when it became the first main sequence star found to be orbited by an exoplanet. It was detected by Michel Mayor and Didier Queloz at the Observatoire de Haute-Provence with the radial velocity technique and confirmed just six days later in the same way by Geoffrey Marcy and Paul Butler working at the Lick Observatory in California. This was just the start. These four astronomers went on to make many of the early exoplanet discoveries.

The rapid confirmation of 51 Pegasi b (Box 12.1) was possible because the exoplanet was orbiting extremely close to its star (0.053 AU) so that its period was very short (4.2 days). With a minimum mass around half that of Jupiter it was clearly a gas giant but with a surface temperature around 1200°C; the first hot Jupiter. This was surprising at the time since theory says that gas giants cannot form so close to a star.

BOX 12.1   NAMING EXOPLANETS

The nomenclature for naming exoplanets is baroque. The first exoplanet dis-covered going round a star is designated "b" (the star itself is "a") and subse-quent ones are "c," "d," "e," ..., etc. This will be prefixed by the name of the star (e.g., 70 Virginis), or its catalogue number (e.g., HD80806), or the name of the discovery instrument followed by a number (e.g., WASP-4). If several exoplanets are discovered orbiting a star at the same time, lowercase letters are given in the order of distance from the star; closest is "b," next is "c," and so on. Hence, f is the most distant in the Kepler-62 multiple system of five exoplanets.

In a clash between these two systems, the order of discovery trumps distance from the planet. Thus, the latest discovery in the 55 Cancri system is termed 55 Cancri f even though it lies between the earlier discovered 55 Cancri c and 55 Cancri d. This means that subsequent discoveries do not require earlier exo-planets to be relettered.

Kepler has yielded a huge number of candidate planets that are designated as KOIs (Kepler Objects of Interest). Thus, Kepler-62f was originally designated KOI-701.04.

EXERCISE 12.1

By assuming that 51 Pegasi is one solar mass show that a reasonable estimate for the period of the orbit of 51 Pegasi b is 4.2 days.

EXERCISE 12.2

How might we now explain the location of hot Jupiter exoplanets?

Astrobiologists are most interested in exoplanets that are Earth-like, and this has been an important driver of search strategies. So, for example, surveys have generally concentrated on sun-like stars, taken to be those of spectral types F, G, or K.

12.2 Some Exoplanets Can Be Imaged Directly

You may think that the most obvious way to discover an exoplanet is to image it using a large aperture telescope. The larger the aperture of a telescope (a) the more light it can collect and the dimmer the object it can see and (b) the greater the spatial resolu-tion—the ability to distinguish two objects that are close together on the sky. But it turns out that direct imaging of exoplanets is extremely hard.

12.2.1 Planets Are Extremely Dim Compared to Their Host Stars

While it is the case that some of the furthest exoplanets (e.g., OGLE-2005-BIG-39 L b, at 28,000 ly) could not be imaged with current technology because they are too dim or too distant to be resolved, the biggest problem for imaging exoplanets is that they reflect something like one billion times less visible light than their host stars.

We can illustrate this by comparing the luminosities of Jupiter and the sun:

$$L_\odot/L_J = 3.84 \times 10^{26} \text{ W}/5.42 \times 10^{17} \text{ W} = 7.1 \times 10^8 \sim 10^9.$$

Hence, Jupiter is of order one billion times fainter than the sun.

The odds can be improved by looking for planets in emitted light. Planets absorb some of the visible light that strikes them, warming them so they emit at longer infrared wavelengths. A sun-like star gives out appreciably less IR than visible light: the planet is now typically only one million times dimmer than the star!

Despite these tremendous difficulties tens of extrasolar planets have been directly imaged, most of them by using two tricks which effectively block out the light from the host star, reducing the problem to one of detecting the very faint planet. The tricks are to use a coronagraph or nulling interferometry.

12.2.2 Coronagraphs Create Artificial Eclipses

A coronagraph eclipses a star with an occulting mask at the telescope's focal plane. Any remaining light (from exoplanets) is then reimaged by a CCD camera. An alternative design has a hole in the primary mirror of a telescope that corresponds to the star, so the image brought to prime focus has no starlight.

Gemini Planet Imager, a coronagraph on the 8-m Gemini South telescope at Cerro Pachón in Chile will detect contrasts of 1 in 10^7 at IR wavelengths. SPHERE, an instrument with three coronagraphs, is being installed at ESO's 8.2-m Very Large Telescope (VLT) at Cerro Paranal in Chile.

12.2.3 Nulling Interferometry Cuts Out Starlight

In nulling interferometry, starlight recorded by two telescopes is combined exactly $180°$ out of phase by introducing a $+\pi$ delay in one of the delay lines (Figure 12.1). This produces destructive interference and the starlight is suppressed. Any exoplanet is offset from the central star and its off-axis light takes a different optical path so that constructive interference occurs between the fringes from the two telescopes.

Provided that the starlight has been sufficiently blocked any exoplanet can be imaged. The depth and breadth of the null (i.e., how effectively the starlight is suppressed) depends on the angle subtended between the exoplanet and its host star, θ. For typical values of θ an interferometer with two telescopes does not give a null large enough but this can be overcome by having more telescopes (three, four, or even more) in the interferometer. The aim is to have a null which is broad and deep enough to cut out the starlight but not too broad to suppress any inner planets.

12.2.4 Direct Imaging Reveals High Mass Planets

As we have seen, in infrared the contrast between a star and an exoplanet will typically be about 1 million-fold, but this is a bit more favorable for exoplanets with a large surface area, plus an internal source of heat to add to the IR emitted as processed starlight. It also helps if the parent star is dim, for example, a red dwarf, and if the planet is not too close.

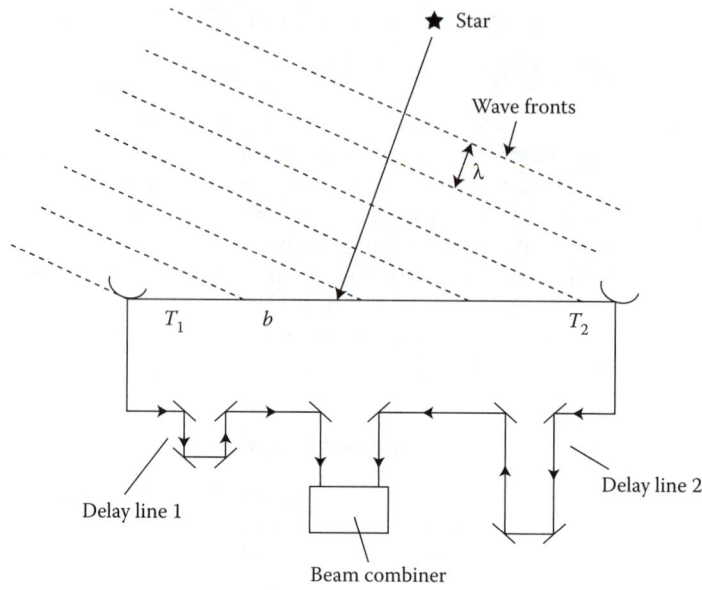

FIGURE 12.1   A two telescope interferometer.

12.3 Astrometry Detects Binary Systems by Stellar "Wobble"

In a binary system, such as an exoplanet and its star, both bodies orbit a common center-of-mass (COM). Because the mass of the planet is so much less than that of the star the planet appears to orbit the star. But the star's orbit around the COM means that it too traces out an orbit, although very much smaller than the planet's. If the system is sufficiently close the tiny shift in the position of the star in the sky can be measured directly. This is termed astrometry. Unfortunately, the positional shifts are so small that only one exoplanet has been discovered by this method to date (Figure 12.2).

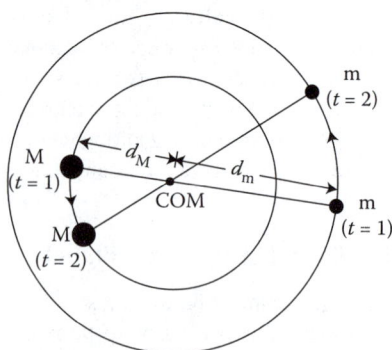

FIGURE 12.2   In a binary system, both objects have elliptical orbits around a common center-of-mass. $r = d_M + d_m$.

ESAs Gaia mission is able to plot the positions of stars to about 20 micro-arcsecond precision. It is expected to find between 10,000 and 50,000 gas giants. However, Gaia will not be able to find Earth-mass planets.

12.4 The Radial Velocity Method Uses the Doppler Effect

The radial velocity method is based on the same idea as astrometry, namely, detecting the tiny motion of a star orbiting the center of mass it shares with one or more exoplanets. But instead of trying to measure the motion directly, provided a component of the star's motion carries it periodically toward us and away from us, this radial velocity can be detected by Doppler shifting of the star's spectral lines (Box 12.2). This is plotted as a radial velocity curve; a graph of v_r versus t. This is the radial velocity (RV) method, sometimes more colloquially referred to as the Doppler wobble method.

As can be seen from Figure 12.3, the RV method provides the period of the exoplanet's orbit. From this, Kepler's third law can be used to calculate the semi-major axis of the exoplanet's orbit. It also gives a lower limit on its mass, though, as we shall see, we cannot find the mass itself. The method is currently sensitive enough to detect

---

BOX 12.2   DOPPLER EFFECT

The observed frequency, f, of a star with a radial velocity v_r, emitting radiation of frequency f_0 in its rest frame is given by:

$$f = (1 - v_r/c)f_0. \tag{12.1}$$

The equivalent relationship exists for the observed wavelength, λ, and the rest frame wavelength, λ_0:

$$\lambda = (1 - v_r/c)\lambda_0. \tag{12.2}$$

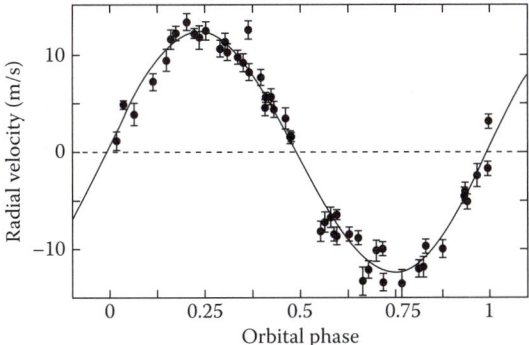

FIGURE 12.3   Gliese 581a is a red dwarf (M3V) star 20.3 ly away. In 2005, a Neptune mass planet (Gliese 581b) was discovered orbiting 0.04 AU from the star (period 5.4 days). Since then, using the radial velocity method, three other planets have been found orbiting this star.

a radial velocity as low as 1 m s⁻¹ and can tease out several exoplanets going around the same star. Not surprisingly it works best for finding high mass planets close to their star since this combination produces the largest stellar "wobble" and is revealed by short observation times. Postgraduate research grants generally run for 3 years. This makes it extremely hard to detect an exoplanet *de novo* with a period great than a year! This selection effect biased our view of the nature of exoplanets early on when most exoplanet discoveries were made by the RV method.

12.4.1 Defining the Radial Velocity

For an exoplanet, orbital inclination, i, is the angle of the plane of the orbit relative to the plane perpendicular to the line-of-sight to its star. So $i = 0°$ is a face-on orbit, and $i = 90°$ is an edge-on orbit.

If we happened to view an exoplanet-star system exactly edge-on, the velocity vector of the star would always be pointing directly toward us or away from us. In this case, the radial velocity, v_r, that we found from the Doppler shift of the star's spectral lines would be the star's total velocity, v. What about a binary system that is exactly face-on? Now the star's velocity vector has no component toward or away from us. In other words, there is no radial velocity we can measure; the RV method cannot be used. Of course, in most instances, the inclination of the star's orbit about the center of mass of the binary system is somewhere between edge-on and face-on, and we find a radial velocity lower than the total velocity by an indeterminate amount. Indeterminate because we usually have no way of knowing what the inclination is.

12.4.1.1 Determining Radial Velocity Is a Challenge

In theory, we can find the radial velocity from the shift in wavelength of the spectral lines:

$$\Delta\lambda = (\lambda - \lambda_0)/\lambda_0 = -v_r/c. \tag{12.3}$$

However, the shifts in wavelength are too small to be detected directly. Instead, the spectrograph transforms wavelength into positions, and it is the positions of spectral lines that are measured. For this to yield a reliable estimate for v_r, the spectrograph must achieve positional stability <5 nm over years, and even then, statistical modeling of the spatial locations of numerous lines (of order a thousand) is required. This constrains the method to relatively cool stars like the sun, because O, B, and A spectral type stars have too few narrow lines. Furthermore, it is necessary to correct for the Earth's orbital motion (up to 30 km s⁻¹) and rotation (0.5 km s⁻¹), and for the rotation of the star. In addition, star spots (cf. sunspots) and stellar oscillations are significant sources of noise. Getting a radial velocity curve is not a trivial task!

12.4.1.2 RV Calculations Must Sidestep the Unknown Inclination

As we have seen, because of the unknown inclination of the star's orbit we do not know how close v_r is to the true velocity. However, from trigonometry we do know that $v_r = v\cos\theta$, where θ is the angle between our line of sight and the star's velocity vector (see Figure 12.4).

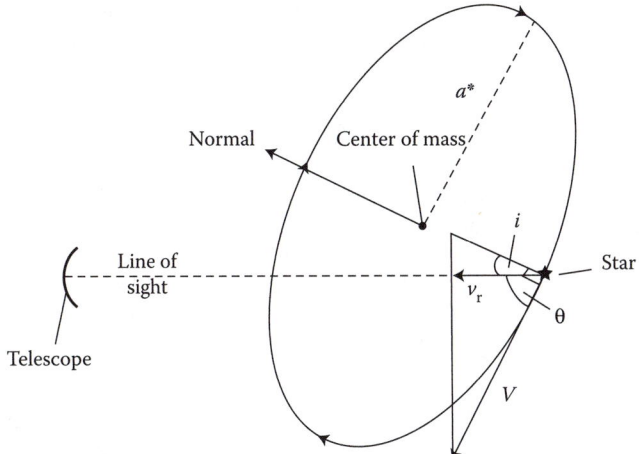

FIGURE 12.4 The orbit of a star around the center-of-mass of its planetary system. i is the inclination, the angle between our line of sight and the normal (i.e., at $90°$) to the plane of the orbit. Note that $\cos\theta = v_r/v$.

Astronomers normally work with the inclination, i, of stars' orbits rather than with θ.

$$\theta = 90 - i.$$

Now because the sine function is out of phase with the cosine function by $90°$ we can write:

$$\cos(90 - i) = \sin i.$$

Hence,

$$v_r = v\cos\theta = v\cos(90 - i) = v\sin i. \tag{12.4}$$

So, although we do not know i or v, we have defined the lower bounds on the star's velocity on the radial velocity curve; $v\sin i$ (star approaching) and $-v\sin i$ (star receding) (see Figure 12.5).

12.4.2 Finding the Period and Semi-Major Axis of an Exoplanet's Orbit

For a star with a single planet, the radial velocity graph is a sine curve from which the period is easily found. Note that the period of the planet orbiting the COM is the same

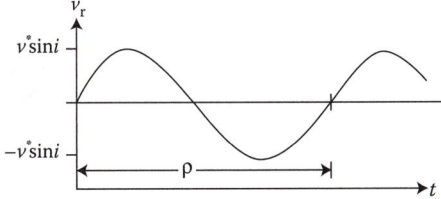

FIGURE 12.5 Radial velocity curve for a star orbited by an exoplanet.

as the period of the star orbiting the COM. Thus the distance of the exoplanet from its host star, a_P, can be found unambiguously from the period using Kepler's third law (K3) which, ignoring the negligible mass of the planet, is given by

$$a_P = (GM^*P^2/4\pi^2)^{1/3}, \tag{12.5}$$

where M^* is the mass of the star. In fact, with the appropriate choice of units (a_p in AU, P in years) K3 can be approximated to

$$a_P^3 \ (M^*/M_\odot)P^2. \tag{12.6}$$

Our next task is to find M^*.

Only if a star is a component of a binary star system can we determine its mass directly. But we can *estimate* the mass of a single star in several ways. One is to use the mass-luminosity relation, which for stars with $0.5M_\odot < M^* < 2M_\odot$ is:

$$L^*/L_\odot = (M^*/M_\odot)^4. \tag{12.7}$$

The luminosity of the star, L^*, can be found by measuring the flux of photons, F, from it:

$$L^* = 4\pi d^2 F, \tag{12.8}$$

Of course, this depends on knowing its distance, d. For distances out to ~500 pc we can currently do this to reasonable accuracy using heliocentric parallax, and with Gaia this method can be extended to $>10^4$ pc.

Once we have the mass of the star, we can obtain a_P from K3.

12.4.3 The RV Method Provides a Lower Bound on Planetary Mass

We can use the RV method to determine a lower bound on the mass of the planet because for a binary system consisting of an exoplanet orbiting a star with masses M_P and M^*, respectively, in circular orbits at distances R_P and R^* from a common COM, we can write:

$$M_P R_P = M^* R^*, \tag{12.9}$$

so,

$$M_P = R^* M^*/R_P. \tag{12.10}$$

We know M^*, but what about R_P and R^*? Let us deal with R_P first.

At any instant, the distance between the planet and the star is

$$a_P = R_P + R^*, \tag{12.11}$$

so,

$$R^* = a_P - R_P.$$

Substituting this into Equation 12.9, we get

$$M_\mathrm{P}R_\mathrm{P} = M^*(a_\mathrm{P} - R_\mathrm{P}) = a_\mathrm{P}M^* - M^*R_\mathrm{P},$$
$$M_\mathrm{P}R_\mathrm{P} + M^*R_\mathrm{P} = a_\mathrm{P}M^*,$$
$$R_\mathrm{P}(M_\mathrm{P} + M^*) = a_\mathrm{P}M^*,$$
$$R_\mathrm{P} = a_\mathrm{P}M^*/(M_\mathrm{P} + M^*).$$

But since $M_\mathrm{P} \ll M^*$ we see that to a good approximation $R_\mathrm{P} = a_\mathrm{P}$, and we can rewrite Equation 12.10 as

$$M_\mathrm{P} = R^*M^*/a_\mathrm{P}. \tag{12.12}$$

We have already worked out M^* and a_P.

Next we find R^*, the distance of the star from COM. We can get this from the period P. The distance, d, traveled by the star in one orbit $= 2\pi R^*$.
Now,

$$v = d/t = d/P, \text{ so } d = Pv.$$

Hence,

$$Pv = 2\pi R^* \quad \text{and so} \quad R^* = Pv/2\pi. \tag{12.13}$$

However, as we have seen, we do not know i or v, we only have $v\sin i$. Hence, we need to rewrite Equation 12.13 as

$$R^*\sin i = v\sin i \times P/2\pi. \tag{12.14}$$

$R^*\sin i$ is the *lower bound* on the distance of a star from the COM of the planetary system it is host to.

Finally, we have all we need to get a lower bound on the mass of the planet from Equation 12.18:

$$M_\mathrm{P} = R^*M^*/a_\mathrm{P},$$

By multiplying both sides by $\sin i$ we get,

$$M_\mathrm{P}\sin i = (R^*\sin i \times M^*)/a_\mathrm{P}. \tag{12.15}$$

12.4.4 The RV Method Is Biased to Detect Massive Planets with Short Periods

The critical constraint for the RV method is the velocity of the star as it orbits the common center-of-mass of the exoplanet–star binary system. For an exoplanet in a circular orbit in which $m_\mathrm{p}\sin i \ll M^*$, it is:

$$v = 28.4 \, [m_\mathrm{p}\sin i/(P^{1/3}M^{*2/3})], \tag{12.16}$$

where $m_\mathrm{p}\sin i$ is in Jupiter masses (M_J), P is in years and M^* is in solar masses.

The current record for stellar velocity is 0.51 m s^{-1}, measured by the HARPS spectrograph on ESOs 3.6-m telescope at La Silla Observatory in Chile. This is about the speed of a crawling baby and was determined for a star 1.3 pc away, Alpha Centauri B, around which an exoplanet has recently been discovered orbiting at a distance of 0.04 AU.

EXERCISE 12.3

Alpha Centauri B has a mass of $0.934M_\odot$. The exoplanet ($m_p\sin i = 1.13M_\oplus$) orbits with a period 3.236 days. Use Equation 12.16 to confirm the velocity of Alpha Centauri B. (Take $M_\odot = 2 \times 10^{30}$ kg, $M_\oplus = 6 \times 10^{24}$ kg, and $M_J = 1900 \times 10^{24}$ kg.)

Since astrobiologists are interested in finding Earth-mass planets in the habitable zone of sun-like stars (Earth twins), we explore the minimum mass that the RV method can detect at present.

EXERCISE 12.4

(a) What is the minimum mass of an exoplanet that can be detected orbiting 1 AU from a solar mass star assuming that a radial velocity of 0.5 m s^{-1} can be measured?

If we were prepared to consider exoplanets in habitable zones of lower mass stars we could detect even lower mass exoplanets. For example, the habitable zone of a red dwarf star with mass $0.5M_\odot$ is 0.23–0.38 AU. Both the lower mass of the star and the smaller distance of the exoplanet contribute to a larger stellar radial velocity. We could consider even lower mass stars, but their habitable zones may turn out not to be very "user friendly" as we shall see in Chapter 13.

EXERCISE 12.5

Calculate the lowest detectable $m_p\sin i$ for a planet in the habitable zone of a $0.5M_\odot$ assuming we can measure a stellar radial velocity of 0.5 m s^{-1}.

Planned third generation spectrographs will be able to measure radial velocities <0.1 m s^{-1}. One such spectrograph is EXPRESSO which will be used with the VLT from about 2016. Even more impressive will be CODEX, a spectrograph intended for the European Extremely Large Telescope (E-ELT), a 39.3-m instrument which will be built on Cerro Armazones in Chile. CODEX will be able to detect radial velocity <0.02 m s^{-1}. It is planned to be operational by the early 2020s.

EXERCISE 12.6

In this question you are going to calculate the lower bound for the mass of the exoplanet HR 18033b from data provided by the radial velocity method. This exoplanet is in orbit around HR 18033a, a G5V spectral class star with $L^* = 1.30L_\odot$ (Take $L_\odot = 3.84 \times 10^{26}$ W, $M_\odot = 1.99 \times 10^{30}$ kg, and $M_J = 1.89 \times 10^{27}$ kg.)

    a. From Figure 12.6 write down: (i) the period, P, for the orbit of HR 18033a, (ii) the maximum radial velocity of HR 18033.
    b. Calculate $a^*\sin i$.
    c. Use the mass–luminosity relation, $L^*/L_\odot = (M^*/M_\odot)^3$ to find the mass of HR 18033a.
    d. Use Kepler's third law to find a_p.
    e. Calculate $m_p\sin i$.
    f. Comment briefly on the likely nature of HR 18033b.

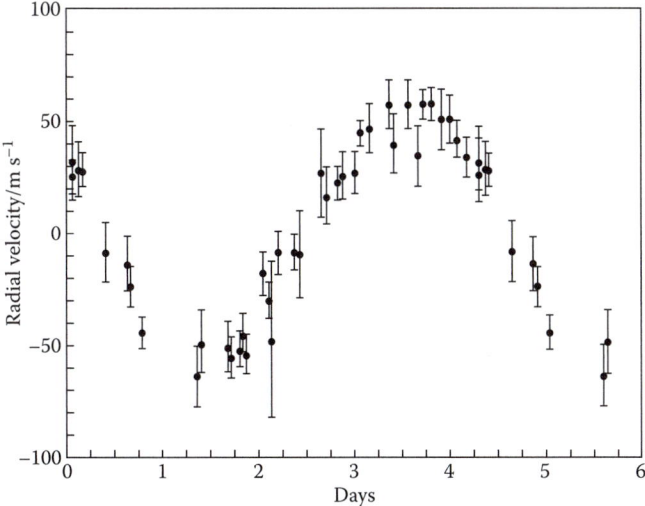

FIGURE 12.6   Radial velocity measurements for HR 18033a over about one orbit.

12.5 Exoplanets Are Revealed When They Transit Their Star

The transit method takes advantage of the fact that some planets periodically pass in front of their host star as viewed from Earth. These events are called transits. Extrasolar planets are far too distant to be seen as discs, but periodic dimming of the star is a telltale signature of a transit (Figure 12.7).

12.5.1 Exoplanet Transits Dim Starlight

The occultation depth, the fractional drop in the light of the star by a transiting planet, $\Delta F_P/F$, is:

$$\Delta F_P/F = (R_P/R*)^2, \tag{12.17}$$

where F is the flux from the star, ΔF_P is the change in flux during the transit; R_P and $R*$ are the radii of the planet and star, respectively. This equation assumes that the stellar disc is of uniform brightness, but this is not true. The brightness of the disc falls off smoothly toward the edge, an effect called limb darkening. It is seen most clearly in the images of the sun. It is caused by the fact that toward the edge of the disc the photons we see are coming from shallower regions in the photosphere—a geometrical line of sight effect—and so are coming from cooler gas. (There is a limit to the depth we can see into the photosphere; see Box 12.3.)

In practice, to find the radius of a planet transit curve are fit to models which take account of the star's mass, radius and limb darkening.

For Jupiter transiting the sun we see that

$$\Delta F_J/F = (R_J/R_\odot)^2 \sim (70{,}000/700{,}000)^2 = 0.1^2 = 0.01, \text{ or } 1\%.$$

This is well within the detection limits of current technology.

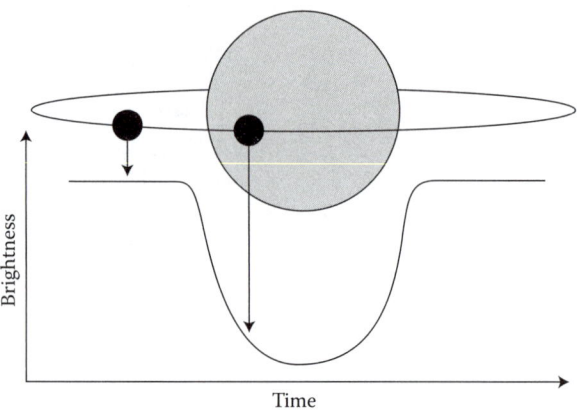

FIGURE 12.7   Light curve of an exoplanet transit.

EXERCISE 12.7

Calculate the percentage change in the sun's brightness caused by the transit of an Earth-sized planet. Take $R_{\oplus} = 6.4 \times 10^3$ km.

For an exoplanet with a thick atmosphere the occultation depth also depends on wavelength. This is because the radial extent of the planet is determined by the height of the atmosphere at which the optical depth reaches 1. This is determined by the structure and composition of the atmosphere and depends on the wavelength.

BOX 12.3   OPTICAL DEPTH

Optical depth, τ, is a measure of the extent to which radiation is attenuated by scattering or absorption as it passes through a medium, such as the photosphere of a star, the interstellar medium or a planetary atmosphere. It is given by

$$I/I_0 = e^{-\tau}, \tag{12.18}$$

where I_0 is the intensity of radiation entering the medium and I is the intensity of the radiation leaving it. The amount of light transmitted drops exponentially as the optical depth increases:

$$I = I_0 e^{-\tau}. \tag{12.19}$$

If the optical depth is low, the medium transmits radiation easily and is said to be optically thin. If the optical depth of a medium is high, it is optically thick.

For an optical depth, τ, of unity, $I = I_0 e^{-1} = 0.37 I_0$; only 37% of the original radiation is transmitted. It is usual to argue that we see to a depth in the atmosphere where optical depth is more or less unity. The justification for this is that for $\tau \ll 1$, $e^{-\tau} \sim (1 - \tau)$, so Equation 12.19 becomes $I = I_0(1 - \tau)$. Obviously, when $\tau = 1$, $I = 0$.

Different frequencies of radiation are attenuated by different amounts; that is, optical depth depends on the wavelength. Thus, the optical depth of the Earth's atmosphere is greater for blue light than for red light because blue light suffers more Rayleigh scattering. That is why the sun on the horizon at dusk is not only dimmer but also redder.

Optical depth is determined by three factors:

1. The absorption coefficient of the material, κ. This depends on the wavelength. For IR light in the Earth's atmosphere, greenhouse gases have the highest coefficients, while N_2 and O_2 have lower coefficients. At UV wavelengths, O_2 has the highest coefficient.
2. The density of the medium, ρ
3. The path length, s

Hence $\tau = \kappa \rho s$.

The transit method only works if the orbital plane of the planet is edge-on with our line of sight. The probability, $P(T)$, of this happening, assuming random orientations of exoplanet orbits and ignoring eccentricity, depends only on the radius of the star and the semi-major axis of the planet's orbit, a_p.

$$P(T) = R*/a_p. \tag{12.20}$$

For Jupiter:

$$P(T) = R_\odot/a_J = 7 \times 10^5 \text{ km}/5.2 \times 1.5 \times 10^8 \text{ km} \sim 9 \times 10^{-4} \sim 0.1\%.$$

EXERCISE 12.8

What is the probability of a transit of a sun-like star by an exoplanet at 1 AU, assuming the orientations of planetary orbits are randomly distributed.

It is easy to see that the smaller the orbit of an exoplanet, the greater the chance of seeing a transit, so the method is biased to detect short period exoplanets.

12.5.2 The Transit Method Allows Several Parameters to Be Deduced

Much can be gleaned about an exoplanet from the transit method that cannot be worked out by other methods. The combination of transit method and radial velocity method applied to a single exoplanet is particularly powerful.

12.5.2.1 The Period Gives the Semi-Major Axis of the Exoplanet's Orbit

Kepler's third law can be used to find the semi-major axis of the exoplanet's orbit, assuming we know the mass of the star. In most cases, the mass of the planet is small in comparison to the star's mass* and can be ignored. In this case, we can use Equation 12.5.

EXERCISE 12.9

HD209458b orbits a G0V star of mass = $1.13M_\odot$ with a period of 3.525 days. The planet has $M = 0.71M_J$ and $R = 1.35R_J$. (a) Calculate the semi-major axis of this planet's orbit. Express your result in AU. (b) Comment on the nature of HD209458b.

12.5.2.2 Transits Allow Mass to Be Refined

The transit method allows the mass of the planet to be estimated with greater precision. The semi-major axis of the exoplanet's orbit can be deduced from its period using Kepler's third law, provided we have an estimate for the mass of the star. Although the orbit may not be *exactly* edge-on to get a transit, so strictly we only have $M_P\sin i$, the inclination can often be deduced from the shape of the light curve so M_P can be calculated. From its mass the surface gravity is given, and this will tell us the sort of atmosphere it is likely to be able to retain.

Finding the inclination of the exoplanet's orbit using the transit method requires the transit duration, T_{dur}, the time for which the occultation depth is maximal and the ingress and egress time, the intervals during which the dimming and brightening occur as the planet encroaches on and leaves the disc of the star. An exoplanet with an exactly edge-on orbit with respect to the Earth will pass across the diameter of the stellar disc. By contrast, an exoplanet in an inclined orbit will transit above or below the diameter so will have a shorter transit but longer ingress and egress times. Hence, the inclination can be calculated from the transit curve.

T_{dur} is derived as follows:

1. The sky-projected distance between the center of the star and the center of the planetary disc at conjunction (when the straight line, star–planet-Earth, can be drawn) is the impact parameter, b.

   For a circular orbit of radius a (Figure 12.8a):

   $$b = a\cos i.$$

2. The sky-projected distance of the transit is 2ℓ. The distance ℓ is given by Pythagoras's theorem (Figure 12.8b):

   $$\ell = [(R^* + R_P)^2 - b^2]^{1/2}. \tag{12.21}$$

3. The distance around the entire orbit (assuming it is circular) is $2\pi a$. During the transit the planet will move along part of this orbit with arc length αa (Figure 12.8c) which corresponds to the sky-projected transit distance 2ℓ.

Hence, $\sin(\alpha/2) = \ell/a$. From Equation 12.21 we get

$$T_{dur} = P\alpha/2\pi = (P/\pi)\sin^{-1}(\ell/a) = (P/\pi)\sin^{-1}\{[(R^* + R_P)^2 - b^2]^{1/2}/a\}. \tag{12.22}$$

If the radial velocity method is used for an exoplanet that is transiting its parent star, then the inclination of the orbit can be deduced from the radial velocity curve.

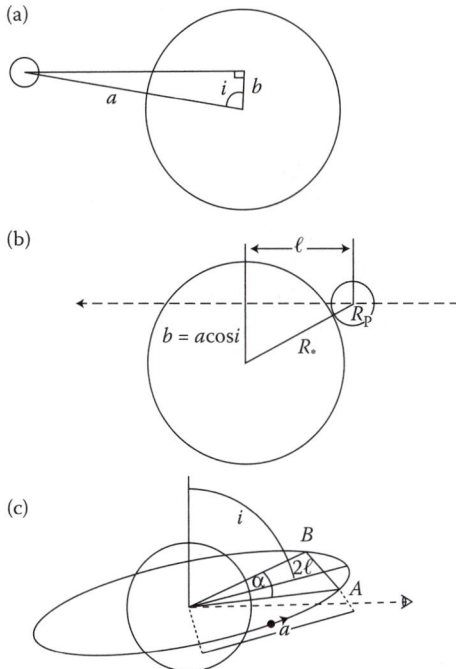

FIGURE 12.8 (a) Impact parameter, b, is a function of the inclination. (b) ℓ is half the sky-projected transit. (c) During the transit, the planet will move through angle α (in radians) carrying it a sky-projected distance of 2ℓ.

This depends on the Rossiter–McLaughlin (RM) effect (see Figure 12.9). Stars rotate. Light from a rotating star is blue-shifted on the approaching side, and red-shifted on the receding side. As a planet with zero inclination transits the star, if it is in a prograde orbit it will occlude first a fraction of the blue-shifted light then the same fraction of red-shifted light. This will produce a symmetrical variation in the radial velocity curve. For a planet in an inclined orbit the fraction of blue-shifted and red-shifted light is not the same, and this produces an asymmetrical variation in the radial

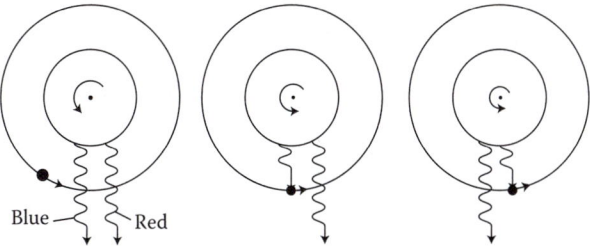

FIGURE 12.9 The transit produces a small blip (the RM effect) on the radial velocity curve, from which the inclination, i, of an exoplanet's orbit can be deduced. The orbiting planet (black disc) blocks out the blue-shifted light before the red-shifted light. When $i = 0°$, the modulation to the radial velocity curve is symmetrical. At $i = 30°$, less blue light is occluded early in the transit and more red light is occluded later in the transit. For $i = 60°$, there is no interruption of blue-shifted light.

velocity curve. The greater the inclination, the larger this asymmetry will be. Of course, the technique will also reveal that a planet is in a retrograde orbit.

12.5.2.3 Exoplanet Radius, Volume, and Bulk Density Can Be Found

The radius of the planet, R_P, can be found from the fractional drop in light (Equation 12.17). The radius of the star will be estimated by stellar modeling based on its spectrum. For a main sequence star, this is easy to appreciate from the H–R diagram (Chapter 1). From R_P we can calculate the planet's volume and with its mass we then can find its bulk density. This allows intelligent guesses to be made about the composition of the planet; that is, whether it is a terrestrial planet, a gas giant, or an ice giant.

12.5.2.4 Exoplanetary Temperature and Atmosphere Composition May Be Revealed

In some instances it may be possible to gain the spectroscopic signature of a planetary atmosphere. When the planet is in transit, a small fraction of the star's light will be transmitted through the atmosphere of the planet. So, in essence, the approach is to subtract the spectrum of the star alone from the spectrum of the star during the transit. The residual signature is the spectrum of the planet's atmosphere. In addition, it is possible to study the radiation from the planet itself. This makes use of the fact that periodically the planet will disappear *behind* the star. This is termed a secondary eclipse. If the starlight during the secondary eclipse is subtracted from the light just before or just after the eclipse only the signature of the planet is left. The radiation from the planet can enable its temperature to be deduced. However, not all transiting exoplanets have a secondary eclipse because their orbits are eccentric and oriented significantly away from edge-on as viewed from Earth.

> **EXERCISE 12.10**
>
> Draw sketches to show the configurations of star and exoplanet orbit as seen from Earth that will give: (i) a transit and a secondary eclipse, (ii) a transit without a secondary eclipse.

12.5.2.5 Exoplanet Albedo Can Be Estimated

At visible wavelengths the starlight reflected from the planetary surface (e.g., by clouds) becomes a dominant contribution to the occultation depth whereby

$$\Delta F_P / F \sim A(R_P/a)^2, \tag{12.23}$$

where A is the geometric albedo. Hence, determining R_P/a can yield the albedo.

12.5.2.6 False Positives Arise with the Transit Method

A downside of the transit method is that it may generate a substantial number of false positives. This is because similar light curves can be obtained from a variety of other phenomena; for example, star spots rotating into and out of view, or eclipsing binary

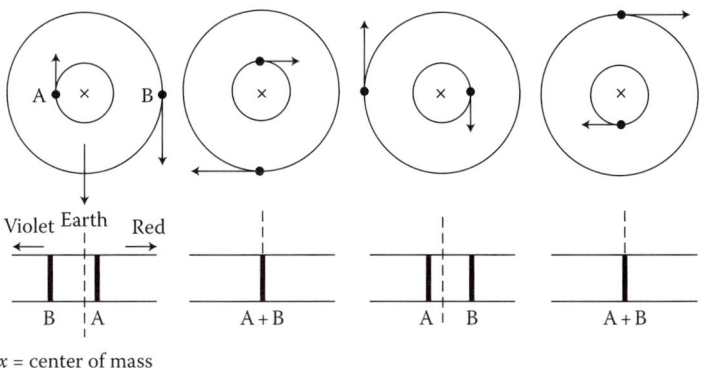

x = center of mass

FIGURE 12.10 Eclipsing binary systems can be detected by spectroscopy.

star systems. Hence, normally three successive transits are required to claim a *candidate* planet, and transit detections must then be confirmed by a different method; in practice, this is generally the radial velocity method. This also provides an independent determination of the planet's mass. Eclipsing binary stars can be distinguished by spectroscopy. As the two stars orbit, the spectral lines of each are alternately Doppler shifted toward blue then toward red as they move radially toward and then away from the Earth. This reveals itself as periodic splitting of the spectral lines from the binary system (Figure 12.10).

12.5.3 Transit Timing Variation Uncovers Multiple and Circumbinary Exoplanets

If several planets orbit a star, each perturbs the orbit of the others. Small variations in the timing of a transit can therefore suggest the presence of another planet, even if it does not itself transit. The technique is called transit timing variation (TTV). In a similar manner, variations in the timing of eclipses of an eclipsing binary system can reveal the presence of a planet that orbits both stars. Several circumbinary exoplanets (exoplanets that orbit binary star systems) have been discovered in this way.

12.5.4 Transits Can Potentially Detect Earth-Mass Planets

Our ability to detect transiting planets depends on how small a dip in the light curve of the star we can detect. From Equation 12.23 a planet's radius is:

$$R_P = (\Delta F_P/F)^{\frac{1}{2}} R^*. \tag{12.24}$$

If we assume that a planet has the same density throughout, then its mass will be proportional to volume and so to the cube of its radius. Thus,

$$m_P = \rho \times 4/3 \times \pi R_P^3 = \pi \times 4/3 \times [(\Delta F_P/F)^{1/2} R^*]^3 = 4/3\pi\rho(\Delta F_P/F)^{3/2} R^{*3}. \tag{12.25}$$

To explore a ballpark minimum mass for a planet to be detectable by the transit method currently, let us assume we can measure f_P of order 0.01% from a space observatory. With a star having radius $R_\odot$ and a planet the same density as Earth $(\rho_\oplus = 5520 \text{ kg m}^{-3})$ this gives a mass:

$$m_P = 4/3\pi \times 5520 \text{ kg m}^{-3} (0.0001)^{3/2} (7 \times 10^8 \text{ m})^3$$
$$= 7.9 \times 10^{24} \text{ kg(cf.} M_\oplus \sim 6 \times 10^{24} \text{ kg).}$$

So, transits can detect Earth-mass planets.

EXERCISE 12.11

KOI-961.03 is a Mars-sized planet $(R_P = 0.57R_\oplus)$, one of three small planets, in orbit around a red-dwarf star with a radius one-sixth that of the sun.

    a. What fraction of the star's light was blocked by the transits of the planet?

    b. Estimate the mass of KOI-961.03 assuming it is a terrestrial planet with an average density of 4000 kg m^{-3}.

    c. The system lies 38.7 pc away. What angle would the orbit of KOI-961.03 around its star (0.0154 AU) subtend on the sky if it were face-on?

    d. Calculate the period P (in years) of the planet's orbit from $P^2 = a^3/M*$ assuming the mass of the red dwarf $M* = 0.13M_\odot$. Convert the period in years to days.

12.5.5 Most Transit Detections Have Been Made from Space

Both ground-based and orbiting telescopes have used the transit method to hunt for exoplanets, but the great majority of discoveries have been made by the orbiting observatories which can detect transits as small as 0.01% deep. Here we focus on Kepler.

12.5.5.1 Kepler Had 961 Confirmed Exoplanets by April 2014

Kepler has been a very successful planet hunter with 3845 candidates (KOIs) awaiting confirmation in April 2014, of which only 5% to 10% are likely to be false positives.

Kepler trails the Earth in its orbit in such a way that sunlight never enters the instrument. It monitors the brightness of 145,000 main sequence stars continuously in a fixed 12° field of view which straddles the constellations Cygnus, Lyra, and Draco. This region, while at the edge of the Milky Way, avoids dust lanes and so minimizes extinction of starlight, and is well above the ecliptic plane, so false positives due to transits by asteroids or Kuiper belt objects are highly unlikely.

Kepler is a Schmidt camera with a 1.4-m primary mirror but an aperture of 0.95-m imposed by the diameter of the Schmidt corrector plate. A 95 megapixel CCD camera at the focal plane acts as a photometer. The system collects far more data than can be relayed back so data from only about 5% of the pixels—mostly those monitoring the stars of interest—is selected.

In May 2013, the failure of the second of four reaction wheels prevented the spacecraft from accurate pointing, but there is still a huge amount of data awaiting analysis; hence, Kepler exoplanets will continue to be discovered long after the mission is terminated.

12.6 Gravitational Lensing Can Unveil Exoplanets

The General Theory of Relativity predicts that light is bent by massive objects. This means that galaxies or galaxy clusters can act as gravitational lenses, bending the light of much more distant light sources, such as quasars or galaxy clusters, so as to produce magnified, though distorted, images of the source. These lensing effects generally last for millions of years because both source and lens are so far away that the alignment is maintained over long time scales. The lenses are massive enough that the source images and the lensing object can be resolved by large aperture telescopes (Figure 12.11).

12.6.1 Gravitational Microlensing Events Are Short-Lived

If the lens has a low mass (e.g., a star) the displacement of light by the lens cannot be resolved, although the source will be magnified by the lens, producing a brightening and dimming of the source. These microlensing events are generally transient, lasting only seconds to months, because the lensing objects are so close (a few kpc) that the alignment occurs only briefly.

For a point source and a single lensing star the light curve produced by the system going into and out of alignment is a symmetrical bell curve (see Figure 12.12). If the lens is a binary system—such as a star with an exoplanet—a much more complex light curve results, the shape and amplitude of which depends on the geometry of the lensing system and the geometry of the binary.

12.6.2 Microlensing Yields Information About Exoplanet Systems

The theory of lensing says that if a point source and single lens star are perfectly aligned with respect to an observer, the image of the source is maximally magnified

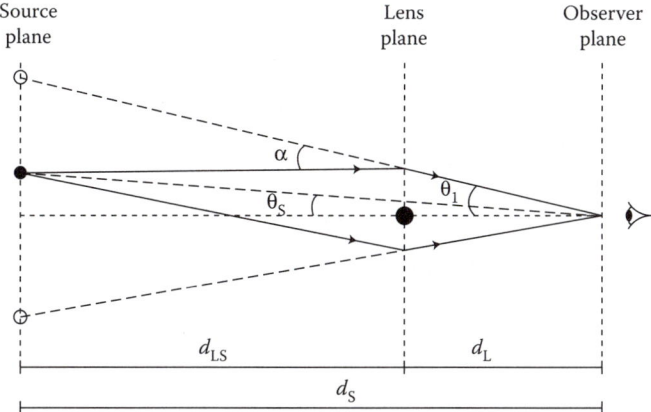

FIGURE 12.11 The principle of gravitational lensing. α is the bending angle, θ_S is the actual angle subtended (without lens effect) by the source at the observer, θ_1 is the observed angle subtended (due to lens effect) by the apparent source at the observer, d_L is the angular distance to the lens, d_S is the angular distance to the source, and d_{LS} is the angular distance between the lens and the source.

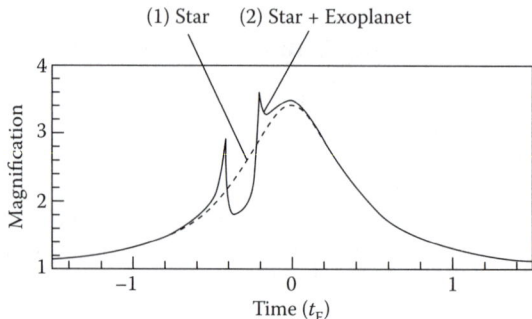

FIGURE 12.12   Microlensing by a star–exoplanet binary.

and stretched into a ring, termed an Einstein ring, with radius θ, the Einstein radius θ_E, in radians:

$$\theta_E = \{(4GM/c^2) \times [(d_S - d_L)/(d_L d_S)]\}^{\frac{1}{2}}, \qquad (12.26)$$

where G is the gravitational constant, M is the mass of the lensing star, d_S is the distance of the source star from the observer and d_L is the distance of the lensing star from the observer. θ_E obviously contains information about the lensing system which we would like to know. The problem is that for microlensing events this ring is too tiny to be resolved.

EXERCISE 12.12

By calculating a value for θ_E (in arcseconds) show that it is too small to be resolved under typical gravitational microlensing conditions; that is, for a solar mass lensing star at 1000 pc and $d_S = 2000$ pc. (Take $G = 6.67 \times 10^{-11}$ N m^2 kg^{-2}, $M_\odot = 2 \times 10^{30}$ kg; 1 pc $= 3.086 \times 10^{16}$ m; 1 rad $= 57.2957795°$.)

The result of Exercise 12.12 shows that θ_E is far too small to be resolved. But its size is, in theory, encoded in the duration of the brightening of the light curve. This is just twice the time it takes for source to cross the Einstein ring. Hence, the Einstein ring radius time, t_E, is given by

$$t_E = \theta_E/\mu \qquad (12.27)$$

The Einstein time t_E depends on the relative proper motion, μ, of the lens with respect to the source. Of course, t_E is still a degenerate combination of parameters. Thus, the proper motion depends on the relative lens–source parallax which depends on the relative lens–source distance. Values for d_S and d_L can be deduced from the spectra if the light of the lens can be isolated from that of the source. This allows μ, θ_E, and the mass of the lensing star to be found.

Of course perfect alignment is extremely rare, but the maximum magnification is a function of the displacement of the lens from the source at the closest approach, and can be read directly from the light curve. For an imperfect alignment, at closest approach there are two arc-like images of the source, one outside the Einstein ring, the major image, and another one on the opposite side and inside the ring. If an exoplanet is close to the major image as it sweeps past during the lensing event then the image will be further perturbed yielding a short blip of duration that is proportional to $t_{E,p}$, the crossing time of the Einstein ring radius of the planet and:

$$t_{E,p} = q^{1/2}t_E \qquad (12.28)$$

where $q = m_p/M^*$, the mass ratio of the exoplanet and its star. If the mass of the star has been deduced from θ_E, then we have the mass of the exoplanet. This is possible in about half of cases. Generally, q will be small, which means that the lensing event will be very brief. A lensing event may last only a few hours for an Earth-mass planet or 4 to 5 days for a Jupiter-mass planet.

From the time of the planetary blip and its magnitude the angular separation between the exoplanet and star on the sky at the time of the lensing event can be worked out, which places a lower limit on the size of the orbit.

In practice, considerable modeling is needed to tease out the geometry of the lensing system and the characteristics of the exoplanet–star binary system. Matters are complicated by the fact that the alignment of lensing system and source star must get close to particular regions within the Einstein ring, known as caustics, for the deviation in the light curve due to the exoplanet to be noticeable. The caustics are only a small area, so the chance of spotting an exoplanet by microlensing is small, even if it is there!

12.6.3 There Are Both Pros and Cons to Lensing

Gravitational microlensing works best for systems 3–6 kpc away, distances that are taxing for other methods. It has the advantage that it is most sensitive to exoplanets 1–5 AU from their host stars. It also has a high sensitivity to low mass exoplanets. It is the only method that can potentially discover an Earth-mass exoplanet from the ground. It is also the only method capable of detecting exoplanets in other galaxies.

However, it has serious disadvantages. Microlensing events are extremely rare because they depend on highly improbable precise alignments between the source and lens that are relatively close to us. This means that millions of stars need to be monitored to catch a handful of microlensing events. The perturbations of lensing light curves by exoplanets are brief and hence easily missed. The alignments are never repeated, so verification, a key step in any scientific investigation, is not possible.

Despite these, since 2005 and as of June 2013, eighteen exoplanets have been found as microlensing events in surveys such as the Optical Gravitational Lensing Experiment (OGLE). These surveys automatically scan millions of stars looking for transient changes in brightness that are achromatic: gravitational lensing is achromatic, that is, light is bent to the same extent regardless of its wavelength.

12.7 Detection Methods Are Biased

Each of the detection methods we have explored is biased to find some types of exoplanet easier than others. For example, the radial velocity (RV) method finds high mass exoplanets close to their star more readily than low mass exoplanets at large distances. So the detection of lots of hot Jupiters by RV need not mean that hot Jupiters are common, it just tells us that the RV method is good at spotting them. In other words, when RV finds numerous hot Jupiters, this may be a selection effect. We would need to take this into account if we wanted to use RV data to estimate the distribution of exoplanet masses or orbital sizes.

EXERCISE 12.13

Comment on the selection bias of the transit method.

12.7.1 Survey Statistics Can Estimate What We Cannot Detect

To estimate how common exoplanets are we need to assess how many we fail to see with our given techniques. In other words, we need to assay the fraction of stars with exoplanets we *could* detect if we looked at N stars, given our current technology. Clearly for microlensing it would be very small, for transits the fraction would be larger but still only a few percent since it will only work for virtually edge-on systems. The radial velocity method would yield a higher fraction because a wider range of inclinations provide a detectable signal, and so on.

It may seem impossible to know how many planets we do not detect, let alone any of their properties, but in fact we can use a variety of statistical techniques, combined with ways of quantifying the limitations of our techniques, to do just that. To take an obvious example: if a technique has a 10% probability of finding a planet and we survey a thousand stars and discover 100 planets, then the implication is that each star has, on average, one exoplanet. This argument can be refined; for example, we can assess the probability that a given technique will find a particular type of exoplanet around a specific type of star. In this way, for every type of exoplanet found, we can infer the number that was not. Perhaps surprisingly, we can infer the existence of a large population of exoplanets from a very small number of discoveries with clever use of statistics. Even if we do not find a single planet, we can learn something from how many stars we looked at. For example, if we calculate a 10% probability of finding a planet, and look at 500 systems and do not find any planet, then it is likely that less than 1 in 50 stars have a planet orbiting around them.

Of course, the statistical modeling is much more complicated than the examples above, but the basic idea is the same. In fact, it tells us that exoplanets are extremely common. We take up the story in Chapter 13.

13

Exoplanetary Systems

13.1 Surveys Probe Exoplanet Properties

As of August 2014, the existence of 1788 exoplanets has been confirmed; there are over 3500 candidate exoplanets awaiting confirmation, and much data is still to be analyzed. This number is now large enough that distributions of properties—such as semi-major axis, eccentricity and inclination—are becoming statistically meaningful. Consequently, we are able to see which exoplanets, and which exoplanetary system architectures are possible and typical. This provides clues to how planetary systems form and places our solar system in context.

A question for astrobiology is what sort of star and exoplanet system architectures can support worlds that are habitable for billions of years. A number of exoplanet surveys focus on sun-like stars for obvious reasons. These are late-F, G, and early-K spectral type stars.

13.1.1 Exoplanet Diversity Is Large

Exoplanets discovered so far range in mass from ~1/80th Earth mass (discovered by pulsar timing) to >$13M_J$, and can be anywhere between 0.001 and 1000 AU from their host star. Intriguingly, a dearth of exoplanets with masses between 30 and $60M_\oplus$ (0.1 to $0.2M_J$) exists across all distances. The reason for this is not obvious. The most massive and distant worlds have generally been discovered by direct imaging or gravitational microlensing. But the great majority of exoplanets have been found with the transit method, followed by the radial velocity (RV) method (Plate 13.1). The former sees mostly exoplanets with semi-major axes <0.1 AU, while RV finds exoplanets out to 20 AU, although there are fewer worlds between ~0.2 and ~1 AU.

13.1.2 Exoplanets Are Classified According to Their Size

A number of systems for classifying exoplanets have been devised. The one which is in most widespread use at present is based on size (see Table 13.1). Earths and super-Earths are almost certainly rock/metal worlds, although a substantial amount of water is also possible. The lower bound on Earth-like planets is because planets with masses less than $0.5M_\oplus$ in a habitable zone are not likely to be able to hold onto a significant atmosphere and probably do not have plate tectonics. A $2.0R_\oplus$ upper bound on super-Earths corresponds to about $10M_\oplus$, massive enough to hold onto hydrogen and helium gas so that they could morph into an ice giant or a gas giant. It is well to note that

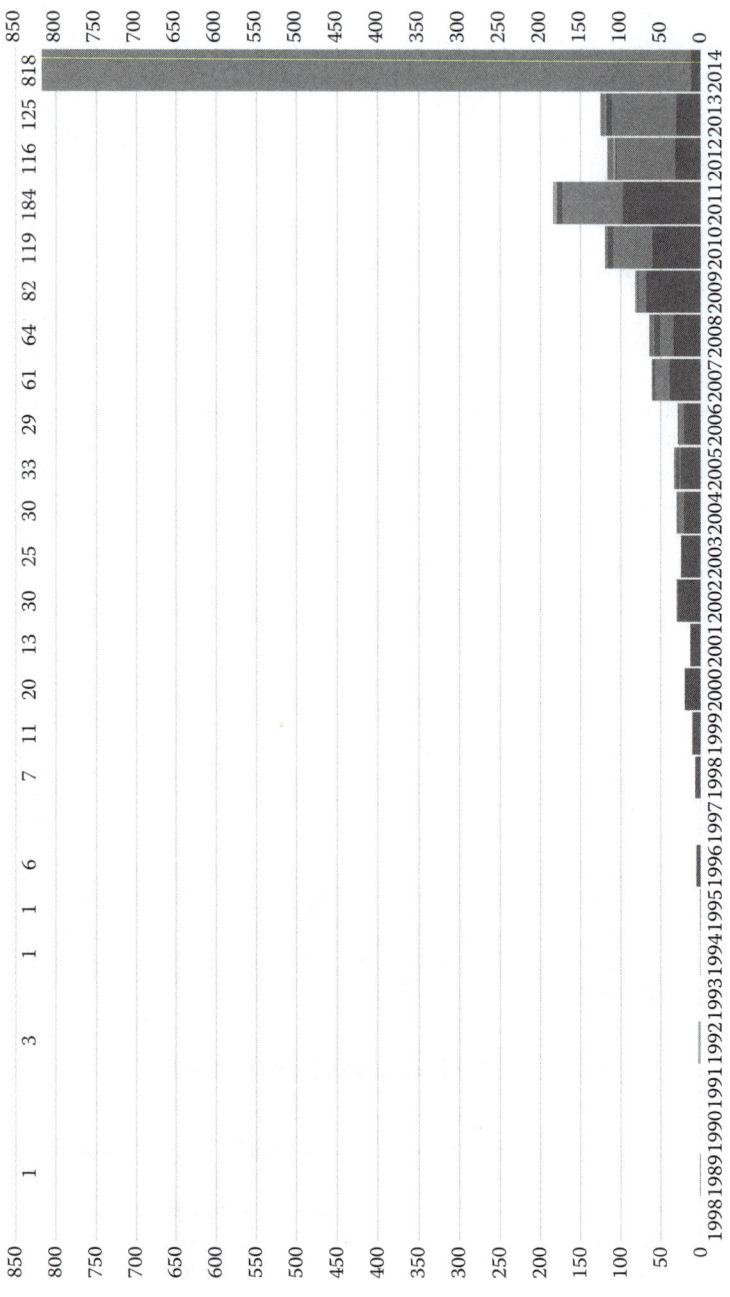

PLATE 13.1 (See color insert.) Exoplanet discoveries by year showing the discovery method by colors: radial velocity (blue); transit (green); timing (yellow); direct imaging (red); microlensing (orange). (Photo credit: Wikipedia "Exoplanets" May 2014.)

TABLE 13.1

Planetary Classification by Size

Class	Size Range ($R_\oplus$)	Mass Range
Earth-sized	0.8–1.25	0.5–2.0$M_\oplus$
Super-Earths	1.25–2.0	2.0–10$M_\oplus$
Neptunes[a]	2.0–6.01	10–30$M_\oplus$
Jupiters[b]	6.0–152	0.1–10M_J
Very large	15–24	>10M_J

[a] Neptune is $3.7R_\oplus$ and $17M_\oplus$.
[b] Jupiter is $11.1R_\oplus$ and $318M_\oplus$.

these boundaries are rather arbitrary, and exoplanet classification is sure to be refined as we find more of them.

13.1.3 Hot Jupiters Were Discovered Early

Many of the first exoplanets to be discovered by the radial velocity (RV) method were hot Jupiters, gas giants with periods less than 10 days. The proximity to their parent star gives them high surface temperatures. For example, HAT-P-2b, tidally locked because its period is just 5 days, has dayside and nightside temperatures of 2400 K and 1200 K, respectively. Atmospheric modeling shows that this huge temperature gradient drives ferocious winds.

The early discovery of hot Jupiters is not surprising given the bias of the RV method. A 2005 RV survey show that they are found around 1.2% of sun-like stars, but Kepler sees a smaller proportion.

Most hot Jupiters are in low eccentricity orbits because the intense tidal forces circularize their orbits. Hot Jupiters could not have formed where they are found. It used to be thought that type II migration explained how hot Jupiters got where they are. However, the high inclinations of most hot Jupiters and retrograde orbits of many them cannot be generated by type II migration. Instead, it is likely that gravitational scattering between several gas giants in a system is responsible for flinging hot Jupiters into their close orbits.

Systems with hot Jupiters have clearly suffered gravitational mayhem and it is unlikely that any terrestrial planet could have survived. However, a couple of simulations have found that post-migration assembly of terrestrial planets from remaining planetesimals and planetary embryos could be feasible.

13.1.4 Small Exoplanets Are Commonest

Extrapolating from a Keck telescope survey of 1000 sun-like stars shows that 20% of them could have gas giants orbiting inside 20 AU and that less massive and more distant gas giants are most common.

That said, size and mass distributions, determined by transit and radial velocity methods, respectively, are broadly similar and show that small, low mass exoplanets are more common than large, high mass ones.

A 2013 survey concluded that 22% of sun-like stars have an Earth-sized (1–2$R_\oplus$ in this context) planet in the habitable zone.

13.1.4.1 The Planet/Star Ratio Is of Order One

The HARPS survey found that at least 50% of stars have one or more planets of any mass orbiting with a period of less than 100 days, that is, within about 0.4 AU. The implication is that the number of exoplanets is, to first order, equal to the number of stars. The PLANET collaboration has made a detailed statistical analysis based on three recent exoplanetary discoveries by microlensing and concludes that each star in the Milky Way may host on average 1.6 planets.

Of course, these estimates are hardly the last word, but the consensus is that planets are the rule, not the exception.

13.1.5 Exoplanet Composition Can Sometimes Be Deduced

For those exoplanets where we know both the mass (from radial velocity) and the radius (from the transit method), we have the bulk density. Except in the most obvious cases, these three parameters will be consistent with many different compositions and detailed modeling is required to home in on the most likely ones. However, just for fun, let us consider a couple of straightforward examples.

> **EXERCISE 13.1**
>
> Suggest the most likely bulk compositions for the following exoplanets: (a) Kepler 10b with $M = 4.6 M_\oplus$, $R = 1.4\ R_\oplus$, and a bulk density of 8800 kg m^{-3}; (b) Kepler 11e with $M = 8.4 M_\oplus$, R $= 4.5\ R_\oplus$ and a bulk density of 500 kg m^{-3}.

Most exoplanets do not yield their secrets so easily. To constrain the structure of exoplanets, it is usual to assume the planet is layered (i.e., differentiated) and derive equations of state for the relevant composition (iron alloys, silicates, and ices) under self-gravity that satisfy the observed mass–radius relation (Box 13.1). This approach is currently being developed by Damian Swift and others working at the Lawrence Livermore National Laboratory. They have come up with new methods for deriving and testing equations of state, working out the mass–radius and mass–pressure relations for materials relevant to planetary interiors. In doing so the researchers have made specific predictions for exoplanets having different compositions, with pressures at the center ranging from just 800 MPa to a whopping 1900 GPa (the pressure at the center of the Earth is reckoned to be about 375 GPa). Of course, determining what exoplanets are made of is a step toward understanding whether or not they could be habitable (Figure 13.1).

13.1.5.1 Compositions Can Be Plotted on a Ternary Diagram

Ignoring atmospheres, differentiated terrestrial planets can be composed of three bulk materials: water, silicate mantle, and iron core. Any of these can be solid or liquid depending on pressure and temperature. They can be mixed in appropriate proportions to determine a planet's radius at a given mass. This can be represented on a ternary diagram (Figure 13.2).

Ternary diagrams are commonly used in geology to plot the composition of a rock with three minerals. Each vertex of the equilateral triangle is 100% of one of the

---

BOX 13.1   EQUATIONS OF STATE

An equation of state describes how the volume, V, or density, ρ, of a system varies as a function of pressure, P and temperature, T. The simplest equation of state (EOS) is that for an ideal gas:

$$PV = nRT, \tag{13.1}$$

where n is the amount of gas in moles and R is the universal gas constant.

This EOS can be applied to relatively low density atmospheres such as that of the Earth, but does not work for the deep high density atmospheres of gas and ice giants since these are not ideal: molecular forces cannot be neglected.

For solids being compressed slowly enough that they are able to lose heat and remain at the same temperature (i.e., isothermal conditions) the simplest equation of state describes how compressible the solid is. On the other hand, solids being heated at constant pressure experience thermal expansion. The simplest equation of state in this situation defines the thermal expansion coefficient, the change in volume with temperature.

Solids in planetary interiors are subjected to increasing pressure, compression, and hence heating the closer they are to the center of the planet, to an extent determined by the mass and size of the planet. Minerals change phase in response. The physics of self-compression and mineral phase changes can be captured in EOS.

Models include equations of state for several materials (e.g., iron, silicate, and water) that can mix in various proportions and determine a planet's radius at a fixed mass.

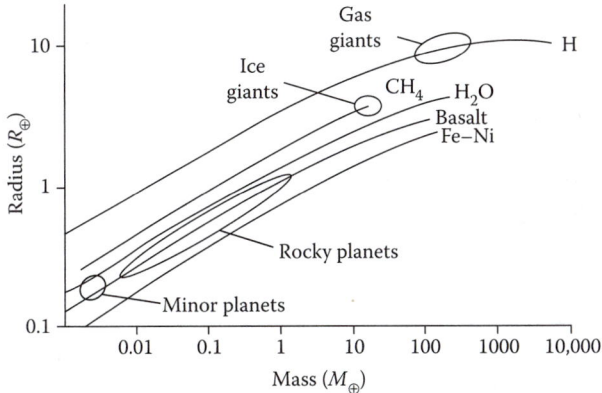

FIGURE 13.1   Mass–radius relations for planets with different compositions.

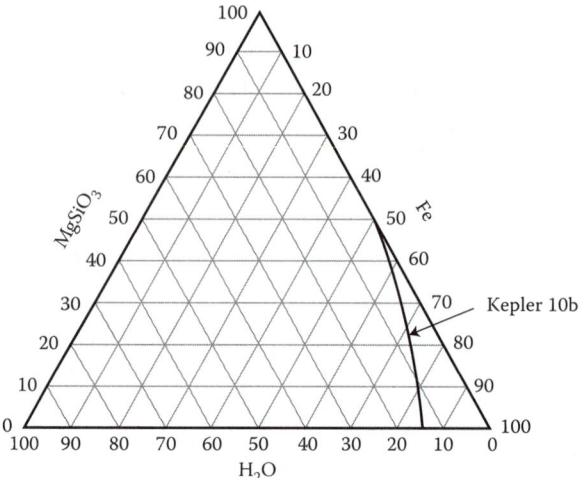

FIGURE 13.2   The Si–H_2O–Fe ternary diagram. Here Si denotes silicate (SiO_4^{2-}) minerals. The curved line is the iso-radius line for Kepler 10b. It joins all compositions that can account for the bulk density of the exoplanet.

minerals. Decreasing proportions of that mineral map to lines parallel to the side of the triangle *opposite* the vertex.

EXERCISE 13.2

a. The Earth is approximated by $Si_{68}Fe_{32}$. (Water can be ignored since it makes up just 0.05–0.1% of the Earth by mass.) Plot this on Figure 13.2.

b. Plot the composition $Si_{60}H_2O_{10}Fe_{30}$ on the ternary diagram.

EXERCISE 13.3

a. Calculate the bulk density for a planet if it is of composition $Si_{20}H_2O_{10}Fe_{70}$. Take the densities of silicate rock, water, and iron at STP as 3000 kg m^{-3}, 1000 kg m^{-3}, and 7900 kg m^{-3}, respectively.

b. This composition lies on the iso-radius line for Kepler 10b which has a bulk density of 8800 kg m^{-3}. Explain the discrepancy between this value and the one you found in (a).

13.1.5.2 Protoplanets with Mass Greater Than $1.5M_\oplus$ May Become Mini-Neptunes

As planets accrete a solid core they may reach a mass at which they can hold onto light gases in a protoplanetary disc and so end up enshrouded in a dense atmosphere of H and He. Helmut Lammer of the Austrian Academy of Sciences, and colleagues has modeled the balance between capture and removal of hydrogen by protoplanets with masses between 0.1 and $5M_\oplus$ in the habitable zone of sun-like stars. They found that cores with masses <$0.5M_\oplus$ and of density similar to that of the Earth failed to capture the gas. Between 0.5 and $1.5M_\oplus$ protoplanets accrete gas but subsequently lose it to the fierce UV fluxes of young stars. But higher mass cores hold onto almost all of the hydrogen. The implication is that most of the worlds currently classed as super-Earths, such as Kepler 62e and Kepler 62f, may actually be more like mini-Neptunes, with extremely high surface pressures that would make habitability unlikely even within a habitable zone.

13.1.5.3 Could Water Worlds Exist?

Some simulations of planet formation predict the accretion of massive ($\sim6M_\oplus$) ice-covered worlds beyond the snowline. These super-Earths have a mantle of ice overlying a metal/silicate core but were not thought massive enough to acquire a H/He envelope and morph into "mini-Neptunes." If they subsequently migrated inward to the habitable zone they would become covered in a deep global ocean, becoming water worlds. They could be distinguished from terrestrial planets by their bulk density. Whether these planets have a steam atmosphere and hence risk a runaway greenhouse, or are cloud covered would depend on their distance from their host star. Cloud cover would be indicated by a high albedo.

13.1.5.4 How Big Must a Planet Be to Have Plate Tectonics?

We saw in Chapter 3 that plate tectonics could be crucial to habitability mainly by operating the silicate–carbonate cycle that stabilizes surface temperature. Astrobiologists are therefore keen to understand what it takes for a rocky planet to develop plate tectonics. Attempts to scale plate tectonics up to that of super-Earth have led to completely contradictory findings from different research groups. Models are simply too poorly constrained currently.

13.1.6 Exoplanet Temperature Can Be Estimated

A first stab at estimating the effective temperature (T_{eff}) of an exoplanet can be made from the flux of radiation it gets from its parent star, which depends on the average distance between the planet and the star and on the spectral type and luminosity class of the star. However, T_{eff} also depends on the albedo of the planet, which is usually not known, so has to be guessed from what its "surface" might be made from; gas, rock, water, ice, and so on. If the exoplanet has an atmosphere then its surface temperature may be somewhat higher than its T_{eff} if there are greenhouse gases present. One way to determine the temperature is to measure the difference in infrared radiation as the planet is eclipsed by its parent star. This has now been done for several exoplanets. If the star's brightness during secondary eclipse is subtracted from its brightness before or after, only the radiation from the planet remains.

13.2 Exoplanet Host Star Properties

Surveys of exoplanet properties are beginning to show us what it takes to build planets and provide datasets for testing models of planet formation.

13.2.1 Metal-Rich Stars Are More Likely to Host Gas Giants

Metals are needed to build planets. This is especially true of gas giants which have to form a massive ($\sim10M_\oplus$) iron and silicate kernel in order to capture their hydrogen/

helium envelopes in the limited time during which the protostellar nebula exists (3–5 Myr). The data so far shows that gas giant-endowed stars on average have a *higher* metallicity, Z, than the sun (i.e., [Fe/H] > 0, see Chapter 1, Box 1.4). It is surprising then that the sun has four gas giants despite having comparatively low metallicity.

13.2.2 Did Terrestrial Planets Form 11 Billion Years Ago?

Planets smaller than Neptune ($<4R_\oplus$) are found around stars with a great range of metallicities. In fact, almost three quarters of stars hosting Earths and super-Earths have metallicity the same as or lower than solar metallicity. Recall that metals are produced by stellar nucleosynthesis, so each round of star formation will increase the metallicity of the interstellar medium from which the next generation of stars is spawned. Given that terrestrial planets can be forged in low metallicity environments, it seems plausible that they formed early in cosmological history. This is supported by studies showing that early (~11 Gyr old) galaxies have metallicity of one-third that of the solar metallicity.

If rocky worlds can form at low metallicities and galaxies make metals early, then maybe Earth-like, habitable planets were forged when the universe was young. Metals include oxygen and carbon which means that water and organic molecules would have been abundant in the early universe, perhaps paving the way for the emergence of life within a couple of billion years of the Big Bang. Because on average it takes higher metallicity to make gas giants than rocky planets (because they have high mass metal–rock cores) might there have been a brief time when stellar systems lacking gas giants were common? Without migrating gas giants to cause havoc, such systems may have been especially likely to have habitable planets in stable orbits.

13.2.3 Most Stars with Planets Have Low Lithium Abundance

Lithium is destroyed by nuclear fusion in the cores of stars but should survive in stellar atmospheres. It was therefore unexpected to find that the sun has 140-fold less lithium than it is predicted to have. Moreover, a number of studies show that the great majority of stars with exoplanets have low lithium abundance. This is not an age correlation. Old stars (>1 Gyr) without exoplanets have the predicted amount of lithium. However, the depletion correlates with the surface temperature of the star. Stars with $T_{eff} > 5860$ K did not show the anomaly.

It seems that exoplanet-hosting stars have greater convective mixing of their atmosphere than barren stars, and this drags lithium into the deep interior of the star where it is destroyed. Higher convection occurs because stars that form protoplanetary nebula are slow rotators. This idea is supported by a HARPS survey of 500 stars which shows that lithium-depleted exoplanet-host stars are indeed slow rotators. What is the link between rotation speed and convection? It seems that slow rotating stars develop a high degree of differential rotation between the radiative core and the convective envelope which does not happen in fast rotators. The strong differential rotation at the base of the convective envelope is responsible for enhanced lithium depletion. Slow rotation results from a long-lasting star–disc interaction; as matter accretes onto the star, angular momentum is transferred to the disc. Modeling suggests that long-lived discs (5 Myr) may be a necessary condition for this scenario. Stars which destroy their discs early have not been spun down, neither have they acquired planets.

13.3 Habitable Exoplanets

13.3.1 $\eta_\oplus$ Is the Proportion of Stars with Habitable Planets

Astrobiologists would like to know the proportion of stars that host habitable worlds. This parameter is termed $\eta_\oplus$ (Eta-sub-Earth or Eta of Earth).

$$\eta_\oplus = \text{number of habitable planets/number of stars} \qquad (13.2)$$

It is usual to define a habitable planet as being a terrestrial planet, although the size and/or mass ranges adopted in surveys or models vary somewhat. It is also usual to specify that the planet is orbiting a sun-like star but M-type stars are increasingly seen as hosts for habitable planets. If these are included, $\eta_\oplus$ increases dramatically because they constitute 75% of all main sequence stars.

A habitable planet must of course be in a habitable zone (HZ) but there are a number of ways to define this. We explore this below, but the most optimistic definition would make Venus, Earth, and Mars habitable, whereas the most conservative just has a habitable Earth. The choice of HZ could make a big difference to the value of $\eta_\oplus$. James Kasting and Chester Harman at Penn State University argue that this difference is important because a smaller value of $\eta_\oplus$ means that to characterize such planets would need larger telescopes.

Clearly $\eta_\oplus$ is a moveable feast and any quoted value has to be seen in the context of how it is arrived at.

13.3.2 An Earth-Sized Planet in the Habitable Zone Has Been Discovered

Kepler has discovered dozens of Earth-sized exoplanets orbiting closer than their stars' habitable zones. It has also found two super-Earths within the HZ of the K2V star Kepler 62a. Kepler 62e and Kepler 62f are 1.6 and 1.4 Earth radius, respectively. Although their masses are not known (their radial velocities could not be measured), their small size precludes them from being mini-Neptunes. Hence, they are likely to have masses between 2 and 4 Earth masses and be solid. Models of the type mentioned in Section 13.1.6 suggest they could have substantial amounts of water. If so, they may be covered by deep oceans, making them water worlds, a kind of planet not seen in our own solar system. Whether they have atmospheres is not known.

EXERCISE 13.4

If Kepler 62f had the same bulk density as the Earth, what would its mass be?

Three other Kepler super-Earths have been proposed as potentially habitable—Kepler 22b, Kepler 61b and Kepler 69c—but the first two are probably not rocky worlds and the last is now thought to be a super-Venus.

Three super-Earths in a 5- or 7-planet system around the M1.5 V star Gliese 667C discovered by the radial velocity technique are also in the HZ. The innermost, Gliese 667Cc, receives 90% of the light Earth does but most of this is in the IR so it has $T_{\text{eff}} = 277$ K. If it has an atmosphere with greenhouse gases it would likely be uncomfortably hot.

In March 2014, the discovery of the first Earth-sized planet in the habitable zone was announced. Kepler 186f orbits a M1 star 151 pc away with a mass and radius about half that of the Earth. The planet has a radius of $1.1R_\oplus$ and orbits 0.356 AU from the star in 129.9 days. It may be tidally locked, but at the very least is expected to rotate very slowly. Its mass is unknown but if it has the same composition as Earth (one-third iron, two-thirds silicate rock) it will have a mass of $1.44M_\oplus$. It is close to the outer edge of the habitable zone, and 1D climate modeling with a N_2–CO_2–H_2O atmosphere suggest that it will have $T > 273$ K only if it has at least 0.5 bar CO_2 in its atmosphere. This is an extremely exciting discovery, although its distance means that there is no possibility that its atmosphere could be characterized even with next generation instruments such as the James Webb Space Telescope.

13.3.3 How to Define a Habitable Zone

Of course, whether an exoplanet is in a habitable zone (HZ) depends on how we define it.

13.3.3.1 Conservative HZ

The most conservative definition goes from moist greenhouse at the inner edge to the maximum CO_2 greenhouse at the outer edge (see Sections 3.3.4.1 and 3.3.4.2). But it is worth noting that Kopparapu's (2013) model assumes that a habitable exoplanet has a carbonate–silicate cycle that is able to regulate atmospheric CO_2 levels and surface temperature. This will widen the HZ. Alternatively, the HZ can be defined by taking runaway greenhouse and CO_2 condensation as the inner and outer boundaries. This may provide even more leeway, though for planets around stars with surface temperatures <5000 K there is very little difference between the moist greenhouse and runaway greenhouse limits. That is because cool stars radiate mostly in the infrared, where water is a particularly good absorber.

13.3.3.2 Optimistic HZ

Some astrobiologists go further and argue for an optimistic or empirical HZ by assuming that Venus and Mars were both once habitable (Table 13.2). How does this work?

The inner limit is fixed by Magellan spacecraft observations of Venus showing that liquid water cannot have flowed on its surface for at least 1 Gyr. At this time, solar luminosity was 92% of its present value. Venus is 0.72 AU from the sun. So, now it gets $(1/0.72)^2 = 1.93$ times the solar flux falling on the Earth. And one billion years ago it would have received $0.92 \times 1.93 = 1.77$-fold greater insolation than the current

TABLE 13.2

Habitable Zone (HZ) Distances (AU) for the Solar System

		Inner HZ		Outer HZ	
Model	Moist Greenhouse	Runaway Greenhouse	Recent Venus	Maximum Greenhouse	Early Mars
Kopparapu et al. (2013)	0.99	0.97	0.75	1.70	1.77
Kasting et al. (1993)	0.95	0.84	0.75	1.67	1.77

Earth. This corresponds to an orbital distance of $(1/1.77)^{1/2} = 0.75$ AU. This "recent Venus" boundary is of course quite arbitrary since we have no idea whether water ever flowed on the surface of Venus or when.

Early Mars is used as a marker for the outer limit of the optimistic HZ. The valleys and channels on its surface were clearly carved by running water, so Mars must have been warm enough for surface liquid water then. The latest time these fluvial features formed is taken to be at 3.8 Gyr, when the sun's luminosity was only 75% of its current value. Mars lies at 1.52 AU from the sun, so it now receives $(1/1.52)^2 = 0.43$ of the insolation of the Earth. This means the average insolation of Mars 3.8 Gyr ago could have been no more than $0.74 \times 0.43 = 0.32$ of the present solar radiation falling on the Earth. This flux corresponds to a distance of 1.77 AU from the sun now.

The obvious difficulty here is that the climate models struggle to keep Mars out of the freezer (see Chapter 10, Section 10.5.3).

Despite the obvious problems, astronomers frequently use the optimistic HZ definition on the grounds that it ensures that all *potentially* habitable planets are recognized in exoplanet surveys. Conservative definitions are used as a guide to ensure that missions to hunt out habitable worlds are built to a high enough specification.

13.3.3.3 HZ Location Varies with Stellar Temperature

The location of the habitable zone will vary with the surface temperature of the star and hence its spectral class. Both the inner and outer boundaries move inward through the standard sequence of spectral classes: OBAFGKM (Plate 3.1).

EXERCISE 13.5

What is the relative probability of finding a habitable planet around an F-type star compared to an M-type star?

13.3.3.4 Desert Worlds Can Be Very Close to Their Host Stars

An arid planet with no oceans and very low atmosphere water content could be habitable much closer to a sun-like star. The inner edge of the habitable zone for such a planet is defined as the temperature and pressure at which water boils. Modeling by Sara Seager and colleagues shows that a world with 1 bar surface pressure, an atmospheric CO_2 mixing ratio of 10^{-4}, and a relative humidity of 1% could be habitable at 0.38 AU from the sun if it had an albedo of 0.8 or at 0.59 AU if it had an albedo comparable to that of the Earth. Dew could form during the night at the poles. If the surface pressure is too low, the water loss timescale is too short to support life. For example, at 0.1 bar, all water would have gone by one billion years. Atmospheric CO_2 has an indirect effect on the water loss timescale because it controls tropopause and stratosphere temperature. A low CO_2 mixing ratio minimizes the greenhouse effect in the troposphere but makes the tropopause and the stratosphere warmer, so more water is transported into the stratosphere where it is dissociated and lost. The reason for this is that CO_2 in the stratosphere is a coolant because the radiation it emits is lost to space.

13.3.3.5 Three-Dimensional Climate Models Are Needed

As we have seen, there is considerable uncertainty in how to define the boundaries for circumstellar habitable zones. This means that our current estimates for $\eta_\oplus$ are provisional at best. One key reason for this is that one-dimensional (1-D) climate models do not capture all of the physics and chemistry atmospheres. Most do not incorporate clouds, for example.

The way to resolve this is to modify the 3-D global climate models (GCMs) that exist for Earth so that they can be applied to exoplanets. GCMs are mathematical simulations used for weather forecasting and predicting climate change. They couple general circulation models for the atmosphere and ocean—the equations that describe fluid mechanics on a rotating sphere—with the thermodynamics of radiation and heat transport, and the more sophisticated of them include representations of biogeochemical cycles. The Earth's atmosphere, oceans, and land surface are treated as a grid of volume elements. Actually, the models are 4-D because the equations are integrated over time. Processes which operate on a scale smaller than the volume elements, such as moist convection, land surface albedo variations, and cloud cover are parameterized. This means that instead of these processes being calculated explicitly from basic physics, a simple approximation is made by an appropriate choice of variables (parameters). These 3-D models could be generalized for different fluxes and spectral makeup of incoming starlight and so used to calculate more realistic HZ boundaries.

The results of a 3-D model, which included clouds and variations in relative humidity, were published in December 2013. Jeremy Leconte and colleagues at the Institute Simon Pierre Laplace in Paris and Robin Wordsworth at the University of Chicago show that the insolation threshold for a runaway greenhouse is 375 W m^{-2}. This is higher than previously thought and confirms 0.95 AU as the inner edge of the habitable zone around a sun-like star. Interestingly, this model predicts that even with warm surface temperatures (e.g., 330 K) the tropopause is cold (120 K) so that most water vapor condenses leaving the stratosphere dry. In other words, there is no moist greenhouse limit to the habitable zone.

13.3.3.6 The HZ Depends on the Planetary Mass

A planet can retain atmospheric gases over geological timescales provided that the average speeds of the particles is less than one-sixth the escape speeds, which depends on the planetary mass and the temperature at the top of its atmosphere, the exobase (see Chapter 3, Section 3.3.5). In theory, Mars has sufficient mass at present to hold onto N_2, O_2, CO_2, H_2O, and CH_4 because it is so cold. If Mars had an Earth-like atmosphere and was in the HZ its exobase temperature would be much higher and it would likely lose its atmosphere. In other words, for a planet to be continuously habitable it needs a mass substantially higher than that of Mars.

Lower mass planets have lower surface gravity. This increases the column depth of the atmosphere (assumed to be N_2–H_2O–CO_2 in HZ climate models) to produce modest increases in the greenhouse effect (at the inner edge) and albedo (at the outer edge). Consequently, the HZ moves out slightly. Super-Earth's have HZs that are marginally closer and wider.

13.3.3.7 Is a Planet in the HZ?

It is not a straightforward matter to say whether an exoplanet is in the HZ, however we define it. The usual strategy is to try to calculate its effective temperature from the flux of radiation it receives from its star. This requires the semi-major axis and eccentricity of the orbit to be calculated, and the uncertainties on these can be high. Unless we have the composition of the planet's surface or atmosphere we have to assume an albedo, or use a climate model and assume the composition of the atmosphere. Even if we can capture the spectrum of a giant exoplanet's atmosphere, we can neither make quantitative estimates of its composition to within the tens of percent level yet, nor can we determine the atmospheric mass and composition of small exoplanets. In most cases, we cannot tell whether there are clouds, or the planet's rotation direction or speed. Consequently, model calculations are done to span a range of plausible values.

13.3.4 $\eta_\oplus$ Has Been Derived from Kepler Data

An estimate of $\eta_\oplus$ has been derived from a 2011 Kepler dataset by Wesley Traub of the Jet Propulsion Laboratory, California Institute of Technology. The brightest F-type, G-type, and K-type stars were selected. These had 278 candidate exoplanets orbiting with periods between 2.5 and 40 days. From this sample, survey bias corrections implied the presence of 12,300 terrestrial planets in the HZ around all 35,896 bright stars in the dataset. Terrestrial planets were defined as having radii: $0.5 < R/R_\oplus < 2.0$. The HZ was defined in three ways, normalized to the current solar system these were as follows: wide (0.72–2.00 AU), nominal (0.80–1.80 AU), and narrow (0.95–1.67 AU). The number of planets in the HZ was an average over all three of these options deduced for each of the three spectral classes of star.

EXERCISE 13.6

Calculate $\eta_\oplus$ for the 2011 Kepler dataset.

In fact, the error bars on this result are ±14%, so there is still considerable uncertainty about $\eta_\oplus$. The calculated value of $\eta_\oplus$ actually corresponded to the number of terrestrial planets expected in the nominal HZ of a G-type star. This is arguably too optimistic. For a more realistic narrow HZ around a G-type star, $\eta_\oplus = 24\%$.

13.3.5 Habitable Planets Around M-Dwarfs?

Because of their large numbers, low mass, cool M-dwarfs may be the most abundant exoplanet hosts. Habitable exoplanets around M-type stars are also easier to detect. Because the HZ is closer to lower mass stars the orbital periods are shorter, the signal is larger, and there is a bigger chance of detecting them than if they orbited a sun-like star.

EXERCISE 13.7

Why is (a) the signal larger and (b) the chance of detecting a transit higher for an exoplanet in the HZ around an M-type star than around a G-type star?

By combining data from two exoplanet European Southern Observatory (ESO) surveys in Chile, HARPS (High Accuracy Radial Velocity Planet Searcher) and UVES (Ultraviolet and Visual Echelle Spectrograph), Mikko Tuomi of the University of Hertfordshire, UK, and colleagues have detected radial velocity signals of eight exoplanets around nearby red dwarfs. Statistical analysis of this result suggests that every red dwarf has at least one planet.

Analysis of 3987 Kepler M-dwarfs suggest that 51% of these cool stars could have Earth-sized planets (0.5–$2R_\oplus$) within a conservative habitable zone. Data from an ESO/HARPS radial velocity survey of 101 M-dwarfs reckon that 41% have Earth-sized planets in the habitable zone. The uncertainties in these figures are large but it does seem that substantial numbers of M-type stars host terrestrial planets in their HZ.

Perhaps surprisingly planets orbiting in the HZ of cooler stars are more likely to be ice-free than those around hotter stars. This is because ice absorbs much of the longer wavelength near-IR light emitted by cool stars. By contrast, the dominant visible and UV light of hotter stars is reflected by ice rather than absorbed.

There are, however, potential hazards to being close to M-dwarfs. We explore these now.

13.3.5.1 Habitable Planets Around M-Dwarfs May Be Tidally Locked

The HZs around M-dwarfs are narrow and extremely close to the star. This guarantees that habitable exoplanets hosted by M-type stars will be in virtually circular orbits and they may be tidally locked also.

The timescale for tidal locking, τ_{TL}, is given by

$$\tau_{TL} = 10^{12} \text{ yr } (a/\text{AU})^6 (M/M_\odot)^{-2}. \tag{13.3}$$

Clearly, in the case of the Earth, tidal locking is not going to happen in the 10 Gyr main sequence lifetime of the sun.

EXERCISE 13.8

Calculate the tidal locking timescale for an exoplanet around an M-dwarf of $M = 0.1 M_\odot$ at semi-major axis of (a) 0.1 AU; (b) 0.23 AU.

Exercise 13.8 shows us that tidal locking will happen swiftly for planets close to the inner edge of the habitable zone of red dwarfs, though at the outer edge it may take many billions of years.

The temperature difference between the dayside and nightside of a tidally locked world will be ameliorated by a dense atmosphere or deep oceans since these will serve to transport heat, though this will set up high winds blowing from dayside to nightside.

Jun Yang, Nicolas Cowen and Dorian Abbot, in collaboration between Northwestern University and the University of Chicago, have used a 3-D global climate model which includes clouds. Earlier models for the HZ have been based on 1-D radiative-convective models which are limited by their inability to predict cloud behavior. Adding clouds into the mix shows that the high stellar flux on the dayside of a tidally locked world produces strong convection. This produces high albedo water clouds near the substellar location that would have a net cooling effect. In fact, the clouds have a negative feedback stabilizing influence on planetary temperature. Substellar clouds also effectively block outgoing radiation from the surface, reducing or even completely reversing the thermal emission contrast between dayside and nightside. In effect, a high altitude cloud deck is cold; in other words, it radiates mostly at visible wavelengths rather than IR. This reversal would be detectable with the James Webb Space Telescope, and if detected would confirm the presence of substellar water clouds. If this is the case, the researchers reckon it would roughly double the frequency of habitable planets orbiting red dwarf stars.

EXERCISE 13.9

Draw a diagram to show the location of an exoplanet with respect to its star, as viewed from the Earth, when the temperature of the planet's (a) dayside and (b) nightside could be measured.

13.3.5.2 M-Dwarf Radiation Could Be Problematic

The cool surface temperatures of M-type stars (2300–3800 K) means that most of their energy is radiated in the IR and their small size makes their luminosity extremely low (~10^{-4} to $0.07L_{\odot}$). These characteristics are by no means insurmountable challenges for life because presumably any photosynthetic biochemistry would have adapted to harness the energy of the low flux of IR photons as efficiently as possible.

M-dwarfs are highly variable stars. They have stellar cycles akin to solar cycles but star spots can cover a much larger proportion of their surface than higher mass, hotter stars. This will cause substantial variations in luminosity.

Many, perhaps the majority, of M-dwarf stars are flare stars. They erupt extremely powerful flares that cause the star to brighten across a broad range of wavelengths, principally in x-ray, UV, and radio frequencies. They can generate as much as 10^4-fold more x-ray photons than a typical solar flare. The close location of the habitable zone means these outbursts could make the surface of a planet inhospitable.

13.3.5.3 Red Dwarf Magnetic Fields Are Problematic

Relatively young red dwarfs have strong magnetic fields. Aline Vidotto of the University of St. Andrews, UK, and colleagues argue that these would compress the magnetospheres of exoplanets in the HZ perhaps even down to the surface of the planet. This would expose the atmosphere to sputtering by the stellar wind, which itself would be intense because of the closeness of the planet to the star.

13.3.5.4 Planets Around M-Dwarfs Are Probably Water-Depleted

Jack Lissauer of the Space Science and Astrobiology Division of NASA's Ames Research Centre has dynamically modeled the accretion of Earth-mass planets in the M-dwarf habitable zone. Temperatures there are higher than those encountered by material that accreted to form the Earth, and collision velocities are higher. The implication is that planets orbiting in the HZ may have water abundances that are substantially less than the Earth. This might compromise habitability.

13.3.6 Eccentric Orbits Influence Exoplanet Habitability

It can be hard to determine the eccentricity of an exoplanet orbit. Indeed, sometimes a single planet in a moderately high eccentricity orbit turns out on closer analysis to be two planets in nearly circular orbits.

High eccentricity orbits pose two different problems for life. One is that a high mass planet in an eccentric orbit could clear the HZ of terrestrial planets by gravitational scattering. The other is that a potentially habitable planet may not be in the HZ for all of the time if its orbit is too eccentric. We look at these in turn.

13.3.6.1 Giant Planets Can Clear Out the Habitable Zone

A massive planet can pump up the orbital eccentricity, e, of a terrestrial planet in the HZ if it gets too close to the HZ. This can result in the terrestrial planet being excited into an orbit that periodically carries it beyond the HZ even though its semi-major axis may not change very much. In extreme cases, $e > 0.4$, the terrestrial planet can be ejected from the HZ altogether.

But how close is too close? It transpires that the distance a giant planet has to be from an Earth-like planet to exert a gravitational effect is $\leq nR_H$, with R_H being the Hill radius of the giant planet and n an empirically derived multiplier that depends on the eccentricity, e_G, of the giant planet's orbit and whether the terrestrial planet is inferior or superior to the giant planet. The Hill radius is given by

$$R_H = (m_G/3M^*)^{1/3} a_G, \qquad (13.4)$$

where m_G and a_G are the mass and semi-major axis of the giant planet, respectively and M^* is the mass of the star. There are two cases to consider (Figure 13.3):

1. *The giant planet is superior to the HZ.* In this situation, the giant has an inward gravitational reach $n_{int}R_H$ interior to its periastron. The periastron distance is $a_G(1 - e_G)$, so the reach is $a_G(1 - e_G) - n_{int}R_H$.

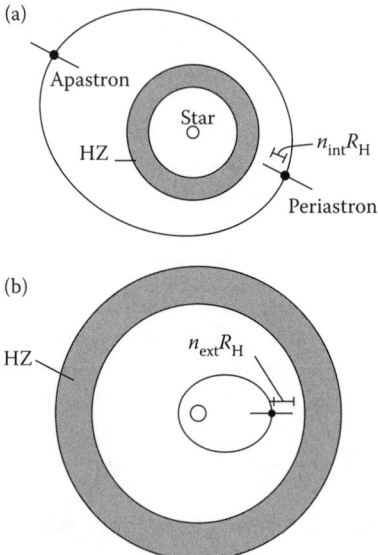

FIGURE 13.3 Periastron and apastron distances for gas giants in superior orbits (a) or inferior orbits (b), respectively are critical for survival of planets in the HZ.

2. *The giant planet is inferior to the HZ.* In this case, the giant has a gravitational reach $n_{ext}R_H$ exterior to its apastron. The apastron distance is $a_G(1 + e_G)$, so the reach is $a_G(1 + e_G) + n_{ext}R_H$.

If the reach extends throughout the habitable zone, there is no safe place for a terrestrial planet to hide within it.

EXERCISE 13.10

A gas giant HD 80606 b (mass = $3.49M_J$) orbits a $0.9M_\odot$ G5V star with a semi-major axis of 0.449 and eccentricity 0.9337. The habitable zone around this star is calculated to be from 0.84 to 1.66 AU. The multiplier for this highly eccentric orbit is $n = 17$. (a) Sketch the system and specify whether the giant planet is inferior or superior to the HZ. (b) Show that a terrestrial planet cannot survive within the habitable zone around this star. (Take $M_J = 1.9 \times 10^{27}$ kg and $M_\odot = 2 \times 10^{30}$ kg.)

A 2010 analysis of 79 transiting exoplanets by Barrie Jones and Nick Sleep at the Open University, UK, which explored the gravitational reach of gas giants into the habitable zones (defined by the Kasting et al. climate model), showed that most system architectures are compatible with terrestrial planets in the HZ (Figure 13.4). The entire HZ is safe if the giants are confined to orbits well inside or outside it. (The second case must be true since our solar system has Jupiter!) If a giant planet is moved closer to the inner or outer edge of the HZ its reach begins to extend into the inner and outer parts of the HZ, respectively. A terrestrial planet can still remain within the HZ provided it occupies the outer or inner regions of the HZ, respectively. It is even possible for an Earth-like planet to survive with a giant inside the HZ provided the giant is in a low eccentricity orbit. In fact, only two of the 79 systems studied by Jones and Sleep do not offer safe haven at the present time. One of these is HD 80606 b. If the requirement is for habitability over the past 1.7 Gyr, then a further 28 systems fall by the wayside; that is, only 49 systems are habitable over such a sustained period.

13.3.6.2 Can Life Survive on Worlds in Eccentric Orbits?

Mars has a relatively high eccentricity. Taking the outer limit of the HZ in the solar system as 1.67 AU, this places Mars right at the edge of the HZ when it is close to aphelion.

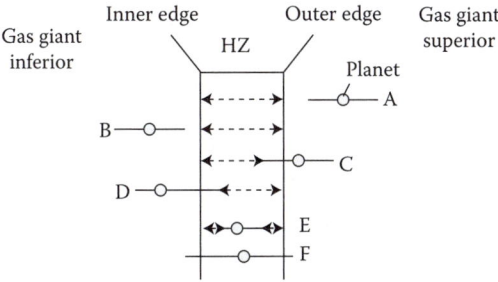

FIGURE 13.4 The gravitational reach of a giant planet determines how much of a habitable zone can be populated by terrestrial planets. Only for F is a habitable planet excluded.

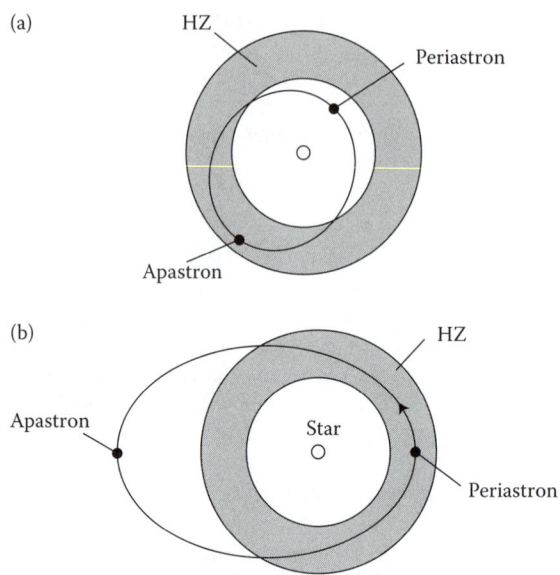

FIGURE 13.5 Periodically a planet in a highly eccentric orbit could be outside the HZ around periastron (a) or apastron (b).

What are the consequences for planets in orbits so eccentric that they are carried outside of the HZ periodically? This depends on many factors. A relatively short period reduces the time in the inhospitable state and so limits the heating and/or cooling. A planet outside the HZ near periastron is much less compromised than one outside at apastron (Figure 13.5).

EXERCISE 13.11

Why is habitability more compromised by an eccentric orbit that carries a planet outside the HZ at apastron than periastron?

Exercise 13.11 suggests that life may be able to survive if its host planet leaves the habitable zone around periastron even if it could not at apastron. The bottom line is that life on an exoplanet in an eccentric orbit that carries it beyond the HZ at apastron must be able to weather prolonged cold periods. This unpleasant situation could be ameliorated by deep oceans, the high thermal inertia of which could buffer extreme swings in temperature. In fact, calculations show that if an Earth-like planet had an orbital eccentricity of 0.2 (compared to the current and maximum values for Earth of 0.017 and 0.061, respectively), provided it kept the same semi-major axis then the ratio of radiation fluxes at periastron and apastron,

$$F_p/F_a = [(1 + e)/(1 - e)]^2, \qquad (13.5)$$

would be comparable to the summer/winter insolation ratio at mid-latitudes on Earth due to the obliquity.

EXERCISE 13.12

Calculate F_p/F_a (a) for Earth with $e = 0.017$ and; (b) for an exoplanet with $e = 0.2$ at a distance of 1 AU from a star with $L = L_\odot$ and $T_{eff} = 5770$ K. (c) Why might this apparently modest variation in flux due to high eccentricity pose a problem for life on an Earth-like world?

A dense atmosphere could act as a thermal blanket for a planet carried periodically across the outer boundary of the habitable zone. Organisms might adapt by a variety of strategies including living at depth or by hibernating.

Planets in the HZ around M-dwarfs, provided they are not harassed by badly behaving gas giants, are unlikely to be in eccentric orbits since the habitable zones are so close to the star that the orbits are circularized.

13.3.7 Planets Exist in Binary Systems

Binary star planetary systems fall into two architectures. A planet can be in orbit around a single component of the binary or around both. These are described as S and P orbits, respectively.

13.3.7.1 S Orbits Are Possible for Close Binaries

Circumstellar discs are required to form planets and young binary star systems have circumstellar discs around the primary star—usually the component with the highest mass. It has been thought that in close binaries (i.e., separated by ≤40 AU) the secondary star could increase the velocity of planetesimals in the disc so that collisions would result in fragmentation rather than accretion (although this could be opposed by gas drag), and would truncate the circumprimary disc, thereby removing potentially planet-forming material. Yet despite these theoretical considerations, discs massive enough to accrete planets—that is, discs ~20 AU across containing $0.06M_\odot$ of material—have been observed in close binary systems. In one instance, separate discs around both primary and secondary stars have been observed.

Radial velocity observations have revealed gas giants in binary systems in which the two stars are as close as 40 AU. An interesting question is whether binary systems can host terrestrial planets in their habitable zones. Given that ~30% of stars have binary companions the number of habitable worlds would be appreciably higher if this were the case.

13.3.7.2 Habitable Planets Probably Exist in Binary Systems

To explore whether S orbits are stable in HZs of binary systems, numerical simulations of discs usually assume that accretion of planets in binary systems happens in the same way as around single stars but looks at how the secondary star might disrupt it. This shows that for a planet to be in a circular stable orbit in a binary system its orbital radius cannot exceed a critical value, a_C, which is a function of the semi-major axis and eccentricity of the secondary star's orbit, and the stellar mass ratio, $\mu = M_2/(M_1 + M_2)$. The bigger the separation and lower the eccentricity, the higher a_C is.

One simulation, by Nadar Haghighipour of the University of Hawaii and Sean Raymond at the Laboratoire d'Astrophysique de Bordeaux, France, investigated the behavior of the disc around a sun-like primary star. The disc contained moon- to Mars-sized planetary embryos, distributed between 0.5 and 4 AU, and a Jupiter mass planet at 5 AU. Simulations varied the characteristics of the secondary star orbit and its mass. One aim of the study was to investigate how easily wet terrestrial planets would form. The model gave planetary embryos beyond the snow line (2 AU) water inventories that corresponded to those of carbonaceous chondrites. These were expected to deliver water to terrestrial planets forming closer in. The HZ was taken to extend from 0.95 to 1.5 AU. The simulations showed that Earth-like planets with substantial amounts of water could form in the HZ around the primary over a range of secondary star masses, distances, and orbital eccentricities.

But it is not a free-for-all situation. For a given binary separation, the water content of the resulting terrestrial planets falls as the eccentricity of the secondary star increased. This is an effect mediated through the gas giant. At higher eccentricity the secondary star gets closer to the gas giant at perihelion (see Figure 13.6). This pumps up the eccentricity of the gas giant so that it disrupts the disc, ejecting most of the water-bearing planetary embryos. In effect the gas giant is acting as a go-between, transferring angular momentum from the secondary star to planetary embryos in the disc so as to alter the water content of any accreting terrestrial planets.

Intriguingly, in the absence of a gas giant, terrestrial planets still accreted (including in the HZ) but it took longer; moreover, the greater the eccentricity of the secondary star (up to $e \sim 0.4$), the larger the planets that formed and higher their water content. But at extreme eccentricities increasing numbers of planetary embryos are ejected from the disc and the efficiency of terrestrial planet formation falls.

Models like this are useful, but they are relatively low resolution and the planetary embryos and gas giants are put in "by hand." We have no idea how the earlier (and less well understood) steps in planetary formation are affected by the presence of a second star, nor the extent to which this might alter the final architectures of binary planetary systems.

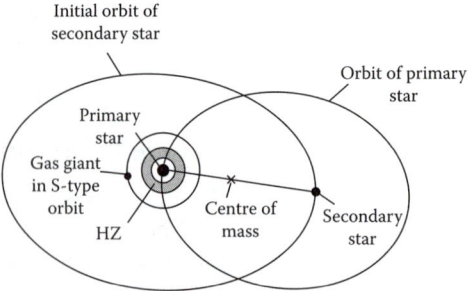

FIGURE 13.6 In an S-type orbit a planet orbits one component of a stellar binary. Here a close binary system has a Jovian-planet orbiting outside the HZ of the primary star. As the eccentricity of the secondary star is racked up, its perihelion gets closer to the gas giant.

13.3.7.3 Circumbinary Planets Have Been Identified

Six systems have been confirmed in which planets orbit two suns. The existence of such circumbinary planets has been postulated for decades and has been popularized in science fiction, most notably the planet Tatooine in the *Star Wars* films.

The first P orbit circumbinary system to be confirmed, Kepler 16, is an eclipsing binary of low mass stars (~0.7 and $0.2M_\oplus$) and its Saturn-sized gas giant reveals itself by transits and by small deviations in the timing of the stellar eclipses. It is a very compact system. The two stars orbit each other in 41 days in a fairly eccentric orbit with a semi-major axis of 0.22 AU, and the planet orbits the binary in a nearly circular orbit at a semi-major axis of ~0.7 AU and a period of 229 days (Figure 13.7).

That the planet's orbit has only a tiny inclination with respect to the plane of the stars' orbits implies that the planet accreted from a circumbinary disc.

The habitable zones of circumbinary systems are more complicated to determine than those around single stars. Stephen Kane and Natalie Hinkel of NASAs Exoplanet Science Institute, Caltech, have calculated stellar flux contours around a variety of binary systems. From this they found the wavelength at which the combined flux peaks and assumed black body radiation so that the Wien displacement law could be used to find the effective temperature. From this polynomial functions of T_{eff} were used to calculate runaway greenhouse and maximum greenhouse habitable zone boundaries. Of course, the HZ boundaries will oscillate periodically as the two stars orbit each other. If this is significant, then the time evolution of these boundaries has to be derived. This can be translated into seasonal changes in radiation flux and hence surface temperature as the planet orbits. Finally, whether a habitable planet can be stable in this dynamic set-up has to be ascertained.

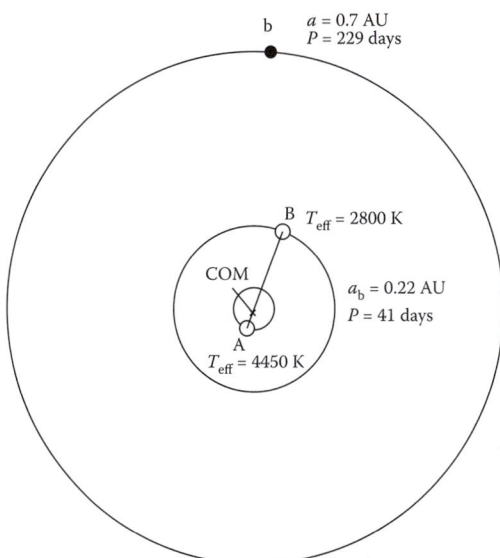

FIGURE 13.7 In circumbinary systems such as Kepler 16 the planet orbits the center-of-mass of the binary star system. This is described as a P-type orbit.

Recall that there is a critical semi-major axis, a_C, the minimum allowed semi-major axis for a planet to be in a stable orbit. If eccentricity is zero, a_C is given by the quadratic:

$$a_C = (1.60 + 4.12\,\mu - 5.09\,\mu^2)a_b, \tag{13.6}$$

where μ is the mass ratio and a_b is the binary separation. Figure 13.8 shows two configurations calculated by Kane and Hinkel. In (a) a G2V-K2V binary with separation $a_b = 0.1$ AU can harbor a planet in its HZ because a_C has a much smaller radius than the inner HZ boundary. By contrast (b) shows an equal mass M0V-M0V binary with

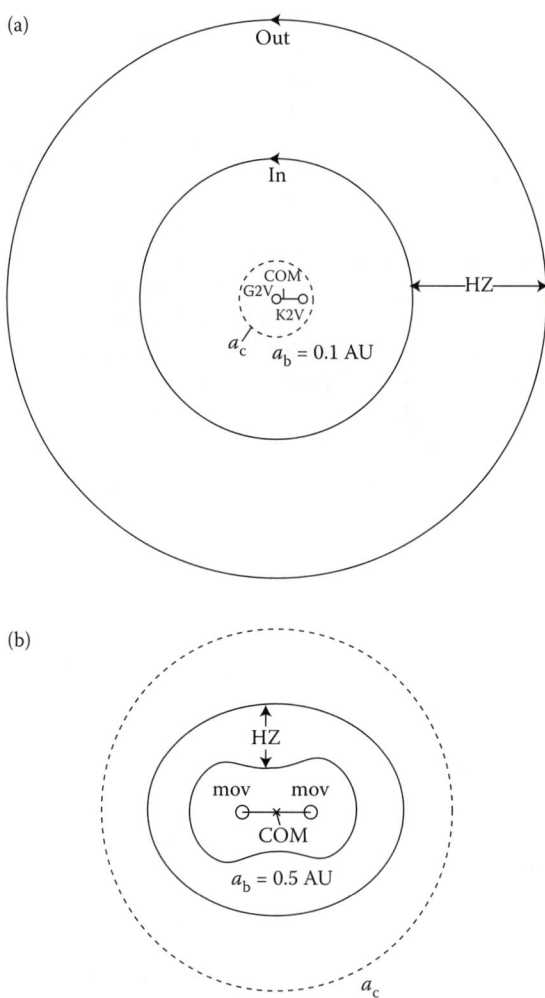

FIGURE 13.8 Circumbinary system HZs may (a) allow or (b) disallow stable planetary orbits depending on whether they are larger or smaller than a critical semi-major axis.

$a_b = 0.5$ AU. Here a_C is beyond the outer boundary of the HZ, so no habitable planet can be accommodated. Note how distorted the shape of the HZ is.

EXERCISE 13.13

Confirm that the M0V-M0V binary (Figure 13.8b) with $a_b = 0.5$ AU cannot have a habitable planet. The long axis of the outer boundary of the HZ of this system has a semi-major axis, $a_{OHZ} = 0.9$ AU.

The asymmetry in HZs in circumbinary systems can have the odd consequence that even if a planet is in a perfectly circular orbit it experiences variations in radiation fluxes and peak wavelengths and so temperature fluctuations over its orbit. This is the case with Kepler 16 which simulations show to be moving in and out of the outer boundary of the HZ during each orbit (Figure 13.9). The average surface temperature oscillates periodically between 200 and 215 K as it does so and in response to the changing alignments of the two stars with respect to the planet.

A further complication to the habitability of circumbinary planets has been explored by Paul Mason at the University of Texas and colleagues. Nascent stars generally rotate more rapidly than middle-aged ones so have more intense magnetic fields. This has two consequences for a planet in the HZ. One is that it can compress the magnetosphere of the planet, making its atmosphere more vulnerable to sputtering. The second is that young stars have many more flares and hence radiate more extreme UV and x-rays that can sterilize the surface of a planet. Binary stars in close orbit can have one or both components tidally locked. This can reduce their rotation rates and hence reduce their magnetic field strengths to below those seen for single stars, as is the case for Kepler 34, Kepler 47 and Kepler 64. This scenario is obviously good for circumbinary planet habitability because it reduces the time during which they are very active, with high fluxes of UV and x-rays from flares and atmosphere-sputtering high energy particles. The bad news is that this is not invariable. Tidal locking of very close, low mass binaries can synchronize the stars into very high rotation rates so that their magnetic fields are particularly intense, making the system appear forever young. This has happened in Kepler 16.

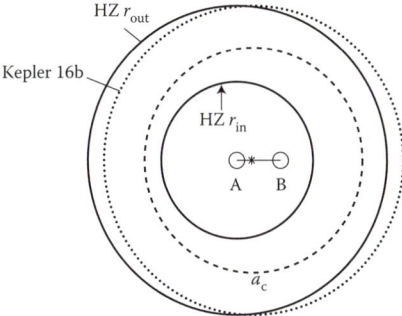

FIGURE 13.9 The HZ of Kepler 16 is offset from the center-of-mass of the binary, so the circular orbit of the planet takes it in and out of the HZ.

13.4 The Galactic and Habitability

13.4.1 The Milky Way Is a Spiral Galaxy

Galaxies are gravitationally bound "cities" of stars. The Milky Way is a barred spiral (SB) galaxy (Figure 13.10). With a mass $\sim 2 \times 10^{12} M_\odot$, it is the second most massive galaxy in a local group of about 50 galaxies.

The entire galaxy is enveloped in a huge halo that contains most of its mass in the form of dark matter. The halo contains globular clusters (see Box 13.2). The central bulge consists mostly of extremely red stars. The stars in the halo, globular clusters, and bulge are extremely old (12.5 Gyr), so called population II stars. Those in the halo, including those in most globular clusters have very low metallicities, but those in the bulge can have high metallicities. This reflects the very high rates of star formation and therefore large numbers of metal-dispersing core collapse supernovae early in the life of the Galaxy.

The bulge is surrounded by a gas-rich disc. Disc stars have Keplerian orbits; the sun orbits the Galaxy once every 230 Myr. As disc stars orbit they encounter rigidly rotating spiral density waves. These are spiral-shaped density enhancements that act rather like moving traffic jams on a motorway. Generally, spiral density waves rotate more slowly than stars. So stars approach the wave from behind, slow down, and after some delay emerge ahead of the wave. Gas is compressed as it ploughs into the wave, triggering star birth in giant molecular clouds. This produces spiral arms, revealed by HII regions, open clusters, OB star associations, and the occasional type II supernova. What is more, it means that the disc is home to young (<10 Gyr), higher metallicity, population I stars, like the sun.

13.4.2 Galactic Chemistry Influences Habitability

Supernovae (SNe) are important contributors to the chemical evolution of the Galaxy. Intermediate mass stars, when they reach the asymptotic giant branch, provide most of the C, N, and some O, while Type II SNe provide O, the r-process elements, and the α elements that are built by successive fusion of He nuclei. The α elements typically measured in stellar spectra are Mg, Si, Ca, and Ti (see Chapter 1, Section 1.4.6).

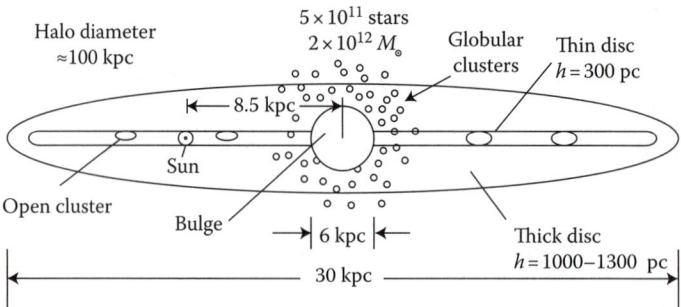

FIGURE 13.10 The Milky Way: h, scale height, the distance above the midplane where the number of stars has decreased by a factor of e.

The primary source of the iron group elements (Fe, Co, Ni) are a completely different class of supernova termed Type Ia. Like all Type I SNe they can be distinguished from Type II by the lack of hydrogen lines in their spectra. Type Ia SNe result from the complete detonation of either a single white dwarf as it accretes material from a binary companion (single degenerate case) or the explosion of two colliding white dwarfs (double degenerate case). In either scenario the trigger for the explosion is when a critical mass, the Chandrasekhar limit of $1.44M_\odot$, is exceeded. At this point, the white dwarf ignites in an all-consuming nuclear conflagration.

Over the lifetime of the Galaxy, as the rate of star formation has declined, there has been a rise in the number of Type Ia SNe relative to Type II. This has resulted in a fall in the O/Fe ratio coupled with a rise in Fe/H.

EXERCISE 13.14

What might you expect to happen to Mg/Fe and Si/Fe ratios over the Galactic lifetime?

Interestingly, it has been suggested that the ratio of α elements to iron is greatest in stars with planets, particularly for Mg/Fe. So even if [Fe/H] is low maybe α elements aid planet formation. If so, the result of the exercise above implies that future generations of stars may be less likely to host planets.

The drop in Mg/Fe and Si/Fe ratios might also mean that later-forming planets have larger core/mantle ratios.

We have argued that a habitable terrestrial planet needs a magnetic field, mantle convection and plate tectonics. Ultimately these are powered by the internal heat provided by long half-life radioisotopes—^{40}K, ^{232}Th, ^{235}U, and ^{238}U—the abundance of which relative to Fe has been declining with time because the rate of star formation in the Galaxy has decreased, so less are being forged in supernovae.

13.4.3 The Galaxy Has a Habitable Zone

It has been suggested that only a relatively narrow torus of the Galaxy can harbor life. This region is termed the Galactic habitable zone (GHZ).

The idea that particular features of the Milky Way could be inimical to habitability had been around for a while but the notion of a GHZ was developed by astronomer Donald Brownlee and paleontologist Peter Ward to support their hypothesis that complex life is rare, made in a popular book *Rare Earth: Why Complex Life Is Uncommon in the Universe* published in 2000. They published a more technical account in collaboration with astrophysicist Guillermo Gonzalez a year later. Their concept is that the GHZ is governed by two factors: metallicity and catastrophic events.

13.4.3.1 Low Metallicity Chokes Planet Accretion

There is a radial gradient in metallicity, Z, as measured by [Fe/H], high near the Galactic center and decreasing through the thick disc and halo and toward the rim of the thin disc. Although, as we have seen, terrestrial planets can form around stars with a wide range of metallicities, there is a minimum Z below which accretion of metal/silicate worlds becomes unlikely. Recent work by Jarrett Johnson and Hui Li suggests this limit is about one-tenth solar metallicity. Below this, and the little dust that was there in a stellar nebula would photo-evaporate before planets could accrete.

EXERCISE 13.15

A star has [Fe/H] = +0.5. By how many times does this star's metallicity differ from that of the sun?

Guillermo Gonzalez derived an empirical linear equation relating [Fe/H] to the mean galactocentric distance (R_m) of a star and to stellar age from observations of nearby stars. He found that the metallicity gradient in the thin disc is −0.07 kpc^{-1} so that,

$$[Fe/H] = -0.01 - 0.07(R_m - R_\odot) - 0.035t, \tag{13.7}$$

where $R_\odot$, the distance of the sun from the galactic center, is taken as 8.5 kpc, t is the age of the star in Gyr with $t_{now} = 0$. Hence, the average metallicity at the location of the sun is given by

$$[Fe/H] = -0.01 - 0.07(8.5 - 8.5) - 0.035 \times 0 = -0.01.$$

There are three caveats to Equation 13.7. It holds for the thin disc only, the spread of metallicities is large, and the linear dependence of [Fe/H] on age is assumed to hold for the entire lifetime of the disc.

EXERCISE 13.16

Use Equation 13.7 to determine whether stars that formed at the rim of the thin disc could host planets. Although the oldest galactic halo star is 13.2 Gyr, the thin disc is thought to have formed much later, 8.8 Gyr ago. Comment on your result.

13.4.3.2 The Galactic Center Is Dangerous

There are plenty of opportunities for habitable planets to meet a sticky end in the Galactic bulge. Stars are more closely packed there than in the disc. This means that tidal disruption of the Oort clouds around bulge stars would increase the risk of cometary impacts on planets.

The chemical evolution of the bulge also has a huge influence. When the Galactic bulge first formed, from ~13 Gyr onward, its star formation rate was very high. Consequently, the number of type II supernovae at this time was high and this drove a rapid increase in the metallicity of the bulge interstellar medium. With abundant metals bulge stars should have plenty of planets but they risked being sterilized by the intense fluxes of γ-rays, x-rays and cosmic rays from the SNe. This effect is enhanced by the higher density of stars in the bulge.

The Galactic bulge metallicity is now very high. Consequently, stellar systems will have made many gas giants and their migration is bad news for planets in habitable zones. That said, the Galactic bulge is now relatively depleted in gas so the current rate of star formation is low. Consequently, the probability that planets are accreting there now is small.

13.4.3.3 *The Boundary of the GHZ Has Been Defined*

Many researchers argue that the large uncertainties in the factors that determine the GHZ mean that its boundaries are stochastic. That has not prevented Charles Lineweaver and colleagues from calculating boundaries for the GHZ at 4 kpc and 10 kpc from the Galactic center. Stars may migrate substantial radial distances over their lifetimes, so it is always possible that they could move in or out of the GHZ.

13.4.4 Are There Habitable Planets in Clusters?

13.4.4.1 *Globular Clusters Are Probably Barren*

There are a couple of theoretical reasons to suspect that globular cluster stars will not be hosts to planets. One is that they have metallicities too low even for terrestrial planets to accrete.

The other has to do with the number density of stars in the cores of GCs which is 10^4 to 10^5 per pc^3, compared to stellar densities in the solar neighborhood of order 0.1 per pc^3. Even if planets did form they are unlikely to remain in stable orbits for more than a few hundred million years because of gravitational disruption by close stellar encounters.

Transit surveys of 47 Tucana and Omega Centauri have failed to identify any exoplanets. At the very least this implies that exoplanets in GCs are rare. In fact, only one has been identified in any GC up to now. PSR B1620–26 b is a $2.5M_J$ circumbinary planet orbiting a pulsar–white dwarf binary in M4. The extreme antiquity of this cluster (12.7 Gyr) probably makes this the oldest exoplanet discovered.

BOX 13.2   STAR CLUSTERS

Galaxies are home to two very different types of star clusters.

Open clusters (OCs) are groups of up to a few thousand very young stars that all formed at the same time in the same molecular cloud. Their youth (millions to tens of millions of years old) means that they are dominated by a few high mass, hot, bright stars. This is clearly seen in one of the most spectacular clusters, M45 (the Pleiades or "Seven Sisters") in Taurus. Open clusters are found in the spiral arms of the thin discs of spiral galaxies such as the Milky Way. They are not permanent structures. Their high mass stars rapidly evolve; exploding as core collapse supernovae and effectively vanishing. In addition, the mutual gravity of stars in an open cluster is not enough to hold the clusters together and over time they evaporate.

Globular clusters (GCs) are spheres of 10^5 to 10^6 stars held together by their mutual gravity. The Milky Way has ~150 GCs and they are distributed around the bulge, with most in the halo but some in the disc and bulge. There are two populations of globular cluster but both are extremely ancient (>12.5 Gyr old). They are dominated by low mass red dwarfs and stellar number densities are very high. The most spectacular globular clusters are 47 Tucana and Omega Centauri, both visible in the southern hemisphere.

13.4.4.2 Open Clusters Likely Harbor Planets

Open clusters (OCs) are extremely young and as they derive from stellar nurseries would be expected to have exoplanets. They have high stellar densities though not as extreme as GCs. Several exoplanets have indeed been found in OCs; for example, Kepler 66 b and Kepler 67 b are both Neptune-sized worlds in the cluster NGC 6811 in Cygnus, and two hot Jupiters have been detected by the radial velocity method in the Beehive cluster, M44, in Cancer.

In summary, we do not expect to see habitable planets in globular clusters but we may in open clusters.

13.4.5 Biosignatures

We have seen how to quantify the number of habitable planets. But how many are actually inhabited is another question, and one that we are unlikely to be able to answer for some time. One strategy to get at this information is to hunt for exoplanet biosignatures. There are two types of biosignatures: those required for life, principally water and carbon dioxide; and the molecules produced by life, for example, oxygen (O_2 and O_3), methane, and nitrous oxide (N_2O).

It is important to be aware that in no instance does the presence of any of these *prove* habitability or the existence of life. Water is the second most common molecule in the universe after H_2, and occurs mostly in environments that are completely inhospitable, so the NASA mantra "Follow the water" must always be seen in context. The atmosphere of Venus is almost pure carbon dioxide yet the surface of Venus is uninhabitable. And there are abiogenic ways of making O_2, O_3, CH_4, and N_2O as we shall see, although a *prima facie* case might be made for biology if the coexistence of two or more showed a marked disequilibrium of the exoplanet (see Chapter 6). Hence, a judgment on whether a planet is inhabited is always going to be based on a balance of probabilities, unless we actually go there.

Until that exciting day we must make the best of remote observations which give us access to the atmosphere and surface of an exoplanet. Repeated observations over time could reveal dayside/nightside differences and seasonal changes—examples of temporal biosignatures.

13.4.5.1 Earth Provides Practice in Acquiring Biosignatures

Spectral lines of biosignature molecules in a planetary atmosphere are found over visible and infrared wavelengths (Figure 13.11).

Earth is obviously a prime test bed for detection of biosignatures. Remote sensing of the Earth's atmosphere shows that the most prominent absorption lines are those of water (5–7 μm and 18 μm), ozone (9.6 μm) and CO_2 (15 μm). An emission feature in the center of the CO_2 band reveals the high temperature of Earth's exobase. This provides clues to gas loss mechanisms at the top of the atmosphere. We can deduce Earth's effective temperature from the black body curve that fits best to the IR spectrum. The spectrum of Earth captured by Carl Sagan and colleagues with the Galileo spacecraft showed the simultaneous presence of O_2 and CH_4, evidence that the Earth system is far from equilibrium.

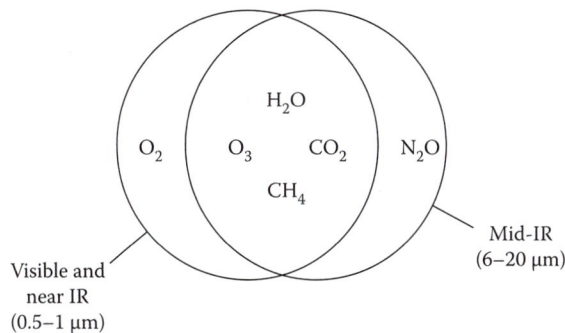

FIGURE 13.11   Venn diagram showing the best windows for detecting biosignature molecules.

We see changes to the spectrum that depend on solar illumination (i.e., differences between dayside and nightside), viewing angle, cloud coverage, and the proportion of land and ocean in the visible hemisphere. Clouds are a problem. In visible and near-IR regions of the spectrum clouds are very reflective with much the same albedo across all wavelengths (Figure 13.12). This is a source of noise. In the infrared, clouds reduce emitted radiation from the troposphere because they are cooler than the surface. In this annoying way, they hide most of the interesting spectral features.

Of course, Earth is an easy target and it is straightforward to show that it is a habitable planet with oceans, a greenhouse atmosphere that boosts the effective temperature, global geochemical cycles, and life (as revealed by the spectral signature of plant photosynthetic pigments) because we can observe it with almost arbitrarily high signal-to-noise ratio (SNR) and spatial spectral resolution. For exoplanets, where SNR and resolution are limited, it becomes much harder. Researchers take terrestrial spectra and process them to reproduce the low SNR and poor resolution that might be expected when observing a distant exoplanet. These degraded data are then used to

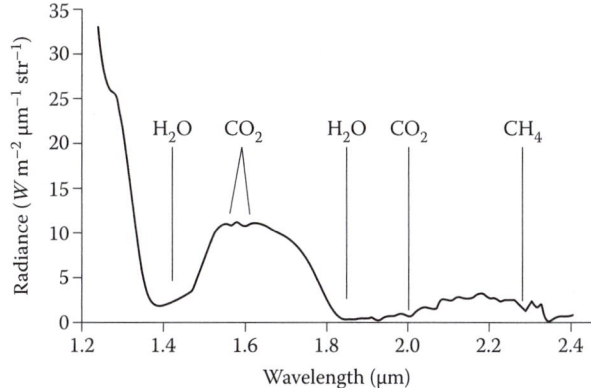

FIGURE 13.12   Near-IR spectrum of Earth's atmosphere is consistent with life on Earth!

test whether our information processing and modeling can recover the biosignatures we are hunting for, and help us define the uncertainties that we expect to work with.

13.4.5.2 Atmosphere Spectra Have Been Obtained

The first spectra of exoplanet atmospheres were published in 2007. They were from transiting hot Jupiters. One (HD 209458b) showed emission features from silicate clouds and another, actually the closest hot Jupiter (HD 189733b), revealed Na, H, and H_2O.

For some years researchers have been studying the HR 8799 system. HR 8799 is an A5V star of $1.5M_\odot$ that is 40 pc away in Pegasus. It has four planets/brown dwarfs (the masses are high but poorly constrained) that are at large distances from their star and can be imaged directly. Remarkably, atmospheric spectra of all four planets have been obtained revealing a variety of molecules, all of which are found in Jupiter (Table 13.3). Intriguingly, the questionable CO_2 signature (it might be hydrogen cyanide) is variable.

Near-IR transmission spectrum of the transiting hot Jupiter HD 189733b which lies 63 ly away in Vulpecula allowed Giovanna Tinetti and colleagues to discover water and methane in its atmosphere. Subsequently, neutral oxygen and carbon monoxide were discovered. With a surface temperature of 1117 K this world has a bloated exosphere of atomic hydrogen that is evaporating at a prodigious rate. But at this high temperature, water and CH_4 should react to form CO, so the detection of methane is a mystery. Perhaps more worrying is the presence of OI and methane together, albeit at low abundances. This illustrates the difficulty of trying to assess whether a planet is habitable from its atmospheric composition.

13.4.5.3 Water Need Not Mean Habitability

We cannot yet see liquid surface water on the surface of an exoplanet, although liquid water polarizes light and this signal might be detectable with future technology. For now we must rely on the detection of water vapor in an exoplanet's atmosphere. But water vapor can exist in an atmosphere even if the pressure and temperature are such that liquid water cannot exist, either because it is too hot to condense, as we have seen for hot Jupiters, or because it is frozen at the surface. So, water can exist over a range of temperature and pressures that are incompatible with life. Some extremophiles seem to be able tolerate very high pressures, so habitability is best

TABLE 13.3

Inventory of Molecules Detected in HR 8799 Planetary/
Brown Dwarf Atmospheres

Planet	CO_2	NH_3	C_2H_2 (Acetylene)	CH_4	H_2O
b	✓	✓	✓		
c	? or HCN	✓	?		✓
d	✓		✓	✓	
e			✓	✓	

judged by the temperature extremes for terrestrial organisms which go from ~255 K to 395 K (−18°C to +122°C). The lower limit is set by growth. Some organisms can survive much colder conditions, depending on how they are frozen, but they do not do anything interesting so do not produce evidence for their existence. There may also be wriggle room to these values, but the important principle is that there are temperature limits to life.

An absence of water is implied by the presence of very water soluble gases in an atmosphere such as SO_2 and H_2SO_4. The Venusian atmosphere contains these compounds. Should we regard them as anti-biosignatures?

13.4.5.4 Some Biosignature Gases Are Produced by Metabolism

Numerous gases are produced by the metabolic activities of organisms but to be useful biosignatures they must be produced in amounts that can be detected. For remote sensing of exoplanets this condition is currently met by O_2, O_3, and is likely to be satisfied by CH_4 and N_2O in the not too distant future. On anoxic worlds, just like Earth earlier than 2.4 Gyr ago, reduced sulfur-containing gases such as methanethiol (CH_3SH) and dimethyl sulfide ($H_3C–S–CH_3$) could be in high concentrations, since these are produced in large amounts by some anaerobic bacteria on Earth.

Biosignatures can be false friends. In a high UV environment, photodissociation of water could produce detectable levels of O_2.

> **EXERCISE 13.17**
>
> State two environments in which a planet in a habitable zone could be exposed to a high UV flux.

Oxygen could also accumulate in a dry CO_2-rich atmosphere in which the rocks are already highly oxidized. Oxygen could only have been detected in Earth's atmosphere for the past 2.2 Gyr or so but life existed here for at least one billion years prior to this, so lack of O_2 in a planetary atmosphere is not evidence for the absence of life.

Detection of ozone (at 9.6 µm) shows the presence of O_2, but has a highly nonlinear relation to how much oxygen there is. That is because as more O_2 is converted into O_3 (increasing the intensity of the 9.6 µm line as you would expect) the stratosphere cools and this *reduces* the intensity of the 9.6 µm line.

Low resolution mid-IR spectroscopy could not detect methane in Earth's atmosphere unless its abundance was appreciably higher. Provided that water was present, substantial amounts of CH_4 can be formed by serpentinization of mantle rock, so its detection is not of itself evidence for life. The coexistence of O_2 and CH_4 shows marked disequilibrium. While this cannot rule out an interesting set of abiogenic processes, it does argue *prime facie* for biology.

Almost all nitrous oxide (N_2O) in Earth's atmosphere comes from the metabolism of denitrifying bacteria. These have a crucial role in the nitrogen cycle (Figure 13.13) in which atmospheric nitrogen is fixed by bacteria such as cyanobacteria and converted into nitrate, NO_3^-. This is assimilated by marine phytoplankton and land plants to build nitrogen-containing biomolecules such as proteins and nucleobases. Atmospheric N_2 is regenerated by anaerobic denitrifying bacteria which use NO_3^- as an electron donor in respiration rather than oxygen. A lifeless world with an active

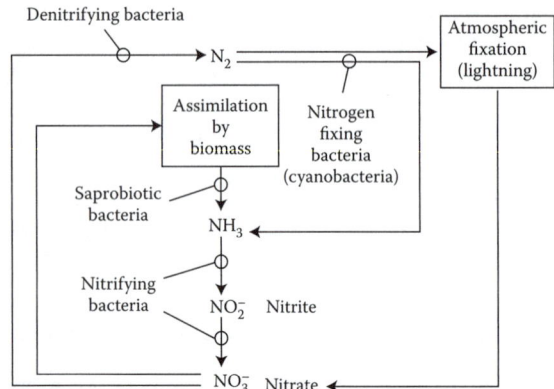

FIGURE 13.13   The nitrogen cycle. The biosignatures N_2O is produced by denitrifying bacteria.

hydrological cycle might fix N_2 by lightning to produce N_2O in the absence of free oxygen in the atmosphere via dissociation of carbon dioxide:

$$CO_2 \rightarrow CO + O^+ + e^-, \tag{13.8a}$$

$$O^+ + N_2 \rightarrow NO^+ + N, \tag{13.8b}$$

$$NO^+ + N \rightarrow N_2O^+, \tag{13.8c}$$

$$N_2O^+ + e^- \rightarrow N_2O. \tag{13.8d}$$

However, N_2O is rapidly hydrolyzed by near UV so is not going to last long in an atmosphere lacking O_2, such as the early Earth.

Even remote detection of N_2O on Earth now would be hard because of its low abundance at the surface (0.3 ppmv), which drops off markedly in the stratosphere.

13.4.5.5 *The Red Edge Is a* Relatively *Unambiguous Biosignature*

Earth's plants have a sharp 10-fold increase in reflectance between approximately 700 and 750 nm wavelength because of their photosynthetic pigments. This strong signal is termed the red edge. It has been detected in the spectrum of Earthshine, the integrated scattered light spectrum of the Earth reflected back to us from the surface of the moon not illuminated by the sun. (It can be seen as a faint ghost of the lunar disc during the waxing crescent phase.)

Extraterrestrial photoautotrophs have no reason to display a red edge over the same wavelengths as photosynthesizers on Earth. Indeed, light harvesting organisms on worlds around other spectral type stars would be expected to absorb across different wavelengths. For example, photoautotrophs on planets orbiting M-type stars would likely harness photons in the near-IR, perhaps to 1.1 μm or longer. They might make up for the low stellar flux by capturing photons across the visible spectrum also, giving a reflectivity drop over 0.4–1.1 μm.

Some minerals have reflectance edges similar in slope and strength to the terrestrial red edge (though at different wavelengths) so an extrasolar planet reflectance edge must be interpreted with care. A red edge signal that varied in intensity over time would be suggestive, as this would be expected of seasonal variations in biomass.

Of course the combination of a red edge and O_2 in the atmosphere would be very exciting, but what about oxygen without the red edge? This might be what is termed cryptic photosynthesis. If photoautotrophs are confined to water, their red edge is much harder to find than land biomass red edge because of the lower reflectivity of water. Plankton biomass would need to be 10-fold higher on Earth before its contribution to the red edge could be detected with the current technology.

13.4.5.6 *Time-Varying Signatures Could Indicate Habitability*

If the integrated IR emission from an exoplanet exhibits variations that correlate with seasons (assuming the planet has a nonzero obliquity), habitability might be entertained. However, dust storms on Mars are seasonal and can be global in coverage, so we must not be too carried away by such a signal from an exoplanet.

The dayside–nightside temperature contrast will be less on a habitable planet with a dense atmosphere and water oceans than on an arid world with a tenuous atmosphere. The average global dayside–nightside temperature difference, ΔT, for Earth is surprisingly small just 2.8 K. As a percentage change, this is $(100 \times \Delta T)/T = (100 \times 2.8)/288 = 1\%$. For Mars $\Delta T = 42$ K so $(100 \times \Delta T)/T = (100 \times 42)/210 = 20\%$. A Mars-size signal might be detectable. Of course Venus would pass this habitability test!

As we have seen, Earth's atmosphere has changed over geological timescales, probably from CO_2-rich to CO_2/CH_4 and then to O_2-rich. Inhabited exoplanets may experience similar secular changes, and to this end some researchers, like Lisa Kaltenegger at Harvard University, are modeling how the spectrum of an Earth-like planet would evolve over 4.5 Gyr.

14

Prospecting for Life

14.1 Rare Earth versus the Principle of Mediocrity

The principle of mediocrity says that there is nothing about us, our planet, our star, or our Galaxy that are special. From this we infer that the universe must be teeming with life and that even intelligent life should be common. The principle of mediocrity underpins the Copernican and Darwinian revolutions and is supported by modern observational astronomy. In stark contrast is the rare Earth hypothesis that argues that life is extremely rare because of the fine-tuning required for its existence. In particular, this includes being in the Galactic habitable zone, being in a circumstellar continuously habitable zone around a star of spectral type F7 to K1 only, having a stable orbit unperturbed by the migration of other planets in the system, being a rocky planet of the right mass with plate tectonics, and a large moon to stabilize obliquity. The rare Earth idea also calls upon the idea that complex life (eukaryotes) arose on Earth only once and took ~1.7 Gyr to do so, implying that it is a low probability event.

In what follows, we explore issues raised by the debate between rare Earth and mediocrity.

14.1.1 Is an Early Origin a Guide to the Probability of Life?

Many astrobiologists argue that since life originated on Earth soon after it became habitable it follows that life in the universe is common. The first part of this argument is hard to assess because we do not know when the Earth became habitable or when life originated. If we except there was liquid water on the Earth's surface as early as 4.4 Gyr ago and life originated 3.5 Gyr ago, then we have 900 Myr. This is rather a long time but is probably about the maximum. There are good arguments to be made that would delay the onset of habitability and we could speculate about an earlier origin of life, thereby shortening the timescale.

The second part of the argument seems intuitively obvious. If life got started here as soon as it could then surely the origin of life must be a high probability event. This follows from the fact that improbable events are infrequent so we would accept to have to wait longer for them to happen. But this common sense view may be misleading and several researchers have developed statistical models to test it.

14.1.1.1 Earth Is Not a Randomly Selected System

If the Earth had been randomly selected for study, this early appearance of life could be interpreted to mean that the origin of life is a high probability event. We might then expect life to be commonplace in the universe. In a randomly selected system, assuming that biogenesis proceeds via a series of discrete steps, the time taken to make a step is related to how difficult that step is and is a measure of the probability of its occurrence. It takes longer to make hard, low probability transitions.

But the Earth is not a randomly selected system; its selection is biased by the fact that it is the only exemplar we know about and this changes the inferences we make. A second selection effect is that it took nearly 4 Gyr for evolution to lead to organisms capable of pondering the probability of life elsewhere in the universe. If this time is necessary, then it would be impossible for us to find ourselves here if life had arisen much later than it did. In other words, if 3.5 Gyr is required for intelligent life capable of asking astrobiological questions, then we had to find ourselves on a planet where life arose relatively early, regardless of the probability of biogenesis.

Several statistical analyses have been derived to estimate how likely it could be that life emerges.

14.1.1.2 If Biogenesis Is a Lottery It Is Likely

An analysis by Charles Lineweaver and Tamara Davies comes from the mathematics of lottery gambling, estimating the chance of winning a lottery by how quickly a lottery winner wins, and conclude that life might indeed be common. They argue that our existence on Earth does not say anything about the probability of biogenesis. It could be infinitesimally small—indeed the Earth could be the only inhabited world in the universe—and necessarily we find ourselves on it. If this were the case, it would imply that the Earth was unique, though this is not supported by the evidence. In this situation, where our existence does not constrain the probability of biogenesis, it could be anywhere between 0 and 1. However, the fact that biogenesis on Earth was fast can be used to constrain the likelihood of biogenesis occurring within a given time window. The mathematical model converts this into constraints on the fraction of habitable planets that actually have life, a parameter termed f_ℓ. Assuming 600 Myr as an upper limit for the time life took to appear on Earth, Lineweaver and Davis calculated that for terrestrial planets older than 1 Gyr, f_ℓ is most probably close to 1 and >0.013 at the 95% confidence level. The model includes a time during which biogenesis is frustrated by sterilizing impacts. It is not valid if the impacts of the late heavy bombardment actually enhance the likelihood of the origin of life, as some researchers have argued. But taken at face value, if habitable planets are common, the implication of this study is that life is common in the universe. A large uncertainty is how close to Earth-like a terrestrial planet has to be and still have the probability of biogenesis close to that of Earth. This is really just another way of asking the question: How special is the Earth?

14.1.1.3 If We Account for Selection Effects Biogenesis Is Rare

Another approach is to use Bayesian statistics, which takes account of the selection effects, to estimate the chance of life originating. Bayes' theorem allows the probability of an event to be calculated given the likelihood of some other event on which it is conditional. For example, we could use Bayes' theorem to calculate the probability that a black vehicle on a London street is a London taxicab, given the prior information that London taxicabs are black.

BOX 14.1   POISSON DISTRIBUTION

A Poisson distribution describes the probability of a given number of events occurring in a fixed time interval (e.g., buses per hour on a given route), given that the events occur with a known average rate (e.g., six buses per hour), and each event occurs independently of the time of the previous one (which might be the case for buses on a journey randomly afflicted by traffic jams). It predicts the spread (variance) of the number of occurrences around the known average number (mean), in a given time. Poisson distributions have the property that the mean and variance are the same. It is the probability distribution used for events that happen relatively rarely (biogenesis), but which have plenty of opportunities to happen (lots of habitable planets).

David Spiegel and Edwin Turner have a model in which biogenesis can only occur during a limited time window when an Earth-like planet is habitable. Early on, a planet is assumed to be inhospitable because of massive impacts, and the upper limit to the interval will be set by the main sequence lifetime of its host star. The probability of biogenesis is evaluated assuming a Poisson distribution (Box 14.1), after taking account of the fact that it happened within the required time and sufficiently early for humans to evolve subsequently—that is the Bayesian bit. The details are complicated but the outcome is instructive. Although the early emergence of life on Earth could mean that extraterrestrial life is common, the model is consistent with life being extremely rare. The "take home" message is that early biogenesis does not necessarily mean that biogenesis has a high probability: it is a selection effect of our existence; the probability of biogenesis may be low.

This approach comes with a health warning. The Poisson distribution is a huge simplification since it ignores the idea that biogenesis is likely to be complex sequence of events that play out over time rather than a single event. That said, some biochemists argue that the *process* of biogenesis, getting the first living system (whatever that was) from abiotic chemistry would be extremely rapid, perhaps just tens or hundreds of thousands of years. Once all necessary conditions were satisfied life would emerge: delay and the opportunity might be lost as environmental perturbations disrupted nascent biochemical networks. Of course, this is speculative. Since we do not know how life originated we do not know how robust or vulnerable prebiotic systems on the cusp of life were.

EXERCISE 14.1

Speculate on the effect that vulnerability of prebiotic systems to environmental perturbations might have on the probability of biogenesis. Statistical models are not required!

The most important way to refine this Bayesian approach would be the discovery of life that has arisen independently on another planet, or evidence for a second genesis on Earth. This would strengthen the idea that life might be common.

14.1.2 Life May Be Rare or Common in the Galaxy

If we know how probable it is that life evolves on a given suitable world, then we can estimate how common life is in the Galaxy. One way to do this is to use the expression:

$$F = f_p \times n_e \times f_\ell, \qquad (14.1)$$

where F is the fraction of stars in the Galaxy today orbited by planets on which life has emerged independently, f_p is the fraction of stars in the Galaxy with planetary systems, n_e is the fraction of planetary systems with a habitable planet, f_ℓ is the fraction of habitable planets on which biogenesis has happened.

All the evidence says that exoplanets are common. Indeed, to a first approximation we can argue that $f_p \approx 1$ so Equation 14.1 becomes:

$$F \approx n_e \times f_\ell.$$

At present, as we have seen, we cannot quantify f_ℓ with any certainty. Statistical analysis of exoplanet surveys can provide a handle on n_e but there is an alternative approach to estimating n_e. If we go with the assumption that the Earth is not special (i.e., that rocky planets with $M \approx M_\oplus$, a global magnetic field and plate tectonics, in continuously habitable zones, are common), then n_e could be of order 1, and F is then determined largely by f_ℓ. A counterargument is to assume that the Earth is a typical habitable planet and that it is also special in several ways. Thus, if we consider just some of the parameters thought to be important for habitability and consider the range of values they have for Earth, then a simple statistical argument gives a much smaller value for n_e. Thus, 95% of stars in neighborhood are less massive than the sun, 90% of planets are smaller than Earth as deduced from Kepler statistics, 96% of planetary orbits are more eccentric than Earth's and 90% of stars lie closer to the Galactic center than the sun. From this we infer that a typical habitable planet is a high mass rocky world in a circular orbit around a star in the upper decile for mass located toward Galactic edge. Using Bayes theorem to work out a habitability likelihood for each of these properties, then Monte Carlo simulations (Box 14.2) to determine the number of habitable planets, Dave Waltham shows that only 10^{-4} randomly

---

BOX 14.2   MONTE CARLO METHODS

Monte Carlo methods are algorithms that use a large numbers of trials to obtain a numerical result for some variable. They are generally used for problems that cannot easily be solved analytically. In particular, they can be used to generate samples from a probability distribution. For example, a Monte Carlo simulation of the behavior of repeatedly tossing a fair coin could be made by drawing numbers randomly from [0.1, 0.2, 0.3, 0.4, 0.5, 0.6, 0.7, 0.8, 0.9, 1.0], and assigning values less than or equal to 0.50 as heads and greater than 0.50 as tails. A single trial is a simulation but not a Monte Carlo simulation: this is a statistical procedure, so requires a large number of trials. As the number of trials increases, the numbers of heads and tails more closely approximates the expected 50% for each. In essence, Monte Carlo methods are similar to games of chance played in a casino; hence their name!

chosen planets will have all four parameters in the right range. For each parameter introduced, the number of habitable planets will drop. In this model F is determined by n_e and f_ℓ, with n_e small.

14.1.3 Several Hard Steps Could Be Needed for Intelligent Observers to Emerge

It took ~4 Gyr for humans to evolve on Earth and our planet will be uninhabitable in about another billion years. In other words, intelligent observers emerged toward the end of the time when Earth is habitable. The implication is that intelligence is rare. This could be because it requires several hard steps. This idea was proposed in 1983 by Brandon Carter. He argued that we would expect life to have appeared early, relative to when intelligent life appeared, if a sequence of n critical, hard steps that took a long time were required for intelligent observers to evolve. In other words, the selection effect of our presence accounts for early biogenesis on Earth by virtue of the necessity for several low probability steps for humans to evolve. Carter's statistical model implied that $n < 10$, with a best guess being $n = 4$.

The argument is that the hard, critical steps happen stochastically with differing but very low probability. Each hard step had to follow previous ones in the right order. Normally, evolution proceeds relatively rapidly when adaptive innovations are easy, high probability events. By contrast, hard steps stall the pace of evolution because they are highly unlikely to occur. In other words, the hard, critical steps determine the timescale of evolution.

In 1998, Robin Hanson of George Mason University returned to this issue. He used Monte Carlo simulations of the times it took to pick locks that could be opened with various levels of difficulty. He showed that if there is a *limited* time for the successful picking of several locks in succession, the time taken to open the hard locks is about the same regardless of how hard they actually are. This sounds counterintuitive, should not locks take longer to pick the harder they are to open? This would be true if time was not restricted, and all attempts lead to success. But that is not how life on Earth has evolved. There has only been a limited period from biogenesis to the emergence of humans, so evolution has successfully negotiated the required n hard steps, but presumably, there are an unknown number of habitable worlds on which the sequence of hard steps stalled at various stages; that is, failures. Earth got lucky. The outcome of the Monte Carlo simulation is that the n hard steps tend to be equally spaced in time. Put another way: for easy steps, the time taken to make the step is a reliable guide to its probability, but for hard steps the step duration says nothing about their probability. It means that the origin of life might be harder than its early appearance on Earth suggests.

14.1.3.1 How Many Steps Are Needed?

There have been several attempts to home-in on the number of critical steps, n, required for the evolution of complex life and intelligent life. The larger n is, the longer the time needed. Hence, n depends on assumptions about how long life has existed on Earth, and how long it will continue to be habitable.

Andrew Watson of the University of East Anglia, UK, has derived probability density functions (how the probability of an event changes with time) for each of the steps in models with different values for n. This allowed him to calculate the probability

TABLE 14.1

Probabilities for the Emergence of Prokaryotes Depend on the Time Available

Earth Habitable from (Gyr)	Prokaryote Origin (Gyr)	Probability
4.4	3.5	<0.01
3.85	3.5	<0.003
3.85	3.7	~0.001

Source: Data from Watson (2002).

that a particular number of steps would happen in a given time. If we consider the set of all habitable planets, on most of them no hard steps will occur, and for those in which the first happens, on only a few will the second step take place. The more steps there are, and the shorter the time available, the lower the probability of getting to a particular stage in evolution. Clearly, the subset in which all *n* steps required intelligent observers to emerge will be extremely small.

To see how this works imagine that it takes three critical steps to give birth to prokaryotes: the shorter the time available the lower the probability that prokaryotes emerge (Table 14.1).

Watson regarded these probabilities as worryingly low. For example, if $n = 8$, the probability that intelligent observers emerged would be vanishingly small. He suggested two solutions. One is that terrestrial life first originated on Mars, over longer timescales (thereby pushing up the probabilities), and was delivered to Earth on a Martian meteorite. As we shall see there are several intriguing arguments for a Mars genesis, though none is compelling. A second solution is that there is only a single critical step involved in the origin of prokaryotes. This automatically increases the likelihood of their emergence. By assuming one step to prokaryotes and only a further three steps to get to humans (i.e., $n = 4$), and taking Earth as being habitable for ~5 Gyr in total (~4 Gyr before humans, and about one billion years left), Watson's model gets a probability for biogenesis not too different from that of Lineweaver and Davis.

14.1.4 What Are the Hard Steps?

The number of critical steps depends on assumptions about how long life has existed on Earth, and how long it will continue to be habitable. For terrestrial life, given its 5 Gyr tenure, an upper value for $n \sim 8$ to get intelligent observers is sensible. Higher values produce unrealistically low probabilities. But which of the many adaptations life made were the hard transitions?

14.1.4.1 Several Hard Step Sequences Have Been Proposed

A number of scientists, from cosmologists to evolutionary biologists, have suggested sequences of hard transitions that range from $n = 1$ to $n = 10$. Below is a heavily triaged list of postulated and plausible contenders:

- Abiotic synthesis of the first replicating system
- Origin of protein synthesis controlled by nucleic acids

- Origin of free-living prokaryotes (acquisition of a cell membrane)
- Oxygenic photosynthesis
- Origin of eukaryotes
- Origin of photosynthetic eukaryotes
- Emergence of intelligence with language; that is, intelligent observers

14.1.4.2 Unique Events Are Presumably the Hardest

A major difficulty is deciding which steps are indeed critical. We know that it is highly unlikely that several critical events will be bunched in time because we expect them to be evenly distributed. One discriminator is to insist that critical steps are those that have such a low probability they happened only once.

14.1.4.3 Some Events May Have Happened Only Once

A sensible contender for a one-off event is the evolution of the photosystem II water-splitting complex that allowed oxygenic photosynthesis. That evolution stumbled across a mechanism to catalyze such a highly thermodynamically unfavorable reaction once seems remarkable, and the likelihood of this happening multiple times must be very low. That this was a unique event is supported by the fact that the complex is very similar in all oxygenic photosynthetic organisms.

Many astrobiologists would argue that the origin of eukaryotes is the most difficult transition made by life on Earth. It represents a step change from simple life to complex life. It certainly took some time. Using biogenesis at 3.5 Gyr and the origin of eukaryotes at 1.8 Gyr gives 1.7 Gyr, with considerable uncertainty.

14.1.4.4 Is Intelligence with Language Sufficient?

As for intelligence with language, whether this was a singular event on Earth depends on how we view the intelligence and sophisticated communication of which other animals—for example, squid, dolphins, or chimpanzees—are capable. Perhaps the discriminator is that humans can communicate much more widely than any other animal by virtue of technology. This has been made possible by the development of the formal languages of mathematics and computing, which can be used to manipulate concepts logically and quantitatively in ways not achieved by natural languages (English, Mandarin Chinese, Arabic, etc.).

14.2 Is Life Seeded from Space?

There seems no doubt that organic molecules synthesized in space have been delivered to Earth, but have living micro-organisms arrived here from interplanetary or even interstellar space? This idea, panspermia (Greek: "all seed"), is that life is propagated from one world to another throughout the Galaxy is an old idea but extraordinarily hard to test.

Recent interest in panspermia has been awakened by the realization that some terrestrial microorganisms can survive in extremely hostile environments. Cyanobacteria in Antarctica, for example, have to endure temperatures below −50°C, and very high levels of UV light beneath the ozone hole.

It is important to recognize that panspermia is not a hypothesis about the origin of life, though it is often assumed that if it happens then life on Earth originated elsewhere. If this were the case, then we do not know where life originated (unless we turn out to be obviously related to Martian life) so our prospects for understanding how life originated in any detail are seriously thwarted. It also invalidates arguments about how common life is that rely on the timing of the emergence of life on Earth.

14.2.1 Survival of Ancient Bacteria Makes Panspermia Plausible

Bacteria can package themselves into a thick, protective outer coat and go into suspended animation in which their metabolism drops to virtually zero, becoming bacterial spores. These are extraordinarily hardy to heat, desiccation, ionizing radiation, and damaging chemicals.

A number of reports suggest that bacterial spores can remain viable for a long time. The record, by a large measure, is 250 million years! This work identified spores in brine inclusions (small isolated "bubbles" of salty water) within salt crystals deposited in the Salado beds, New Mexico, during the Permian period. The age was established both by the nature of invertebrate fossils in the vicinity and by radioisotope dating of minerals. Fluid removed from the inclusions, when cultured, grew living *Bacillus* bacteria.

Of course, all claims for resuscitation of ancient microorganisms are open to the criticism of modern contamination but this study went to extraordinary lengths to reduce the risk of contamination, at most, to one part in a billion. Consistent with the idea that the organisms were genuinely old was that a DNA sequence of one of the strains revealed it to be a previously unknown variety of *Bacillus*.

If these bacteria are authentically Permian in age, then long timescales are not a bar to panspermia *per se*.

14.2.2 Life May Be Uncommon Despite Panspermia

If interstellar distances are traversable, given the age of the Galaxy, life could be widely distributed. However, this does not mean that life is common. For this to be the case, panspermia would need to be effective; that is, it would deliver to a high number of habitable planets and have a high success rate in establishing itself. It does not matter that organisms are widely disseminated to hundreds of worlds if few manage to establish themselves in the new environments. In other words, panspermia could be ineffective in two ways. It may not propagate well or it may not establish well.

For panspermia to be plausible micro-organisms, or at least their spores, must be sufficiently robust and/or protected to survive a variety of hazards during the journey. The most obvious way in which panspermia could happen is if organisms are carried by a small solar system body such as an asteroid, meteorite, or the nucleus of a comet (so-called lithopanspermia). Alternatively, stellar radiation pressure could provide motive power, but this would only work for particles up to a few microns across: naked DNA, individual bacteria, at most a small clump of bacteria.

14.2.3 Radiation Is a Major Hazard

Cosmic radiation and UV radiation are serious hazards faced by micro-organisms in space because they have sufficient energy to damage DNA. Cosmic radiation and short wavelength UV ($\lambda < 125$ nm) are ionizing radiation because they carry enough kinetic energy to ionize any atom or molecule they collide with. Ionizing radiation forms free radicals such as the superoxide anion O_2^- or the hydroxyl radical OH in the presence of water or oxygen molecules and it is these free radicals that are responsible for much of the DNA damage caused by ionizing radiation. Hence, in the dry, oxygen-free environment of space, relatively few free radicals would be generated, and the extreme cold would limit the diffusion of any that were made. We might therefore expect exposure to space to attenuate the damaging effect of radiation.

EXERCISE 14.2

a. Calculate the energy of a photon with $\lambda = 125$ nm.
b. Does the value you calculate represent the lower bound on the energy of a photon that can ionize a molecule or atom?

14.2.3.1 Some Bacteria Have High Radioresistance

To be able to assess the likelihood of bacteria surviving the radiation gauntlet it would be useful to see how tolerant they can be. *Deinococcus radiodurans* is one of the most radiation tolerant organism known (Table 14.2), by virtue of having multiple copies of its genome and an ability to repair DNA damage rapidly. It can withstand an acute dose of 5000 Gy (Box 14.3), enough to inflict hundreds of double stranded breaks in its DNA, with little loss of viability. It is also able to survive the vacuum, desiccation, and cold of space. This impressive toughness has earned *D. radiodurans* the epithet "Conan the bacterium," although other organisms, for example, the Archaean *Thermococcus gammatolerans*, have similar radioresistance.

TABLE 14.2

Some Absorbed Radiation Doses

Condition	Absorbed Radiation Dose[a] (Gy)
Medical x-ray	$<10^{-3}$
Low level radiation sickness	0.5
Human lethal	5
E. coli lethal	200–800
Tardigrade (animal) lethal	4000
D. radiodurans lethal	15,000
Limit for radiation-hardened electronics	10^9

[a] Acute doses except in the case of the electronics.

BOX 14.3   RADIATION DOSE

The SI unit of absorbed radiation dose is the gray (Gy). One gray is the absorption of one joule of energy, in the form of ionizing radiation per kilogram of matter. $1\ Gy = 1\ J/kg = 1\ m^2\ s^{-2}$. The equivalent unit in the cgs system is the rad, equivalent to 0.01 Gy.

The biological effect of absorbed radiation depends on the nature of the radiation, and this is expressed as equivalent dose, the SI unit of which is the sievert (Sv). $1\ Sv = 1\ J/kg$. The relationship between absorbed radiation dose, D, and equivalent dose, H, is:

$$H = QD,$$

where Q is a factor that depends on the nature of the radiation. It is 1 for ionizing electromagnetic radiation and β particles, 10 for fast neutrons and protons, and 20 for α particles.

EXERCISE 14.3

Natural background radiation averages ~0.4 mGy with the highest natural background on Earth, in Iran, being 260 mGy. In the light of this, comment on the evolution of highly radioresistant organisms such as *D. radiodurans*.

In fact, evidence suggests that radioresistance in micro-organisms is a "side effect" of the adaptations that have evolved to survive desiccation. In other words, *D. radiodurans* and its cousins happen to be resistant to radiation not because they evolved where ambient radiation dose was high but because their environment was sometimes arid.

14.2.3.2 *Experiments Have Revealed Limits to Survival in Space*

A number of experiments done in low Earth orbit (~300 km) to test the effects of exposing micro-organisms to space have been conducted from 1966 onward, including most recently on the International Space Station. ESA's 1992 Exobiology Radiation Assembly (ERA) experiments examined the effect of UV on *Bacillus subtilis* spores over 11 months. These showed that ultraviolet radiation is a big killer and that the drop in the viability in organisms directly correlates with the amount of damage to the DNA. Even exposure to the vacuum of space damages DNA and this is synergistic with that caused by UV. In other words, the combined effect of vacuum and UV on DNA is more than the linear addition of the two agents acting separately. These results run counter to the idea that dry, anoxic, and cold environments attenuate DNA damage by UV. This may be because longer wavelengths of UV are not very ionizing and do not do so much of their harm via free radicals.

Cosmic radiation posed problems as well. Even hardy *D. radiodurans* failed to survive 7 months unprotected in space, having suffered terminal DNA damage.

ESA's experiments between 1992 and 2007 using the BIOPAN facility located on the external surface of the Russian Foton re-entry capsule allowed exposure to space over 10 to 17 days to be investigated. One study revealed that shielding from UV by a

10-μm thickness of rock dust was needed to protect a monolayer of *B. subtilis* spores. This seems to rule out the hypothesis that radiation pressure could act as a vehicle for panspermia.

14.2.4 Can Micro-Organisms Survive Lithopanspermia?

Encasing micro-organisms in a substantial thickness of silicate rock or a rock/ice mix affords protection against radiation, so meteorites and comets could be vehicles for long journeys through space in relative safety. This scenario is lithopanspermia.

For lithopanspermia to work the vehicle has to deliver its life to the surface of a habitable world. This means frictional aerobreaking ("re-entry") in an atmosphere followed by a relatively low velocity impact. In addition, a meteorite will have been launched from the surface of another world at a velocity greater than escape velocity at the start of the journey. Remarkably, there are terrestrial organisms that could withstand the stresses.

14.2.4.1 Extremophiles Can Survive High Accelerations

Meteorites or comets have velocities of 10–70 km s^{-1} on entering Earth's atmosphere and experience huge deceleration forces and ram pressure heating as they enter an atmosphere. This often causes them to explode several kilometers above the surface. Such an airburst was almost certainly responsible for the Tunguska event in Siberia in 1908. Any impactor fragments that survive the fireball are likely to land with relatively low velocity—about 500 km per hour. Stony meteorites get heated above the melting point of rock but the ablation of rock liquid and vapor carries away much of the heat. In addition, rock is a poor conductor of heat. For these reasons, the fusion crust on the surface of meteorites (where liquid rock has refrozen) is very thin (~1 mm), and it is easy to show that the inside of a meteorite is barely warmed by its passage through the atmosphere. In summary, it is perfectly plausible for micro-organisms to survive landing if cocooned by a few centimeters of rock.

The challenge posed to an organism of being launched by a high velocity impact to greater than escape velocity is greater than a modest meteorite landing. However, simulations show that ejecta at the edge of an impact crater can be accelerated to escape velocity without being excessively heated or shocked. *D. radiodurans* and *B. subtilis* have been demonstrated to survive accelerations as high as 33,000*g*.

14.2.4.2 Are We All Martians?

If early Mars was habitable, it has been argued that there are several reasons to expect life might have originated more readily on Mars than Earth. We consider each of these in turn.

1. Mars was not as heavily bombarded as Earth because of its lower gravity. Some astrobiologists argue that impacts actually helped create habitable environments on Mars so a lower impactor flux is not necessarily positive. For example, phosphorus could be provided in a biologically accessible form as phosphite derived from a phosphide mineral in iron–nickel meteorites.

2. The effect of a large impact on Mars is less than on Earth because of the lack of an ocean to vaporize and send a lethal heat pulse into the crust. However, the warm, wet Mars model both enhances habitability and makes a Northern ocean on Mars more likely, so the efficacy of this mitigation is not clear.

3. Being smaller, Mars would have cooled more rapidly and hence a deep subsurface habitable zone for hyperthermophiles could have formed earlier, where nascent life would have been protected from UV radiation.

4. The low surface gravity of Mars would have meant hydrodynamic escape of hydrogen from the photodissociation of water vapor, and hence the formation of redox gradients would have been earlier.

5. Steven Benner of the Westheimer Institute for Science and Technology, USA, argues that early Mars could have had borates and molybdates that were lacking on Archaean Earth. These would facilitate the synthesis of organic molecules useful to biochemistry, in particular ribose and then RNA. Without them, the products of organic reactions driven by heat or UV light are tars; complex mixtures of polycyclic aromatic hydrocarbons (PAHs) and heterocyclic compounds (those with more than one element in their rings) that are intractable to biochemistry. Mars has borates and molybdates because elemental boron and molybdenum were oxidized by oxygen produced by the photodissociation of water. Earth would have lacked them until 2.4 Gyr ago. In addition, the boron abundance on Mars is much higher than on Earth. Benner also maintains that the more arid conditions on Mars, compared with water-covered Earth, would limit the hydrolysis of ribose and hence make RNA synthesis easier. However, synthesis of nucleotides without ribose as a reactant has now been demonstrated under plausible aqueous conditions.

6. Low prebiotic availability of phosphate has been cited as a limiting factor in terrestrial prebiotic chemistry. Bio-available phosphate now comes from biological recycling when organisms die, but initially it had to be derived from the breakdown of phosphate minerals, mainly apatite, during water–rock interactions. Experiments show that chlorine-bearing apatite in Martian meteorites are considerably more water-soluble than terrestrial hydroxyl- and fluorine-apatites. Phosphate concentrations of early wet Martian environments could be more than twice that of Earth.

As we can see, some of the physicochemical arguments for Martian genesis being earlier than a terrestrial one are not that convincing. Starting from the premise that life is a freak accident, Paul Davies of Arizona State University has developed a statistical argument that it is far more likely that life started on Mars. If we accept this, it is not impossible that lithopanspermia transported Martian micro-organisms to Earth. For dynamical reasons, transfer in the reverse direction is far less likely. It follows that if they survived, and no separate genesis occurred on Earth or left descendants, then we are all Martians.

EXERCISE 14.4

Why is panspermia from Earth to Mars more difficult than from Mars to Earth?

In light of this, if past or extant life is discovered on Mars, there are two possible outcomes, both profoundly interesting. One is that it has a biochemistry that is obviously like that of Earth. If this is the case, either terrestrial life was derived from Mars or (with lower probability) Earth life infected Mars. The other is that the biochemistry on Mars is very different from that on Earth. This would imply that Martian life is a separate genesis. This would reset the statistical arguments about how common life is in the universe.

14.3 Metrics for Extraterrestrials

The possibility that we might be able to detect radio transmissions broadcast by technological civilizations on worlds around other stars was first proposed by Guiseppe Cocconi and Philip Morrison in a paper published in 1959 in *Nature*. An artificial signal, unlike a natural one, would be expected to have a narrow bandwidth (i.e., a narrow frequency range; see Box 14.4). Noise is proportional to bandwidth, so the narrower the bandwidth the higher is the signal-to-noise ratio. They suggested that the best frequency to search would be 1420 MHz—the 21 cm radio emission line of neutral hydrogen—since extraterrestrials attempting interstellar radio communication could reasonably expect that receivers tuned to this wavelength would be developed early in radio astronomy, which has been the case on Earth. They also pointed out that background noise from the atmosphere and Galaxy is minimal between 1 and 10 GHz. Cocconi and Morrison suggested listening for signals from sun-like stars within 15 light years. Of these, four stars were in regions of the sky with low Galactic background 21 cm emission.

14.3.1 Is the Drake Equation More Than a Guess?

In 1960, Frank Drake used the 26-m Tatel dish of the National Radio Astronomy Observatory at Green Bank, West Virginia to hunt for 21 cm signals from two such stars Epsilon Eridani and Tau Ceti over 4 months. Although unsuccessful, Drake subsequently hosted a search for extraterrestrial intelligence (SETI) meeting at Green Bank. In preparation for this, he wrote the Drake equation, an amalgam of the factors that would determine the number, N, of detectable civilizations in the Galaxy. The civilizations are detectable by virtue of their radio communications, at least as originally envisaged. The Drake equation is:

$$N = R^* \times f_p \times n_e \times f_\ell \times f_i \times f_c \times L, \tag{14.2}$$

where R^* is the average number of stars formed in our Galaxy per year, f_p is the fraction of those stars that have planets, n_e is the fraction of those planets that are habitable, f_ℓ is the fraction of habitable planets that actually develop life, f_i is the fraction that develop intelligent life, f_c is the fraction on which intelligent life develops technology that broadcasts its signals into space, and L is the length of time over which the broadcasts occur/years.

Note that since the f terms are ratios the unit of N is R^*L = number per yr × yr = number, as required. Our ability to evaluate each of these terms, and our confidence in the values we give them, drops as we go to the right.

The first three terms in the equation are currently amenable to direct observation. The most recent estimate puts the star formation rate, $R*$, in the Milky Way as 7 per year. To a first approximation, exoplanet statistics suggests that each star is host to one planet, and the Kepler team have argued from estimates of $\eta_\oplus$ that 5.4% of all stars may have a habitable zone, although if red dwarfs cannot host habitable worlds this figure is too large by a factor of ten. This figure does not take account of any Galactic habitable zone, which could also reduce $\eta_\oplus$ by a factor of 10.

The number of habitable planets on which life actually emerges, f_ℓ, is not yet knowable, and statistical modeling has thrown up estimates for biogenesis probability from close to 1 (>0.13) to 0.001. Brownlee and Ward's original rare Earth hypothesis gives the combination of $n_e \times f_\ell$ as 10^{-11}; in other words, Earth is more or less the only inhabited planet in the Galaxy.

The final three terms in the Drake equation are even harder to evaluate. We might argue that the f_i term at least requires a definition of intelligence. On Earth, do we include all or some of the great apes, or whales, or squid? If we confine intelligence to hominids how far back into our prehistory do we go? But this may not be so important; whichever we choose, intelligent life emerged late on Earth, some 3.5 Gyr after the origin of life, a time that is within a factor of two of the main sequence lifetime of a sun-like star. We cannot be sure that the late development of intelligent life means that it is a low probability event. However, the evolutionary biologist Ernst Mayr has put $f_i = 10^{-9}$ on the grounds that billions of species have come and gone over the lifetime of the Earth but only one developed intelligence. The counterargument is that as life has become increasingly complex, the development of intelligence becomes ever more likely.

The ability of an organism to communicate across space, as instantiated in f_c, depends not just on its intelligence but also on whether it is physically capable of building the technology. The nature of a planet can constrain the evolutionary paths available to an organism. For example, the size and shape of organisms are constrained by gas diffusion, heat transfer, power/weight ratios, and so on, which depend on planetary variables such as atmospheric pressure and composition, temperature, and acceleration due to gravity. This might make it impossible for tool-making species to evolve. Even on Earth where toolmakers have emerged, this has only been the case for a few species. Dolphins are clearly intelligent but they do not have the anatomy or physiology needed to fashion tools in stone, metal, or silicon, as we have. There is also the possibility that intelligent organisms with the physical attributes to build technology might not have the behavioral imperative to do so: conditions on their world might be so pleasant there is no requirement for them to do so.

The lifetime of a communicating civilization is the Drake equation parameter that is arguably the hardest to pin down. Such a civilization may occur several times on a single world, with episodes of radio-silence of variable durations intervening as a result of natural or societal catastrophes. The total number of civilizations on such a world would be given by the reappearance factor $(1 + n_r)$ that is added to the equation, where n_r is the average number of times a civilization reappears on a planet that has experienced one previously that has vanished. Each of these would have its associated lifetime.

A single civilization may colonize several star systems, spreading out from its home world with a given expansion speed. Each of these colonies would survive for its own lifetime. This scenario requires a set of three equations.

Even the length of a single spatially fixed civilization is currently indeterminate. We can construct two end-members. Earth has had a communicating civilization for about 100 years (though whether our radio leakage could be detected out to a sphere of radius 100 light years is questionable, see Section 14.4.1) but during this time, nuclear weapons with the capability of destroying that civilization have been developed. The pessimistic inference is that L could be measured in a few tens or hundreds of years. The other, optimistic, end-member is a communicating civilization that survives as long as Earth remains habitable: sometime between 0.5 and 1 Gyr. The difference this makes to N is of course enormous.

The original estimate of the 1961 Green Bank meeting was that $N \sim L$ and was between 10^3 and 10^8 civilizations (see Table 14.3).

The only parameters for which we have reasonably confident values now are $R* = 7$ and $f_p \sim 1$. For the remainder it is anyone's guess, although we know that N is at least 1! One wit has described the Drake equation as packing the greatest amount of ignorance into the smallest space. Does this mean that it is a waste of time? No. The original intention was to focus minds, not come up with a definitive number, and in that, it has succeeded.

14.3.2 Alternatives to the Drake Equation Have Been Developed

Sara Seager, a planetary scientist at Massachusetts Institute of Technology, motivated by the impossibility of evaluating many of the terms in the Drake equation, has put together a reduced form with the more modest goal of estimating how many planets in a realistic-sized star survey would have detectable biosignature gases.

$$N = N* \times F_Q \times F_{HZ} \times F_O \times F_L \times F_S \qquad (14.3)$$

Here, N is the number of planets with detectable biosignature gases, $N*$ is the number of M-type stars within the sample, F_Q is the fraction of quiet stars, F_{HZ} is the fraction with rocky planets in the habitable zone, F_O is the fraction of observable systems, F_L is the fraction with life, F_S is the fraction with detectable spectroscopic signatures.

Seager reckons that our best chance of detecting biosignature gases is of planets orbiting the M-type stars since they dominate the stellar population. Many M-type stars are flare stars, which could expose planets in the habitable zone to potentially lethal fluxes of radiation, hence the requirement to look at quiet stars. Only for a fraction of the planets will it be possible to obtain the spectral signature of the atmosphere, hence the F_O term.

TABLE 14.3

The Original "Solution" to the Drake Equation

Parameter	Value
$R*(\text{yr}^{-1})$	1
f_p	0.2–0.5
n_e	1–5
f_ℓ	1
f_i	1
f_c	0.1–0.2
L (yr)	10^3–10^8
N	$\sim 10^3$–10^8

TABLE 14.4

Estimate for the Number of Planets with
Detectable Atmospheric Biosignatures
in a Sample of 30,000 M-Type Stars

Parameter	Value
N^*	30,000
F_Q	0.2
F_{HZ}	0.15
F_O	0.001
F_L	1
F_S	0.5
N	2

Table 14.4 gives Seager's own estimates for the parameters in her equation. N^* is defined, the next two terms are amenable to observation, and F_O can be modeled without recourse to controversial assumptions. This leaves only two parameters that are currently guesses, though F_S could be determined from observation.

14.4 SETI and the Fermi Paradox

The search for radio signals from Epsilon Eridani and Tau Ceti—Project Ozma—was the start of the Search for Extraterrestrial Intelligence (SETI). Project Ozma used the 26 m Tatel telescope at Green Bank and a low noise single channel amplifier to scan the 400 kHz band around the 1420.4 MHz neutral hydrogen line. This part of the microwave spectrum is dubbed the "water hole" because of its proximity to the hydrogen and hydroxyl radical lines (Figure 14.1). The signals were recorded on tape

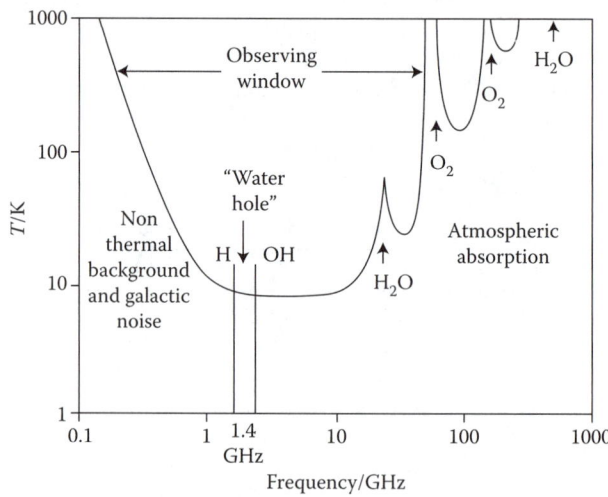

FIGURE 14.1 The terrestrial microwave window.

for later "off-line" analysis but nothing unusual was detected except for the then top secret US U2 spy plane!

A nice feature about broadcasting at radio frequencies is that they are not much broadened by their journey through the interstellar medium, and digital processing and Fourier analysis allows the frequency components of a radio signal to be nicely resolved. However, if the extraterrestrial source and Earth are moving relative to each other, then signals will be Doppler shifted. This is why a range of frequencies is scanned rather than just a single frequency.

A relative speed of $\pm v$ km s^{-1} between the source and the Earth will produce a Doppler shift given by

$$[(\lambda_e - \lambda_0)/\lambda_0] = v/c$$

so,

$$\lambda_e/\lambda_0 = (1 + v/c), \qquad (14.4)$$

where λ_e is the emitted wavelength and λ_0 is the observed wavelength.

EXERCISE 14.5

Rewrite Equation 14.4 in terms of frequency and calculate the frequency range a SETI search should be conducted to detect a 1420 MHz signal if the extraterrestrial source and Earth are moving with a relative speed of ± 300 km s^{-1}.

14.4.1 Radio SETI

The longest running program, the Ohio State University SETI program (OSU SETI), operational from 1963 to 1997, used a Kraus-type radio telescope in which a flat primary collector reflected radio waves to a parabolic reflector. In 1977, it recorded a 72-s high signal-to-noise ratio extraterrestrial signal at 1420 MHz. This "Wow!" signal has never been seen again and there is no explanation for it.

Despite a plethora of SETI programs using ever more advanced technology, such as the SERENDIP program of the University of California, Berkeley, which currently uses the 305-m fixed Arecibo radio telescope no convincing signal has been received.

14.4.2 Optical SETI

In 1961, just a year after Frank Drake's first radio search, Charles Downes, a coinventor of the laser, proposed that aliens might attempt communication by nanosecond pulses of laser light. We would detect this as a very powerful emission line in the optical spectrum of a star. In principle, information could be conveyed by variations in the frequency with which the pulses arrived.

There are a number of advantages of optical SETI over radio SETI. Because of the higher frequency of light, optical telescopes are much smaller than their radio counterparts for the same sensitivity and resolution. The resolution of a telescope, θ, whether an optical or a radio telescope is:

$$\theta \sim \lambda/D.$$

BOX 14.4   BANDWIDTH AND INFORMATION TRANSMISSION

·Bandwidth, B, is the difference between the highest and lowest frequency in a signal:

$$B = v_{hi} - v_{lo} = c/\lambda_{lo} - c/\lambda_{hi}.$$

With $v_{hi} - v_{lo} = \Delta v$ and $\lambda_{hi} - \lambda_{lo} = \Delta\lambda$, and after some algebra and neglecting very small terms it can be shown that:

$$\Delta v \sim c\Delta\lambda/\lambda^2.$$

In addition, because of the higher frequency of light, optical telescopes can transmit over greater bandwidths (see Box 14.4) than microwave/radio devices, so have a bigger capacity for information; that is, they can send information faster. We will now see how this works.

EXAMPLE 14.1

For a signal with a central wavelength of 1000 nm and $\Delta\lambda = 1$ nm

$$\Delta v = (3 \times 10^8 \text{ m s}^{-1} \times 1 \times 10^{-9} \text{ m})/(10^{-6} \text{ m})^2 = 3 \times 10^{11} \text{ s}^{-1} \text{ or } 300 \text{ GHz}.$$

EXERCISE 14.6

Show that the bandwidth for the higher frequency signal with central wavelength 100 nm and $\Delta\lambda = 1$ nm is 100-fold bigger.

Shannon–Hartley theorem, a central tenet of information theory, states the channel capacity C, the theoretical upper bound on the rate that error-free data can be transmitted by a signal of power S in the presence of (white Gaussian) noise of power N, is:

$$C = B\log_2(1 + S/N).$$

Hence, for a given signal-to-noise ratio, the rate at which information can be sent is directly related to the bandwidth. B is higher for higher frequencies so optical SETI is intrinsically a faster channel for conveying information than radio SETI.

Other advantages of optical SETI are: (1) there is less interference of visible light signals by natural sources than is the case for microwave/radio signals and (2) there appears to be no astrophysical process that generates nanosecond pulses of light.

Although there is no natural choice for wavelengths in optical SETI as there is for radio SETI, optical telescopes are broadband instruments so transmissions are likely to be detected. In any case it is more difficult to pin down the precise frequencies at which an optical signal is broadcast, compared with a radio signal, since the process to do this is intrinsically noisy at visible wavelengths and visible light suffers greater frequency broadening and absolute Doppler shifts than radio wavelengths.

The idea was largely ignored in 1961 because the power needed to generate a laser that could outshine the light of a star seemed incredibly high. However, by the late 1990s lasers had been developed that could produce picosecond pulses of 1000 TW:

thousands of times brighter than the sun. Given the rapid improvement in our laser technology (a doubling of power every year for more than 3 decades) we could expect the lasers of more advanced extraterrestrials to be orders of magnitude brighter. This prompted Paul Horowitz to begin an Optical SETI (OSETI) search by piggybacking on the 1.5-m telescope at Oak Ridge Observatory, Harvard, Massachusetts. This instrument was being used to hunt for exoplanets, and Horowitz had a beam splitter collect one-third of the starlight whenever the telescope was in use.

Extremely short pulses (3 ns for Horowitz OSETI) has the advantage for us as receivers that during this time the number of photons from the host star of the extraterrestrial civilization is negligible. A G2V star at 300 pc delivers $\sim 3 \times 10^5$ photons m^{-2} s^{-1} to a telescope. So during a 3- ns pulse this will be $\sim 10^{-3}$ photons. The timescale of the pulses also matches the fast response times of photomultiplier tubes (PMTs) used to detect the photons.

There are currently a couple of optical SETI programs. That at Harvard has now upgraded to a 1.8-m telescope and SETI at Berkeley uses a 0.76-m telescope. No signal has yet been detected.

14.4.3 How Far Away Would Passive Radiation Reveal Our Presence?

Guglielmo Marconi made the first *verifiable* long distance (Transatlantic) radio broadcast in 1902, but this should not be taken to mean that the radiosphere—the sphere of space centered on Earth with radius the furthest distance our radio/microwave broadcasts have reached—is currently at ~ 112 ly. Early radio signals were weak, most of their power would have been reflected back to Earth by the ionosphere, and any that did escape has been attenuated by the inverse square law to a level indistinguishable from ambient radio noise. Indeed, all amplitude modulated (AM) signals are reflected from the ionosphere, and frequency modulated (FM) radio and TV broadcasts, which do escape, were not transmitted commercially before 1936. What is more AM, FM, and TV are broadband signals so their power is spread over a large range of frequencies. At any given frequency, the signal is weak and would not be detected beyond Pluto even by a dish as big as Arecibo. By contrast, narrowband signals such as radar, used by the military or by astronomers to map asteroids, could be detected by Arecibo over distances of 1000 ly. The difficulty here is that radar is beamed and unless the receiver is pointed directly toward the beam and tuned to the right frequencies, it would be missed.

In general, the distance R at which a radio signal could be detected is given by

$$S/N = P_t G_t A_r/(4\pi R^2 BkT), \tag{14.5}$$

where S/N is the signal-to-noise ratio (SNR), P_t and G_t are the power and gain of the transmitter respectively, A_r is the effective area of the receiver, B is the bandwidth of the signal, k is the Boltzmann constant and T is the antenna temperature.

A couple of terms here need some explanation. The gain of a transmitter describes how well it converts electrical power into radio waves in a *specific direction*. A high gain transmitter puts most of its energy into a narrow beam, so is highly directional, whereas a low gain one has a wide beam. The receiver mounted at the focus of a radio telescope, often called the antenna, generates some noise. It also picks up noise from its surroundings, such as radiation scattered from the ground, synchrotron radiation from electrons in the Galaxy, absorption by carbon dioxide and oxygen in the atmosphere, and cosmic microwave background radiation. All of this noise is typically given as the power of black body radiation of a given temperature, the antenna temperature, which is approximately constant over small regions of the spectrum. The antenna temperature is a measure of the sensitivity of the receiver. The noise power that any signal will have to compete with is proportional to the bandwidth, so narrowband signals are the easiest to detect.

We can make a few assumptions to examine the practicality of signal detection. First we adopt a SNR of 25, comparable to that of radio SETI programs. The argument is that once the SNR gets lower than this, it becomes hard to hear the signal above the noise. The effective area of a radio telescope is:

$$A_r = \varepsilon \pi D^2, \tag{14.6}$$

where ε is the fractional efficiency and D is the aperture (diameter) of the dish. If we adopt Arecibo as the receiver with $D = 305$ m and 50% efficiency then the effective area is:

$$A_r = 0.5\pi \times 305^2 = 3.65 \times 10^4 \text{ m}^2.$$

Finally, with $k = 1.381 \times 10^{-23}$ J K^{-1} we can rewrite Equation 14.5 as

$$R = 2.9 \times 10^{12} \, (P_t G_t / BT)^{1/2} \text{ m or } 3.07 \times 10^{-4} \, (P_t G_t / BT)^{1/2} \text{ ly.}$$

EXERCISE 14.7

Calculate the distance, R, at which S/N ratio of the following radio signals fall below 25 and hence can be assumed undetectable: (a) VHF TV with $P_t G_t = 5$ MW, $B = 6$ MHz, $T = 50$ K. (b) FM radio with $P_t G_t = 5$ MW, $B = 150$ kHz, $T = 430$ K. (c) Astronomical radar with $P_t G_t = 22$ TW, $B = 0.1$ Hz, $T = 40$ K.

In fact, except for a few narrowband radar signals, which an alien would need to be favorably placed to receive, Earth is invisible beyond a few AU. If we consider narrow band radar signals then Earth's radiosphere has a radius of perhaps 30 ly or ~10 pc. This is a volume of 4189 pc^3. The number density of stars in the solar neighborhood is estimated to be 0.11 pc^{-3}, so we might expect the radiosphere to encompass about 420 stars. As of 2013, a total of 409 stars, including brown dwarfs and white dwarfs have been identified within 10 pc. If there are any technologically advanced civilizations on planets around these stars, they will not have seen any of our TV programs, but they may have detected a message transmitted by the Arecibo dish in 1979.

14.4.4 Where Is Everybody?

The original Drake equation estimate for the number of communicating civilizations in the Galaxy was 10^3–10^8, yet despite 50 years of searching no extraterrestrial signal has been spotted.

In addition, we could expect many of these civilizations to have developed interstellar travel before us, given the relative youth of the sun, and hence to have visited us, either in person or via robotic space probes, yet there is no compelling evidence for this.

This contradiction between the expectation of a high number of communicating civilizations and lack of evidence for them prompted the Italian nuclear physicist Enrico Fermi in 1950 to ask, "Where is everybody?," so is known as the Fermi paradox. Many explanations have been proposed.

14.4.4.1 SETI Has Not Been Long or Far-Reaching Enough

Of course SETI may have failed because we have not looked for long enough or/and our search has not surveyed a sufficiently large part of the sky. Alternatively, maybe our search strategy is flawed: we are looking at the wrong part of the electromagnetic spectrum. That we have not apparently been visited, or detected any alien artifacts, implies that if there are other extraterrestrial civilizations they are too separated in time and/or space.

14.4.4.2 Extraterrestrial Civilizations May Be Widely Separated in Time

Separation in time might be because communicating civilizations have a short lifetime; that is, L in the Drake equation is small. There are several possibilities here. Civilizations such as ours are very vulnerable to catastrophic events. A large asteroid/comet impact, the eruption of a supervolcano, or a nearby gamma ray burst, would produce planet-wide climate change that could disrupt a civilization for long periods or even destroy it. The catastrophe may be self-inflicted. Climate change generated by our need for energy and weapons of mass destruction are serious threats. Maybe civilizations are intrinsically self-limiting.

14.4.4.3 Extraterrestrial Civilizations Are Widely Separated in Space

Separation in space would be large if communicating civilizations were sparsely distributed through the Galactic disc so that signals take a long time to reach us, although the upper limit is only 100,000 years. Large distances would also explain why we have not been visited if interstellar spaceflight is impossible, although no rules of physics forbid it as far as we know. However, calculations show that for a 10-cm thick aluminum spacecraft hull, at $0.5c$ relativistic boosting of the energy of protons and electrons in the ISM would pose a fatal radiation hazard for living systems and radiation-hardened electronic circuits. In practice, this would constrain spaceflight speed to much less than $0.5c$. Does this mean there has not been time for colonizing civilizations to have reached us yet?

Simple calculations show this is unlikely. Diffusion models for the spread of civilizations between neighboring stellar systems show that with reasonable assumptions, the Galaxy could be colonized within 10^6 to 10^8 years. For example, Ian Crawford of Birkbeck, University of London, postulates that the speed of a colonization wavefront, V_{col}, is given by

$$V_{col} = d/(t_{travel} + t_{consolidation}),$$

where d is the average distance between colonies, t_{travel} is the time it takes to travel from one colony to the next, and $t_{consolidation}$ is the time it takes for a colony to become sufficiently well established to launch a colonizing mission itself. With $d = 10$ ly, interstellar spacecraft traveling at $0.1c$ and $t_{consolidation} = 400$ yr then V_{col} would be 0.02 ly per yr. Just 10,000 jumps like this could span the Galactic disc in 5 Myr.

This model is rather optimistic in its high velocity of space travel and the lack of an attrition factor to account for the inevitable loss of colonies through disasters. However, diffusion models are relatively insensitive to the velocity of interstellar spaceflight; the greatest uncertainty is in the consolidation time. However, even if the speed of interstellar spaceflight is 0.01c and $t_{consolidation}$ is 5000 years, the speed of Galactic colonization is only reduced by an order of magnitude and we might still expect to have been visited. Yet there is no evidence for this, wacky and unsubstantiated claims of UFOs and alien abduction notwithstanding!

14.4.4.4 Are There Resource Constraints on Interstellar Travel?

If interstellar spaceflight is possible and our estimates of colonization rates are realistic, it might be that extraterrestrial civilizations cannot undertake it because the resources required are too large. This might seem implausible, particularly if the civilization is older and more technologically developed than ours and hence would be expected to harness energy and materials more effectively than we can. After all, we are close to being able to generate energy sustainably from nuclear fusion, and it is not inconceivable that we could capture an asteroid, place it in Earth orbit, and harvest it for elements, such as rare Earth elements, that we risk running out of, within a few decades. But we need to reckon with the energy costs of interstellar spaceflight. The current record for the fastest human spacecraft, Voyager 1, is a paltry 17.2 km s^{-1}. To accelerate 1000 kg to $0.1c$ requires $\sim 5 \times 10^{17}$ J, about one-thousandth the Earth's annual energy consumption. This energy has to be supplied from on-board fuel, the ISM, or projected from the home stellar system in some way. Many space scientists think this is the single biggest difficulty for journeying between the stars.

14.4.4.5 Are Extraterrestrial Civilizations Intrinsically Rare?

Perhaps the solution to the Fermi paradox is the most obvious one, that we are the most advanced life forms in the Galaxy. The sun is only 4.5 Gyr old, and stars that are 10 Gyr old have sufficient metallicity to form planets. The implication is that one or more of the hard steps toward intelligent communicating life forms (see Section 14.1.3) are actually so difficult that we are the only ones to have made it. Maybe the Galaxy has not yet been colonized because a future step, one we have yet to attempt, is the limiting factor. Could it be that interstellar spaceflight is so hard that it has defeated all other intelligent extraterrestrials?

Bibliography

Adcock CT, Hausrath EM, Forster PM. Readily available phosphate from minerals in early aqueous environments on Mars. *Nat Geosci* 2013;6:824–7.

Allen JF, Martin W. Evolutionary biology—Out of thin air. *Nature* 2007;445:610.

Allwood AC, Walter MR, Kamber BS, Marshall CP, Burch IW. Stromatolite reef from the early Archaean era of Australia. *Nature* 2006;441:714–8.

Anbar AD, Duan Y, Lyons TW, Arnold GL, Kendall B, Creaser RA et al. A whiff of oxygen before the great oxidation event. *Nature* 2007;317:1903–6.

Andrews-Hanna JC, Zuber MT, Banerdt WB. The borealis basin and the origin of the Martian crustal dichotomy. *Nature* 2008;453:1212–5.

Atreya SK, Adams EY, Niemann HB, Demick-Montelara JE, Owen TC, Fulchignoni M et al. Titan's methane cycle. *Planet Space Sci* 2006;54:1177–87.

Bada JL, Lazcano A. Some like it hot, but not the first biomolecules. *Science* 2002; 296:1982–3.

Belbruno E, Gott JR. Where did the moon come from? *Astron J* 2005;129:1724–45.

Benner SA, Ricardo A, Carrigan MA. Is there a common chemical model for life in the universe? *Curr Opin Chem Biol* 2004;8:672–89.

Bibring J-P, Langevin Y, Mustard JF, Poulet F, Arvidson R, Gendrin A et al. The OMEGA team. Global mineralogical and aqueous Mars history derived from OMEGA/Mars express data. *Science* 2006;312:400–4.

Bishop JL, Noe Dobrea EZ, McKeown NK, Parente M, Ehlmann BL, Michalski JR et al. Phyllosilicate diversity and past aqueous activity revealed at Mawrthvallis, Mars. *Science* 2008;321:830–3.

Blake RE, Chang SJ, Lepland A. Phosphate oxygen isotopic evidence for a temperate and biologically active Archaean ocean. *Nature* 2009;464:1029–32.

Bontognali TRR, Sessions AL, Allwood AC, Fischer WW, Grotzinger JP, Summons RE et al. Sulfur isotopes of organic matter preserved in 3.45-billion-year-old stromatolites reveal microbial metabolism. *PNAS* 2012;109:15146–51.

Boyd ES, Peters JW. New insights into the evolutionary history of biological nitrogen fixation. *Front Microbiol* 2013;4:1–12.

Brandes JA, Hazen RM, Yoder HS. Inorganic nitrogen reduction and stability under simulated hydrothermal conditions. *Astrobiology* 2008;8:1113–26.

Brasier M, Mcloughlin N, Green O, Wacey D. A fresh look at the evidence for early Archaean cellular life. *Phil Trans R Soc B* 2006;361:889–902.

Buick R. Early life: Ancient acritarchs. *Nature* 2010;463:885–6.

Buick R. When did oxygenic photosynthesis evolve? *Phil Trans R Soc B* 2008;363:2731–43.

Burbidge EM, Burbidge GR, Fowler WA, Hoyle F. Synthesis of the elements in stars. *Rev Mod Phys* 1957;29:547–655.

Caldeira K, Kasting JF. The life span of the biosphere revisited. *Nature* 1992;360:721–3.

Campbell IH, O'Neil H St. C. Evidence against a chondritic Earth. *Nature* 2012;483:553–8.

Canup R. Forming a moon with an Earth-like composition via a giant impact. *Science* 2012;338:1052–5.

Canup R, Asphaug E. Origin of the moon in a giant impact near the end of the Earth's formation. *Nature* 2001;412:708–12.

Cassan A, Kubas D, Beaulieu J-P, Dominil M, Horne K, Greenhill J et al. One or more bound planets per Milky Way star from microlensing observations. *Nature* 2012;481:167–9.

Chyba CF, Phillips CB. Possible ecosystems and the search for life on Europa. *PNAS* 2001;98:801–4.

Clifford SM, Parker TJ. The evolution of the Martian hydrosphere: Implications for the fate of a primordial ocean and the current state of the northern plains. *Icarus* 2001;154:40–79.

Cocconi G, Morrison P. Searching for interstellar communication. *Nature* 1959;184:844–6.

Connelly JN, Bizzaro M, Krot AN, Nordlund A, Wielandt D, Ivanova MA. The absolute chronology and thermal processing of solids in the solar protoplanetary disk. *Science* 2012;338:651–5.

Crowe SA, Døssing LN, Beukes NJ, Bau M, Kruger SJ, Frei R et al. Atmospheric oxygen three billion years ago. *Nature* 2013;501:535–8.

Cuzzi JN, Alexander CMO'D. Chondrule formation in particle-rich nebular regions at least hundreds of kilometres across. *Nature* 2006;441:483–5.

Dale CW, Burton KW, Greenwood RC, Gannoun A, Wade J, Wood BJ et al. Late accretion on the earliest planetesimals revealed by the highly siderophile elements. *Science* 2012;336:72–5.

Dauphas N, Pourmand A. Hf-W-Th evidence for rapid growth of Mars and its status as a planetary embryo *Nature* 2011;473:489–92.

David LA, Alm EJ. Rapid evolutionary innovation during an Archaean genetic expansion. *Nature* 2011;469:93–6.

Dhuime B, Hawkesworth CJ, Cawood PA, Storey CG. A change in the geodynamics of continental growth 3 billion years ago. *Science* 2012;335:1334–6.

Domagal-Goldman SD, Kasting JF, Johnston DT, Farquhar J. Organic haze, glaciations and multiple sulfur isotopes in the Mid-Archean Era. *Earth Planet Sci Lett* 2008;269:29–40.

Doyle LR, Carter JA, Fabrycky DC, Slawson RW, Howell SB, Winn JN et al. Kepler-16: A transiting circumbinary planet. *Science* 2011;333:1602–6.

Drake MJ, Righter K. Determining the composition of the Earth. *Nature* 2002;416:39–44.

Dreibus G, Palme H. Cosmochemical constrains on the sulfur content of the Earth's core. *Geochim Cosmochim Acta* 1996;60:1125–30.

Dreibus G, Wanke H. Mars: A volatile-rich planet. *Meteoritics* 1985;20:367–82.

Edelstein WA, Edelstein A. Speed kills: Highly relativistic spaceflight would be fatal for passengers and instruments. *Nat Sci* 2012;4:749–54.

Elkins-Tanton LT. Linked magma ocean solidification and atmospheric growth for Earth and Mars. *Earth Planet Sci Lett* 2008;271:181.

Fairén AG, Fernández-Remolar D, Dohm JM, Baker VR, Amils R. Inhibition of carbonate synthesis in acidic oceans on early Mars. *Nature* 2004;431:423–6.

Farquhar J, Wing BA. Multiple sulfur isotopes and the evolution of the atmosphere. *Earth Planet Sci Lett* 2003;213:1–13.

Fedo CM, Whitehouse MJ, Kamber BS. Geological constraints on detecting the first life on Earth: A perspective form the early Archaean (older than 3.7 Gyr) of southwest Greenland. *Phil Trans Roy Soc B* 2006;361:851–67.

Forget F, Pierrehumbert RT. Warming early Mars with carbon dioxide clouds that scatter infrared radiation. *Science* 1997;278:1273–6.

Forget F, Wordsworth R, Millour E, Madelaine J-B, Kerber L, Leconte J et al. 3D modelling of the earlier Martian climate under a denser CO_2 atmosphere: Temperature and CO_2 ice clouds. *Icarus* 2013;222:81–99.

Freeland SJ, Hurst LD. The genetic code is one in a million. *J Mol Evol* 1998;47:238–48.

Fru EC, Ivarsson M, Kilias SP, Bengston S, Belivanova V, Marone F et al. Fossilized iron bacteria reveal a pathway to the biological origin of banded iron formation. *Nat Commun* 2013;4:2050.

Früh-Green GL, Kelley DS, Bernasconi SM, Karson JA, Ludwig KA, Butterfield DA et al. 30,000 years of hydrothermal activity at the lost city vent field. *Science* 2003;301:495–8.

Gaidos EJ, Nealson KH, Kirschvink JL. Life in ice-covered oceans. *Science* 1999;284:1631–2.

Gomes R, Levison HF, Tsiganis K, Morbidelli A. Origin of the cataclysmic late heavy bombardment period of the terrestrial planets. *Nature* 2005;435:466–9.

Greenwood JP, Itoh S, Sakamoto N, Warren P, Taylor L, Yurimoto H. Hydrogen isotope ratios in lunar rocks indicate delivery of cometary water to the moon. *Nat Geosci* 2011;4:87–92.

Haghighipour N, Raymond SN. Habitable planet formation in binary planetary systems *ApJ* 2007;666:436–46.

Halliday AN. A young moon-forming giant impact at 70–110 million years accompanied by late-stage mixing, core formation and degassing of the Earth. *Phil Trans R Soc B* 2008;366:4163–81.

Hanson R. Must early life be easy? The rhythm of major evolutionary transitions. 1998. http://hanson.gmu.edu/hardstep.pdf (Accessed on October 13, 2014).

Haqq-Misra JD, Domagal-Goldman SD, Kasting PJ, Kasting JF. A revised hazy methane greenhouse for the Archaean Earth. *Astrobiology* 2008;8:1127–37.

Hartmann WK, Quantin C, Mangold N. Possible long-term decline in impact rates. 2: Lunar impact-melt data regarding impact history. *Icarus* 2007;186:11–23.

Hartogh P, Lis DC, Bockelée-Morvan D, de Val-Borro M, Biver N, Küppers M et al. Ocean-like water in the Jupiter-family comet 103P/Hartley. *Nature* 2011;478:218–20.

Haskin LA, Wang A, Jolliff BL, McSween HY, Clark BC, Marais DJD et al. Water alteration of rocks and soils on Mars at the spirit rover site in Gusev crater. *Nature* 2005;436:66–9.

Hauri EH, Weinreich T, Saal AE, Rutherford MC, Van Orman JA. High pre-eruptive water content preserved in lunar melt inclusions. *Science* 330;333:213–5.

Hedrich S, Schlömann M, Johnson DB. The iron-oxidizing proteobacteria. *Microbiology* 2011;157:1551–64.

Hemmingway D, Nimmo F, Zebker H, Less L. A rigid and weathered ice shell on Titan. *Nature* 2013;500:550–2.

Hörst SM, Yelle RV, Buch A, Carrasco N, Cernogora G, Dutuit O et al. Formation of amino acids and nucleotide bases in a Titan atmosphere simulation experiment. *Astrobiology* 2012;12:809–17.

Howard AW. Observed properties of extrasolar planets. *Science* 2013;340:572–6.

Hren MT, Tice MM, Chamberlain CP. Oxygen and hydrogen isotope evidence for a temperate climate 3.42 billion years ago. *Nature* 2010;462:205–8.

Huynen, MA, Dandekar T, Bork P. Variation and evolution of the citric acid cycle: A genomic perspective. *Trends Mircrobiol* 1999;7:281–291.

Israelian G, Delgado Mena E, Santos NC, Sousa SG, Mayor M, Udry S et al. Enhanced lithium depletion in Sun-like stars with orbiting planets. *Nature* 2009;462:189–91.

Javaux E, Marshall C, Bekker A. Organic-walled microfossils in 3.2-billion-year-old shallow-marine siliciclastic deposits. *Nature* 2010;463:934–8.

Jones BW, Sleep PN. Habitability of exoplanetary systems with planets observed in transit. *MNRAS* 2010;407:1259–67.

Kaltenegger L, Segura A, Mohanty S. Characterizing model spectra of the first potentially habitable super earth Gl581d. *ApJ* 2011;733:35–47.

Kaltenegger L, Traub WA, Jacks K. Spectral evolution of an Earth-like planet. *ApJ* 2007;658:598.

Kane SR, Gelino DM. The habitable zone and extreme planetary orbits. *Astrobiology* 2012;12:940–5.

Kane SR, Hinkel NR. On the habitable zones of circumbinary planetary systems. *ApJ* 203;726:7.

Kargel JS, Kaye JZ, Head JW, Marion GM, Sassen R, Crowley JK et al. Europa's crust and ocean: Origin, composition, and the prospects for life. *Icarus* 2000;148:226–65.

Kasting JF. Runaway and moist greenhouse atmospheres and the evolution of Earth and Venus. *Icarus* 1988;74:472–94.

Kasting JF, Howard MT. Atmospheric composition and climate on early Earth. *Phil Trans R Soc B* 2006;361:1733–42.

Kasting JF, Whitmire DP, Reynolds RT. Habitable zones around main sequence stars. *Icarus* 1993;101:108–28.

Kaufman A, Johnston DT, Farquar J, Masterson AL, Lyons TW, Bates S et al. Late Archaeanbiospheric oxygenation and atmospheric evolution. *Nature* 2007;317:1900–3.

Kelley DS, Karson JA, Früh-Green GL, Yoerger DR, Shank TM, Butterfield DA et al. A serpentinite-hosted ecosystem: The Lost City hydrothermal field. *Science* 2005;307:1428–34.

Kerr RA. Rovers, dust and a not-so-wet Mars. *Science* 2005;308:192.

Kite ES, Williams J-P, Lucas A, Aharonson O. Paleopressure of Mars' atmosphere from small ancient craters. arXiv:1304.4043

Knauth LP, Burt DM, Wohletz KH. Impact origin of sediments at the opportunity landing site on Mars. *Nature* 2005;438:1123–8.

Knauth LP, Lowe DR. High Archean climatic temperature inferred from oxygen isotope geochemistry of cherts in the 3.5 Ga Swaziland Supergroup, South Africa. *Geol Soc Am Bull* 2003;115:566–80.

Konhauser KO, Lalonde SV, Planavsky NJ, Pecoits E, Lyons TW, Mojzsis SJ et al. Aerobic bacterial pyrite oxidation and acid rock drainage during the great oxidation event. *Nature* 2011;478:369.

Konhauser KO, Hamade T, Raiswell R, Morris RC, Ferris FG, Southam G et al. Could bacteria have formed the Precambrian banded iron formations? *Geology* 2002;30:1079–82.

Kopp RE, Kirschvink JL, Hilburn IA, Nash CZ. The Paleoproterozoic snowball Earth: A climate disaster triggered by the evolution of oxygenic photosynthesis. *PNAS* 2005;103:11131–6.

Kopparapu RK. A revised extimate of the occurrence rate of terrestrial planets in the habitable zones around Kepler M-dwarfs. *ApJ Lett* 2013;767:L8.

Kopparapu RK, Ramirez R, Kasting JF, Eymet V, Robinson TD, Mahadevan S et al. Habitable zones around main-sequence stars: New estimates. *ApJ* 2013;765:131.

Lane N, Martin WF. The origin of membrane bioenergetics. *Cell* 2012;151:1406–16.

Lathe R. Early tides: Response to Varga et al. *Icarus* 2006;180:277–80.

Lazar C, McCollom TM, Manning CE. Abiotic methanogenesis during experimental komatiiteserpentinization. Implicatios for the evolution of the early Precambrian atmosphere. *J Chem Geol* 2012;326–7:102–12.

Leconte J, Forget F, Charnay B, Wordsworth R, Pottier A. Increased insolation threshold for runaway greenhouse processes on Earth-like planets. *Nature* 2013;504:268–71.

Lineweaver CH, Davies TM. Does the rapid appearance of life on Earth suggest life is common in the Universe? *Astrobiology* 2002;2:293–304.

Lissauer J. Planets formed in habitable zones of M dwarf stars probably are deficient in volatiles. *ApJ Lett* 2007;660:L141-52.

Lorenz, LD, Niemann HB, Harpold DN, Way SH, Zarnecki JC. Titan's damp ground: Constraints on Titan surface thermal properties from the temperature evolution of the Huygens GCMS inlet. *Meteoritics Planet Sci* 2006;41:1705–14.

Luhmann JG, Johnson RE, Zhang MHG. Evolutionary impact of sputtering of the Martian atmosphere by O^+ pick-up ions. *J Geophys Res* 1992;19:2152–4.

Manga M, Patel A, Dufek J, Kite ES. Wet surface and dense atmosphere on early Mars suggested by the bomb sag at home plate, Mars. *Geophys Res Lett* 2012;39:192.

Marinova MM, Aharonson O, Asphaug E. Mega-impact formation of the Mars hemispheric dichotomy. *Nature* 2008;453:1216.

Martin W, Muller M. The hydrogen hypothesis for the first eukaryote. *Nature* 1998;329: 37–41.

Marty B, Zimmermann L, Pujol M, Burgess R, Philippot P. Nitrogen isotopic composition and density of the Archean atmosphere. *Science* 2013;342:101–4.

Mayor M, Queloz D. A Jupiter-mass companion to a solar-type star. *Nature* 1995;378: 355–9.

McCollom TM. Methanogenesis as a potential source of chemical energy for primary biomass production by autotrophic organisms in hydrothermal systems on Europa. *J Geophys Res* 1999;104:30729–42.

McCollom TM, Hynek BM. A volcanic environment for bedrock diagenesis at Meridiani Planum on Mars. *Nature* 2005;438:1129–31.

McKay CP, Smith HD. Possibilities for methanogenic life in liquid methane on the surface of Titan. *Icarus* 2005;178:274–6.

McKay DS, Gibson EK, Thomas-Keptra KL, Vali H, Romanek CS, Clemett SJ et al. Search for past life on Mars: Possible relic biogenic activity in Martian meteorite ALH84001. *Science* 1996;273:924–30.

McKeegan KD, Kallio APA, Heber VS, Jarzebinski G, Mao PH, Coath CD et al. The oxygen isotopic composition of the sun inferred from captured solar wind. *Science* 2011;332:1528–32.

McMahon S, Parnell J, Ponicka J, Hole M, Boyce A. The habitability of vesicles in Martian basalt. *Astron Geophys* 2013;54:1.17–21.

McSween Jr HY, Taylor GJ, Wyatt MB. Elemental composition of the Martian crust. *Science* 2009;324:736–9.

Middleton C, Gilmore J. Anthropic selection of a solar system with a high $^{26}Al/^{27}Al$ ratio: Implications and a possible mechanism. *Icarus* 2009;201:821–3.

Minton DA, Malhotra R. A record of planet migration in the main asteroid belt. *Nature* 2009;457:1109–11.

Mouginot J, Pommerol A, Beck P, Kofman W, Clifford SM. Dielectric map of the Martian northern hemisphere and the nature of plain filling materials. *Geophys Res Lett* 2012;39:L02202.

Mustard JF. Hydrated silicate minerals on Mars observed by the Mars reconnaissance orbiter CRISM instrument. *Nature* 2008;454:305.

Neish CD, Somogyi A, Smith MA. Titan's primordial soup: Formation of amino acids via low-temperature hydrolysis of tholins. *Astrobiology* 2010;10:337–47.

Nemchin AA, Pidgeon RT, Whitehouse MJ. Re-evaluation of the origin and evolution of >4.2 Ga zircons from the Jack Hills metasedimentary rocks. *Earth Planet Sci Lett* 2006;244:218–33.

Niemann HB, Atreya SK, Bauer SJ, Carignan GR, Demick JE, Frost RL et al. The abundances of constituents of Titan's atmosphere from the GCMS instrument on the Huygens probe. *Nature* 2005;438:779–84.

Nixon SL, Cousins CR, Cockell CS. Plausible microbial metabolisms on Mars. *Astron Geophys* 2013;54:1.13–16.

Norman MD, Borg LE, Nyquist LE, Bogard DD. Chronology, geochemistry, and petrology of a ferroan noritican orthosite clast from Descartes breccia 67215: Clues to the age, origin, structure, and impact history of the lunar crust. *Meteorit Planet Sci* 2003;38:645–61.

Olgin JG, Smith-Konter BR, Pappalardo RT. Limits of Enceladus's ice shell thickness from tidally driven tiger stripe shear failure. *Geophys Res Lett* 2011;38:L02201.

Orgel LE. *The Origin of Life: Molecules and Natural Selection.* John Wiley & Sons Inc, 1973.

Osinski GR, Tornabene LL, Banerjee NR, Cockell CS, Flemming R, Izawa MRM et al. Impact-generated hydrothermal systems on Earth and Mars. *Icarus* 2013;224:347–63.

Paniello RC, Day JMD, Moynier F. Zinc isotope evidence for the origin of the moon. *Nature* 2012;490:376–9.

Pasek MA, Mousis O, Lunine JI. Phosphorus chemistry on Titan. *Icarus* 2011;212:751–61.

Paulton SW, Fralick PW, Canfield DE. The transition to a sulfidic ocean ~1.84 Gyr ago. *Nature* 2004;431:173–7.

Planavsky NJ, Asael D, Hofmann A, Reinhard CT, Lalonde SV, Knudsen A et al. Evidence for oxygenic photosynthesis half a billion years before the Great Oxidation. *Nat Geosci* 2014;7:283–6.

Planavsky NJ, McGoldrick P, Scott CT, Li C, Reinhard CT, Kelly AE et al. Widespread iron-rich conditions in the mid-Proterozoic ocean. *Nature* 2011;477:448–51.

Planavsky NJ, Bekker A, Hofmann A, Owens JD, Lyons TW. Sulfur record of rising and falling marine oxygen and sulfate levels during the Lomagundi event. *PNAS* 2012;109:18300–5.

Powner MW, Gerland B, Sutherland JD. Synthesis of activated pyrimidine nucleotides in prebiotically plausible conditions. *Nature* 2009;459:23942.

Proskurowsky G, Lilley MD, Seewald JS, Fruh-Green GL, Olsen EJ, Lupton JE et al. Abiogenic hydrocarbon production at lost city hydrothermal field. *Science* 2008;319:604–7.

Ramirez RM, Kopparapu R, Zugger ME, Robinson TD, Freedman R, Kasting JF. Warming early Mars with CO_2 and H_2. *Nat Geosci* 2013;7:59–63.

Raymond SN, Barnes R, Kaib NA. Predicting planets in known extra-solar planetary systems. III: Forming terrestrial planets. *ApJ* 2006;644:1223–31.

Raymond SN, Quinn T, Lunine JI. High-resolution simulations of the final assembly of Earth-like planets. 1: Terrestrial accretion and dynamics. *Icarus* 2006;183:265–82.

Reinhard CT, Raiswell R, Scott C, Anbar AD, Lyons TW. A late Archaean sulfidic sea stimulated early oxidative weathering of the continents. *Science* 2009:326:713–6.

Reinhard CT, Planavsky NJ, Robbins LJ, Partin CA, Gill BC, Lalonde SV et al. Proterozoic ocean redox and biogeochemical status. *PNAS* 2013;110: 5357–62.

Rivera MC, Lake JA. The ring of life provides evidence for a genome fusion origin of eukaryotes. *Nature* 2004:431:152–5.

Robert F. Planetary science: A distinct source for lunar water. *Nat Geosci* 2011;4:74–5.

Rosing MT. ^{13}C-depleted carbon microparticles in >3700-Ma sea-floor sedimentary rocks from west Greenland. *Science* 1999;283:674–8.

Rosing MT, Bird DK, Sleep NH, Bjerrum CJ. No climate paradox under the faint young sun. *Nature* 2010;464:744–7.

Saal AE, Hauri EH, Van orman JA, Rutherford MJ. Hydrogen isotopes in lunar volcanic glasses and melt inclusions reveal a carbonaceous chondrite heritage. *Science* 2013;340:1317–20.

Sagan C, Thompson WR, Carlson R, Gurnett D, Hord C. A search for life on earth from the Galileo spacecraft. *Nature* 1993;365:715.

Sagan L. On the origin of mitosing cells. *J Theor Biol* 1967;14:255–74.

Schopf JW. Microfossils of the early Archean Apex chert: New evidence of the antiquity of life. *Science* 1993;260:640–6.

Schulze-Makuch D, Grinspoon DH. Biologically enhanced energy and carbon cycling on Titan? *Astrobiology* 2005;5:560–4.

Seager S. Exoplanet habitability. *Science* 2013;340:577–81.

Seager S, Turner EL, Schafer J, Ford EB. Vegetation's red edge: A possible spectroscopic biosignature of extraterrestrial plants. *Astrobiology* 2005;5:372–90.

Segura TA, McKay CP, Toon OB. An impact-induced, stable, runaway climate on Mars. *Icarus* 2012;220:144–8.

Segura TJ, Toon OB, Colaprete A. Modeling the environmental effects of moderate-sized impacts on Mars. *J Geophys Res* 2008;113:E11.

Segura TL, Toon OB, Colaprete A, Zahnle K. Environmental effects of large impacts on Mars. *Science* 2002;298:1977.

Shen Y, Farquhar J, Masterson A, Kaufman AJ, Buick R. Evaluating the role of microbial sulfate reduction in the early Archaean using quadrupole isotope systematics. *Earth Planet Sci Lett* 2009;279:383–91.

Shirey SB, Richardson SH. Start of the Wilson cycle at 3 Ga shown by diamonds from subcontinental mantle. *Science* 2011;333:434–6.

Som SS, Catling DC, Harnmeijer JP, Polivka PM, Buick R. Air density 2.7 billion years ago limited to less than twice modern levels by fossil raindrops imprints. *Nature* 2012;484:359–62.

Spiegel DS, Turner EL. Bayesian analysis of the astrobiological implications of life's early emergence on Earth. *PNAS* 2012;109:395–400.

Squyres SW, Kasting JF. Early Mars: How warm and how wet? *Science* 1994;265: 744–9.

Stanley S, Elkins-Tanton L, Zuber MT, Parmentier EM. Mars' palaeomagnetic field as a result of a single-hemisphere dynamo. *Science* 2008;321:1822–5.

Stewart AJ, Schmidt MN, van Westrenen W, Liebske C. Mars: A new core crystallization regime. *Science* 2007;316:1323–5.

Sugita S, Schultz P. Efficient cyanide formation due to impacts of carbonaceous bodies on a planet with a nitrogen-rich atmosphere. *Geophys Res Lett* 2009;36:L20204.

Susanti D, Mukhopadhyay B. An intertwined evolutionary history of methanogenic Archaea and sulfate reduction. *PLoS One* 2012;7:e45313.

Swain MR, Vasisht G, Tinetti G. The presence of methane in the atmosphere of an extrasolar planet. *Nature* 2008;452:329–31.

Swift DC, Eggert JH, Hicks DG, Hamel S, Caspersen K, Schwegler E et al. Mass-radius relationship for planets. *ApJ* 2012;744:59–69.

Szostak J. Origins of life: Systems chemistry on early Earth. *Nature* 2009;459:171–2.

Takami H, Noguchi H, Takaki Y, Uchiyama I, Toyoda A, Nishi S et al. A deeply branching thermophilic bacterium with an ancient acetyl-CoA pathway dominates a subsurface ecosystem. *PLoS One* 2012;7:e30559.

Tarduno JA, Cottrell RD, Nimmo F, Hopkins J, Voronov J, Erickson A et al. Evidence for a dynamo in the main group pallasite parent body. *Science* 2012;338:939–42.

Thomas-Keprta KL, Clement SJ, McKay DS, Gibson EK, Wentworth SJ. Origin of the magnetite crystals in ALH84001. *Geochim Cosmochim Acta* 2009;73:6631–77.

Tsiganis K, Gomes R, Morbidelli HF. Origin of the orbital architecture of the giant planets of the solar system. *Nature* 2005;435:459–61.

Vreeland RH, Rosenzweig WD, Powers DW. Isolation of a 250 million-year-old halotolerant bacterium from a primary salt crystal. *Nature* 2000;407:897–900.

Wacey D, Kilburn MR, Saunders M, Cliff J, Brasier MD. Microfossils of sulphur-metabolizing cells in 3.4-billion-year-old rocks of Western Australia. *Nat Geosci* 2011;4:698–702.

Waite JH, Young DT, Cravens TE, Coates AJ, Crary FJ, Magee B et al. The process of tholins formation in Titan's upper atmosphere. *Science* 2007;316:870–5.

Walsh KJ, Morbidelli A, Raymond SN, O'Brien DP, Mandell AM. A low mass for Mars driven from Jupiter's early gas-driven migration. *Nature* 2011;475:206–9.

Waltham D. Anthropic selection and the habitability of planets orbiting M and K dwarfs. *Icarus* 2011;215:518–21.

Watson A. Implications of an anthropic model of evolution for the emergence of complex life and intelligence. *Astrobiology* 2008;8:175–85.

Weber RC, Lin P-Y, Garnero EJ, Williams Q, Lognonne P. Seismic detection of lunar core. *Science* 2011;331:309–12.

Williams TA, Foster PG, Cox CJ, Embley TM. An archaeal origin of eukaryotes supports only two primary domains of life. *Nature* 2013;504:231–6.

Withers P, Neumann GA. Enigmatic northern plains of Mars. *Nature* 2001;410:651.

Wochner A, Attwater J, Coulson A, Holliger P. Ribozyme-catalyzed transcription of an active ribozyme. *Science* 2011;322:209–12.

Wolf ET, Toon OB. Hospitable Archaean climates stimulated by a general circulation model. *Astrobiology* 2013;13:656–76.

Wolszczan A, Frail DA. A planetary system around the millisecond pulsar PSR1257 + 12. *Nature* 1992;355:145–7.

Wordsworth R, Forget F, Millour E, Head JW, Madelaine J-B, Charney B. Global modeling of the early Martian climate under a denser CO_2 atmosphere: Water cycle and ice evolution. *Icarus* 2013;222:1–19.

Wray JJ, Hansen ST, Dufek J, Swayze GA, Murchie SL et al. Prolonged magmatic activity on Mars inferred from the detection of felsic rocks. *Nat Geosci* 2013;6:1013–7.

Yang J, Cowan NB, Abbot DS. Stabilizing cloud feedbacks dramatically expand the habitable zone of tidally-locked planets. *ApJ Lett* 2013;771:L45.

Zahnle K, Arndt N, Cockell C, Halliday A, Nisbet E, Selsis F et al. Emergence of a habitable planet. *Space Sci Rev* 2007;129:35–78.

Zuber M. Planetary science: Mars at the tipping point. *Nature* 2007;447:785–6.

Index

A

α-elements, 366–367
α-particle x-ray spectrometry (APXS), 261
α-proteobacteria, 249, 251
Absorption, 37
Absorption lines, 6, 8
Absorption spectrum, 10
Acasta gneiss, Canada, 118, 211
Acetate, 196
Acetogens, 207, 230
Acetothermus autotrophicum, 207
Acetyl CoA, 196–197, 227
Acetyl phosphate synthesis, 227, 231–232,
Acetyl thioester, 227, 232
Acetylene
    Titan, 302–303
    Titan atmosphere, 299
Achondrites, 96
Acidic hydrothermal vents (Black smokers), 226
Acritarchs, 247–248
Activation energy, 42
Activation energy barrier, 194
Adenine (A), 178
Adenosine triphosphate (ATP), 196–197, 200
    synthesis, 205
Adiabat, 124
Aerobic respiration, 196, 198
Albedo, 71–72, 146–147, 152
    Enceladus, 305
ALH84001, 256–257, 279
    history, 286
    life, 286
Alkaline vents, 227
    proton gradients, 231
Alternative replicators, 222
Aluminium isotope (26Al), 99, 102–103
Amazonian age, 260
Amino acids, 175, 216
    polymerization, 177
    structure, 176
Ammonia, 38, 144–145
    chemical properties, 172
Ammonia formation, 60
Ammonia-water binary system, Titan, 295, 297
Ammonia-water binary system, phase
        diagram, 294
Anabolism, 182

Anaerobic organisms, 236
Anaerobic respiration, 196, 19, 202, 207, 226
Anorthosite, lunar, 106–107
Anoxygenic photosynthesis, 240, 246
Antenna temperature, 396
Anthracine, 38
Anticodon, 192
Antiparticles, 2
Archaea, 180
Archaea, 186–187
Archaean
    ocean, 137
    surface temperature, 137
    temperature, 151
Archaean atmosphere
    carbon dioxide, 149
    carbon dioxide concentration, 151–152
    hydrogen, 150
    methane, 152–153, 155
    methane/carbon dioxide ratio, 153
    nitrogen, 150
    organic haze, 153
    pressure, 150
Archaean eon, surface temperature, 135,
        139, 154
Archaean raindrops, 151
Arecibo radiotelescope, 395
Arrhenius equation, 42, 297
Asteroids, 67, 115, 133–135
Astrochemistry
    density range, 41
    rate, 40
    rate, 41
    temperature range, 40–41
Astrometry, 324
Asymtotic giant branch, 25–26
Athenosphere, 83
Atmosphere,
    Archaean, 235
    carbon dioxide concentration, 254
    CH4/CO2 ratio, 235
    column mass, 76
    early Mars, 276
    Hadean, 235
Atom, electronic transition, 8–9
Atomic (neutral) hydrogen, 35
    hyperfine atomic transition, 35–36
Atomic number, 2